Harnessing Phytoconstituents for Wound Healing

Geeta Deswal
https://orcid.org/0000-0003-3648-0643
Guru Gobind Singh College of Pharmacy, Yamuna Nagar, India

Ajmer Singh Grewal
https://orcid.org/0000-0002-2418-9616
Guru Gobind Singh College of Pharmacy, Yamuna Nagar, India

Kumar Guarve
Guru Gobind Singh College of Pharmacy, Yamuna Nagar, India

IGI Global
Scientific Publishing
Publishing Tomorrow's Research Today

Vice President of Editorial	Melissa Wagner
Director of Acquisitions	Mikaela Felty
Director of Book Development	Jocelynn Hessler
Production Manager	Mike Brehm
Cover Design	Jose Rosado

Published in the United States of America by
IGI Global Scientific Publishing
701 East Chocolate Avenue
Hershey, PA, 17033, USA
Tel: 717-533-8845 | Fax: 717-533-7115
Website: https://www.igi-global.com E-mail: cust@igi-global.com

Library of Congress Cataloging-in-Publication Data

LCCN: 2025026751 (CIP Data Pending)
ISBN13: 9798337329086
Isbn13Softcover: 9798337329093
EISBN13: 9798337329109
British Cataloguing in Publication Data
A Cataloguing in Publication record for this book is available from the British Library.

Table of Contents

Detailed Table of Contents

The process of wound healing is also a very complex biological phenomenon which is necessary to reestablish the structure and functionality of tissue following an injury. The paper has given a detailed analysis of wounds dividing them into different types, such as acute and chronic wounds, and open and closed wounds. It further explores the four major phases of wound healing that include hemostasis, inflammation, proliferation and remodeling. The complex cellular and molecular processes that occur in each stage are in sync to heal the injured tissue. Outside the fundamental science, the discussion covers the challenges to wound care, which include infection treatment, the protracted healing of non-healing wounds, and issues that are associated with the systemic conditions, such as diabetes and immunodeficiency. A good understanding of these aspects is an essential part of designing efficient treatment strategies and expanding the limits of clinical results in wound care.

of wound healing and critically examines preclinical and emerging clinical evidence supporting phytochemical-based interventions. It further discusses current challenges, translational limitations, and prospects, emphasising the role of systems biology, nanotechnology, and personalised medicine in advancing safe, effective, and sustainable wound-healing strategies.

Section 2
Phytoconstituents in Wound Management

Abraham Olufemi Asuku, Bioresources Development Centre, National Biotechnology Research and Development Agency, Nigeria
Ekundayo Temitope Ola, Bioresources Development Centre, National Biotechnology Research and Development Agency, Nigeria

In both traditional and modern medicine, the bioactive chemicals found in plants, also known as phytoconstituents, have long been recognized for their medicinal properties. These comprise substances with a wide range of pharmacological characteristics, like anti-inflammatory, antibacterial, antioxidant, and anticancer effects, such as alkaloids, terpenoids, glycosides, flavonoids, and polyphenols. With an emphasis on their mechanisms of action, therapeutic value, and potential for drug development, this chapter explores the various roles that medicinal plants and their phytoconstituents play in wound prevention and care. The chapter also covers advancements in phytochemical research, including bioavailability enhancements and pharmaceutical inclusion. Notwithstanding their enormous therapeutic potential, restrictions including safety, standardization, and regulatory issues are still necessary for their widespread use. By understanding and applying nature's abundance of medicinal compounds, phytoconstituents offer a practical and effective way to improve global health.

Diabetes and obesity are the main causes of wounds which are regarded as a major issue that impacts the healthcare industry in many nations. Unhealthy lifestyle choices and behaviors exacerbate wounds. After an injury, wound healing is a complex physiological process that is necessary to restore the epithelial barrier. Because of their well-known anti-inflammatory, angiogenesis, re-epithelialization, and antioxidant actions, polyphenols and flavonoids have been shown in numerous studies to have wound-healing capabilities. They have potent anti-inflammatory, antibacterial, and antioxidant qualities and are mostly found in fruits, vegetables, tea, and wine. The complex molecular processes by which polyphenols and flavonoids influence wound healing include immune response modulation, collagen synthesis enhancement, oxidative stress reduction, and angiogenesis promotion. Additionally, the NF-κB and MAPK pathways, two important signaling pathways involved in tissue repair and inflammatory regulation, are influenced by these chemicals. According to research, flavonoids and polyphenols may also support angiogenesis, the process of forming new blood vessels, and extracellular matrix remodeling, both of which are essential for proper wound closure and healing. This chapter provides a comprehensive review of the most current studies on the role of flavonoids and polyphenols in wound healing, with a focus on the potential therapeutic applications of these substances to enhance wound care and prevent chronic wounds.To determine the usefulness of polyphenols and flavonoids in therapeutic settings, more clinical research is necessary to examine their pharmacological characteristics and safety profiles.

Manju Jhakar, School of Pharmacy and Emerging Sciences, Baddi University of Emerging Sciences and Technology, Baddi, India
Nidhi Rani, Chitkara College of Pharmacy, Chitkara University, India
Ashish Kashyap, Bhojia Dental College and Hospital, Baddi, India
Thakur Gurjeet Singh, Chitkara College of Pharmacy, Chitkara University, India

Plant-derived bioactive compounds such as terpenoids and essential oils (EOs) are increasingly recognised for their therapeutic value in wound healing and tissue regeneration. There are many biological activities, e.g., anti-inflammatory, antioxidant, antibacterial, and wound-healing activity, which are exhibited by these volatile oils. Current science validates their roles in current medical practices, which have traditionally been the cornerstone of herbal medicine. This chapter explores the multifaceted functions of terpenoids and essential oils in healing tissue, with a specific interest in their mode of action in a range of therapeutic settings. Their role in repairing and healing joints is highlighted, particularly by antioxidant and anti-inflammatory mechanisms that mitigate oxidative stress and enhance tissue regeneration. Neuroprotective activity, especially in the treatment of neurodegenerative diseases, is also examined. Additionally, the chapter explores their application in stimulating tissue development and collagen synthesis. Lavender, chamomile, rosemary, and thyme are essential oils that have also shown potential in bone and wound healing, which can be used along with scaffolds. Advances in extraction methods such as hydrodistillation and supercritical fluid extraction have enhanced the therapeutic properties of these compounds. Importantly, although the preclinical data are compelling for many agents of interest, there is a significant translational gap: with few exceptions, such therapies have not been adopted in routine clinical practice due to unmet needs related to standardization, bioavailability of delivery engineering, and regulatory validation. This chapter therefore reframes the discussion from cataloguing preclinical promise to critically analysing the translational barriers facing terpenoids and EOs, and proposes a structured four-stage research strategy to guide their validation as clinically approved wound-healing therapeutics.

Wound closure is a multifaceted biological phenomenon through which hemostasis, inflammatory phase, proliferative or re-epithelialization and maturation are different stages of wound healing. As natural astringents, tannins are available as phytoconstituents in several herbal plants and have become particularly remarkable pharmacological potential for wound closure benefits. Tannins exhibited strong astringent, antimicrobial, anti-inflammatory, and antioxidant characteristics, making them effective in enhancing tissue repair. Recent developments in biomaterials have resulted to the creation of tannin-based wound dressings, hydrogels, and bioactive scaffolds, which allow for a controlled release of tannins, enhancing their stability as well as bioavailability while maintaining moisture balance at the wound site. Tannin-infused nanofibers and hydrocolloid dressings have exhibited promise in supporting faster wound closure by creating an optimal healing environment. With their diverse therapeutic benefits, natural based tannins may provide a valuable herbal alternative in comparison to synthetic agents for wound healing, offering safer and more effective approaches for the management of wound.This chapter highlighted pathophysiology of wound healing, sources and classification of tannins, therapeutic mechanism of tannins, and applications of several naturally derived tannins in wound care and closure as well as the challenges aspects and future perspectives of tannins as natural astringents for the management of wound closure.

The foundation for early medicine was established by ancient herbal practices, which still influence healthcare today. Ayurveda, Unani, and Traditional Chinese Medicine are examples of systems that have developed over centuries of spiritual, scientific, and cultural integration. Textual and archeobotanical records from Egypt, Greece, India, and Mesopotamia demonstrate the systematic usage of therapeutic plants. More than 80% of people still use herbal remedies today, particularly in rural regions. The preventive and holistic aspects of herbal therapy are reviving interest and bolstering traditional treatments. With an emphasis on clinical efficacy, illness management, prevention, and historical context, this study investigates the integration of traditional systems into contemporary care. Safety, regulation, standardisation, validation, and ethics are also examined. China, India, and Germany's integration models demonstrate the possibility of interdisciplinary, policy-driven cooperation.

Foreword

Wound healing using different types of plants and their extracts has been a common practice since ancient times. Phytoconstituents have therefore been widely utilized for the treatment of injuries, burns, and infections across various ancient civilizations. Nowadays, modern science has also recognized the significant role of phytoconstituents in wound healing. However, there is a growing global demand for safe, cost-effective, and biologically based alternative therapies in the global healthcare sector.

The present book entitled "Harnessing Phytoconstituents for Wound Healing" serves as an excellent resource for stakeholders in the global healthcare industry. It illustrates the integration of traditional knowledge with modern scientific approaches to explore the potential of plant-based compounds and to promote the herbal sector. Several bioactive molecules, such as flavonoids, alkaloids, terpenoids, tannins, and phenolic acids, are commonly used in the treatment of various diseases because of their unique properties, including antimicrobial, anti-inflammatory, antioxidant, and tissue regenerative activities.

The editors and contributing researchers involved in this volume possess comprehensive and up-to-date knowledge of the subject, aiming to bridge existing scientific gaps. The book thoroughly explores the process of wound healing through herbal approaches using plant-based materials and their derivatives. It also discusses formulation strategies, safety considerations, quality standards, limitations, and emerging techniques to ensure the efficacy and stability of phytoconstituents.

Importantly, this book will be valuable for laboratory professionals, budding researchers, clinicians, pharmacognosists, and industrial innovators. With the increasing prevalence of chronic wound problems worldwide, this volume serves as a scholarly resource and an inspiring guide for readers to explore nature's molecular diversity as a valuable asset in advancing the science of healing.

I sincerely commend the editors for bringing this important issue to the forefront and for highlighting the immense potential of plants in promoting natural and effective wound healing.

Arun Lal Srivastav
Chitkara University, India

Preface

Wound healing is a highly complex and dynamic biological process that is vital to preserving the structural and functional integrity of living tissues. Ranging from minor abrasions to chronic non-healing ulcers, wounds constitute a significant clinical and socioeconomic burden worldwide. The increasing prevalence of chronic wounds associated with diabetes mellitus, vascular insufficiencies, infections, traumatic injuries, and aging populations has increased the demand for therapeutic strategies that are safe, effective, and economically accessible. In recent years, natural products and phytoconstituents have garnered significant scientific interest owing to their diverse pharmacological activities, broad availability, and relatively low incidence of adverse effects. The edited volume "*Harnessing Phytoconstituents for Wound Healing,*" has been developed as a comprehensive scholarly resource aimed at examining the therapeutic potential of phytochemicals and herbal interventions in contemporary wound management.

OVERVIEW OF THE SUBJECT MATTER

The field of wound healing integrates principles from biology, pharmacology, biotechnology, medicinal chemistry, and clinical medicine. Wound repair involves a coordinated cascade of cellular and molecular events including inflammation, proliferation, extracellular matrix remodeling, and tissue regeneration. Disruption in any of these phases can result in delayed healing or chronic wound formation. Modern medicine has made significant progress in wound care through synthetic drugs, advanced dressings, and biomaterials; however, limitations such as high cost, side effects, antimicrobial resistance, and incomplete healing outcomes persist. Phytoconstituents offer a promising complementary and alternative approach. Polyphenols, flavonoids, terpenoids, tannins, saponins, and other plant-derived molecules exhibit antioxidant, anti-inflammatory, antimicrobial, angiogenic, and collagen-promoting properties that are important for effective wound repair. Traditional systems of medicine such as Ayurveda, Traditional Chinese Medicine, and ethnomedicine have long utilized plant-based formulations for wound treatment. Contemporary research now seeks to validate these practices through scientific investigation, molecular studies, and computational approaches. This book provides a structured exploration of wound healing from fundamental biological mechanisms to advanced computational studies on phytochemicals. It bridges traditional knowledge with modern scientific methodologies, highlighting the therapeutic potential of plant-based compounds in both acute and chronic wound management.

RELEVANCE OF THE TOPIC IN TODAY'S WORLD

In the present global healthcare landscape, wound management has become a pressing concern. The increasing prevalence of diabetes and associated complications such as diabetic foot ulcers has significantly escalated the incidence of chronic wounds. According to global health estimates, millions of patients suffer from non-healing wounds that often lead to infections, amputations, and decreased quality of life. The economic burden on healthcare systems is substantial, particularly in developing countries where access to advanced wound care technologies may be limited. Phytoconstituents present an economical and sustainable alternative for wound treatment. Their multi-targeted mechanisms of action make them suitable for addressing complex wound environments characterized by oxidative stress, microbial infection, and impaired tissue regeneration. Also, the integration of phytochemicals into modern formulations such as gels, ointments, nanocarriers, and bioengineered dressings is opening new avenues in translational research and pharmaceutical development. Advances in molecular biology, computational chemistry, and bioinformatics have enabled deeper insights into the interaction of phytochemicals with biological targets involved in wound healing. Molecular docking and in silico studies are now widely used to predict the efficacy and mechanism of plant-derived compounds, thereby accelerating drug discovery and development processes. This book situates itself at the intersection of traditional herbal knowledge and modern scientific innovation, reflecting the evolving paradigm of integrative wound care.

TARGET AUDIENCE

This book is designed to serve a wide spectrum of readers including:

- Researchers and academicians in pharmaceutical sciences, pharmacognosy, medicinal chemistry, biotechnology, and biomedical sciences

- Undergraduate, postgraduate, and doctoral students in pharmacy, life sciences, and allied health disciplines
- Clinicians and healthcare professionals involved in wound management and dermatological care
- Scientists engaged in natural product research, herbal drug development, and translational medicine
- Professionals in pharmaceutical and biotechnology industries focusing on herbal formulations and novel therapeutics

By presenting both fundamental concepts and advanced research insights, this book aims to function as a valuable reference for academic study, research exploration, and practical application.

CHAPTER OVERVIEW

Chapter 1: Understanding Wounds: Types, Stages, and Challenges

This chapter introduces the fundamental concepts of wound classification, stages of healing, and challenges associated with acute and chronic wounds. It provides a foundational understanding of wound etiology, pathophysiology, and complications, establishing the basis for subsequent discussions on therapeutic interventions.

Chapter 2: The Science of Wound Healing: Biological Processes and Key Mediators

This chapter elaborates on the biological processes involved in wound repair, including inflammation, proliferation, and remodeling. It highlights the roles of cytokines, growth factors, enzymes, and signaling pathways that orchestrate tissue regeneration, offering a detailed scientific perspective on healing mechanisms.

Chapter 3: Wound Healing Demystified: Cellular and Molecular Insights

Focusing on cellular and molecular aspects, this chapter explores the contributions of fibroblasts, keratinocytes, macrophages, and endothelial cells in wound healing. It provides insights into molecular signaling networks and genetic regulation that influence tissue repair and regeneration.

Chapter 4: Phytoconstituents: Nature's Arsenal for Healing

This chapter presents an overview of plant-derived bioactive compounds and their therapeutic relevance in wound management. It discusses the diversity of phytoconstituents and their pharmacological properties.

Chapter 5: Polyphenols and Flavonoids: Antioxidants at the Forefront

Polyphenols and flavonoids are widely recognized for their antioxidant and anti-inflammatory activities. This chapter examines their mechanisms in reducing oxidative stress, promoting collagen synthesis, and enhancing angiogenesis, emphasizing their significance in accelerating wound healing.

Chapter 6: Terpenoids and Essential Oils: Aromatic Agents in Tissue Repair

This chapter explores the role of terpenoids and essential oils in wound care, focusing on their antimicrobial, anti-inflammatory, and tissue-regenerative properties.

Chapter 7: Tannins: Natural Astringents for Wound Closure

Tannins are known for their astringent and antimicrobial properties. This chapter highlights their ability to promote wound contraction, reduce exudation, and facilitate tissue repair through protein precipitation and antioxidant mechanisms.

Chapter 8: Healing Through the Ages: Traditional Herbal Remedies

This chapter traces the historical use of medicinal plants in wound management across various traditional systems of medicine. It emphasizes ethnopharmacological knowledge and its relevance in contemporary therapeutic research.

Chapter 9: Phytoconstituents: A Promising Avenue for Diabetic Foot Ulcer Management

This chapter focuses on the role of phytochemicals in managing diabetic foot ulcers. It discusses pathophysiology, therapeutic targets, and plant-based interventions for improving healing outcomes in diabetic patients.

Chapter 10: Wound Healing by Phytochemicals: Insights from Molecular Docking and Computational Analysis

The final chapter explores modern computational approaches in studying phytochemicals for wound healing. Molecular docking, in silico screening, and predictive modeling are discussed as tools for identifying potential therapeutic candidates and understanding their mechanisms of action.

CONCLUSION

Harnessing Phytoconstituents for Wound Healing represents a collective effort to integrate traditional herbal wisdom with contemporary scientific research. By bringing together contributions from experts across diverse disciplines and geographical regions, this volume provides a holistic understanding of phytoconstituents and their role in wound management.

The book underscores the importance of interdisciplinary collaboration in advancing wound care research and highlights the potential of plant-derived compounds in developing safe, effective, and accessible therapeutic solutions. It is our sincere hope that this volume will inspire further research, innovation, and clinical translation in the field of natural product-based wound healing.

We extend our heartfelt gratitude to all contributing authors for their scholarly efforts and to the publishers for their support in bringing this work to fruition. We also acknowledge the readers and researchers whose continued interest in this field drives scientific progress and innovation.

Geeta Deswal
Guru Gobind Singh College of Pharmacy, Yamuna Nagar, India

Ajmer Singh Grewal
Guru Gobind Singh College of Pharmacy, Yamuna Nagar, India

Kumar Guarve
Guru Gobind Singh College of Pharmacy, Yamuna Nagar, India

Section 1
Fundamentals of Wound Healing

Chapter 1
Understanding Wounds:
Types, Stages, and Challenges

Pranjal Sachan
http://orcid.org/0009-0004-8844-5940
Sanskriti College of Higher Education and Studies, Bhoganipur, India

ABSTRACT

The process of wound healing is also a very complex biological phenomenon which is necessary to reestablish the structure and functionality of tissue following an injury. The paper has given a detailed analysis of wounds dividing them into different types, such as acute and chronic wounds, and open and closed wounds. It further explores the four major phases of wound healing that include hemostasis, inflammation, proliferation and remodeling. The complex cellular and molecular processes that occur in each stage are in sync to heal the injured tissue. Outside the fundamental science, the discussion covers the challenges to wound care, which include infection treatment, the protracted healing of non-healing wounds, and issues that are associated with the systemic conditions, such as diabetes and immunodeficiency. A good understanding of these aspects is an essential part of designing efficient treatment strategies and expanding the limits of clinical results in wound care.

INTRODUCTION

Wounds, which can be defined as the breakages in the continuity of body tissues, are some of the most prevalent clinical manifestations that most people present with in medical facilities around the world. They can be caused by physical, thermal, chemical, microbial, or immunological injuries of the skin, and both have severe physiological and psychological effects on the subjects. Where most wounds have a predictable pattern of healing, some do not and instead, they end up being chronic,

DOI: 10.4018/979-8-3373-2908-6.ch001

resulting in prolonged suffering, disability and in extreme situations, causing life-threatening complications. Since they are common and affect the socioeconomic status of patients, it is vital to comprehend the principles of wound formation, its types, and the healing process, as well as deal with the challenges to enhance the quality of patient care and increase therapeutic interventions (Malone & Schultz, 2022).

The human skin is the biggest organ in the human body and plays a very important role of shielding the internal body tissues against external damages. It controls the heat and stops microbial entry and helps in the immune patrol. Violation of this barrier leads to the initiation of a cascade of cellular and molecular events in the process of repairing tissue integrity. Although such a process is usually effective, it depends on a large number of local and systemic factors, such as infection, comorbid diseases, nutritional status, age, and medications. When these factors come into play, the wounds may turn out to be chronic and their treatment may be complicated and expensive. Wound care has an astronomical economic cost.

In developed nations with high income, like the United States, chronic wounds are a problem that is impacting more than 6.5 million people per year, and the bill is estimated to be more than 25 billion a year. The situation is further exacerbated in lower and middle-income nations, where healthcare infrastructure might be reduced, subsequent diagnosis, and lack of care access as well as high rates of infection and amputation. This burden is a large number of chronic wounds including diabetic foot ulcers, pressure ulcers, and venous leg ulcers. They are connected to a long hospital stay, high readmission rates, and significant changes in quality of life (Harding, 2015).

The growth in the rates of non-healing wounds is because of the growing commonality of chronic diseases like diabetes mellitus, peripheral vascular disease, and obesity. As an illustration, the global incidence of diabetes is expected to be up to 643 million by 2030 (IDF Diabetes Atlas, 2021), and diabetic foot ulcers are predicted to strike up to 25 percent of patients with diabetes in their lifetime. Not only are these ulcers painful and debilitating, they are also linked with high risk of infection, sepsis and amputation of lower limbs. Complications of diabetic foot constitute the largest number of hospital admissions of any complication of diabetes and are a serious risk factor to mortality.

The necessity of efficient, affordable and available wound management measures is more pressing than it was previously. Although the traditional wound care strategies of debridement, dressings, antibiotics, and surgery are still the primary treatment options, they are not always effective in the context of multidrug resistance, comorbidities, and delayed healing. Moreover, excessive antibiotic use in wound treatment has led to the disastrous increase in antimicrobial resistance (AMR), thus, the world is looking for an alternative and adjunctive treatment that can enhance healing without accelerating this situation (Frykberg & Banks, 2015). The application of phytoconstituents, which is one of the potential areas of research, could be considered a potential source of therapeutic agents with antimicrobial, anti-inflammatory, antioxidant, and pro-healing effects that are derived using medicinal plants. The flavonoids, alkaloids, tannins, and terpenoids and saponins are phytochemicals which have been used in ancient systems of traditional medicine all over the world, such as Ayurveda, Traditional Chinese Medicine (TCM), and African ethnomedicine. Over the past few decades scientific studies have commenced in authenticating most of these traditional applications and have shown the molecular basis of the beneficial action of plant-derived compounds on wound healing (Ousey et al., 2018). As an example, flavonoids have been demonstrated to mediate pathways associated with inflammation, angiogenesis and collagen production, and terpenoids and alkaloids are also potent antimicrobials against pathogens that tend to cause wound infections. Another group of plant-derived compounds are polyphenols, which are very powerful antioxidants which alleviate the effect of oxidative stress, one of the key factors in slowing wound healing.

These results indicate that phytoconstituents can be used as useful complements or replacements to synthetic medications in wound treatment, especially in the context where other treatments are not available, or patients cannot afford them. Nonetheless, there are numerous obstacles that prevent the clinical implementation of phytochemicals into the mainstream wound care, among them being variation in plant composition, un-standardized formulations, regulatory limitations and scanty clinical trial evidence. In order to address these barriers, it is necessary to have a better insight into the pathophysiology of wounds and the challenges that hamper healing and affect that process. The knowledge can not only be used to design specific phytotherapeutic interventions but also to empower healthcare providers to do tailored treatment to a specific patient.

This chapter focuses on the principles of wound science, and the first part is about classifying wounds according to etiology, duration, depth, and extent of contamination. Then we get into the twists and turns of normal wound healing processes such as hemostasis, inflammation, proliferation, and remodelling as embedded in the complicated biological mechanisms. Knowledge of these phases is essential in the detection of deviations which define the chronic wounds and also in the development of interventions which may correct or offset the deviations. We look also into the myriad factors which hinder the successful healing of wounds, both the ones inside the patient and the ones outside the environment.

These are microbial infections, biofilm formation, poor vascularity, recurrent trauma and systemic complications like diabetes, malnutrition and immunosuppression. All of these may extend the inflammatory period, decrease tissue perfusion and impair repair cellular functions. More to the point, we emphasize the interplay between these challenges in the dynamic and, very often, self-perpetuating cycle that predisposes chronic wounds, rendering them so hard to treat. Lastly, we ponder over the application of this information on the phytoconstituent-based therapies. We discuss the existing studies on medicinal plants and their active compounds as wound-healing agents that have been demonstrated to have, examine the mechanisms through which they act and discuss the opportunities and challenges of using these agents in clinical practice. Thus, we will fill the gap between conventional knowledge and modern science to give evidence-based phytotherapy in wound treatment a framework.

This chapter is not just aimed at teaching, but also at encouraging more research and innovation in the area of wound management. With the increasing burden of chronic wounding in the XXI century, interdisciplinary mechanisms integrating the knowledge of molecular biology, pharmacognosy, clinical medicine, and population health will become necessary. The awareness of wound healing types, phases, and difficulties preconditions us the creation of more efficient, sustainable and patient-centered solutions by nature and technologies alike. To conclude, the science of wounds is an ancient and modern, practical and philosophical, clinical and scientific one. It addresses some of the deepest issues of human biology, health systems and philosophy of therapy Figure. 1 (Doughty et al., 2006).

The recent demand to renew the interest in phytoconstituents presents a window of opportunity to re-discover ancient remedies in the light of a contemporary research and create new formulations that are valid at the same time both scientifically and culturally sustainable. In that regard, this chapter is one that is timely and indispensable to the area of integrative wound care, which is on the rise. In the future, several key trends will probably determine the future of wound care. To begin with, accurate diagnostics such as verified biomarker panel and point-of-care molecular tests will enable earlier wound stratification as healing-competent or healing-impaired wounds, which would enable clinicians to implement more specific interventions than use standardized protocols.

Second, biomaterial innovations, including bioengineered scaffolds, cells in matrices, and controlled-release therapeutic delivery systems, will be useful in improving the processes of repair through the delivery of specific growth factors or antimicrobial agents in the microenvironment of the wound. Third, regenerative treatments (e.g., stem-cell-based therapy, extracellular vesicles, and immunomodulators) promise to revert dysfunctional healing repercussions, particularly in non-responsive, chronic wounds. Moreover, the digital health technologies incorporation such as wound imaging, AI-powered wound assessment application, wearable biosensors, and telemedicine platforms are likely to enhance the remote monitoring process, timely complication detection, and prompt changes to the therapeutic plan. Application of omics-based personalized medicine in which genetic, proteomic, and metabolomic analyses are used as determinants of personalized care pathways is likely to change prognostication and therapy selection.

Figure 1. Four-year wound care strategy with emphasis on accuracy diagnostics, biomaterials, regenerative therapy, electronic monitoring, and patient-centred care.

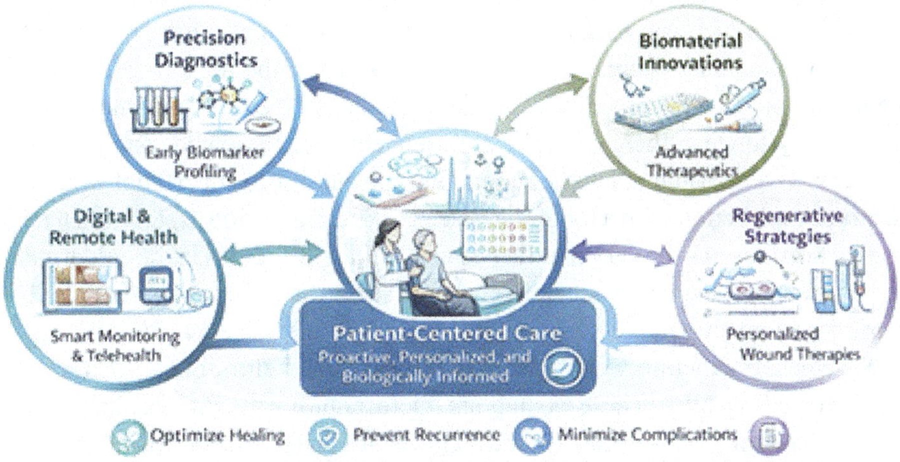

CLASSIFICATION OF WOUNDS

The classification of wounds forms a basis of clinical practice and research and forms a methodical basis of identifying the nature and severity of the wound. Correct classification will enable correct treatment planning and prognostication as well as exchange of information among health care providers. It is also essential in the design of the clinical trials, policymaking, and design of the advanced therapeutic regimens such as the use of phytoconstituent-based treatment. Segmentation of wounds may be done through a number of criteria including etiology, duration, depth, level of contamination and tissue involvement.

Both systems of classification present individual information on the wound environment and healing potential. Classification according to Etiology. Etiology is a term used to describe the cause of the wound and it is one of the chief factors which are taken into account in the process of initial clinical examination. There are broadly categorized wounds which include mechanical/traumatic, surgical, pathological, burn-related, and iatrogenic (Mahmoudi & Gould, 2020). Traumatic Wounds Traumatic wounds are a heterogeneous category of injuries caused by mechanical forces applied abruptly on the body, the cause of which is usually an accident, blunt collision, sharp objects, or external objects. Such injuries are not planned and they usually happen in emergency situations like road accidents, workplace accidents, sport injuries, falls and violent cases. The nature, severity and direction of the external force may create a traumatic wound that may be as simple as superficial abrasion or as severe and life threatening as a disruption in the tissue.

A jagged or irregular tear in the skin and the underlying tissues, made by a sharp object, which can be glass, metal, or tool or by blunt injury, overstretching the skin to the point of breaking it, is known as one of the most frequent types as the laceration. Lacerations can be excessive and can be easily contaminated, so it is necessary to clean them carefully and frequently stitch, which can ensure good recovery. The abrasion is another common type and it is a superficial wound that occurs due to friction with a rough surface e.g. skin scraping on asphalt when taking a fall. Abrasions may be very painful because the nerve endings in the epidermis are very thick and, though they are unlikely to cause excessive bleeding, they are likely to trigger an infection unless cleansed adequately (LeBlanc et al., 2025).

Severer are puncture wounds, caused by sharp and slim things such as nails, needles or animal bites. These wounds are microscopic on the surface although their sub-surface ramifications may extend significantly in the tissues posing the risk of anaerobic bacterial infection, such as Clostridium tetani (tetanus) and Pseudomonas species, because the wound channel is not sufficiently aerated. Puncture wounds are usually underestimated due to depth and lack of visibility and may not be given proper debridement or prophylaxis. The most destructive type of traumatic wound is the avulsion whereby a part of the skin and the tissue behind are forcefully removed off the body. Avulsions are normally the outcomes of high impact accidents like explosions, entrapments of machinery, or animal attacks, and may include muscular, bone, and vascular tissue. The wounds are generally characterized by a large tissue loss, a disrupted blood supply and a high rate of necrosis and infection.

To replace the form and function, surgery, grafting or reconstructive surgeries are frequently required. Together, traumatic wounds can be challenging to manage clinically because they are unpredictable and may be contaminated, besides the fact that the involvement of tissues may be different. Timely evaluation, good cleaning, proper antimicrobial cover, and close observation are critical elements of good traumatic wound care. Their control is even further complicated in resource constrained environments or where patients possess underlying chronic health conditions which inhibit wound healing as is the case of diabetes or immunity suppression (Whittam et al., 2016). It is important to understand the comparatively unique forms and processes of traumatic wounds in order to customize response during emergencies and subsequent long-term treatment interventions, including the possibility of using phytoconstituent-based interventions with antimicrobial, anti-inflammatory, and tissue-repair properties.

Surgical Wounds

Surgical wounds are willful incisions or excisions that are made through a scalpel or surgical tool during an operation under a sterile environment. Contrary to traumatic wounds, surgical wounds are well thought out, done with precision and usually handled under controlled environments in order to reduce the chances of infection, and to give the best healing experience. They constitute an essential element of contemporary medicine, which is required in various types of surgeries, starting with small dermatological removals and ending with organ transplants and complicated reconstructions.

Surgical wounds may be of vastly different sizes, depths, anatomical sites, and uses. Others are shallow skin cuts and others can be deep cuts through skin and may include a muscle, fascia, bone or even internal organs. These wounds are generally categorized according to the extent of contamination into four types namely clean, clean-contaminated, contaminated and dirty-infected with the extent of postoperative infection risk being higher in the progressive categories. The clean wounds are made in a sterile state and never getting access to any organ system, which is harbouring microbes, like the respiratory or the gastrointestinal tracts.

These have the least infection rate and normally healed by primary intention where pericutaneous edges are estimated using sutures or staples to allow quick and low scarring of the wound. Clean-contaminated wounds are those that are entered into a colonized space (e.g. bowel surgery, bladder surgery) under the controlled circumstances and without any extra contamination, whereas contaminated wounds are those where there is a significant violation of aseptic technique or where there is a fresh spillage of bowel or gastrointestinal contents. Infected or dirty wounds can be either current or traumatic wounds not managed within a certain period of time, and at risk of developing complications the most.

Much relies on the general health of the patient, nutritional condition, comorbid conditions (e.g., diabetes mellitus, immunosuppressive therapy or peripheral vascular disease), competence and technique of the surgeon, and postoperative wound care practice in terms of healing surgical wounds. Surgical wounds are not resistant to development of complications (wound dehiscence or reopening of the wound), surgical site infections (SSIs), hematoma, seroma, and hypertrophic scars and keloids despite the fact that their healing process is relatively predictable (Rodrigues et al., 2018). There are also possible risks of incisional hernia especially in the abdominal surgery where the tissue pushes through the weakened surgical area. Prophylactic antibiotics, use of the right suture materials, sterile dressing, and postoperative wound monitoring are some of the critical aspects used to reduce these risks.

Other surgical wound closure measures like the absorbable sutures, tissue adhesives, negative pressure wound therapy, and the incorporation of the antimicrobial-impregnated dressing have also enhanced the healing outcomes. Interest in the application of phytoconstituent-based therapies in surgical wound care has increased over the last few years. A number of plant-derived compounds such as flavonoids, alkaloids, tannins and terpenoids have demonstrated good results in the improvement of collagen production, inflammation reduction, angiogenesis and antimicrobial effects, all of which are essential in surgical wound repair.

Such natural agents may be applied as wound dressings, gels, sprays, and they provide an alternative to traditional post-surgical care, particularly in patients who experience a slow rate of wound healing, or in those areas where synthetic drugs are not readily available. With the increasing need of a safer, cost-effective, and bioactive wound healing agent, the phytoconstituents provide a potential research and clinical implementation area. The knowledge of the nature, types, and risks of surgical wound enable clinicians to make decisions on both traditional and integrative approaches to treatment with the aim of achieving improved patient outcomes and mitigating the global impact of surgical wound complications.

Pathological Wounds

Pathological wounds are wounds that are non-healing or chronic wounds and they occur when there is underlying processes of systemic or local disease which often is manifestation of a disturbance of the normal body processes of preserving the integrity of tissues and/or starting normal healing processes. Various to traumatic or surgical wounds, which are usually triggered by some identifiable external injury or intervention, pathological wounds develop in an insidious way and are highly associated with chronic diseases like diabetes mellitus, vascular problems, auto-immune disorders, malignancies, or neurological disabilities.

Such wounds may be challenging to treat, because they occur in a complex pathologic environment that is high in impaired perfusion, poor cellular response, chronic inflammation, and microbial colonization, and systemic metabolic dysfunction. The diabetic foot ulcer (DFU) is one of the best-known examples of pathological wounds and is a debilitating and recurring wound, which is caused by the direct combination of peripheral neuropathy, poor glycemic control, and peripheral arterial disease. DFU patients lose sensation in feet hence they are oblivious to simple injuries that develop into deep-infected ulcers (Liu et al., 2025). Inhibition in the youthful healing of diabetic wounds occurs through impaired leukocyte functioning, decreased angiogenesis, diminished collagen production, which requires long-term medical treatment and can result in surgical debridement or amputation, in the worst case scenario.

Likewise, the other significant subtypes of pathological wounds are the venous leg ulcers (VLUs) and arterial ulcers, which are related to chronic vascular insufficiency.

VLUs are caused by venous hypertension caused by faulty valves or the blockage of the veins resulting in tissue edema, hypoxia, and skin breakdown, especially in the area around the medial malleolus of the lower leg. Lipodermatosclerosis, hemodidion deposition, and eczema usually accompany these ulcers and in many cases they recur very often unless the underlying venous pathology is repaired. Arterial ulcer on the other hand is a result of poor blood flow either through atherosclerosis or thrombus and is usually accompanied by a punched-out appearance and severe pain, especially in the distal extremities, such as toes or heels.

Poor perfusion causes the delay of healing, and even simple injuries may quickly degenerate to necrosis and gangrene without revascularization. Pressure ulcers (synonymous: decubitus ulcers, bedsores) are a different form of pathological wound, and they develop as a result of prolonged pressure on the skin and tissue underlying it, in most cases, around bony prominences, such as the sacrum, heels, and hips. They occur frequently in non-ambulant, bedridden, or impaired patients who are also compounded by incontinence, malnutrition, and a lack of sensation. The pathogenesis includes localized ischemia, tissue anoxia, inflammation, cell death, and stages of pathogenesis ranging between erythema and complete loss of tissue tissue including muscle and bone.

To prevent complications, it is necessary to detect and intervene early with the help of pressure relief strategies, nutritional help, and wound care. Besides these, cancerous wounds, particularly when they are at an advanced stage of the disease, soft tissue sarcomas or metastasized tumors are all considered pathological wounds. Inflammation in the tissue, tissue necrosis, bleeding, and malodor characterize these lesions, which may cause severe distress to the patient and create problems in the palliative care of the wound. Vasculitis, pyoderma gangrenosum, as well as lupus ulcers might also lead to chronic and non-healing wounds highly resistant to traditional treatment and controlled by immune-modulating agents.

The pathology of the underlying wounds involves improper activation of the immune system, release of inflammatory cytokines and a general destruction of the tissue, which provides an unfavorable environment to promote the healing process of the tissue. Treatment of pathological wounds is a multidisciplinary event that involves treatment of the wound and the underlying medical condition (Manisha et al., 2025). Debridement, infection control, advanced wound dressings, offloading, revascularization procedures, and systemic disease management are also all aspects of care. In this regard, phytoconstituents have recently been receiving more attention on their capability of controlling wound healing pathways.

Curcumin, resveratrol, aloe vera polysaccharides and Centella asiatica triterpenoids are examples of compounds that have anti-inflammatory, antioxidant and pro-angiogenic effects that have the potential to complement traditional treatments, particularly among patients with chronic and recurring pathological wounds. These natural agents can potentially help to decrease oxidative stress, stimulate fibroblast activity, and recover microvascular activity which is potentially promising where the synthetic use of pharmacological therapy has not been successful. As they are complex in their derivation and occur in large rates of frequency, pathological wounds continue to be one of the most difficult types of wounds to treat and of significant concern to future research and therapeutic development in wound care, especially the phytotherapeutic area.

Burns and Radiation Wounds Burns and radiation wounds are also serious types of tissue damage, whose pathophysiology is complex, and the damage can be both deep and long-lasting. Causal factors Exposure to thermal (heat or flame), chemical (acidic or alkaline substances), or electrical causes are usually the cause of burn injuries and they are classified by depth into first-degree (superficial), second-degree (partial-thickness), and third-degree (full-thickness) burns. The most frequent are the thermal burns, they happen when there is contact with hot liquids (scalds), steam, flames, or hot objects, and cause lesions in the skin, and in a more severe case, in the underlying tissue, such as muscle and bone.

Deep tissue necrosis may develop after brief exposure to corrosive substances, which may cause further tissue necrosis even when the chemical is neutralized (chemical burns). Electric burns are singularly hazardous because they may result in internal, profound tissue harm and arrhythmia of the heart even though the actual surface damage may be minor. The pathways of healing of burn wounds are anchored on the area of the burn (calculated by TBSA) and the depth of the involvement. Superficial burns can be treated by simple treatment yet deep or large-scale burns can be treated and necessitate surgical excision, grafting and rehabilitation. The complications that do occur include fluid loss, infection, hypertrophic scarring and contractures which are life threatening in extreme cases (Noor et al., 2017).

On the contrary, radiation wounds are caused by the exposure to ionizing radiation applied in medical practice or oil spills. These injuries occur as radiation destroys the DNA and cell architecture of the skin and deeper tissues resulting into delayed cell death, vascular constriction, fibrosis, and ineffective wound healing. In contrast to thermal burns, radiation injury may have delayed symptoms such as erythema, desquamation, ulceration and necrosis, and such injuries may take months or even years out of time to develop. The process of healing is usually long-term and erratic, as it is marked with continuous inflammation, vascular injury, and the failure of cells to regenerate.

The radiation wounds are also prone to secondary infections and may lead to radiation induced malignancy of the skin like squamous cell carcinoma in chronically damaged skin. Effective management will need advanced wound care involving hyperbaric oxygen therapy, topical growth factors and skin grafting. The newer studies have shown promise of phytoconstituent-based therapies in the treatment of burn and radiation wounds. Anti-inflammatory, antioxidant and antimicrobial, natural products, including aloe vera, calendula, turmeric and honey, were demonstrated to enhance the healing process of burn wounds in terms of epithelialization, decreasing bacterial colonization, and increasing collagen remodelling. Compounds such as curcumin, resveratrol and centella extracts have been shown to decrease oxidative DNA damage, stimulate vascular repair and fibrosis in radiation injuries.

Introduction of these natural agents in topical use, hydrogels and dressing is a developing edge of the burn and radiation wound care, particularly in low resource environments where advanced medical care is scarce. On the whole, burns and radiation wounds necessitate a complex and multi-modal healing process that entails the use of instant injury reduction, infection mitigation, pain management, and long-term functional and aesthetic recovery. Iatrogenic Wounds Iatrogenic wounds are wounds, which are the result of medical or surgical intervention as collateral damage.

Such types of wounds are of a paradoxical nature since they are obtained as a result of a procedure meant to heal or diagnose the condition of a patient. They may occur in many forms such as pressure injuries due to immobility over a long period of time or insufficient padding during surgery, extravasation injuries due to intravenous (IV) therapy, skin tears during dressing changes, radiation burns due to therapy or injuries caused by antiseptics or chemotherapeutic extravasation. Among the most widespread examples, one should find pressure ulcers, which are formed during the process of surgery because of a prolonged period of operation without proper repositioning or support. The other prevalent type is extravasation injuries, in which the IV-introduced fluids, including chemotherapeutic agents or hyperosmolar solutions, leak out into the surrounding tissues to cause chemical burns, necrosis, or compartment syndrome (Fui et al., 2019).

Even though people tend to consider iatrogenic wounds as easily preventable, they bring a huge burden to the healthcare systems because of their chronicity, infection, and medicolegal possibilities. Some of the causes of these wounds include; poor technique, inadequate training, inadequate monitoring and institutional protocols. Timely wound healing is reliant on the timely diagnosis, the timely identification and timely interventions in iatrogenic wounds. Regrettably, these injuries are underreported particularly in minor injuries or initially asymptomatic and the result can be further complications, delayed healing or even litigation when the care is not taken promptly.

There is also the interest of using phytochemicals in the management of iatrogenic wounds as a safer and supportive method of treatment. Topical use of herbal preparations may help decrease the inflammation, regenerate epithelium, and cover the microbes. An example of this is the use of plant-based antiseptics and antimicrobial oils including tea tree oils and neem as a complementary procedure in the treatment of IV-related infection or minor burns. Others with strong antioxidant activity include flavonoids and polyphenols, which could be able to counter the oxidative stress associated with extravasation-generated tissue damage or radiation-generated wounds. Guidelines such as incorporation of such natural agents in the prevention and treatment of iatrogenic injuries could help to improve patient safety particularly in settings where conventional interventions might be constrained.

Physiology and Wound Stages Healing. Healing of wounds is a complicated yet systematic procedure that occurs in a number of distinct phases. The first one is the Hemostasis Phase, which is a direct reaction of the body to the injury. The blood vessels tighten, and the clotting process starts in this stage, preventing bleeding fast. The platelets accumulate in the place of injury and create a clot to cover the injury and release messages that draw the attention of other healing cells. This is followed by the Inflammatory Phase during which the body sends the white blood cells to the wound site to eliminate any infection and destroy dead or damaged tissue. This is necessary to clean up the wound but causes swelling, redness, heat and pain. Once inflamed, the body enters into the Proliferative Phase where new tissue begins to develop. At this phase, cells are engaged in developing new blood vessels (angiogenesis), collagen, and skin cells, which envelop the wound.

This is the period whereby the wound begins to close physically. The next is Remodeling or Maturation Phase which may take weeks to months. In this case, the new tissue develops greater power and structure because the collagen is rearranged, and extra blood vessels are eliminated. All these phases can be affected by a number of Factors Modulating Each Phase, including age, nutrition, oxygen level, infection, underlying diseases (including diabetes, medications, stress), etc. Interruption and Failure of Healing Phases may occur when any of these factors disrupt the normal healing process resulting in wounds needing a prolonged time to heal or in an inappropriate manner Figure. 2 (Kumar et al., 2024).

This explains why proper care and attention should be taken at all the stages of wound healing to guarantee complete healing. Systemic and local factors, including age, nutrition, oxygen supply, infection, diabetes, ischemia, and mechanical stress, have a strong impact on the efficiency of this physiological sequence by slowing down the healing process and causing chronic wounds. Thus, it is necessary to comprehend physiological processes as well as sequential phases of wound healing, both to clinically manage them and to design medicines that would improve tissue re-epithelialization and healing Figure. 3.

Figure 2. How wounds are caused and initiated at the physiological level.

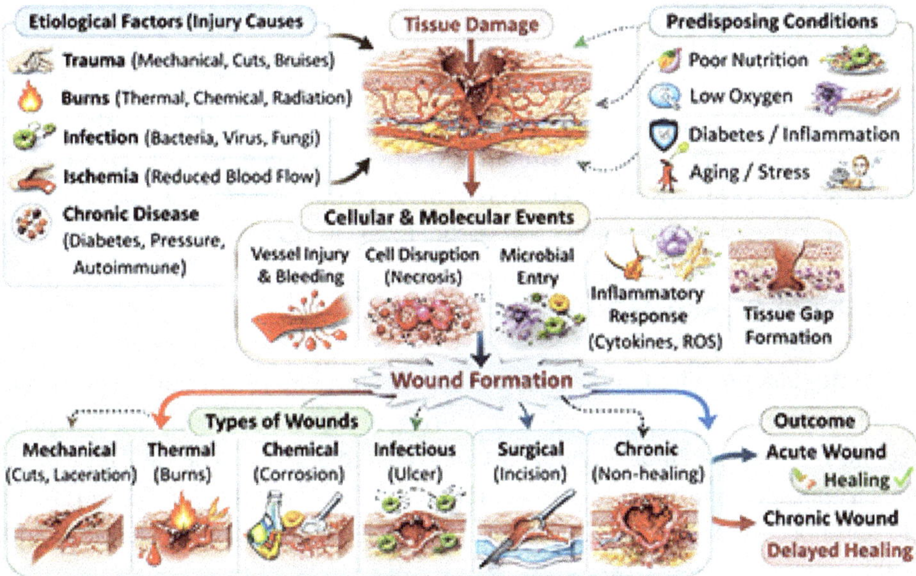

Hemostasis Phase

The hemostasis phase is the first step in the wound healing process. It begins immediately after an injury occurs. The main goal of this phase is to stop the bleeding. Blood vessels constrict to reduce blood flow, and platelets rush to the injury site to form a blood clot. This clot not only helps seal the wound but also releases chemical signals that attract other cells needed for healing. This phase is quick and usually lasts a few minutes to a few hours.

Inflammatory Phase

The inflammatory phase follows hemostasis and is the body's natural response to injury. It usually lasts from 1 to 5 days. During this phase, white blood cells such as neutrophils and macrophages move into the wound site (Porel et al., 2025). They help fight infection, clear out dead cells, and prepare the area for new tissue to grow. This phase is often marked by redness, heat, swelling, and pain, which are normal signs that the immune system is active.

Proliferative Phase

In this phase, the body starts to rebuild the wound with new tissue. It usually begins a few days after the injury and can last for several weeks. Cells such as fibroblasts produce collagen, a protein that gives the skin strength and structure (Rautela et al., 2024). New blood vessels form through a process called angiogenesis, and the wound starts to close as new skin cells grow over the area. This is also the phase where granulation tissue, a pink, soft tissue, forms to fill in the wound.

Remodeling (Maturation) Phase

The remodeling phase is the final stage of wound healing and can last for several months, sometimes even up to a year. During this time, the newly formed tissue becomes stronger and more organized. Collagen fibers are realigned and remodeled, and any unnecessary blood vessels are removed. The scar tissue formed during this phase is not as strong as the original skin but helps restore function and protect the wound area.

Figure 3. Overview of the normal wound-healing process from clot formation to final tissue remodeling.

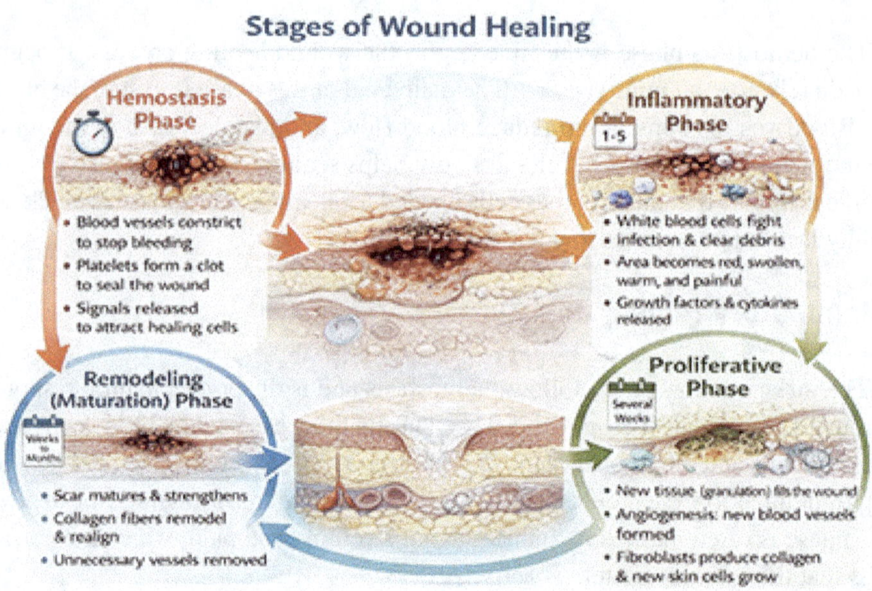

FACTORS MODULATING EACH PHASE

Several internal and external factors can affect how well and how fast each healing phase occurs. These include the patient's age, overall health, nutrition, and hydration status. Conditions like diabetes, infections, and poor blood circulation can delay healing. Certain medications, such as steroids and chemotherapy, may also interfere with the healing process. Proper wound care, oxygen supply, and cleanliness are essential to support every stage of healing.

Interruption and Failure of Healing Phases

When one or more phases of wound healing are delayed or disrupted, the wound may not heal properly. This can happen due to infections, repeated trauma to the wound site, chronic diseases (like diabetes), or poor immune function. A wound that fails to heal within the normal time frame may become chronic and require medical attention. Failure in healing can also result in excessive scarring, poor tissue strength, or an open wound that becomes more prone to further complications (Ko et al., 2021).

Physiology of Wound Healing

Wound healing is a complex and highly orchestrated physiological process that restores the integrity of damaged tissue following injury. It involves a cascade of cellular and molecular events aimed at halting bleeding, clearing pathogens and debris, repairing tissue architecture, and re-establishing functional capacity. The process is generally described in four overlapping but distinct phases: hemostasis, inflammation, proliferation, and remodeling (or maturation). Each phase engages multiple cell types, signaling pathways, and bioactive molecules including cytokines, chemokines, growth factors, and enzymes all functioning in a synchronized manner to achieve successful tissue repair. The efficiency of these stages depends on both systemic and local factors such as age, nutrition, infection, ischemia, and chronic diseases, which can delay or impair healing. Importantly, this physiological sequence also represents a therapeutic window for the application of phytoconstituents, which may modulate inflammation, stimulate fibroblast activity, promote angiogenesis, or act as antioxidants to neutralize excessive reactive oxygen species (ROS).

The first phase, hemostasis, begins immediately following injury to limit blood loss and establish a temporary matrix for cell migration. Vascular injury leads to the rapid constriction of blood vessels (vasoconstriction), platelet aggregation, and activation of the coagulation cascade. Platelets adhere to the exposed subendothelial matrix, primarily via von Willebrand factor, and release a host of granules containing adenosine diphosphate (ADP), serotonin, thromboxane A2, and platelet-derived growth factor (PDGF). These substances promote further platelet aggregation, stimulate local vasoconstriction, and recruit inflammatory cells. Simultaneously, the coagulation cascade generates fibrin from fibrinogen, forming a stable clot that serves not only as a hemostatic plug but also as a provisional extracellular matrix (ECM) for migrating keratinocytes and inflammatory cells. The clot itself is rich in signaling molecules like transforming growth factor-beta (TGF-β) and epidermal growth factor (EGF), which prime the tissue for the next phase (Ko et al., 2021).

Following hemostasis, the inflammatory phase is triggered, typically within hours of injury, and may last for several days. It is a critical step that orchestrates immune surveillance, microbial clearance, and the removal of cellular and extracellular debris. Neutrophils are the first immune cells to infiltrate the wound site, responding to chemotactic gradients created by damaged tissue and activated platelets. They produce proteolytic enzymes such as elastase and matrix metalloproteinases (MMPs), as well as reactive oxygen species (ROS), which are essential for microbial killing but can also damage host tissues if uncontrolled. Neutrophils also secrete cytokines such as interleukin-1β (IL-1β) and tumor necrosis factor-alpha (TNF-α), which amplify the inflammatory response and recruit additional immune cells, including monocytes and lymphocytes. As neutrophils undergo apoptosis, monocytes infiltrate the wound and differentiate into macrophages a pivotal cell type in wound healing. Macrophages exhibit remarkable plasticity, transitioning from a pro-inflammatory (M1) phenotype in early stages to an anti-inflammatory and tissue-repairing (M2) phenotype in later stages. M2 macrophages secrete a diverse array of growth factors such as vascular endothelial growth factor (VEGF), insulin-like growth factor-1 (IGF-1), and TGF-β, which not only suppress inflammation but also initiate the proliferative phase. Importantly, macrophages are involved in angiogenesis, ECM remodeling, and the activation of fibroblasts hallmark processes of the subsequent healing phase.

The proliferative phase typically begins within 3–5 days after injury and is characterized by the formation of granulation tissue, re-epithelialization, neovascularization, and the deposition of new ECM. Fibroblasts, attracted by PDGF and TGF-β, migrate into the wound and begin synthesizing collagen (initially type III), fibronectin, and other matrix proteins essential for structural integrity. Fibroblasts also contribute to wound contraction by differentiating into myofibroblasts, which express α-smooth muscle actin (α-SMA) and exert contractile forces that bring wound edges closer together. Simultaneously, endothelial cells respond to angiogenic signals such as VEGF and basic fibroblast growth factor (bFGF), proliferating and forming new capillary loops through a process called sprouting angiogenesis. This neovascular network ensures adequate oxygen and nutrient supply to the regenerating tissue. Keratinocytes from the wound margins proliferate and migrate across the wound bed, covering the wound through a process known as re-epithelialization (Dwivedi et al., 2024). This process is tightly regulated by signals such as EGF, keratinocyte growth factor (KGF), and TGF-α. The new epithelium is initially immature and lacks full barrier function, but as the phase progresses, differentiation and stratification restore its protective properties. Granulation tissue, the hallmark of this phase, is characterized histologically by abundant capillaries, proliferating fibroblasts, and loose connective tissue, giving it a red, granular appearance. Importantly, this tissue is metabolically active and susceptible to external insults, requiring protection from mechanical disruption and infection.

The final stage, remodeling or maturation, may span weeks to months depending on the wound size and patient-specific factors. During this phase, the newly formed ECM undergoes remodeling to restore tensile strength and functional integrity. Collagen III is gradually replaced by collagen I, which is more resistant to mechanical stress and contributes to the tensile strength of healed tissue.

The activity of MMPs and their endogenous inhibitors, tissue inhibitors of metalloproteinases (TIMPs), becomes critical in regulating ECM turnover and preventing pathological scarring or fibrosis. Myofibroblasts play a central role during this stage, continuing to contract the wound and depositing ECM components before undergoing apoptosis as healing concludes. Neovascularization also regresses, with many capillaries undergoing apoptosis and the vasculature returning to baseline. The balance between collagen synthesis and degradation is essential in determining the final scar architecture. In ideal circumstances, the healed tissue achieves up to 80% of its original tensile strength, although structural differences in the ECM and reduced vascularity often render scar tissue more vulnerable to reinjury than uninjured skin.

Numerous systemic and local factors can affect the physiology of wound healing. Systemically, conditions such as diabetes mellitus, malnutrition, anemia, and immunosuppression can delay healing or predispose wounds to infection. For example, hyperglycemia in diabetes impairs neutrophil function, reduces nitric oxide availability for vasodilation, and promotes the formation of advanced glycation end-products (AGEs), all of which hinder effective tissue repair. Locally, factors such as wound infection, ischemia, excessive exudate, or repeated mechanical stress can perpetuate inflammation and prevent progression to proliferative and remodeling phases. Chronic wounds are often arrested in the inflammatory phase due to a persistently hostile microenvironment characterized by elevated protease activity, low oxygen tension, and high microbial burden. Such wounds benefit from advanced interventions such as debridement, negative pressure therapy, bioengineered skin substitutes, and increasingly, bioactive phytoconstituents that target specific phases of healing.

Phytoconstituents bioactive compounds derived from medicinal plants can significantly influence the wound healing cascade at various stages. For instance, polyphenols such as quercetin and curcumin possess potent anti-inflammatory and antioxidant properties that may help resolve chronic inflammation by downregulating NF-κB signaling, scavenging ROS, and inhibiting pro-inflammatory cytokines. Alkaloids and flavonoids can stimulate fibroblast proliferation and collagen synthesis, while terpenoids and tannins exhibit antimicrobial effects that reduce the risk of infection and support re-epithelialization.

Certain plant extracts have been shown to enhance angiogenesis, modulate macrophage polarization, and upregulate the expression of growth factors like VEGF and TGF-β. These properties suggest that phytoconstituents may serve not only as supportive agents but also as primary therapeutics in the management of acute and chronic wounds. Importantly, these agents often exhibit low cytotoxicity, high biocompatibility, and multifactorial mechanisms of action, making them ideal candidates for integrated wound healing strategies. Traditional medicines across cultures such as *Centella asiatica*, *Aloe vera*, *Calendula officinalis*, *Azadirachta indica*, and *Curcuma longa* have long been used in wound care, and modern pharmacological investigations continue to validate their efficacy and safety. Incorporating such phytochemicals into topical formulations, hydrogels, wound dressings, or nanocarriers offers promising avenues for clinical translation (Dwivedi et al., 2025).

In conclusion, the physiology of wound healing represents a dynamic interplay of hemostatic, inflammatory, proliferative, and remodeling processes that are tightly regulated by an array of cellular and molecular mediators. A comprehensive understanding of these mechanisms is critical for developing effective therapeutic strategies, especially in the context of chronic wounds or patients with comorbidities. The application of phytoconstituents presents a biologically plausible and therapeutically promising approach to enhance healing by targeting multiple pathways simultaneously. By leveraging the physiological principles of wound repair, researchers and clinicians can design more effective, holistic, and personalized interventions that improve outcomes, reduce complications, and restore tissue function with minimal scarring.

Factors Influencing Wound Healing

Wound healing is a dynamic biological process influenced by a variety of internal and external factors that can either accelerate or hinder tissue regeneration. Understanding these influencing factors is essential in tailoring wound care strategies, especially when integrating phytoconstituent-based therapies. These factors can be broadly classified into systemic (affecting the entire body) and local (specific to the wound site), with several behavioral and environmental contributors also playing a pivotal role (Dwivedi et al., 2024). Each factor has the potential to disrupt the tightly regulated phases of wound healing hemostasis, inflammation, proliferation, and remodeling leading to delayed healing, chronic wounds, or excessive scarring. Below is a comprehensive exploration of the multifaceted influences on wound healing.

Systemic Factors Affecting Wound Healing: The Whole-Body Impact

Systemic factors originate from within the patient and influence healing across all body tissues. Age is one of the most fundamental determinants, with older individuals typically experiencing slower wound healing due to diminished immune responses, reduced collagen synthesis, and lower angiogenic potential. In aging skin, fibroblast activity is diminished, epithelial cell turnover is slower, and stem cell reserves are reduced, all contributing to delayed granulation and epithelialization. Similarly, nutritional status plays a vital role; deficiencies in proteins, vitamins (particularly A, C, and E), and minerals such as zinc and iron compromise the immune response, collagen formation, and enzymatic antioxidant defenses. Malnourished patients often exhibit thin granulation tissue and poor tensile strength in the healing scar (Dwivedi et al., 2025).

Chronic diseases are another major class of systemic factors. Diabetes mellitus is perhaps the most impactful, associated with impaired leukocyte chemotaxis and phagocytosis, increased susceptibility to infection, and poor angiogenesis due to endothelial dysfunction. Additionally, advanced glycation end products (AGEs) and oxidative stress in diabetes alter ECM dynamics and increase inflammation, often arresting wounds in the inflammatory phase. Cardiovascular diseases also impact healing by reducing tissue perfusion and oxygenation, while renal failure and hepatic dysfunction affect metabolic detoxification and nutrient utilization, further complicating wound repair (Cotterell et al., 2024).

Immunosuppression, whether due to diseases such as HIV/AIDS or medications like corticosteroids and chemotherapy agents, diminishes the immune system's ability to clear pathogens and regulate inflammation. Moreover, genetic disorders such as Ehlers-Danlos syndrome and epidermolysis bullosa directly interfere with collagen integrity and skin barrier function, leading to fragile skin and chronic wound formation. These systemic influences underscore the need for a holistic approach to wound management that addresses the patient's overall health, not just the wound site.

Local Factors at the Wound Site: The Immediate Environment

Local factors pertain specifically to the wound environment and have a direct effect on healing progression. One of the most critical local factors is tissue oxygenation. Oxygen is essential for ATP production, fibroblast function, collagen crosslinking, and the respiratory burst used by neutrophils and macrophages to destroy pathogens. Hypoxic wounds exhibit impaired collagen synthesis, poor neovascularization, and increased infection risk. Wounds with ischemia often due to pressure (as in pressure ulcers), vascular disease, or trauma tend to heal slowly or not at all. This makes oxygen therapy and pressure offloading crucial elements in localized wound care.

Another major local factor is infection, which perpetuates inflammation and prevents the progression to the proliferative phase. Pathogens such as *Staphylococcus aureus*, *Pseudomonas aeruginosa*, and *Escherichia coli* not only elicit intense immune responses but also produce toxins and biofilms that protect them from both the host immune system and antimicrobial agents. Biofilms are particularly challenging as they create a hostile, protease-rich microenvironment that degrades ECM and inhibits growth factor signaling. Frequent debridement, antimicrobial dressings, and phytochemical-based antimicrobials are effective in counteracting local infections (Xiong et al., 2025).

Mechanical stress and wound tension also influence healing. Excessive movement, especially in joints, can reopen partially healed wounds or prevent epithelial cell migration. Similarly, foreign bodies such as sutures, debris, or necrotic tissue act as niduses for infection and inflammation, delaying repair. Moisture balance is another key aspect; a dry wound bed desiccates cells and inhibits migration, while excess exudate can macerate surrounding tissue. Modern moist wound healing techniques aim to maintain an optimal level of hydration, allowing epithelial and immune cells to function efficiently. Additionally, pH levels and temperature of the wound bed modulate enzyme activity and cell viability; slightly acidic environments and normothermic temperatures are generally conducive to healing (Percival, 2002).

Behavioral and Lifestyle Factors: Personal Choices with Biological Consequences

Patient behaviors and lifestyle choices can significantly influence the wound healing process. Smoking is one of the most detrimental habits, as nicotine causes vasoconstriction and carbon monoxide reduces oxygen-carrying capacity, resulting in local hypoxia. Smoking also impairs leukocyte function and fibroblast activity, severely limiting tissue regeneration and increasing infection risks. Studies have shown that even temporary cessation of smoking prior to surgery can significantly improve healing outcomes.

Alcohol consumption similarly suppresses immune function and reduces nutrient absorption, particularly of vitamins critical to wound healing like vitamin C and zinc. Chronic alcoholism is associated with liver damage and coagulopathy, further complicating tissue repair. Physical activity, when appropriate, can enhance wound healing by improving circulation and metabolic activity. However, excessive or inappropriate movement especially post-surgery can cause mechanical stress and dehiscence (Lazarus et al., 1994).

Psychological stress and mental health conditions are increasingly recognized as important modulators of wound healing. Chronic stress elevates cortisol levels, which in turn suppress immune responses, reduce fibroblast proliferation, and impair angiogenesis. Patients with depression or anxiety may also be less compliant with wound care regimens, nutritional recommendations, or medical appointments, further complicating their recovery. Thus, wound healing should be approached as a biopsychosocial issue, and interventions should incorporate psychological support where necessary (Powers et al., 2016).

Pharmacological Interference and Iatrogenic Factors: Medications That Impede Healing

Several commonly used medications can interfere with normal wound healing pathways. Corticosteroids, for instance, are potent anti-inflammatory agents but can suppress immune function, inhibit fibroblast proliferation, and reduce collagen synthesis. Their use is often associated with delayed wound healing and increased infection risk, especially in patients undergoing surgery or trauma care. Nonsteroidal anti-inflammatory drugs (NSAIDs) can reduce prostaglandin-mediated vasodilation and inflammation, which are necessary for initial wound signaling and tissue perfusion. While useful for pain control, chronic or high-dose NSAID use may impair angiogenesis and matrix remodeling (Morton & Phillips, 2016).

Chemotherapy agents and radiation therapy, used in cancer treatment, are particularly detrimental. They target rapidly dividing cells and thus impair epithelial regeneration, angiogenesis, and immune cell proliferation. Patients receiving these treatments often present with delayed healing, poor granulation, and necrotic tissue. Anticoagulants such as warfarin or heparin, while necessary for cardiovascular conditions, may interfere with clot stability and initial hemostasis, increasing bleeding and hematoma formation at the wound site.

Moreover, iatrogenic factors those introduced unintentionally during medical care can also affect wound outcomes. Surgical errors, improper wound closure techniques, use of incompatible dressings, and poor aseptic technique can all result in wound complications. Additionally, polypharmacy in elderly or chronically ill patients may lead to unforeseen drug interactions that impair healing or cause adverse effects (Whitney, 2005). A thorough review of the patient's medication history is essential to avoid pharmacological barriers to wound repair.

Environmental and Socioeconomic Factors: The Context of Healing

Environmental and socioeconomic contexts also shape wound healing outcomes. Living conditions, including housing quality, sanitation, and access to clean water, directly influence wound hygiene and the likelihood of infection. Climate can affect skin hydration and barrier function; for instance, cold and dry environments may desiccate wounds, while hot and humid climates may foster bacterial growth. Access to healthcare is a profound determinant patients in rural or low-income settings may lack the resources to obtain advanced wound care products, antibiotics, or follow-up services, leading to the progression of acute wounds into chronic conditions.

Educational background and health literacy are similarly important. Patients who understand the basics of wound care, infection prevention, and the importance of nutrition are more likely to engage in positive healing behaviors. Conversely, lack of awareness or misconceptions, such as overuse of traditional remedies without clinical support, can lead to poor outcomes. Occupational exposure also plays a role, as workers in agriculture, construction, or manual labor may face repeated trauma, chemical exposure, or contaminated wounds that complicate the healing process (Kujath & Michelsen, 2008).

Finally, cultural beliefs and practices may influence wound care choices. In many cultures, herbal remedies and traditional medicines are the first line of treatment, and while some phytoconstituents have demonstrated healing benefits, others may delay healing or introduce pathogens. Integrating validated traditional knowledge with modern medical care is essential for culturally competent wound management.

TRANSLATIONAL CHALLENGES IN WOUND HEALING

Wound healing can face many challenges that slow down or stop the process altogether. One common issue is chronic wounds and non-healing ulcers, which are wounds that do not close within the normal time because of factors like poor blood flow or diabetes. Another major challenge is multidrug-resistant infections, where bacteria become resistant to antibiotics, making infections difficult to treat and leading to longer healing times. Biofilm formation also complicates healing; bacteria form protective layers on wounds that shield them from medicine and the immune system (Negut et al., 2018).

Sometimes wounds may reopen after closing, a problem known as wound dehiscence, or get hurt again through re-injury, both of which delay recovery. Additionally, some people develop excessive scarring and keloid formation, where the scar tissue grows too much, causing discomfort or cosmetic concerns. Finally, healthcare access and disparities play a significant role, as many people do not get timely or proper wound care due to factors like location, income, or healthcare availability, which can worsen outcomes. Addressing these challenges is essential to improving wound healing for all patients.

Despite substantial progress in experimental wound-healing research, translating innovative therapeutic strategies from laboratory settings into routine clinical practice remains a significant challenge. Numerous emerging approaches such as bioactive phytoconstituents, nanotechnology-enabled drug delivery systems, recombinant growth factors, stem-cell therapies, and tissue-engineered skin substitutes demonstrate promising outcomes under controlled preclinical conditions. However, their clinical performance is often influenced by patient-specific variability, including age, metabolic disorders, vascular insufficiency, immune dysfunction, infection status, and wound chronicity. These real-world complexities frequently result in heterogeneous treatment responses, underscoring the gap between experimental efficacy and clinical effectiveness (Tatarusanu et al., 2023).

Standardization and reproducibility constitute additional major barriers to translation. Natural product–based therapies may exhibit variability in phytochemical composition due to differences in plant origin, harvesting conditions, extraction techniques, and formulation stability, complicating dose uniformity and quality control. Likewise, advanced biomaterials and nanocarrier systems demand precise and scalable manufacturing processes to maintain consistent physicochemical properties, therapeutic loading, and safety profiles. Regulatory approval pathways for such complex or combination products integrating biological, material, and pharmacological components are often lengthy and resource-intensive, which can delay or limit their clinical availability despite encouraging scientific evidence.

Economic, infrastructural, and safety considerations further influence translational success. Many advanced wound-care technologies require specialized production facilities, controlled storage conditions, and trained healthcare professionals, leading to elevated treatment costs and restricted accessibility, particularly in resource-limited settings where chronic wounds are highly prevalent Figure. 4 (Percival et al., 2015). Moreover, comprehensive long-term safety data for regenerative therapies, biomaterial scaffolds, and nanotechnology-based interventions are still emerging, necessitating continued evaluation of potential cytotoxic, immunological, and systemic effects following prolonged use. Addressing these translational constraints through improved standardization, cost-effective manufacturing, robust clinical trials, and clear regulatory frameworks will be essential for ensuring that next-generation wound therapies achieve meaningful and equitable clinical impact.

Figure 4. Schematic representation of the translational pathway from preclinical research to clinical implementation in wound healing.

Chronic Wounds and Non-Healing Ulcers

Chronic wounds are injuries that fail to progress through the normal stages of healing and remain open for more than 4 to 6 weeks. Common types include diabetic foot ulcers, pressure ulcers (bedsores), and venous leg ulcers. These wounds are often caused by underlying health conditions such as diabetes, poor circulation, or immobility. Because the normal healing response is disrupted, chronic wounds require special care and sometimes advanced treatments to heal properly.

Multidrug-Resistant Infections

One of the major challenges in wound care today is the rise of multidrug-resistant (MDR) bacteria. These are bacteria that no longer respond to commonly used antibiotics. When these resistant strains infect a wound, they can delay healing, increase the risk of complications, and require stronger or more toxic medications. MDR infections are especially dangerous in hospitals and for patients with weakened immune systems (Dow et al., 1999).

Biofilm Formation

Biofilms are clusters of bacteria that stick together and form a protective layer over the wound surface. This layer shields the bacteria from antibiotics and the immune system, making infections harder to treat. Biofilms can keep a wound in a constant state of inflammation and prevent healing. Special treatments are often needed to break down the biofilm and allow the wound to heal.

Wound Dehiscence and Re-Injury

Wound dehiscence occurs when a previously closed wound reopens, either partially or completely. This can happen due to physical stress, infection, or poor surgical technique. Re-injury is also a concern, especially if the area is not fully healed or if the patient resumes activity too soon. Both issues can slow down the healing process and increase the risk of complications like infection or scarring.

Excessive Scarring and Keloid Formation

While some scarring is a normal part of wound healing, excessive scarring can be problematic. Hypertrophic scars stay within the wound boundary but are raised and firm, while keloids grow beyond the wound's original edges (Sen et al., 2009). These types of scars may cause discomfort, itching, or even limit movement depending on their location. People with darker skin tones are more likely to develop keloids. Managing and preventing such scarring often requires special treatments.

Healthcare Access and Disparities

Not everyone has the same access to quality wound care. Factors such as income level, geographic location, availability of healthcare services, and insurance coverage can greatly affect wound healing outcomes. People in underserved or rural areas may not receive timely treatment, leading to complications or chronic wounds. Addressing these disparities is crucial to ensuring better outcomes for all patients Table. 1.

Table 1. Translational challenges in advanced wound healing and potential solutions.

Domain	Major Challenges	Clinical Implications	Possible Solutions
Preclinical–Clinical Gap	Variable patient physiology, infection, chronic disease	Inconsistent healing outcomes	Patient-specific treatment design, improved clinical models
Standardization	Phytochemical variability, unstable formulations	Dose inconsistency, reduced reliability	Quality control and validated formulation methods
Regulatory Approval	Complex pathways for biologics, biomaterials, nanotherapies	Delayed clinical adoption	Harmonized guidelines and robust safety trials

continued on following page

Table 1. Continued

Domain	Major Challenges	Clinical Implications	Possible Solutions
Manufacturing & Scalability	High production cost, technical complexity	Limited widespread availability	Scalable and cost-effective fabrication technologies
Healthcare Accessibility	Infrastructure and trained personnel requirements	Inequity in treatment access	Simplified delivery systems and public health integration
Long-Term Safety	Limited chronic toxicity and immunogenicity data	Uncertain risk–benefit profile	Extended clinical monitoring and multicenter trials

COMPARATIVE OVERVIEW OF CONVENTIONAL AND MODERN WOUND CARE STRATEGIES

Effective wound management has evolved considerably over time, transitioning from traditional supportive care toward advanced, mechanism-driven therapeutic approaches. Conventional wound care remains the clinical foundation in many healthcare settings and primarily focuses on essential principles such as wound cleansing, surgical or mechanical debridement, infection control through topical or systemic antimicrobials, moisture balance using standard dressings (e.g., gauze, bandages, and antiseptic-impregnated materials), and protection of the wound from further trauma. These approaches are widely accessible, cost-effective, and suitable for managing uncomplicated acute wounds. However, their effectiveness is often limited in chronic or complex wounds, where impaired perfusion, persistent inflammation, biofilm formation, metabolic disorders such as diabetes, and antimicrobial resistance significantly delay healing (Moore et al., 2014). In such conditions, conventional therapies frequently provide symptomatic control rather than actively stimulating tissue regeneration.

Modern wound care strategies, in contrast, are designed to directly influence the biological mechanisms underlying tissue repair. These approaches include advanced wound dressings (hydrogels, hydrocolloids, alginates, foam dressings, and bioactive scaffolds), negative pressure wound therapy, delivery of recombinant growth factors, stem-cell-based therapies, skin substitutes, and nanotechnology-enabled drug delivery systems. Such innovations aim to create an optimal microenvironment that promotes angiogenesis, regulates inflammation, enhances fibroblast proliferation, supports extracellular matrix remodeling, and accelerates re-epithelialization. Additionally, the integration of phytoconstituent-based formulations and biomaterial-assisted delivery platforms represents a growing area of translational research, combining natural bioactive compounds with modern biomedical engineering to achieve synergistic healing effects. These contemporary strategies not only improve healing rates but also reduce complications such as infection, excessive scarring, and recurrence of chronic ulcers (Atkin, 2019).

Despite their advantages, modern therapies may involve higher costs, technical expertise, and infrastructure requirements, which can limit accessibility in low-resource settings. Therefore, an integrative approach that combines the reliability and affordability of conventional care with the regenerative potential of modern technologies is increasingly recognized as the most practical and clinically effective pathway. Such a balanced strategy supports personalized wound management, improves patient outcomes, and aligns with current trends in precision and regenerative medicine.

Another important distinction between conventional and modern wound care lies in their approach to infection control and inflammation modulation. Conventional management typically relies on systemic or topical antibiotics and antiseptics to reduce microbial burden. While effective in acute infections, this strategy is increasingly challenged by antimicrobial resistance, biofilm-forming pathogens, and recurrent infections that prevent wounds from progressing beyond the inflammatory phase. Modern wound care, however, adopts a more targeted and biologically responsive approach. Advanced antimicrobial dressings incorporating silver nanoparticles, iodine complexes, antimicrobial peptides, or plant-derived bioactive compounds provide sustained local activity while minimizing systemic toxicity. Simultaneously, contemporary therapies aim to regulate excessive inflammation through growth factors, cytokine modulation, and antioxidant-rich biomaterials, thereby restoring the balance required for normal tissue repair rather than merely suppressing symptoms.

Furthermore, modern wound care emphasizes regeneration and functional tissue restoration rather than simple wound closure. Traditional practices often culminate in scar formation with limited tensile strength and reduced physiological function. In contrast, emerging regenerative strategies including tissue-engineered skin substitutes, stem-cell-based therapies, extracellular matrix scaffolds, and nanostructured delivery systems seek to recreate the native architecture of skin and supporting tissues (Fife & Carter, 2012). These innovations promote organized collagen deposition, vascular network formation, and improved epithelial integrity, ultimately enhancing long-term clinical outcomes and quality of life for patients with chronic or complex wounds. As research continues to advance, the convergence of regenerative medicine, biomaterials science, and phytoconstituent-based therapeutics is expected to redefine future standards of wound management, moving the field from repair toward true tissue regeneration.

Table 2. Comparison between conventional and modern wound care approaches, illustrating the transition from supportive symptom-based management toward regenerative, mechanism-driven therapeutic strategies for improved healing outcomes.

Aspect	Conventional Wound Care	Modern Wound Care
Primary Goal	Protection and basic healing	Active regeneration and tissue restoration
Core Methods	Cleaning, debridement, antibiotics, gauze dressings	Advanced dressings, NPWT, growth factors, stem cells, nanotechnology
Biological Action	Mainly supportive and protective	Direct modulation of cellular and molecular healing pathways
Infection Control	Systemic/topical antibiotics and antiseptics	Sustained-release antimicrobials, nanoparticles, bioactive compounds
Inflammation Management	Limited control	Targeted cytokine regulation and antioxidant therapy
Healing Outcome	Scar formation, slower healing in chronic wounds	Faster healing, improved tissue architecture, reduced recurrence
Effectiveness in Chronic Wounds	Often limited	Significantly improved outcomes
Cost and Accessibility	Low cost, widely available	Higher cost, requires specialized facilities
Clinical Trend	Symptom-focused care	Regenerative and precision-based therapy
Best Current Approach	—	Integration of both conventional and modern strategies

The comparative features summarized in Table 2 highlight the evolving paradigm of wound management from conventional supportive care toward biologically active and regenerative therapeutic strategies. Conventional wound care remains essential due to its affordability, accessibility, and effectiveness in uncomplicated acute wounds; however, its limited ability to address chronic inflammation, impaired vascularization, biofilm persistence, and antimicrobial resistance restricts outcomes in complex wounds. In contrast, modern wound care approaches are designed to actively modulate the cellular and molecular processes of healing through advanced dressings, targeted antimicrobial delivery, growth factor stimulation, stem-cell-based regeneration, and nanotechnology-enabled systems. These innovations not only accelerate tissue repair but also improve structural and functional restoration while reducing recurrence and complications. Nevertheless, higher costs and infrastructure requirements remain significant barriers, particularly in low-resource settings. Therefore, the integration of conventional practicality with modern regenerative capability represents the most balanced and clinically effective strategy for comprehensive wound management (Werdin et al., 2009).

In addition to therapeutic effectiveness, the comparison presented in Table 2 underscores the broader clinical and translational implications of integrating emerging technologies into routine wound care. Modern strategies increasingly emphasize precision-based intervention, where treatment selection is guided by wound type, patient comorbidities, microbial profile, and local tissue physiology. Such personalization enhances healing efficiency while minimizing unnecessary antibiotic exposure and treatment burden. Furthermore, advances in biomaterials, controlled drug delivery, and bioactive phytoconstituent incorporation are expanding the scope of minimally invasive yet highly effective wound therapies. These developments are particularly relevant for managing chronic and non-healing wounds, which represent a growing global healthcare challenge. Consequently, the future direction of wound management is likely to depend on interdisciplinary collaboration that combines conventional clinical expertise with innovations from regenerative medicine, nanotechnology, and pharmacognosy to achieve safer, faster, and more sustainable healing outcomes.

ASSESSMENT AND DIAGNOSIS OF WOUNDS

Accurate assessment and diagnosis are essential steps in effective wound management. A thorough evaluation helps healthcare providers understand the wound's cause, current condition, and factors that may affect healing. This understanding is critical for developing an appropriate treatment plan and improving patient outcomes (Mustoe, 2004).

The process begins with a clinical evaluation, where the healthcare professional conducts a detailed physical examination of the wound. This includes measuring the wound's size—its length, width, and depth and inspecting the wound bed to identify healthy tissue, dead tissue, or signs of infection. The type and amount of wound drainage (exudate), its color and odor, and the condition of the surrounding skin are carefully noted. These observations provide clues about the stage of healing and any complications present. Additionally, the clinician takes a detailed history, asking about how the wound happened, how long it has been present, any previous treatments, and the patient's overall health. Medical conditions such as diabetes, poor circulation, or immune problems can greatly influence wound healing, so this information is critical.

Beyond the clinical exam, laboratory and imaging tests are often used to gather more information. Laboratory tests like blood work can reveal infections or inflammation and assess other health factors that impact healing. Wound swabs or biopsies can identify the specific bacteria infecting the wound and help select the most effective antibiotics. Imaging techniques such as X-rays, ultrasound, CT scans, or MRIs allow visualization of deeper tissues beneath the wound, helping to detect hidden infections, bone involvement, or other complications. These tools provide valuable insights that cannot be gained from physical examination alone.

Another important aspect of wound assessment is accurate measurement and documentation. Measuring the wound consistently over time tracking length, width, and depth helps monitor healing progress or detect worsening.

Photographs are often taken to provide a visual record. Detailed notes are kept on the wound's appearance, changes over time, treatments applied, and patient symptoms such as pain. Well-organized documentation supports communication between healthcare providers and helps guide treatment decisions.

Finally, healthcare professionals often use risk prediction and prognostic scoring systems to estimate how likely a wound is to heal successfully or develop complications. These scoring tools take into account various factors such as the patient's age, health status, wound size, and type. By identifying wounds at higher risk, providers can tailor treatment plans to address specific needs and prevent problems like infection or chronic wounds. These tools also help set realistic expectations for healing timelines and outcomes (Owens et al., 2007).

In summary, a comprehensive and systematic assessment and diagnosis of wounds are vital to guiding effective treatment and improving healing outcomes. Combining clinical evaluation, laboratory and imaging support, precise measurement, and risk prediction allows healthcare providers to make informed decisions and provide the best possible care.

Clinical Evaluation Techniques

The clinical evaluation of wounds is a fundamental step in understanding the wound's nature, severity, and the appropriate treatment approach. This evaluation starts with a thorough physical examination of the wound itself. The healthcare provider carefully inspects the wound's size by measuring its length, width, and depth. The color and condition of the wound bed are noted to assess whether there is healthy granulation tissue, slough (yellow or white dead tissue), or necrosis (black dead tissue). The presence of exudate (fluid discharge) is evaluated for amount, color, and odor, which can indicate infection or other complications. The surrounding skin is also examined to check for signs of inflammation, such as redness, swelling, warmth, or maceration (softening due to moisture).

It is equally important to assess the wound margins for undermining or tunneling, which are hidden spaces beneath the skin's surface that can trap bacteria and delay healing. The clinician collects a detailed patient history including the cause of the wound (e.g., trauma, surgery, pressure), duration, previous treatments, and any underlying medical conditions such as diabetes or vascular disease. This comprehensive clinical evaluation guides the initial diagnosis and influences the treatment strategy (Ibrahim et al., 2018).

Laboratory and Imaging Support

Laboratory and imaging investigations play a critical role in complementing the clinical evaluation by providing deeper insights into the wound and any associated systemic conditions. Laboratory tests often include complete blood counts to check for signs of infection or anemia, blood glucose levels especially in diabetic patients, and markers of inflammation such as C-reactive protein (CRP) and erythrocyte sedimentation rate (ESR). Wound swabs or biopsies may be taken to identify bacterial colonization or infection, including antibiotic sensitivities to guide targeted therapy.

In addition, imaging studies are essential when wounds extend beyond the surface skin, involve deeper tissues, or when osteomyelitis (bone infection) is suspected. X-rays can reveal bone abnormalities or foreign bodies, while ultrasound imaging can assess soft tissue involvement and fluid collections. More advanced imaging techniques such as magnetic resonance imaging (MRI) or computed tomography (CT) scans provide detailed cross-sectional views to evaluate the extent of tissue damage, abscess formation, or vascular supply to the area. These investigations help in planning surgical intervention if needed and monitoring response to treatment (Criollo-Mendoza et al., 2023).

Wound Measurement and Documentation

Accurate measurement and documentation are vital components of wound management, as they provide objective data to monitor healing progress or deterioration over time. The wound's dimensions length, width, and depth are measured using rulers, calipers, or specialized wound measurement tools. Some clinicians also calculate wound volume by multiplying these dimensions, which is especially useful for irregularly shaped wounds. Digital photography is commonly used to visually document the wound and provides a valuable record that can be compared during follow-up visits.

Consistent documentation should include a detailed description of the wound bed, edges, surrounding skin, and any changes since the last assessment. Recording the amount and characteristics of exudate, pain levels, and patient-reported symptoms is also important. This thorough documentation ensures clear communication between healthcare providers and supports clinical decision-making. Additionally, using standardized wound assessment forms or electronic health records with wound care modules improves the consistency and quality of care.

Risk Prediction and Prognostic Scoring

Risk prediction and prognostic scoring systems are tools designed to estimate the likelihood of wound healing and to identify wounds at high risk of complications such as infection, chronicity, or recurrence. These models integrate multiple factors including patient age, comorbidities (such as diabetes, peripheral vascular disease), wound type and size, nutritional status, and previous wound history (Moses et al., 2023).

For example, the Bates-Jensen Wound Assessment Tool and the Pressure Ulcer Scale for Healing (PUSH) score are commonly used in clinical practice to provide a numerical value that correlates with healing potential. These scores help clinicians stratify patients into different risk categories, enabling personalized treatment plans. High-risk wounds may require more intensive monitoring, aggressive infection control, or advanced therapies such as negative pressure wound therapy or skin grafting. Additionally, prognostic tools aid in counseling patients and caregivers regarding expected healing time and possible outcomes, which is crucial for setting realistic expectations and improving adherence to treatment.

ROLE OF PHYTOCONSTITUENTS IN WOUND HEALING

The process of wound healing is complex and involves multiple biological stages that must proceed in a carefully coordinated manner. Despite the body's remarkable ability to repair itself, many wounds experience delayed or impaired healing due to infection, chronic conditions, or other factors. This has driven the search for effective and safe therapeutic agents that can support or enhance the natural healing process.

Phytoconstituents bioactive compounds derived from plants have garnered significant attention in recent years for their potential to promote wound healing. These natural compounds include a diverse range of substances such as flavonoids, alkaloids, tannins, saponins, terpenoids, and phenolic acids. They are known to possess various pharmacological properties including anti-inflammatory, antioxidant, antimicrobial, and collagen-stimulating effects, all of which are critical in different phases of wound repair.

For instance, the anti-inflammatory activity of certain phytoconstituents helps modulate the inflammatory phase, reducing excessive swelling and preventing chronic inflammation that can hinder healing. The antioxidant properties neutralize harmful free radicals generated at the injury site, protecting cells from oxidative stress and supporting tissue regeneration. Additionally, many plant compounds exhibit antimicrobial effects, helping to prevent or control wound infections a common cause of delayed healing. Furthermore, some phytoconstituents stimulate the production of collagen and other extracellular matrix components, which are essential for tissue strength and integrity during the proliferative and remodeling phases.

Traditional medicine systems worldwide have long utilized plant-based remedies for wound care, and modern scientific research continues to validate these practices through rigorous studies. Incorporating phytoconstituents into wound treatment offers the promise of safer, more affordable, and effective therapies, especially in areas where access to advanced medical care is limited (Hajialyani et al., 2018).

In the chapters that follow, we will explore the specific types of phytoconstituents that influence wound healing, their mechanisms of action, and the challenges and opportunities in developing phytochemical-based wound therapies. Understanding the role of these natural compounds provides a valuable bridge between the biological processes of wound healing and innovative approaches to enhancing tissue repair.

Traditional Herbal Approaches

For centuries, traditional medicine systems around the world have used herbal remedies to treat wounds and promote healing. Many cultures have relied on plants known for their healing properties, applying extracts, poultices, or pastes directly to wounds. These traditional approaches are often based on empirical knowledge passed down through generations. Plants such as *Aloe vera*, *Centella asiatica*, and *Calendula officinalis* have been widely used for their soothing, anti-inflammatory, and antimicrobial effects. Although these methods originated before modern scientific understanding, recent research has begun to validate and explain the biological mechanisms behind their effectiveness. Traditional herbal approaches remain a valuable resource in modern wound care, especially in regions where access to advanced medical treatments is limited.

Phytochemicals with Anti-inflammatory and Antioxidant Effects

Many phytoconstituents possess strong anti-inflammatory and antioxidant properties, which are crucial for effective wound healing. During the inflammatory phase of healing, these compounds help regulate the immune response by reducing excessive inflammation that can damage tissues. Antioxidants neutralize free radicals unstable molecules generated during injury that can delay healing by causing oxidative stress. Common phytochemicals like flavonoids, phenolic acids, and tannins act as natural antioxidants and anti-inflammatory agents. By controlling inflammation and protecting cells from oxidative damage, these compounds support faster tissue repair and reduce the risk of chronic wounds.

Phytoconstituent-Enhanced Wound Dressings

Advances in wound care have led to the development of wound dressings infused with phytoconstituents to enhance healing. These specialized dressings can deliver plant-derived bioactive compounds directly to the wound site, providing sustained antimicrobial, anti-inflammatory, and antioxidant effects. For example, dressings containing extracts of honey, curcumin, or tea tree oil are used to prevent infection and promote tissue regeneration. These dressings improve the local wound environment, maintain moisture balance, and protect against bacterial colonization. The integration of phytoconstituents into wound dressings represents a promising strategy that combines the benefits of traditional herbal medicine with modern biomedical technology.

Synergy with Conventional Therapies

Phytoconstituents can work alongside conventional wound treatments to improve healing outcomes. When combined with antibiotics, surgical debridement, or advanced therapies like negative pressure wound therapy, plant-based compounds may enhance the effectiveness of these treatments. They can reduce side effects, lower infection rates, and support tissue regeneration. This synergistic approach leverages the natural properties of phytochemicals to complement medical interventions, offering a more holistic and effective wound care strategy. Researchers continue to explore optimal combinations to maximize benefits while minimizing risks.

Evidence-Based Clinical Applications

Modern clinical research has provided growing evidence supporting the use of phytoconstituents in wound healing. Numerous studies have demonstrated the safety and efficacy of specific plant extracts and isolated compounds in accelerating wound closure, reducing inflammation, and preventing infection. Clinical trials have examined herbal ointments, gels, and dressings, providing guidelines for their practical use in healthcare settings. Evidence-based applications of phytoconstituents ensure that traditional knowledge is supported by scientific validation, helping integrate these natural therapies into standard wound care protocols Figure. 5 (Agyare et al., 2019). This approach enhances patient confidence and promotes wider acceptance of phytochemical-based treatments in both conventional and complementary medicine.

Figure 5. Phytoconstituent-based mechanisms supporting tissue repair and wound regeneration.

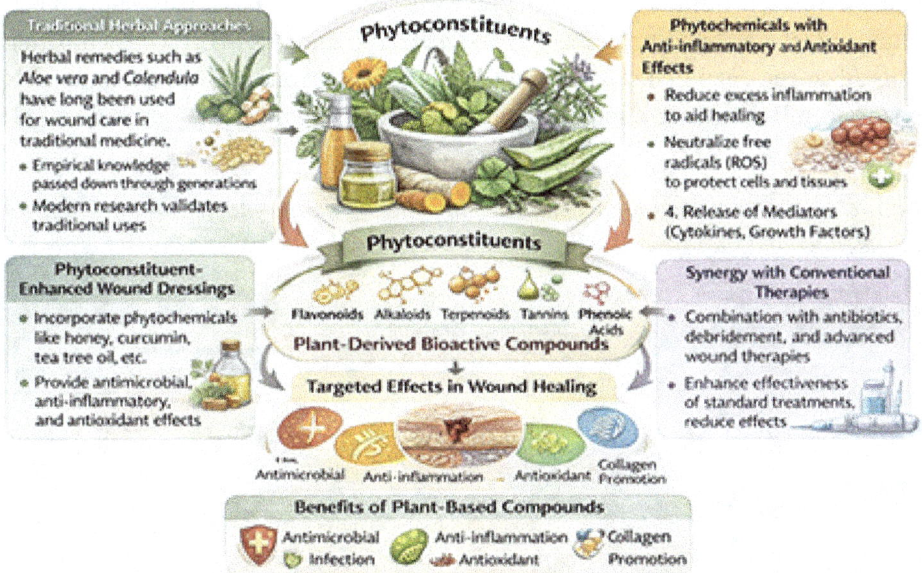

CLINICAL IMPORTANCE OF BIOMARKER MONITORING

Reliable identification and proper validation of biomarkers are becoming very important in modern wound care. Biomarkers are measurable biological signals that help doctors understand what is happening inside a wound. They provide clear and objective information about the condition of the tissue, the level of inflammation, infection risk, and the progress of healing. Because of this, biomarkers can help in predicting whether a wound will heal quickly or may become chronic. They also allow healthcare professionals to monitor how well a treatment is working and make timely changes in therapy when needed.

In recent years, research on wound biomarkers has increased greatly, and many promising indicators have been discovered. However, their use in routine clinical practice is still limited. One major reason is the variation in testing methods used in different laboratories and hospitals, which makes results difficult to compare. In addition, there is a lack of universally accepted standard values that clearly define normal and abnormal biomarker levels. Another important challenge is the shortage of large clinical studies that confirm the reliability and usefulness of these biomarkers in diverse patient populations Figure. 6.

Therefore, improving biomarker research and creating standardized testing systems are essential steps for the future of wound management. Strong and validated biomarker frameworks can improve the accuracy of prognosis, support early clinical decisions, and help design personalized treatment plans for each patient. This advancement will ultimately lead to faster healing, reduced complications, and better overall patient care in both acute and chronic wounds (Pacyga et al., 2024).

Figure 6. Biomarker-guided precision approach for diagnosis, monitoring, and targeted wound therapy.

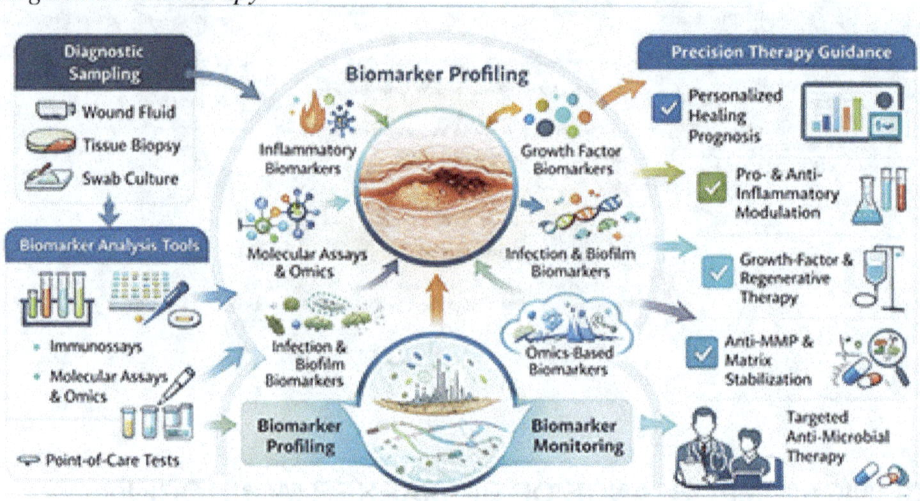

INFLAMMATORY BIOMARKERS

Inflammation plays a very important role in deciding whether a wound heals properly or becomes chronic. During the early stage of healing, a controlled inflammatory response helps remove damaged tissue, fight infection, and prepare the wound for new tissue formation. However, when inflammation continues for a long time or becomes excessive, it can delay healing and lead to chronic, non-healing wounds.

Some key inflammatory biomarkers help in understanding this process. Pro-inflammatory cytokines such as IL-1β, IL-6, and TNF-α increase during active inflammation and are often found at high levels in chronic wounds. In contrast, anti-inflammatory mediators, including IL-10 and other transforming growth signals, help reduce inflammation and support the transition toward the healing phase. Oxidative stress indicators, especially products related to reactive oxygen species, also provide useful information because excessive oxidative stress can damage cells and slow tissue repair.

When pro-inflammatory cytokines remain elevated for a long time, wounds are more likely to remain non-healing or chronic. On the other hand, a balanced inflammatory response supports normal healing and movement toward the proliferation stage, where new tissue forms. Therefore, measuring inflammatory mediators in wound fluid or tissue samples can help doctors identify early whether a wound is healing properly or becoming stalled, allowing timely and more effective treatment decisions.

Growth Factor and Repair-Associated Biomarkers

Proper wound healing and tissue regeneration depend on well-coordinated signals from different growth factors. These biological signals control important healing events such as the formation of new blood vessels (angiogenesis), activation of fibroblast cells that produce connective tissue, and movement of epithelial cells that cover the wound surface. When these processes occur in a balanced and timely manner, normal tissue repair becomes possible.

Several important biomarkers are linked with these repair mechanisms. Vascular endothelial growth factor (VEGF) reflects the development of new blood vessels and is essential for supplying oxygen and nutrients to healing tissue. Platelet-derived growth factor (PDGF) supports the recruitment of fibroblasts and helps in forming the extracellular matrix needed for structural repair. Transforming growth factor-β (TGF-β) plays a major role in collagen production, tissue remodeling, and strengthening of the healed area.

In conditions such as diabetic or ischemic wounds, the expression of these growth factors is often reduced or poorly regulated. This imbalance contributes to slow or incomplete healing. Therefore, monitoring growth-factor levels can help clinicians understand the healing status of a wound, guide the use of regenerative or growth-factor-based therapies, and assess how well a treatment is working over time (Patel et al., 2016).

Extracellular Matrix Remodeling Biomarkers

For a wound to heal properly, the body must maintain a careful balance between the formation of new structural tissue and the breakdown of damaged material. This process is known as extracellular matrix (ECM) remodeling and is essential for restoring strength, stability, and normal function to the repaired area. If this balance is disturbed, healing may slow down or fail completely.

Several important biomarkers help in understanding ECM remodeling. Matrix metalloproteinases (MMP-2 and MMP-9) are enzymes responsible for breaking down damaged matrix components so that new tissue can form. Tissue inhibitors of metalloproteinases (TIMPs) regulate the activity of MMPs and prevent excessive tissue destruction. Another useful indicator is the collagen type I to type III ratio, which reflects the maturity and tensile strength of the repaired tissue. A higher proportion of collagen type I usually indicates stronger and more stable healing.

In chronic wounds, this balance is often disrupted. MMP activity becomes excessively high, while TIMP levels decrease, resulting in continuous breakdown of the matrix instead of proper repair. Because of this, granulation tissue does not develop effectively, and the remodeling phase is delayed. Therefore, analyzing ECM-related biomarkers can help predict poor healing outcomes and guide the selection of treatments that stabilize the matrix and support successful tissue regeneration.

Infection- and Biofilm-Associated Biomarkers

Infection is one of the most important factors that can delay wound healing and make treatment more difficult. When harmful microorganisms grow inside a wound, they increase inflammation, damage healthy tissue, and may form protective biofilms that resist antibiotics and the body's immune response. Because of this, early identification of infection is essential for proper wound management.

Several biomarkers help detect infection and understand its severity. Procalcitonin and C-reactive protein are commonly used indicators that rise in response to bacterial infection and systemic inflammation. Increased bacterial load and the presence of biofilm-related molecular signals within the wound also suggest persistent microbial activity that can prevent healing. In addition, elevated protease activity in wound exudate reflects tissue damage and ongoing inflammatory processes associated with infection.

Detecting these infection-related biomarkers at an early stage allows healthcare professionals to begin appropriate antimicrobial treatment on time, prevent the spread of infection to other parts of the body, and reduce complications. As a result, timely monitoring of these biomarkers plays a key role in improving healing outcomes and supporting effective, targeted wound care.

Emerging Molecular and Omics-Based Biomarkers

Recent progress in molecular biology and modern analytical technologies has led to the development of new and advanced types of wound biomarkers. These next-generation approaches provide deeper insight into the biological processes occurring inside a wound and may allow more accurate prediction of healing outcomes.

Some important emerging biomarkers include microRNA expression patterns, which regulate key processes such as inflammation control and formation of new blood vessels. Proteomic profiles of wound exudate help identify specific proteins that reflect the current stage of healing. In addition, metabolomic and lipidomic markers provide information about cellular energy use, membrane function, and overall metabolic balance within the wound environment. Another promising development is AI-assisted imaging biomarkers, which analyze wound images to estimate healing progress and predict whether a wound is likely to recover or become chronic (Lindley et al., 2016).

These advanced technologies offer very high sensitivity and the possibility of personalized or precision-based wound treatment. However, their routine clinical use is still limited because large-scale clinical validation studies are needed, and the cost of implementation must become more affordable. Even so, these emerging biomarker platforms represent an important future direction in improving wound diagnosis, monitoring, and treatment outcomes.

CLINICAL UTILITY AND FUTURE VALIDATION NEEDS

For wound biomarkers to become part of routine clinical practice, several important requirements must be fulfilled. First, standardized methods for sample collection and laboratory analysis are necessary so that results are reliable and comparable across different hospitals and research centers. Without uniform procedures, biomarker measurements may vary, making clinical interpretation difficult Table. 3.

Second, clear diagnostic thresholds and normal reference ranges must be established. Clinicians need well-defined cutoff values to determine whether a wound is healing normally, becoming infected, or progressing toward chronicity. Third, prospective multicenter clinical studies are essential to confirm the accuracy, safety, and real-world usefulness of these biomarkers in diverse patient populations and healthcare settings.

Another key requirement is the integration of biomarkers with digital health technologies and point-of-care diagnostic tools. Rapid bedside testing and data-driven monitoring systems can help clinicians make timely treatment decisions and track healing progress more effectively.

Strengthening all of these areas will allow wound biomarkers to move beyond research laboratories and become practical clinical decision-support tools. This transition has the potential to improve prognosis, guide appropriate therapy selection, reduce complications, and support truly personalized care for patients with both acute and chronic wounds.

Table 3. Key biomarkers involved in wound healing and their clinical significance.

Biomarker Category	Key Examples	Biological Role	Clinical Significance
Inflammatory Biomarkers	IL-1β, IL-6, TNF-α, IL-10	Regulate inflammation and immune response	Differentiate healing vs. chronic wounds
Growth Factor Biomarkers	VEGF, PDGF, TGF-β	Promote angiogenesis, fibroblast activity, tissue repair	Assess regenerative potential and therapy response
ECM Remodeling Biomarkers	MMP-2, MMP-9, TIMPs, Collagen I/III ratio	Control matrix degradation and remodeling	Predict granulation failure and scar quality
Infection-Related Biomarkers	Procalcitonin, CRP, microbial load markers	Indicate infection and biofilm presence	Guide antimicrobial therapy decisions
Molecular/Omics Biomarkers	microRNA, proteomic and metabolomic signatures	Reflect gene regulation and metabolic status	Enable precision and personalized wound care

FUTURE DIRECTIONS AND INNOVATIONS

The future of wound healing research and treatment is rapidly evolving, driven by advances in technology and a growing understanding of biology, which together offer exciting new possibilities. One of the most promising areas is the integration of nanotechnology with plant-based biomaterials. Nanoparticles derived from phytoconstituents can be engineered to deliver therapeutic compounds precisely to the wound site, enhancing their effectiveness while minimizing side effects. These plant-based nanomaterials can promote faster healing through improved antimicrobial action, controlled release of bioactive agents, and better interaction with the body's cells (Goto & Saligan, 2020). Alongside these innovations, personalized wound management strategies are emerging, where treatments are tailored to the individual patient's unique needs, wound characteristics, and underlying health conditions.

By using detailed patient data, clinicians can select the most suitable therapies, including specific phytochemicals, to optimize healing outcomes. Another cutting-edge tool transforming wound care is artificial intelligence (AI), which is increasingly used to analyze wound images, predict healing trajectories, and assist in clinical decision-making. AI algorithms can process vast amounts of data quickly and accurately, providing insights that help healthcare providers intervene earlier and more effectively (Sapienza et al., 2019).

Finally, there is a growing emphasis on integrative approaches that combine phytotherapy with conventional medical treatments, blending the best of natural and modern medicine. This holistic strategy acknowledges the benefits of plant-derived compounds while leveraging advanced technologies and evidence-based practices to enhance wound healing. Together, these future directions promise to revolutionize wound care, making treatments more effective, personalized, and accessible for patients worldwide (Bodas & Shinde, 2021).

CONCLUSION

In summary, wound healing is a multifaceted biological process involving carefully orchestrated phases that can be influenced by numerous internal and external factors. Understanding the stages of healing, the challenges involved, and the critical role of accurate assessment and diagnosis provides a solid foundation for effective wound care. The growing interest in phytoconstituents highlights the potential of natural plant-based compounds as valuable adjuncts or alternatives to conventional therapies, offering anti-inflammatory, antioxidant, and antimicrobial benefits that can enhance healing.

For clinical practice, these insights underscore the importance of integrating traditional knowledge with modern scientific evidence to develop personalized, effective treatment plans. From a research perspective, continued exploration into novel therapeutic agents, including phytochemicals, along with advanced technologies like nanotechnology and artificial intelligence, promises to improve outcomes and reduce complications. Moving forward, embracing a holistic and integrative approach to wound management one that combines conventional medicine, phytotherapy, patient-centered care, and cutting-edge innovations will be key to advancing wound care practices and improving quality of life for patients across diverse settings.

REFERENCES

Agyare, C., Akindele, A. J., & Steenkamp, V. (2019). Natural products and/or isolated compounds on wound healing. *Evidence-Based Complementary and Alternative Medicine : eCAM, 4594965*, 1–3. Advance online publication. DOI: 10.1155/2019/4594965 PMID: 31186659

Atkin, L. (2019). Chronic wounds: The challenges of appropriate management. *British Journal of Community Nursing, 24*(Sup9, Suppl. 9), S26–S32. DOI: 10.12968/bjcn.2019.24.Sup9.S26 PMID: 31479336

Bodas, K., & Shinde, V. (2021). Healing of wounds: A detailed review on models, biomarkers, biochemical and other wound assessment parameters. *Inflammation, 9*(3), 2069–2085.

Cotterell, A., Griffin, M., Downer, M. A., Parker, J. B., Wan, D., & Longaker, M. T. (2024). Understanding wound healing in obesity. *World Journal of Experimental Medicine, 14*(1), 86898. DOI: 10.5493/wjem.v14.i1.86898 PMID: 38590299

Criollo-Mendoza, M. S., Contreras-Angulo, L. A., Leyva-López, N., Gutiérrez-Grijalva, E. P., Jiménez-Ortega, L. A., & Heredia, J. B. (2023). Wound healing properties of natural products: Mechanisms of action. *Molecules (Basel, Switzerland), 28*(2), 598. DOI: 10.3390/molecules28020598 PMID: 36677659

Doughty, D., Ramundo, J., Bonham, P., Beitz, J., Erwin-Toth, P., Anderson, R., & Rolstad, B. S. (2006). Issues and challenges in staging of pressure ulcers. *Journal of Wound, Ostomy, and Continence Nursing, 33*(2), 125–130. DOI: 10.1097/00152192-200603000-00004 PMID: 16572010

Dow, G., Browne, A., & Sibbald, R. G. (1999). Infection in chronic wounds: Controversies in diagnosis and treatment. *Ostomy/Wound Management, 45*(8), 23–42. PMID: 10655866

Dwivedi, J., Sachan, P., Wal, P., Wal, A., & Rai, A. K. (2024). Current state and future perspective of diabetic wound healing treatment: Present evidence from clinical trials. *Current Diabetes Reviews, 20*(5), 45–65. DOI: 10.2174/157339982 0666230828091708 PMID: 37641999

Dwivedi, J., Sachan, P., Wal, P., Wal, A., Vippamakula, S., Shankara, R. K., & Sharma, M. C. (2025). Clinical evidence of traditional medicines in modulating the immune response and diabetic wound healing. *Current Topics in Medicinal Chemistry, 25*. Advance online publication. DOI: 10.2174/011568026635934925 0707102203 PMID: 40676793

Fife, C. E., & Carter, M. J. (2012). Wound care outcomes and associated cost among patients treated in US outpatient wound centers: Data from the US wound registry. *Wounds : a Compendium of Clinical Research and Practice*, *24*(1), 10–17. PMID: 25875947

Frykberg, R. G., & Banks, J. (2015). Challenges in the treatment of chronic wounds. *Advances in Wound Care*, *4*(9), 560–582. DOI: 10.1089/wound.2015.0635 PMID: 26339534

Fui, L. W., Lok, M. P., Govindasamy, V., Yong, T. K., Lek, T. K., & Das, A. K. (2019). Understanding the multifaceted mechanisms of diabetic wound healing and therapeutic application of stem cells conditioned medium in the healing process. *Journal of Tissue Engineering and Regenerative Medicine*, *13*(12), 2218–2233. DOI: 10.1002/term.2966 PMID: 31648415

Goto, T., & Saligan, L. N. (2020). Wound pain and wound healing biomarkers from wound exudate: A scoping review. *Journal of Wound, Ostomy, and Continence Nursing*, *47*(6), 559–568. DOI: 10.1097/WON.0000000000000703 PMID: 33201141

Hajialyani, M., Tewari, D., Sobarzo-Sánchez, E., Nabavi, S. M., Farzaei, M. H., & Abdollahi, M. (2018). Natural product-based nanomedicines for wound healing purposes: Therapeutic targets and drug delivery systems. *International Journal of Nanomedicine*, *13*, 5023–5043. DOI: 10.2147/IJN.S174072 PMID: 30214204

Harding, K. (2015). Wounds and wound healing: New insights, fresh challenges. *British Journal of Dermatology*, *173*(2), 318–319. DOI: 10.1111/bjd.14036 PMID: 26346074

Ibrahim, N. I., Wong, S. K., Mohamed, I. N., Mohamed, N., Chin, K. Y., Ima-Nirwana, S., & Shuid, A. N. (2018). Wound healing properties of selected natural products. *International Journal of Environmental Research and Public Health*, *15*(11), 2360. DOI: 10.3390/ijerph15112360 PMID: 30366427

Ko, K. I., Sculean, A., & Graves, D. T. (2021). Diabetic wound healing in soft and hard oral tissues. *Translational Research; the Journal of Laboratory and Clinical Medicine*, *236*, 72–86. DOI: 10.1016/j.trsl.2021.05.001 PMID: 33992825

Kujath, P., & Michelsen, A. (2008). Wounds—From physiology to wound dressing. *Deutsches Ärzteblatt International*, *105*(13), 239. DOI: 10.3238/arztebl.2008.0239 PMID: 19629204

. Kumar, P., Hasan, F., Kumar, V., Chawla, R., & Goyal, S. K. (2024). Diabetic wound healing: Navigating physiology, advancements and research frontiers. Journal of Diabetes Research Reviews & Reports, 2–11.

Lazarus, G. S., Cooper, D. M., Knighton, D. R., Margolis, D. J., Pecoraro, R. E., Rodeheaver, G., & Robson, M. C. (1994). Definitions and guidelines for assessment of wounds and evaluation of healing. *Wound Repair and Regeneration, 2*(3), 165–170. DOI: 10.1046/j.1524-475X.1994.20305.x PMID: 17156107

LeBlanc, K., Woo, K., Gethin, G., Beeckman, D., Heerschap, C., Reddy, M., Parker, C., MacAndrew, M., White-Chu, F., & Wiesenfeld, S. (2025). Adults living with dementia and chronic wounds, wound types, care challenges, and impact across dementia stages: A scoping review protocol. *Journal ofTissue Viability, 34*(1), 100845. Advance online publication. DOI: 10.1016/j.jtv.2024.12.009 PMID: 39724737

Lindley, L. E., Stojadinovic, O., Pastar, I., & Tomic-Canic, M. (2016). Biology and biomarkers for wound healing. *Plastic and Reconstructive Surgery, 138*(3, Suppl.), 18S–28S. DOI: 10.1097/PRS.0000000000002682 PMID: 27556760

Liu, H., Sun, W., Cai, W., Luo, K., Lu, C., Jin, A., Zhang, J., & Liu, Y. (2025). Current status, challenges, and prospects of artificial intelligence applications in wound repair theranostics. *Theranostics, 15*(5), 1662–1688. DOI: 10.7150/thno.105109 PMID: 39897550

Mahmoudi, M., & Gould, L. J. (2020). Opportunities and challenges of the management of chronic wounds: A multidisciplinary viewpoint. *Chronic Wound Care Management and Research, 7*, 27–36. DOI: 10.2147/CWCMR.S260136

Malone, M., & Schultz, G. (2022). Challenges in the diagnosis and management of wound infection. *British Journal of Dermatology, 187*(2), 159–166. DOI: 10.1111/bjd.21612 PMID: 35587707

Manisha, N., Niharika, , Gaur, P., Goel, R., Lata, K., & Mishra, R. (2025, March). Gaur, P., Goel, R., Lata, K., & Mishra, R. (2024). Understanding diabetic wounds: A review of mechanisms, pathophysiology, and multimodal management strategies. *Current Reviews in Clinical and Experimental Pharmacology, 20*(3), 207–228. Advance online publication. DOI: 10.2174/0127724328326480240927065600 PMID: 40452447

Moore, Z., Butcher, G., Corbett, L. Q., McGuiness, W., Snyder, R. J., & van Acker, K. (2014). Exploring the concept of a team approach to wound care: Managing wounds as a team. *Journal of Wound Care, 23*(Sup5b, Suppl. 5b), S1–S38. DOI: 10.12968/jowc.2014.23.Sup5b.S1 PMID: 25191792

Morton, L. M., & Phillips, T. J. (2016). Wound healing and treating wounds: Differential diagnosis and evaluation of chronic wounds. *Journal of the American Academy of Dermatology, 74*(4), 589–605. DOI: 10.1016/j.jaad.2015.08.068 PMID: 26979352

Moses, R. L., Prescott, T. A., Mas-Claret, E., Steadman, R., Moseley, R., & Sloan, A. J. (2023). Evidence for natural products as alternative wound-healing therapies. *Biomolecules, 13*(3), 444. DOI: 10.3390/biom13030444 PMID: 36979379

Mustoe, T. (2004). Understanding chronic wounds: A unifying hypothesis on their pathogenesis and implications for therapy. *American Journal of Surgery, 187*(5, Suppl.), S65–S70. DOI: 10.1016/S0002-9610(03)00306-4 PMID: 15147994

Negut, I., Grumezescu, V., & Grumezescu, A. M. (2018). Treatment strategies for infected wounds. *Molecules (Basel, Switzerland), 23*(9), 2392. DOI: 10.3390/molecules23092392 PMID: 30231567

Noor, S., Khan, R. U., & Ahmad, J. (2017). Understanding diabetic foot infection and its management. *Diabetes & Metabolic Syndrome, 11*(2), 149–156. DOI: 10.1016/j.dsx.2016.06.023 PMID: 27377687

Ousey, K., Gilchrist, B., & James, H. (2018). Understanding clinical practice challenges: A survey performed with wound care clinicians to explore wound assessment frameworks. *Wounds International, 9*(4), 58–62.

Owens, B. D., Kragh, J. F.Jr, Macaitis, J., Svoboda, S. J., & Wenke, J. C. (2007). Characterization of extremity wounds in Operation Iraqi Freedom and Operation Enduring Freedom. *Journal of Orthopaedic Trauma, 21*(4), 254–257. DOI: 10.1097/BOT.0b013e31802f78fb PMID: 17414553

Pacyga, K., Pacyga, P., Topola, E., Viscardi, S., & Duda-Madej, A. (2024). Bioactive compounds from plant origin as natural antimicrobial agents for the treatment of wound infections. *International Journal of Molecular Sciences, 25*(4), 2100. DOI: 10.3390/ijms25042100 PMID: 38396777

Patel, S., Maheshwari, A., & Chandra, A. (2016). Biomarkers for wound healing and their evaluation. *Journal of Wound Care, 25*(1), 46–55. DOI: 10.12968/jowc.2016.25.1.46 PMID: 26762498

Percival, N. J. (2002). Classification of wounds and their management. *Surgery (Oxford), 20*(5), 114–117. DOI: 10.1383/surg.20.5.114.14626

Percival, S. L., McCarty, S. M., & Lipsky, B. (2015). Biofilms and wounds: An overview of the evidence. *Advances in Wound Care, 4*(7), 373–381. DOI: 10.1089/wound.2014.0557 PMID: 26155379

Porel, P., Kaur, M., Sharma, V., & Aran, K. R. (2025). Understanding molecular mechanism of diabetic wound healing: Addressing recent advancements in therapeutic managements. *Journal of Diabetes and Metabolic Disorders, 24*(1), 76. DOI: 10.1007/s40200-025-01588-7 PMID: 40060271

Powers, J. G., Higham, C., Broussard, K., & Phillips, T. J. (2016). Wound healing and treating wounds: Chronic wound care and management. *Journal of the American Academy of Dermatology*, *74*(4), 607–625. DOI: 10.1016/j.jaad.2015.08.070 PMID: 26979353

Rautela, D. S., Das, S., Mazumder, A., Sharma, B., Khan, F., & Chaitanya, M. V. (2024). Understanding the influence of diabetes on wound healing and exploring effective management study. *Current Science*, *126*(12), 1438. Advance online publication. DOI: 10.18520/cs/v126/i12/1438-1445

Rodrigues, M., Kosaric, N., Bonham, C. A., & Gurtner, G. C. (2018). Wound healing: A cellular perspective. *Physiological Reviews*. Advance online publication. DOI: 10.1152/physrev.00067.2017 PMID: 30475656

Sapienza, P., Mingoli, A., Borrelli, V., Brachini, G., Biacchi, D., Sterpetti, A. V., Grande, R., Serra, R., & Tartaglia, E. (2019). Inflammatory biomarkers, vascular procedures of lower limbs, and wound healing. *International Wound Journal*, *16*(3), 716–723. DOI: 10.1111/iwj.13086 PMID: 30773823

Sen, C. K., Gordillo, G. M., Roy, S., Kirsner, R., Lambert, L., Hunt, T. K., Gottrup, F., Gurtner, G. C., & Longaker, M. T. (2009). Human skin wounds: A major and snowballing threat to public health and the economy. *Wound Repair and Regeneration*, *17*(6), 763–771. DOI: 10.1111/j.1524-475X.2009.00543.x PMID: 19903300

Tatarusanu, S. M., Lupascu, F. G., Profire, B. S., Szilagyi, A., Gardikiotis, I., Iacob, A. T., Caluian, I., Herciu, L., Giscă, T. C., Baican, M. C., Crivoi, F., & Profire, L. (2023). Modern approaches in wounds management. *Polymers*, *15*(17), 3648. DOI: 10.3390/polym15173648 PMID: 37688274

Werdin, F., Tennenhaus, M., Schaller, H. E., & Rennekampff, H. O. (2009). Evidence-based management strategies for treatment of chronic wounds. *Eplasty*, *9*, e19. PMID: 19578487

Whitney, J. D. (2005). Overview: Acute and chronic wounds. *Nursing Clinics*, *40*(2), 191–205. PMID: 15924889

Whittam, A. J., Maan, Z. N., Duscher, D., Wong, V. W., Barrera, J. A., Januszyk, M., & Gurtner, G. C. (2016). Challenges and opportunities in drug delivery for wound healing. *Advances in Wound Care*, *5*(2), 79–88. DOI: 10.1089/wound.2014.0600 PMID: 26862465

Xiong, Y., Knoedler, S., Alfertshofer, M., Kim, B. S., Jiang, D., Liu, G., Rinkevich, Y., & Mi, B. (2025). Mechanisms and therapeutic opportunities in metabolic aberrations of diabetic wounds: A narrative review. *Cell Death & Disease*, *16*(1), 341. DOI: 10.1038/s41419-025-07583-3 PMID: 40280905

Chapter 2
The Science of Wound Healing:
Biological Processes and Key Mediators

Narinderpal Kaur
http://orcid.org/0009-0004-0081-9793
Chitkara University School of Pharmacy, Chitkara University, India

Shiwali Sharma
Chitkara University School of Pharmacy, Chitkara University, India

Sameer Goyal
Chitkara University School of Pharmacy, Chitkara University, India

Juhi Kumari
Chitkara University School of Pharmacy, Chitkara University, India

Dharmendra Kumar Khatri
NIMS Institute of Pharmacy, NIMS University, Jaipur, India

ABSTRACT

Wound healing is a complex and dynamic process crucial for tissue repair and survival. This abstract provides an overview of wound understanding, encompassing types, stages, and associated challenges. Wound types are diverse, ranging from acute, such as surgical incisions and lacerations, to chronic, including pressure ulcers, diabetic foot ulcers, and venous stasis ulcers. Each type presents unique characteristics and healing trajectories. The healing process itself unfolds in four overlapping stages: hemostasis, inflammation, proliferation, and remodeling. Disruptions in any stage

DOI: 10.4018/979-8-3373-2908-6.ch002

can lead to delayed or impaired healing. Challenges in wound management are multifaceted. These include factors like infection, ischemia, malnutrition, and underlying diseases, which can hinder healing. Furthermore, accurate diagnosis, appropriate wound bed preparation, selection of effective dressings and therapies, and patient compliance are critical determinants of successful outcomes. This understanding, coupled with advances in wound care, including growth factors, negative pressure wound therapy, and advanced dressings, promises improved outcomes and quality of life for individuals with wounds. Further research is needed to personalize treatment strategies and address the increasing prevalence of chronic wounds.

INTRODUCTION

Our skin has created as a result of hundreds of evolutions. Skin is the largest organ of the body which expose to certain challenging factor like physical damage, chemical damage and ultraviolet radiation often result in injury to the skin. This multi-functional and complex organ however as no surprise possesses certain mechanism with the help of the mediators, cells of several types, cytokines and vascular system to heal the injury quickly and efficiently (Olsson et al.,2019). The mechanism to recover from the tissue injury is known as wound healing. Wound Healing is a natural physiological complex reaction to tissue injury. It is complex as it involves the interplay between several cells, cytokines, mediator and vascular system. The process of stoppage of bleeding is designed in a manner with the initial vasoconstriction of blood vessel then platelet aggregation. Then the influx of inflammatory cells takes place, primarily the neutrophil releases, then variety of mediator and cytokines release which in turn promote angiogenesis, Thrombosis and, re-epithelialization (Wallace et al.,2023). A superficial portion of epidermal layer heals quickly and regenerate from adnexal appendages. Adnexal appendages like hair follicle, sweat and sebaceous gland. This process is comparatively faster than other (Grubbs and Manna,2023). Wound healing classifies into two types- acute or chronic. Many people not able to understand what a normal wound healing is, until serious complication occur, like chronic wound or delayed wound. It helps as a physiological response to the disrupt integrity of the skin as well as healing of the organic of the individual. However, the evolutionary advancement, the understanding of wound healing is improving continuously (Sorg and Sorr,2023). The wound healing can occur intentionally or we can say primary intention while it can also occur unintentionally or secondary intention. Primary intention wound is non-infected, non-draining with the wound occur at the surface or edge of skin with close approximation with quick healing. Best example of intentional wound is wound cause during surgical process. While the unintentional wound is infectious, hypoxic or immune dysfunctional and occur may be deeply into the skin with separation of skin with long approximation. It is healing comparatively slower. It results in the loss of skin's tensile strength. Best example is injury by accident (Arif et al.,2024; Gantwerker & Hom .,2011).

ANATOMY AND PHYSIOLOGY OF SKIN

The skin, weighing about 15%of the total body weight, is the largest organ in the human body and plays a crucial role as in interface between the internal and external environment (Sen,2019).

STRUCTURAL LAYERS OF THE SKIN

The skin consists epidermis, dermis and hypodermis (subcutaneous tissue) as their major layers. All of these layers have their own specific, indispensable function in supporting skin integrity and adapting to injury.

Epidermis

The outermost vascular skin layer named epidermis, which is formed as the result of keratinocytes in stratified squamous epithelium. Keratinocytes in stratified squamous epithelium made up the epidermis which is outermost vascular skin layer other cell types are melanocytes, Langerhans cell, (Merkel Badylak,2002). The epidermis consists of a few sublayers: Stratum basal: Basal layer in which keratinocytes grow and begin upward differentiation. Stratum spinosum: Contributes structural strength through desmosomes links. Stratum Lucidum: The bordering barrier consisting of dead, flat corneocytes inter pressed in a lipid matrix. The epidermis regenerating ability is achieved through stem cells present in the basal layer that divide and move upward continuously to populate the other layer (Badylak,2002).

Dermis

Under the epidermis is the dermis, which is a layer rich in connective tissue that gives the skin elasticity and mechanical support. It comprises fibroblasts, mast cell, immune cells, blood vessels, lymphatics, nerve ending, and appendages like further divided into: Papillary dermis: The superficial layer with loose connective tissue and capillaries. Reticular dermis: The more compact, denser area with high collagen and elastic Fiber content. Dermal fibroblasts are the main source of the cellular metrix, such as type I and III collagen, glycosaminoglycan, and fibronectin, which plays a key role in wound healing and tissue remodeling (Friedman,2008; Choi & Diehl,2009).

Hypodermis (Subcutaneous Tissue)

The hypodermis is made up of mainly of adipocytes packed in lobules that are divided by fibrous septate. It exists reservoir of energy, fives insulation, and shield deeper structures from injury. During healing of wounds, the hypodermis also contributes to regulating inflammation and facilitating revascularization (Badylak,2002).

Vascular and Neural Supply

The skin is well supplied with blood, especially in the dermis, where there is a plexus vessels carrying oxygen, nutrients and immune cells to the tissue. Arterioles, capillaries, and venules are disposed in two horizontal plexuses: One at the junction dermal-subcutaneous. These vascular plexuses are crucial for thermoregulation and very much engaged in the inflammation and proliferative stages of healing and cure of wounds (Vestita, M et al.,2022). Nerve fibers in the skin facilitates sensory inputs (e.g., pain, temperature, touch) and autonomic regulation of blood flow and sweet glands. Neuropeptides release from nerve ending, like substance P and calcitonin gene-related peptide (CGRP), have the ability to affect inflammation and repair of tissue (Badylak,2002).

Skin Immune System

The skin contains a dedicated immune system called the Cutaneous. Immune System, which consists of a resident immune cell like Langerhans calls in the epidermis, dendritic cells, macrophages, and mast calls in the dermis. These cells recognize pathogens and tissue injury and trigger relevant immune responses (Badylak,2002; Finnson et al.,2013).

The Extracellular Matrix (ECM)

The ECM is a sophisticated network of protein and polysaccharides that offers structural support and biochemical cues to adjust cells. It governs cell adhesion, migration, proliferation, and differentiation, which are vital events during wound healing (Friedman,2008 and Choi and Diehl,2009). In healthy skin, the ECM have a balance of synthesis and breakdown. After injury, ECM composition dynamically changes to ensure cell recruitment and angiogenesis. Collagen, particularly type I and III, fibronectin, hyaluronic acid and proteoglycans are key to these processes (Choi and Diehl,2009; Freedman et al. 2023).

Appendages and Reservoirs of Steam Cells

Hair's follicles, sweat glands, and sebaceous glands, although underappreciated, are also important for repair in the skin (Pase et al.,2012). Specially, bulge area stem cells within the hair follicles and interfollicular epidermal steam cells are responsible for reepithelialisation during the proliferative phase of healing (Pase et al.,2012). Knowledge of skin anatomy and physiology gives an underlying structure to study of how tissue reacts to injury and heals. Dysfunction or disrupt of any of these structural or functional elements- by trauma, infection, or systemic illness- can impair healing (Grubbs and Manna,2023).

CLASSIFICATION OF WOUND

The wound is classified as acute or chronic. Generally, wound is healed within 4-6 weeks. If the wound healing is done within 12 weeks, it is classifying as acute wound. It is a normal wound condition without severe condition. The world is considered as chronic if it,s healing takes longer than 12 weeks' time period or even not healing properly at all (Sorg and Sorg,2023).

Significantly the aged and the diabetic person is more prone to chronic wound (Grubbs and Manna,2023). Improper care of chronic wound can result as limb amputations and mortality (Rodrigues et al.,2019). Categorization of wounds id important for diagnosis, care, and prediction, since various types of wounds have specific healing patterns and complication risks. Proper knowledge of the types of the wounds guides proper clinical action, ranging from the selection of dressing to the need for systemic therapy (Wang et al.,2018; Pisarik,2016).

Classification according to healing mechanism

One of the most commonly employed frameworks organizes wounds based on their type of healing:

Healing by Primary Intention

Wounds that close by primary intension are those where the edges of the wound are bought directly together, normally by suturing or surgical closure. They have little tissue loss, low risk of infection, and tend to heal quickly with little scarring (Grubbs & Manna,2023).

Healing by Secondary Intention

In contrast, secondary intension wounds are closed and not closed and heal spontaneously, Wounds which include extensive tissue loss, ragged edges, and are at risk of infection are typically example of such wounds. Wounds healing is slower and entails much granulation tissue development, contraction, and epithelialization and usually leads to scarring which is apparent (Grubbs & Manna,2023).

Tertiary Intension healing (Delayed primary closure)

Delayed primary closure or tertiary intension refers to the practice of leaving a wound initially open because it is contaminated or a at risk of infection and then closing it surgically once inflammation has subsided. It is employed to weight the advantages of wound closure against the employed to infection and aiding healing to be maximized (Grubbs and Manna,2023).

Classification according to Time frame

Wounds are also divided according to their time frame: Acute Wounds: These go through the usual healing phases within the proper timeframe and generally heal between 4-6 weeks (Arif et al.,2024). Chronic Wounds: These do not heal within the expected timeframe, usually getting stuck at inflammatory stage. Chronic Wounds comprise diabetic foot ulcer, pressure injuries and venous (legs Wallace et al.,2023; Sorg and Sorg 2023).

Classification according to Etiology

Wounds may be result from varied causative factors, which have a veery important bearing on treatment strategy and outcome.

Mechanical Wounds

These are due to physical injury cuts, abrasion, laceration, or blunt trauma. Mechanical injuries are generally acute and can heal uneventfully with little treatment if not confounded by infection or comorbidities (Sorg and Sorg,2023).

Thermal Wound

Thermal wounds consist of burn (due to heat), frostbite (due to cold), and radiation injury. Burns are also divided by depth- superficial, partial-thickness and full-thickness—and present special problems because of loss of fluid, risk of infection, and the possibility of systemic inflammatory responses (Arif et al.,2024).

Chemical Wounds

Exposure to corrosive substance like alkalis or acids can lead to tissue necrosis and ulceration. Corrosive injuries need specializes decontamination and tent to cause deep tissue damage (Arif et al.,2024).

Surgical Wounds

These are Iatrogenic wounds that are inflicted by design during surgery. Healing is contingent or tissue handling, aseptic technique, and the patient's overall health (Grubbs and Manna,2023).

Infectious Wounds

Infection may be cause and a complication of wounds. Infected wounds usually become chronic and are complicated by delayed healing, necrosis, purulence, and systemic manifestations of inflammation (Arif et al.,2024; Tomasek et al.,2002).

Depth-based classification

Wounds depth will affect dynamics of healing and intervention: Superficial Wound: Include only epidermis. These tends to heal quickly with minimal scaring. Partial Wounds: Include the epidermis and part of the dermis. Reepithelialization and ECM remodeling facilitate healing. (Badylak,2002). Full- thickness Wounds: Penetrate through the dermis and can extent into subcutaneous tissue, muscle, or bone. Such wounds necessitate formation of granulation tissue and could require surgical treatment (Arif et al.,2024).

Classification Based on Cleanliness (Surgical Wounds)

The CDC surgical wounds classification divides wounds by the presence or absence of contamination: Normal: Uninfected operative wounds without inflammation. Normal Contamination: Operative wounds where the respiratory, GIT, or genitourinary tract is entered under controlled condition(Badylak,2002). Contaminated: Open, fresh accidental wounds or major breaks in sterile technique. Dry-infected: Old traumatic with retained devitalized tissue and clinical infection (Arif et al.,2024).

Clinical significance

Refurbishment of injured tissue of the skin consider as, principal function of the wound healing. The epithelial layer of the skinis act as the barrier to creat a protective exposed region from contamination and infection which can caused by microbes and other environmental harmful agents around wound. Wound healing also reliefs to regain the tensile strength during injury (Grubbs and Manna,2023). It can be scrubbed with 0.9% saline solution. However, the concentration of saline solution can be varied from region to region as per different pharmacopeia. Pluronic polyols in combination with saline solution can also be used as an alternative for cleansing of wound. Sometime the clean tap water for cleaning from seeking the medical attention (Pisarik, 2016). Further times, the complications such as hypertrophic scars and keloids create as a result of may be overactive and excess scar tissue formed during wound healing. Both hypertrophic scars and keloids may be associated with itching, burning, or tenderness (Grubbs and Manna,2023).

Cellular response during wound healing

Many authors describe the wound healing process by divide into three to four phases or even more but commonly the wound healing process by studied the 4 phases for relevant study (Sorg and Sorg,2023). These all stages proceed in overlapping manner where one step helps and leads to next step (Landén et al.,2016). Each step has determined time period based on approximation. Each stage has its own significant role. These 4 steps categorize ashemostasis stage (0–several hours after injury), inflammation stage (1– 3 days), proliferation stage (4–21 days) and remodeling stage (21 days– 1 year) (Reinke & Sorg,2012). Deregulation in any of the stages of wound healing cause healing impairment. Example is excessive scarring. This can cause the burden on economic and health status of the society (Landén et al.,2016; Mustoe et al.,2006). Identification of new therapeutic targets and development of more effective treatments are emerging with continuation till date (Sun et al.,2014).

Figure 1. Phases of wound healing

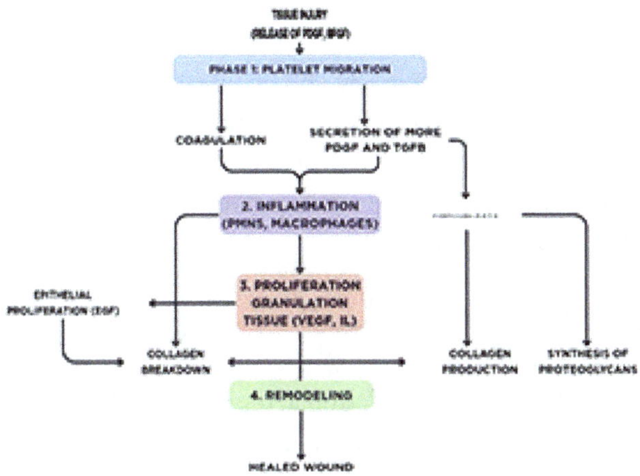

Haemostasis phase

This phase is primarily occurred immediately after the injury. It all starts with the initial three-step phase of coagulation, primary and secondary hemostasis immediately after trauma (Sorg and Sorg,2023). It starts with the rapid contraction of the damaged blood vessel and formation of the blood clot to prevent exsanguination (i.e. severe loss of blood from the entire body or a significant portion of it within a short period of time) as it even lead to death within first 24-hour period of time (Velnar et al.,2009). In this phase, once the skin gets injured, the collagen gets exposed that initiate the cascade of the extrinsic and intrinsic clotting. The vasoconstriction of the damaged blood vessel triggered by the aggregation of thrombocytes that reduce the blood loss and also cause hypoxic condition (Mussbacher, M, et al, 2024). The principle contributor to homoeostasis and coagulation is platelets which are activated when they encounter the vascular sub endothelial matrix. The principle matrix component is fibrinogen (Ozgok Kangal and Regan, 2025). During normal condition, in an uninjured skin, the platelets do not attach to the vessel, and aggregate with each other. The fibrinogen (factor 1) is produced by the hepatocytes and circulate in the blood (Rodrigues et al.,2019). It is also present within the platelets but not cleaved into fibrin fibers (an essential component of blood clot). As we already discuss, there are three sub-phases of haemostatic phase. The primary response is the vasoconstriction that occur immediately after injury to stop blood loss. It proceeds further to primary hemostatic phase that involve platelet plug formation and platelets aggregation. Then last sub- phase, secondary homoeostatic involves activation of coagulation process, where the insoluble fibrinogen release from hepatocytes is convert into insoluble strand that make up the fibrin mesh. The fibrin mesh and platelet plug from secondary and primary hemostasis phase respectively combine to from thrombus to stop bleeding (Rodrigues et al.,2019). Also in this phase, a number of cytokine and growth factors are release, helps in mediating the communication and regulate the activity of different cell component in wound healing process (Landén et al.,2016).

Vasoconstriction

This starts with the vessel constriction at the injured skin part to stop blood loss. The neuronal reflex mechanism because of blood vessel injury causes the blood vessel to contract by vascular smooth muscle cells in the tunica muscularis (Godo and Shimokawa,2017). The reflexive contraction of vascular smooth muscle around injured part elicited by the vasoconstrictor receptor such as endothelin from the damaged endothelium. Additionally, vasoconstriction also regulate by circulating catecholamine, epinephrine, norepinephrine, and prostaglandins released from injured cells (Godo and Shimokawa,2017). Initial reflexive contraction only temporarily reduces bleeding for short time period (like few minutes), as increasing hypoxic condition and following successive acidosis of the wound results in passive relaxation of the muscle with vasodilation (Sorg and Sorg,2023;Rodrigues et al.,2019;Landén et al.,2016). Mediators like bradykinin, fibrinopeptide, serotonin, and thromboxane A_2 supports to regulate the activation of subsequent activation of coagulation cascade as to resolve the issue of bleeding for a longer duration (Teller and White,2009).

Primary haemostasis

In this, platelets plug formation takes place.This phase involve platelet plug formation and platelets aggregation. Schultze was also the first that notice that platelets involve in Thrombus formation. Platelets bud off from megakaryocytes (Brewer, 2006). One of the important components for platelets activity is alpha granule (secretory granule in platelets). Nitric oxide, prostacyclin, and negatively charged heparin-like glycosaminoglycan was produced from monolayer endothelial layer thus exhibit anti- thrombotic property and prevent platelet activation, attachment, and aggregation (Golebiewska and Poole,2015). During wound injury, the blood vessel breach which makes the thrombogenic sub-endothelial matrix wide-open to the external environment. Platelets bind to the receptor present on the matrix throng g protein coupled receptor mechanism which stimulate integrin activation. This increases the attachment of platelets to other platelets and surrounding extracellular matrix. A2β1 is second most widely distributed integrin on the platelets and mediate attachment of platelets with collagen (Rodrigues et al.,2019). Filamentous actin is the most abundant protein in the platelets with 40% of the protein in resting state that surge to approximately 70% during active state (Rodrigues et al.,2019). The attachment of platelet to the ECM assists as the result of conformational change in the shape of actin, contracts, and mechanically seals the blood vessel (Sorrentino et al.,2015). The dead layer over skin after injury (known as eschar) perform several secondary functions. Few function such as protection from bacterial invasion, provide a scaffold for immune cells and harbouring reservoir of growth factor and cytokine to dictate the function of wound cells in early stage of repair (Sorg and Sorg,2023). Transforming growth factor-β (TGF-β), cytokines such as PDGF and Growth factors, insulin growth factor (IGF), and epidermal growth factor (EGF) are vital cell mediator to help in subsequent phases of wound healing. It was release by platelets that forming plug (Rodrigues et al.,2019). The release of the platelets factor is more intense during the initial time period (first few hours) of platelet activation; however, it releases platelet factor for up to 7 days to show its effect of wound healing (Sorg and Sorg,2023). However, platelets are essential in primary hemostasis phases, but their absence does not cause critical defect in wound healing. Thrombocytopenia or low platelets count do not have profound affect the wound healing process was indicated by murine experiments. Mice with thrombocytopenia have relatively more macrophages, and T- cell but do not have any other effect. In fact, growth factor release, rate of wound re-epithelialization, collagen synthesis, or angiogenesis remain similar as normal condition (mice Szpaderska et al.,2003).

Secondary hemostasis

Figure 2. Mechanism of wound healing

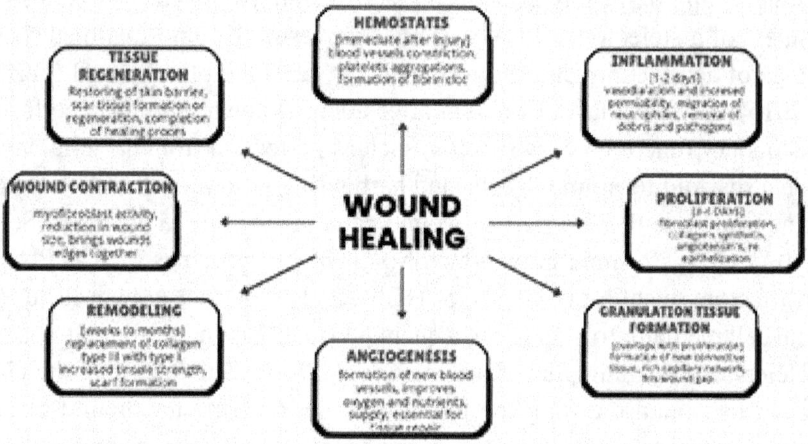

This phase marks the termination of hemostasis phase and beginning of the inflammation phase. In this phase, the coagulation complex gets activated and assemble on the surface of the platelets. Due to the exposure of the endothelial matrix, the coagulation pathways (intrinsic and extrinsic) get activated, followed by activation of factor X (Landén et al.,2016). In this, prothrombin gets convert to thrombin which leads to the breakdown of fibrinogen into fibrins. This fibrin crosslink covalently with factor XII, with which platelets aggregate plug forming secondary hemostasis plug to seal and prevent the blood loss (Rodrigues et al.,2019). Once the sufficient clotting is done, the coagulation process get over and excessive thrombosis get inhibited. In this, the platelet aggregation gets stopped by prostacyclin, thrombin gets inhibit by antithrombin III, and coagulation factors degrade by protein C in its activated form (Sorg and Sorg,2023).

Inflammation phase

This stage starts within hours of trauma and usually continues for 2 to 5 days in acute wounds. It is marked by neutrophil and monocyte recruitments to the wounds, which act against pathogens and remove debris. In this phase, once platelets activated the inflammatory cells migrate toward the site where wound is present on the skin. Tissue injury and late recovery can be caused by uncontrolled and excessive inflammation (Sorg and Sorg, 2023). In this phase, the amplification and recruitment of cells was stimulated for digestion of non- viable tissue, foreign agent (which may be harmful) and bacteria (Sorg and Sorg,2023). The inflammatory phases help to recover the wound bed in clean condition for the further repair processes (Grubbs and Manna,2023). As wound with bacterial imbalance will not heal, the phagocytic activity is critical for the continued progress of wound healing (Sorg and Sorg,2023). Some resident skin cells like keratinocytes, macrophages, dendritic cells and mast cells are wide-open to danger signals due to injury. These danger signal, in general can classify into two types- (a) DAMPs (damage- associate molecular patterns, molecules released by stress cell which go through necrosis. Molecules such as intracellular proteins, DNA and RNA. (b) PAMPs, i.e. which are specific to certain pathogen molecules not found in the host. Example is bacterial essential polysaccharides and polynucleotides (Strbo et al.,2014). Recognition of danger signal was done by pattern recognition receptors, among various receptors toll-like receptor are the characterized one. As a stimulation of TLRs occur, it induces the activation of intracellular signaling pathways that leads to expression of a large number of genes, like cytokines, chemokine and antimicrobial peptides (Takeuchi and Akira,2010). These expressions help to initiate and perpetuate the inflammatory response (Landén et al.,2016).

Action of neutrophil in inflammation phase: -

Among many components primarily, neutrophils predominate. It can be described by its recruitment process, cell adherence diapedesis through the tissue, and reached the wound site at last. It reaches the wound site by attracted by the bacterial product at the site of injury. It possesses engulfment of bacteria along with dead tissue forming the pus visible in skin injured part within period of 2-3 days (Sorg and Sorg,2023; Grubbs and Manna,2023). Neutrophiles are the initial immune cells to migrate to the injury, following chemotactic cues like 1L-8 and complement protein. Their function is phagocytosis of bacteria, deposition of proteolytic enzyme (e.g., elastase, myeloperoxidase), and production of reactive oxygen species (ROS) (Midwood et al.,2004; Landén et al.,2016). The neutrophils reached the wound site where the chemokine is released, once outside the blood vessel the neutrophils get expose to chemokine gradient within the skin. Chemokine released as the complement and by-products of bacterial degradation (Vestweber,2015). DAMPs, H_2O_2, and chemo attractants (CXCL4, CXCL8, CXCL10, CXCL12) helps the neutrophils to react to signals which develop a gradient. This gradient helps the neutrophils to reach the wound site (Rodrigues et al.,2019). The neutrophils move toward the higher concentration. It the noticeable that leucocytes also follow same mechanism for localization. If the wound remain infection free, then commonly neutrophil remain for 2-5 days (Godo and Shimokawa,2017; Teller and White,2009). Once the invasion of microbes was prevented by neutrophil. At the same moment, (G-CSF) granulocyte colony stimulating factor and CXC chemokine are released and reached the bone marrow with the help of systemic circulation, which allow the mature neutrophil to reach peripheral pool and eventually the wound site. Thereafter neutrophil reach the necrotic zone via formyl peptide gradient, in which the formyl-peptide receptor dependent prevails signaling and help the neutrophil to migrate. Neutrophil release toxic granules containing, e.g. lysozyme, cathepsin G, and protease to counteract the infectious threat (Wilgus et al.,2013). Neutrophils exit via special cell death called NETosis. This NETs are formed from the DNA of neutrophils, beads of proteases and from histones. This NETs supports neutrophil capturing the pathogen. While the NETs trapped pathogen, other things are also present like electrical charge differences to prevent spreading of pathogen and anti-microbial protein i.e. neutrophil that kills the pathogen (Fuchs et al.,2007; Tan et al.,2021). NETosis are of two forms- suicidal and vital. Both process are different in the course of time period and end result. The suicidal NETosis initiated by various microbes or cytokine and last for few hours, while the vital NETosis initiate by bacterial lipopolysaccharides, other products of bacteria and last for few minutes. In suicidal NETosis, cell death occurs. Meanwhile, in vital NETosis, cell keeps in function (Fuchs et al.,2007; Tan et al.,2021). The extrusion and apoptosis of neutrophil is crucial as it is must to remove the neutrophil to initiate the next phase. The extrusion product on wound surface called as extrusion slough (Sorg and Sorg, 2023).

Action of macrophages in inflammation phase: -

After injury within approximately 3 days of injury, circulating monocyte reaches the wound site, where they differentiate into macrophages, in response to local milieu. Various chemo-attractive agents help in the differentiation of monocyte into macrophages (chemo-attractive agents like SDF1 and CXCL12) (Rodrigues et al., 2019). The function like debride the wound, clearance of the matrix, and other cell debris like fibrin perform by macrophagesMacrophage are crucial for transition of phase between inflammatory and proliferative phase (Landén et al., 2016). It was also important for overall wound repair process. Thus, inhibition of macrophage and its function cause the delayed wound healing process. Macrophage exhibit differential behaviour and morphological alteration in response to cytokine (Sorg and Sorg, 2023). Monocytes infilter the wounds and differentiate into macrophages, which serves a bimodal function: pro-inflammatory (M1 phenotype) and anti-inflammatory (M2 phenotype) (Barman and Koh, 2020).M1 macrophage secrete inflammatory cytokines L-β, TNF-α, and nitric oxide to maintain infection. With changes in thew environment, macrophages convert to the M phenotype, secreting anti-inflammatory signals and facilitating repair of tissue (Renata Novaes et al., 2022). Macrophage play a key role in moving from the inflammatory to the proliferative phase (Finnson et al., 2013). Wound macrophages are conventionally divided into two main subsets: M1–stimulated and M2-stimulated, but it is an outdated classification. For simplicity, the macrophages divided into category- classically activated (inflammatory) and alternatively activated (anti-inflammatory) (Sorg and Sorg, 2023). The induction of classically activated macrophage is done by proinflammatory stimuli, such as LPS and interferon gamma (IFN- γ), and promote inflammation by releasing ROS (reactive oxygen species), inflammatory cytokines e.g. IL-2, IL-6, and TNF-α) and growth factors like endothelial growth factor, VEGF and PDGF. These macrophages replacing phagocytized apoptotic neutrophils as the main inflammatory mediator. Neo-differentiation of recruited monocytes or switching of existing macrophages stimulate the transition to alternative activation which is a characteristic property of later stage of inflammation (Schultz and Wysocki, 2009). In normal uninterrupted wound healing process, the inflammation commonly takes 2-5 days for recovery. Once all the harmful stimuli have been removed, the wound healing process terminates. Even though the immune response continues throughout the whole process (Landén et al., 2016).

Action of T cells, dendritic cells and mast cells in inflammation phase: -

Dendritic epidermal T cells (DETCs) play a crucial role in epithelialization process. DETCs change its The shape of DETCs change from dendritic to round within the period of 1-2 days of injury. This activates the DETCs to release keratinocyte growth factor (KGF-1), (KGF-2) and (IGF-1). This led to proliferation of keratinocytes throughout the wound. As per depletion studies, the alteration in the amount of DETCs cause the significant closure in wound (Nosbaum et al., 2016). On the other arm of immunity, which is adaptive immunity, B and T cells carried out a more specific and relatively slower response. IL-1 attracts the T lymphocyte cell type to migrate into the wound or site of injury. B cells not only secret antibodies, but also put an influence on immune response. B cell happens to present to wound tissue and help in healing. T lymphocyte helps as growth factors-producing cells as well as immunological effector cells. Deficiency of CD4-CD8 effects the inflammatory activity and cytokine expression. A prolong and increase presence of T cell with a changed CD4-CD8 ratio have observe in case of human chronic wound (Landén et al., 2016). Several cells, which bridge in innate and adaptive immunity play a key according to their types in skin wound healing. Plasmacytoid dendritic cells (PDCs) along with neutronphil infilter the skin wound at the same time. Langerhans cells (LCs) are a subset of epidermal dendritic cells contributing to epidermal immune monitoring (Gantwerker and Hom, 2011). Apart from γ and δ subunits composed of γδT cells is another subset of T cells. Dendritic epidermal T cells (DETC) are the γδT cells present as a subpopulation in the epidermal layer (Mustoe et al., 2006). γδT also eliminates the damaged keratinocytes, releasing growth factor (Schultz and Wysocki, 2009). Mast cells also play a role in inflammatory phase of wound healing as it releases the vasoactive cytokines such as prostaglandins and histamine which enhances capillary permeability and stimulate local dilation to aid the migratory process (Grubbs and Manna, 2023).

Figure 3. Recovery of wound healing

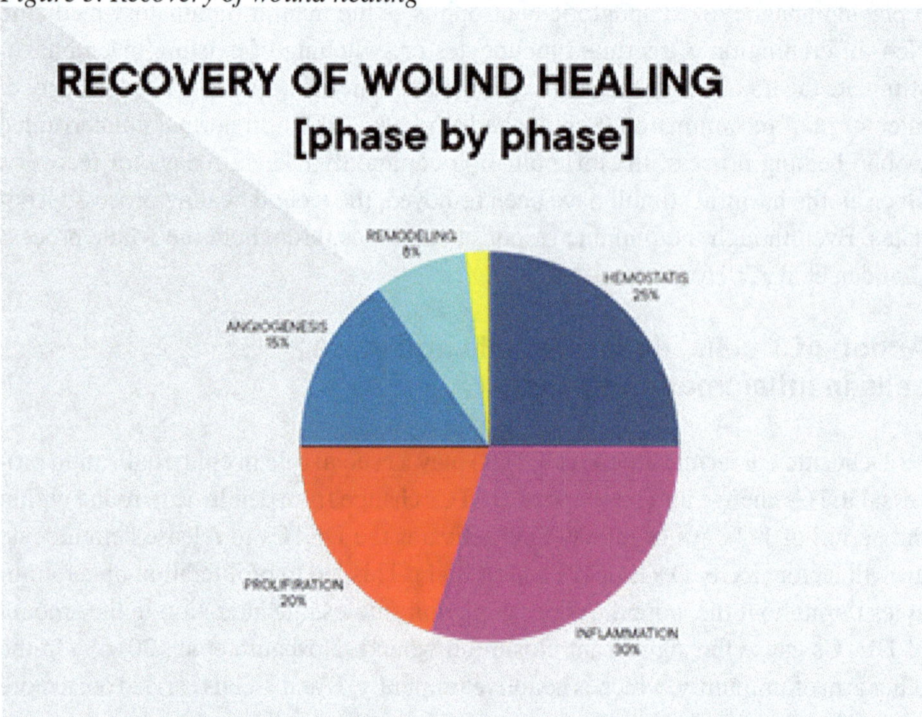

Proliferation phase

It targets on covering would surface (i.e. re-epithelialization), restoring the vascular network and granulation of tissue (Landén et al., 2016). This phase starts approximately after 3-10 days post-injury and it may require days or weeks to complete the wound healing process. During this phase, formation of granulation tissue, and other processes of healing like re-epithelialization, neo-vascularization and immunomodulation occur concurrently (Ozgok Kangal and Regan, 2025).

Granulation Tissue Formation

In this phase, the wound matrix formed in the course of haemostasis phase is replaced by granular tissues. In this phase, wounded skin gets partially recover on the structural and function basis as it consists large number of components like fibroblasts, granulocytes, macrophages, blood vessels, in complex with collagen bundles (Li et al., 2003). In this, fibroblast and endothelial cells are primary key mediators. Fibroblast contribute to angiogenesis also (Ozgok Kangal and Regan, 2025). Fibroblast produce MMPs that degrade the pre-existing fibrin clot. Then a provisional matrix containing collagen type III, glycosaminoglycan, and fibronectin is synthesized. In this manner the pre-existing fibrin clot gets interchanged by new provisional matrix supporting keratinocyte or re-epithelialization. (Ozgok Kangal & Regan, 2025; Wang et al., 2018).

Angiogenesis and vasculogenesis

Endothelial cells migrate, proliferate, and align into tubules, re-establishing oxygen and nutrient delivery to regenerating tissue (Reinke and Sorg, 2012). Nutrients and oxygen is must for the restoration of network of blood vessel as the vascular bundle restoration is very important during wound repair (Landén et al., 2016). There are two mechanisms through which the new vessels (i.e. angiogenesis and vasculogenesis) are formed. Angiogenesis is a sprouting process in which the endothelial cells from the nearby mature vascular network or existing blood vessel aids to grow new vessel into the avascular site. Whereas vasculogenesis is a de novo process during which new vessels form without sprouting with the help of differentiation of progenitor stem cells. These stem cells characteristically found in bone marrow (Schultz et al., 2011). The growth factors, e.g. vascular endothelial growth factor (VEGF), platelet-derived growth factor (PDGF), basic fibroblast growth factor (bFGF), and the serine protease thrombin in the wounds initiated the angiogenesis, which activate the endothelial cells of existing vessels (Wu W.-S, et al, 2025). The administration of anti-angiogenic medications such as bevacizumab may interrupt this phase and lead to chronic wound formation. Neovascularization is critical for efficient wound healing in infant (Ozgok Kangal and Regan, 2025). After secreting proteolytic enzyme for dissolution of basal lamina, the endothelial cells leave the existing vessel, migrate and proliferate toward angiogenic stimulus. These sprouts differentiate into arteries and veins and mature with the help of pericytes and cells of smooth muscles. Under TGF- β stimulation, fibroblast transform into myofibroblasts, which express alpha-smooth muscle actin (α -SMA), allowing them to contract and close the wound (Tomasek et al., 2002). Fibroblast, attracted by PDGF and TGF-β, immigrate into the wounds bed and start producing ECM protein like collage (mainly type III), fibronectin, and glycosaminoglycans (Tomasek et al., 2013). Wound healing also promotes by circulating fibrocyte. Fibrocyte not only supports by contribute to subset of fibroblasts, but also by endorsing the production of cytokines, chemokine, and growth factors, serving as antigen presenting cells as well as enhancing angiogenesis (Pereira et al., 2023).

Wound Contraction

Myofibroblast is key mediators in the wound contraction phase. One of the main features of proliferative phase is wound contraction that cause the reduction in surface area that facilitates the re-epithelialization (Tomasek et al., 2013). During this process, collagen fibres organize in a manner around wounds edges as to enhance the mechanical strength of the tissue (Tomasek et al., 2013). The fibroblasts migrating towards the centre of wound transform into α-SMA positive myofibroblasts due to the alteration in the stiffness of the wound environment. Myofibroblast deposit ECM and exhibit contractile smooth muscle. Post-injury, local fibroblast residing in the dermis layer of uninjured area and sub-cutaneous are considered as the chief source of myofibroblast but there are possible odds that the myofibroblast subsets derive from other cell types including from fibrocytes, MSCs, pericytes, and epithelial cells (Rodrigues et al., 2019; Velnar et al., 2009). The myofibroblast synthesize from differentiated α-SMA occur in response of fibroblast in response with EDA fibronectin, an alternatively spliced segment of fibronectin in the wound bed via $\alpha5\beta1$ and $\alpha v\beta3$ integrin and stimulation by TGF-β. At the day 3, the rise of myofibroblasts in murine wounds initiates and peaks at day 7 following wounding where production of ECM happens. ECM include collagen type I and III and α-SMA (Lau et al., 2009). The liver is a model organ for the study of fibrotic wound healing. Hepatic fibrosis, caused by chronic inflammation, leads to excessive deposition of the ECM and scarring, which is mainly mediated through activated stellate cells (Pisarik, 2016). Integrin assists the polymerization of fibronectin and collagen fibrils and myofibroblast attachment and pulls the fibril perpendicular to wound edge via antirich cytoskeleton, permit wound contraction (Lau et al., 2009). Inflammatory mediators including TNF-α can hinder differentiation, suggesting that myofibroblast differentiation may be postpone and wound-healing process can be delayed because prolonged inflammation in diseased states (Demidova-Rice et al., 2012).

Re-epithelialization

Appendage and wound edge keratinocytes proliferate and migrate over the bed of the wound, depositing a new epithelial layer upon it. (Midwood et al., 2004;Reinke and Sorg, 2012). MicroRNAs are pivotal in controlling EMT. miR-221, for example, promotes TGF-β1-induced EMT in bladder cancer cells by targeting STMN1, associating cancer development with wound healing pathways (Swann, 2010). ERBB2 in prostate cancer was shown to enhance metastatic capacity by mechanisms of EMT significance, reiterating the role of EMT in cellular plasticity (Nussbaum et al., 2018). Re-epithelialization take place by migration and proliferation of keratinocytes. Throughout a course of few hours to 1-day post-injury, the migration of the existing keratinocyte present at the wound edge begins (Nussbaum et al., 2018). Keratinocyte by the wound edge and epithelial stem cell from nearby hair follicle or sweat gland (Demidova-Rice et al., 2012). The body due to the presence of epidermal layer is sheltered from microorganisms, ultraviolet radiation, water loss, and extreme temperature and withstand mechanical stress (Godo and Shimokawa, 2017). Migration is triggered as the result of loss of contact inhibition and physical tension at cell junctions i.e. desmosome and hemi desmosome (Wallace et al., 2023). These junctions are present within the multi-layer present around skin. It consists multilayer epithelium at which keratinocyte is present and these are connected to the nearby keratinocyte by cell-cell junction called desmosome (Wallace et al., 2023). The lowermost basal layer attach to ECM called basement membrane through hemi desmosome (Landén et al., 2016). Additionally, Keratinocytes gets activated by the changes in mechanical tension and electrical gradients, and exposure of H_2O_2, pathogen, etc. This activation cause keratinocyte to undergo epithelial-mesenchymal transition, where they develop migratory phenotype. This interchanges the polarity from front-rear to top-bottom which allow the keratinocyte to migrate across wound laterally (Sorg and Sorg, 2023). Fibroblasts found in granulation tissue and pro-repair macrophages encourages the occurrence of re-epithelization (Woo et al., 2004;Ozgok Kangal and Regan, 2025). Endothelial progenitor cell recruitment into the circulation begins immediately post-injury. NO, VEGF, and MMPs— primarily MMP-9—all monitoring endothelial progenitor cell for mobilization. Likewise, stromal-derived factor 1-α is the primary signal guiding endothelial progenitor cells to gather to areas of ischemia (Schultz et al., 2011;Tomasek et al., 2002).

Remodeling phase

It is the ultimate phase of wound healing, which initiates on day 21 and may continue to proceed further for up to 1 year. The equilibrium between the synthesis and degradation of new tissue is crucial for this phase. Any disturbance causes the development of chronic wound (Godo and Shimokawa, 2017). The end of wound point of marked by the closure of the wounds, but wound remodeling and tissue maturation can still continue for several years (Rodrigues et al., 2019). This phase start at the end of granulation tissue development, regression of the neo vasculature and the maturation of wound. Myofibroblast undergo apoptosis when healing is complete (Mustoe et al., 2006). Various components are stimulated by various signals to form a long-lasting and more organized extracellular matrix (Ozgok Kangal and Regan, 2025). Collagen synthesis occurs in this phase, which last for up to 4 weeks. However, collagen in the wounded area is never recover as organized as healthy skin as it is before the injury (Ozgok Kangal and Regan, 2025). Glycoproteins, and proteoglycans, myofibroblasts also contribute in wound contraction along with the synthesized major extracellular matrix proteins such as collagen types I to VI and XVIII. . During collagen synthesis, hydroxylases enzyme performs major role with the help of oxygen and vitamin C. Thus, the strength of the repaired tissue can distress by hypoxia and vitamin C deficiency. (Ozgok Kangal and Regan, 2025). At this phase, collagen I substitute the rapidly synthesized collagen III in the ECM. Collagen I has a somewhat greater tensile strength than collagen III but takes longer time to deposit. Numerous fresh blood vessel and blood flow decline. Thus, a mature avascular and acellular environment is formed (Pisarik, 2016) Macrophage also play a vital role during wound remodelling phase where they take on fibrolytic phenotype, excessive ECM get break down and ECM debris and apoptotic gets engulfed (Teller and White, 2009). Moreover, Wnt and Notch pathways control important genes like c-Myc and Hes1, regulating proliferation and differentiation in wound healing contexts (Shi et al., 2015). Another key ECM component, elastin must reform elastin fiber to skin to retain skin elasticity. Fascinatingly, elastin fragments, or elastokines release as result of result of normal dermal matrix. Elastokines act as a signaling molecules (Gantwerker & Hom, 2011). Skin components such as hair follicles and sweat glands are permanently lost after serious injury (Landén et al., 2016). After the period of 3 months, only 80% of the wound strength get recover (Teller and White, 2009).). The angiogenic process terminate at the end of the entire process, and the blood flow shrink to pre-injury level. The complete closure of injured tissue site and mechanical strength of the wound is re-established in this process. At the end of wound healing, the inflammation significantly triggers the formation of scar. The process of formation of scar is scarring. The defects like the wound strength can never catch up with the normal skin strength. Scar tissue also lacks rete pegs, which allow for a tight connection between the epidermis and dermis (Teller and White, 2009). Typically, the scar stays for 6 to 9 months and then begin to soften, flatten and become paler (Wallace et al., 2023).

FACTORS AFFECTING WOUND HEALING

Intrinsic Factors

Wound healing is determined by wide range of intrinsic and extrinsic variables that may facilitates or inhibits tissue repair. It is crucial to understand these factors for maximizing clinical acre and enhancing healing (DeRouen and Oro, 2009; Shi et al., 2015)

Age

Increased age harms healing of the wounds through decreased cell proliferation, compromised angiogenesis, and diminished immunity. Older people tend to have delayed re-epithelialization and a higher risk for chronic wounds (Choi and Diehl, 2009).

Nutritional Status

Protein, vitamin C, zinc, and iron nutrient is essential for collagen synthesis, immune response, and cell proliferation. Malnutrition delays healing and raises the risk of infection (Choi and Diehl, 2009).

Comorbidities

Diabetes Mellitus:

With its associated neuropathy, micro vascular disease, and immune dysfunction, diabetes disables all phases of healing, resulting in chronic wounds (Pastar et al., 2010; Ozgok Kangal and Regan, 2021).

Vascular Disease:

Arterial insufficiency limits oxygen supply; venous insufficiency produces edema and inflammation, both disabling repairs (Pastar et al., 2010).

Immunosuppression:

Medical conditions or drugs that decreases immune activity slow down removal of pathogens and cellular waste (Shi et al., 2015).

Extrinsic Factors

Infection

Bacterial colonization and biofilm formation perpetuate inflammation and tissue damage. Infection is a principal cause of decayed healing and wound chronicity (Tome-Garcia et al., 2014; Chhabra et al., 2017).

Mechanical Stress

Excessive pressure, shear, or movement on the wounds disrupts granulation tissue and epithelial migration, particularly in pressure ulcer (Campos et al., 2010).

Smoking

Certain drugs, including corticosteroids and chemotherapies, suppress inflammation and cell proliferation, negative impacting repair (Pastar et al., 2010).

Medications

Certain drugs, include corticosteroids and chemotherapeutic agents, suppress inflammation and cell growth, negatively impacting repair (Pastar et al., 2010).

Environmental Factors

Oxygenation:

Hypoxia compromises cellular metabolism and angiogenesis. Hyperbaric oxygen treatment may enhance results in some wounds (Mace et al., 2005).

Moisture:

Keeping the wound moist enhanced cell migration and minimizes scarring (Shi et al., 2015).

Psychological Stress

Chronic stress stimulates cortisol production, which inhibits immune function and delays wounds healing (Shi et al., 2015).

Genetic Factors

Genetic differences in regular genes for inflammation, ECM production, and growth factors expression may affect healing response in different individual (Niethammer et al., 2009).

Personalized wound care strategies require a holistic approach taking these aspects into account.

Chronic wound

Generally, wound healing process complete within 4 weeks. If somehow the wound is failed to heal in the duration of 4 weeks, then it is characterised as chronic wound. However, it is preferable to hold for 3 months before diagnosing a chronic wound (Swann, 2010; Nussbaum et al., 2018). Model species such as Drosophila melanogaster have identified critical genes for repair of epithelium with the aid of high-throughput genetic screen methods (Wilkinson and Hardman, 2020). Myeloperoxidase from neutrophils, released upon injury, regulates the oxidative burst post-injury, limiting excessive damage to tissues (Mirastschijski et al., 2010). These results play a key role in the design of antioxidant-based wound therapies. Skin regeneration has also been explored in adult zebrafish models due to their optical transparency and regenerative properties (Mirastschijski et al., 2010).

In any of the stages of wound healing, deregulation can cause healing impairment. Example is excessive scarring. This can cause the burden on economic and health status of the society and cause chronic wound (Choi and Diehl, 2009; Al-Khamis et al., 2010). Chronic wound occurs if there may be failure to proceed through orderly and timely reparative process of wound healing to produce anatomic and functional integrity of the injured site (Richardson et al., 2013). In developed countries, it has been estimated that that 1 to 2% of the population will suffer from a chronic wound during their lifetime. In the United States alone, 6.5 million patients suffer from chronic wounds. Histological patterns reflect dysfunctional fibroblast function and continued inflammation (DeRouen and Oro, 2009). Such wounds frequently need specialized interventions such as surgical debridement. Molecular markers may direct these decisions to remove viable tissue (Brem et al., 2007).The primary factor that causes chronic wound formation include age, immune status, malnutrition, infection, insufficient oxygenation or perfusion, smoking, diseases, medications, radiation, and chemotherapy. Most common chronic wound example include vascular ulcers (venous or arterial ulcers), diabetic ulcers, and pressure ulcers (Nussbaum et al., 2018). Chronic wounds are rarely seen in individuals who are otherwise healthy. Patients generally suffer from "highly branded" diseases such as diabetes and obesity have a high risk of chronic wound after injury (Chhabra et al., 2017). From a funding standpoint it is seen only in parts, i.e. diabetes causes a complication like foot ulcers or of spinal cord injuries causes a complication like pressure ulcers (Chhabra et al., 2017).

The most common chronic wounds have several delineating nuances. Discoloration and pain from microvascular injury alone (stage 1), ulceration and skin breakdown (stage 2), to extension to underlying fat (stage 3) or deeper structures (stage 4) initiates progressive, increasing depth of tissue necrosis, which typically followed by pressure ulcers. Diabetic ulcers are accompanied with altered sensorium, and this paraesthesia/anaesthesia eliminate the protective afferent feedback (pain) that prevents soft tissue injury leading to initial ulceration and unnoticed progression. Venous stasis wounds are typically associated with peri wound skin discoloration, pain, variable levels of granulation tissue along with significant exudate. Arterial ulcers lead to acute ischemia often accompanied with significant pain and Eschar (Midwood et al., 2004; Charles et al., 2008).

PATHOPHYSIOLOGICAL CHARACTERISTICS OF CHRONIC WOUNDS

Chronic Inflammation

In contrast to acute wounds in which inflammation is resolved within days, chronic wounds involve ongoing infiltration to macrophages and neutrophils. This chronic inflammatory state leads to prolonged released of proteases, ROS, and pro-inflammatory cytokines, which cause damage to healthy tissue and cleave growth factors and ECM proteins (Hong YK et al., 2023; Szpaderska et al., 2003).

Cellular Dysfunction and Senescence

Fibroblasts and keratinocytes of chronic wounds tend to be senescent, proliferate less, and possess a defective migratory capability. These cells are non-functional and cannot help form granulation tissue and re-epithelialize (Finnson et al., 2013).

Microbial Biofilms

Biofilms made pf groups of bacteria embedded in protective extracellular matrix are responsible for chronicity due to their resistance to antimicrobial therapy and evasion of host immune responses (Badylak, 2002).

Compromised Angiogenesis

Chronic wounds also commonly present with hypoxia as result of inadequate neovascularization, in addition to vascular disease. This hinders nutrient transport and oxygenation important for healing (Rodrigues et al., 2019).

Types of Chronic Wounds

Diabetic Foot Ulcers (DFUs)

These are due to diabetic patient neuropathy, ischemia, and immune dysfunction. DFUs are prone to infection and amputation and are notoriously challenging to heal (Landén et al., 2016; Vestweber, 2015). The diabetic foot constitutes a tremendous challenge for patients, caregivers and the health care system. Up to 25% of individuals with diabetes will develop a foot ulcer during their lifetime (Sorrentino et al., 2015).

Venous Leg Ulcers (VLUs)

VLUs are due to venous hypertension and insufficient, causing tissue edema, inflammation, and subsequent skin breakdown. VLUs are recurrent and need long term compression therapy (Mirastschijski et al., 2010; Li et al., 2003)

Pressure Injuries (pressure ulcers)

Also called decubitus ulcer, these are caused by extended pressure over prominences of bone or muscle and are very challenging to treat (Strbo et al., 2014).

HUMAN AND ECONOMIC COST OF WOUNDS

Wound have remained a challenge throughout the history to our society and continue to represent as an extraordinary burden to the health care system. In 2014 in US, wounds affected more than 8 million people costing an estimated $30 billion. With an increasing aging and obese population high risk comorbidities commensurately increase, growing wound closure product market size of 21.4 billion in 2022 and compound annual growth rate of 4.15% from 2023 to 2030. Increasing surgical procedure almost contribute to the large wound affected population. Careful surgical technique material remains important, as wound dehiscence can lead to a 9.6%

increase in mortality, an additional 9.4 days of hospitalization, and up to $40,000 in hospital charges (Strbo et al., 2014).

Patient Care

Current studies on wound healing focus on detecting molecular level target genes that can be improved to accelerate natural wound healing. The hedgehog signalling pathway has preform a role in epithelial-mesenchymal interaction in wound healing, which create a scope to use this pathway in multiple studies (Schultz and Wysocki, 2009). The understanding of delicate immunological balance is important as it play an important active role throughout the whole wound healing process. The depth knowledge of immunological studies helps for research of wound healing (Nosbaum et al., 2016). In 2009, United States Food and Drug Administration (FDA)-approves Two methods for isolating PRP to treat patients (Szpaderska et al., 2003). Dense network of sensory and autonomous nerve fibres has been present around skin which allow sensation and movement. Nerve fibre regeneration is therefore essential post-injury process. Substance P influence cellular processes like proliferation and angiogenesis substance P is reduced in delayed healing in diabetic wounds, where topical restoration restores healing. (Pisarik, 2016).

Therapeutic Approaches and Innovations in Wound Healing

Optimal wound care incorporates a range of approaches designed to enhance the environment of the wounds, manage infection, and promote tissue regeneration. Biomedical innovations have stimulated the creation of novel therapeutic modalities that address the multicausal reasons behind wounds that fail to heal, particularly chronic and complex wounds (Ozgok Kangal and Regan, 2025).

TRADITIONAL WOUND CARE

Cleansing and Debridement

Necrotic tissue and contaminants must be removed to decrease microbial burden and enhance healing. Surgical, mechanical, enzymatic, and autolytic debridement techniques are used (Brewer, 2006).

Infection Management

Topical and systemic antimicrobial therapy prevents or treat infection. Biofilm presence requires novel antimicrobial approaches (Brewer, 2006).

Moisture Wound Healing

An optimal moisture balance with suitable dressing aids cell migration and pain relief (Berndt et al., 2001).

Advanced Therapeutics

Growth Factor Therapy

Topical use of recombinant growth factors like PDGF and EGF has been utilized to induce cellular activity and enhance healing in chronic wounds (Olsson et al., 2019).

CELLULAR THERAPIES

Stem Cell Therapy:

Mesenchymal stem cells (MSCs) and adipose derived stem cell induce regeneration through differentiation and paracrine signaling (Mousaei Ghasroldasht et al, 2022).

Platelets rich plasma (PRP):

Platelets concentration provides growth factors directly to wound, enhancing repair (Sen, 2019).

Negative Pressure Wound Therapy (NPWT)

Application of sub-atmospheric pressure dehydrates exudate, decreases edema, and provokes granulation tissue development (Sen, 2019).

NEW TECHNOLOGIES

Gene Therapy

Targeting growth factor production, ECM remodeling, and inflammatory genes offers hope for improving repair, especially in ischemic and diabetic wounds (Zheng SY et al., 2023).

Nanotechnology

Nanoparticles and nanofibrous scaffolds enhance drug delivery, antimicrobial function, and stimulate ECM characteristics (Jiang, Z et al., 2023).

Exosome-Based Therapies

Cell free strategies using stem cell derived exosomes leverage paracrine signaling to regulate inflammation and stimulate regeneration (Mustoe et al., 2006).

Biomaterials and Intelligent Dressings

Hydrogels, bioactive dressing, and responsible sensor to the wound's environment are being developed to maximize healing and track progress (Mustoe et al., 2006).

Multidisciplinary Care

Ideal wound healing in most cases demands synergy among endocrinologist, dermatologists, surgeons, nutritionists, and physical therapists tackle systemic issue and deliver holistic care (Teller and White, 2009).

CHALLENGES AND FUTURE DIRECTIONS

There is still a hurdle of high expense, limited availability, and imperfectly comprehension of intricately complex wound biology. Personalized medicine strategies and incorporation of omics technologies can possibly open doors to precision wounds care (Wang et al., 2018; Tomasek et al., 2002).

Future Directions in Wound Healing Research

The wound healing is a field of transforming quickly with the help of advances in molecular biology, bioengineering, and regenerative medicine. New research focuses on fulfilling the unmet needs in chronic and complex wound management by creating more efficacious, personalized, and less invasive treatments (Sun et al., 2014).

Precision Medicine and Biomarker Discovery

Advances in genomics, proteomics, and metabolomics allow biomarkers predicting wound healing ability and therapeutic outcomes to be identified. This allows for personalizes interventions based on patients' individuality (Sorrentino et al., 2015).

Stem Cell and Regenerative Therapies

Stem cell biology advances, such as induced pluripotent stem cells (iPSCs) and gene-edited cells, provide promising routes for regenerating advanced tissue and eliminating the limitations of existing cell therapies (Sorrentino et al., 2015).

Bioengineered Skin and 3D Bio printing

The creation of skin equivalents with a combination of several cell type, vascular network, and ECM components through the use of 3D bio printing technologies seeks to generate functional grafts that closely resemble native skin architecture Takeuchi and Akira, 2010; Midwood et al., 2004).

Modulation of Immune Responses

Knowledge of the immune microenvironment of wounds creates new avenue for modulation of inflammation to support repair and avoid fibrosis and avoid fibrosis and chronicity (Vestweber, 2015).

New Drug Delivery Systems

Nanoparticles, liposomes, and responsive biomaterials are being engineered for controlled and targeted delivery of therapeutics, minimizing side effects and enhancing efficacy (Vestweber, 2015).

Artificial Intelligence and Digital Health

Machine learning and wearable sensors are increasingly being used in wound assessment and management to provide real-time tracking and predictive insights into clinical decision-making (Fuchs et al., 2007).

Microbiome and Wound Healing

New evidence on the role of the skin and wound microbiome in healing mechanisms is emerging. Microbial community manipulation may offer new approaches to infection prevention and regeneration promotion (Brewer, 2006) .

Challenges and Ethical Considerations

With the advent technologies, concerns related to safety, cost-effectiveness, regulatory clearance, and access for all will have to be tackle in order to convert innovations into practice (Takeuchi and Akira, 2010).

CONCLUSION

Wound repair is an intricate, multifunctional biological process requiring the coordination of the activities of cells, signaling molecules, and the extracellular matrix. Improved understanding of this process guided novel therapeutic strategies, revolutionizing the treatment of both acute and chronic wounds. Research continues to identify additional healing outcomes and patient quality of life. Integrated, multidisciplinary treatment continues to be necessary to treat the many factors impacting repair of the wound and to apply scientific knowledge to useful clinical practice (Al-Khamis et al., 2010; Pastar et al., 2010; Ozgok Kangal and Regan, 2021).

REFERENCES

Al-Khamis, A., McCallum, I., King, P. M., & Bruce, J. (2010). Healing by primary versus secondary intention after surgical treatment for pilonidal sinus. *Cochrane Database of Systematic Reviews*, *2010*(1), CD006213. Advance online publication. DOI: 10.1002/14651858.CD006213.pub3 PMID: 20091589

Arif, N., Baumann, L., & Felcht, M. (2024). Combination of secondary intention healing and primary closure to reconstruct large facial defects. *Journal der Deutschen Dermatologischen Gesellschaft*, *22*(10), 1344–1349. DOI: 10.1111/ddg.15481 PMID: 39167560

Badylak, S. F. (2002). The extracellular matrix as a scaffold for tissue reconstruction. *Seminars in Cell & Developmental Biology*, *13*(5), 377–383. DOI: 10.1016/S1084952102000940 PMID: 12324220

Barman, P. K., & Koh, T. J. (2020). Macrophage dysregulation and impaired skin wound healing in diabetes. *Frontiers in Cell and Developmental Biology*, *8*, 528. Advance online publication. DOI: 10.3389/fcell.2020.00528 PMID: 32671072

Brem, H., Stojadinovic, O., Diegelmann, R. F., Entero, H., Lee, B., Pastar, I., Golinko, M., Rosenberg, H., & Tomic-Canic, M. (2007). Molecular markers in patients with chronic wounds to guide surgical debridement. *Molecular Medicine (Cambridge, Mass.)*, *13*(1–2), 30–39. DOI: 10.2119/2006-00054.Brem PMID: 17515955

Brewer, D. B. (2006). Max Schultze (1865), G. Bizzozero (1882) and the discovery of the platelet. *British Journal of Haematology*, *133*(3), 251–258. DOI: 10.1111/j.1365-2141.2006.06036.x PMID: 16643426

Campos, I., Geiger, J. A., Santos, A. C., Carlos, V., & Jacinto, A. (2010). Genetic screen in Drosophila melanogaster uncovers a novel set of genes required for embryonic epithelial repair. *Genetics*, *184*(1), 129–140. DOI: 10.1534/genetics.109.110288 PMID: 19884309

Charles, C. A., Tomic-Canic, M., Vincek, V., Nassiri, M., Stojadinovic, O., Eaglstein, W. H., & Kirsner, R. S. (2008). A gene signature of nonhealing venous ulcers: Potential diagnostic markers. *Journal of the American Academy of Dermatology*, *59*(5), 758–771. DOI: 10.1016/j.jaad.2008.07.018 PMID: 18718692

Chhabra, S., Chhabra, N., Kaur, A., & Gupta, N. (2017). Wound healing concepts in clinical practice of OMFS. *Journal of Maxillofacial and Oral Surgery*, *16*(4), 403–423. DOI: 10.1007/s12663-016-0880-z PMID: 29038623

Choi, S. S., & Diehl, A. M. (2009). Epithelial-to-mesenchymal transitions in liver. *Hepatology (Baltimore, Md.)*, *50*(6), 2007–2013. DOI: 10.1002/hep.23196 PMID: 19824076

Demidova-Rice, T. N., Hamblin, M. R., & Herman, I. M. (2012). Acute and impaired wound healing: Pathophysiology and current methods for drug delivery, part 1: Normal and chronic wounds: Biology, causes, and approaches to care. *Advances in Skin & Wound Care*, *25*(7), 304–314. https://pubmed.ncbi.nlm.nih.gov/22713781/ . DOI: 10.1097/01.ASW.0000416006.55218.d0 PMID: 22713781

DeRouen, M. C., & Oro, A. E. (2009). The primary cilium: A small yet mighty organelle. *The Journal of Investigative Dermatology*, *129*(2), 264–265. DOI: 10.1038/jid.2008.404 PMID: 19148215

Finnson, K. W., McLean, S., Di Guglielmo, G. M., & Philip, A. (2013). Dynamics of transforming growth factor beta signaling in wound healing and scarring. *Advances in Wound Care*, *2*(5), 195–214. DOI: 10.1089/wound.2013.0429 PMID: 24527343

Freedman, B. R., Hwang, C., Talbot, S., Hibler, B., Matoori, S., & Mooney, D. J. (2023). Breakthrough treatments for accelerated wound healing. *Science Advances*, *9*(20), eade7007. Advance online publication. DOI: 10.1126/sciadv.ade7007 PMID: 37196080

Friedman, S. L. (2008). Mechanisms of hepatic fibrogenesis. *Gastroenterology*, *134*(6), 1655–1669. DOI: 10.1053/j.gastro.2008.03.003 PMID: 18471545

Fuchs, T. A., Abed, U., Goosmann, C., Hurwitz, R., Schulze, I., Wahn, V., Weinrauch, Y., Brinkmann, V., & Zychlinsky, A. (2007). Novel cell death program leads to neutrophil extracellular traps. *The Journal of Cell Biology*, *176*(2), 231–241. DOI: 10.1083/jcb.200606027 PMID: 17210947

Gantwerker, E. A., & Hom, D. B. (2011). Skin: Histology and physiology of wound healing. *Facial Plastic Surgery Clinics of North America*, *19*(3), 441–453. DOI: 10.1016/j.fsc.2011.06.009 PMID: 21856533

Godo, S., & Shimokawa, H. (2017). Endothelial functions. *Arteriosclerosis, Thrombosis, and Vascular Biology*, *37*(9), e108–e114. DOI: 10.1161/ATVBAHA.117.309813 PMID: 28835487

Golebiewska, E. M., & Poole, A. W. (2015). Platelet secretion: From haemostasis to wound healing and beyond. *Blood Reviews*, *29*(3), 153–162. DOI: 10.1016/j.blre.2014.10.003 PMID: 25468720

Grubbs, H., & Manna, B. (2023). Wound physiology. In *StatPearls*. StatPearls Publishing., https://www.ncbi.nlm.nih.gov/books/NBK518964/

Hong, Y.-K., Chang, Y.-H., Lin, Y.-C., Chen, B., Guevara, B. E. K., & Hsu, C.-K. (2023). Inflammation in wound healing and pathological scarring. *Advances in Wound Care, 12*(5), 288–300. DOI: 10.1089/wound.2021.0161 PMID: 36541356

Jiang, Z., Zheng, Z., Yu, S., Gao, Y., Ma, J., Huang, L., & Yang, L. (2023). Nanofiber scaffolds as drug delivery systems promoting wound healing. *Pharmaceutics, 15*(7), 1829. Advance online publication. DOI: 10.3390/pharmaceutics15071829 PMID: 37514015

Landén, N. X., Li, D., & Ståhle, M. (2016). Transition from inflammation to proliferation: A critical step during wound healing. *Cellular and Molecular Life Sciences : CMLS, 73*(20), 3861–3885. DOI: 10.1007/s00018-016-2268-0 PMID: 27180275

Lau, K., Paus, R., Tiede, S., Day, P., & Bayat, A. (2009). Exploring the role of stem cells in cutaneous wound healing. *Experimental Dermatology, 18*(11), 921–933. DOI: 10.1111/j.1600-0625.2009.00942.x PMID: 19719838

Li, J., Zhang, Y.-P., & Kirsner, R. S. (2003). Angiogenesis in wound repair: Angiogenic growth factors and the extracellular matrix. *Microscopy Research and Technique, 60*(1), 107–114. DOI: 10.1002/jemt.10249 PMID: 12500267

Mace, K. A., Pearson, J. C., & McGinnis, W. (2005). An epidermal barrier wound repair pathway in Drosophila is mediated by grainy head. *Science, 308*(5720), 381–385. DOI: 10.1126/science.1107573 PMID: 15831751

Midwood, K. S., Williams, L. V., & Schwarzbauer, J. E. (2004). Tissue repair and the dynamics of the extracellular matrix. *The International Journal of Biochemistry & Cell Biology, 36*(6), 1031–1037. DOI: 10.1016/j.biocel.2003.12.003 PMID: 15094118

Mirastschijski, U., Schnabel, R., Claes, J., Schneider, W., Agren, M. S., Haaksma, C., & Tomasek, J. J. (2010). Matrix metalloproteinase inhibition delays wound healing and blocks the latent transforming growth factor-beta1-promoted myofibroblast formation and function. *Wound Repair and Regeneration, 18*(2), 223–234. DOI: 10.1111/j.1524-475X.2010.00574.x PMID: 20409148

Mousaei Ghasroldasht, M., Seok, J., Park, H.-S., Liakath Ali, F. B., & Al-Hendy, A. (2022). Stem cell therapy: From idea to clinical practice. *International Journal of Molecular Sciences, 23*(5), 2850. Advance online publication. DOI: 10.3390/ijms23052850 PMID: 35269990

Mussbacher, M., Kral-Pointner, J. B., Salzmann, M., Schrottmaier, W. C., & Assinger, A. (2024). Mechanisms of hemostasis: Contributions of platelets, coagulation factors, and the vessel wall. In Geiger, M. (Ed.), *Fundamentals of vascular biology*. Springer., DOI: 10.1007/978-3-031-64591-4_8

Mustoe, T. A., O'Shaughnessy, K., & Kloeters, O. (2006). Chronic wound pathogenesis and current treatment strategies: A unifying hypothesis. *Plastic and Reconstructive Surgery, 117*(7, Suppl), 35S–41S. DOI: 10.1097/01.prs.0000225431.63010.1b PMID: 16799373

Niethammer, P., Grabher, C., Look, A. T., & Mitchison, T. J. (2009). A tissue-scale gradient of hydrogen peroxide mediates rapid wound detection in zebrafish. *Nature, 459*(7249), 996–999. DOI: 10.1038/nature08119 PMID: 19494811

Nosbaum, A., Prevel, N., Truong, H.-A., Mehta, P., Ettinger, M., Scharschmidt, T. C., Ali, N. H., Pauli, M. L., Abbas, A. K., & Rosenblum, M. D. (2016). Cutting edge: Regulatory T cells facilitate cutaneous wound healing. *The Journal of Immunology : Official Journal of the American Association of Immunologists, 196*(5), 2010–2014. DOI: 10.4049/jimmunol.1502139 PMID: 26826250

Novaes, R., Costa, T. F. R., Goundry, A. L., Verçoza, B. R. F., Rodrigues, J. C. F., Godinho, J. L. P., Reis, F. C. G., Morrot, A., & Lima, A. P. C. A. (2022). Bone marrow granulocytes downregulate IL-1β and TNF production and the microbicidal activity of inflammatory macrophages. *Biochemistry and Cell Biology, 100*(3), 246–265. DOI: 10.1139/bcb-2021-0460 PMID: 35443139

Nussbaum, S. R., Carter, M. J., Fife, C. E., DaVanzo, J., Haught, R., Nusgart, M., & Cartwright, D. (2018). An economic evaluation of the impact, cost, and Medicare policy implications of chronic nonhealing wounds. *Value in Health, 21*(1), 27–32. DOI: 10.1016/j.jval.2017.07.007 PMID: 29304937

Olsson, M., Järbrink, K., Divakar, U., Bajpai, R., Upton, Z., Schmidtchen, A., & Car, J. (2019). The humanistic and economic burden of chronic wounds: A systematic review. *Wound Repair and Regeneration, 27*(1), 114–125. DOI: 10.1111/wrr.12683 PMID: 30362646

Ozgok Kangal, M. K., & Regan, J. P. (2025). Wound healing. In *StatPearls*. StatPearls Publishing., https://www.ncbi.nlm.nih.gov/books/NBK535406/

Pase, L., Layton, J. E., Wittmann, C., Ellett, F., Nowell, C. J., Reyes-Aldasoro, C. C., Varma, S., Rogers, K. L., Hall, C. J., Keightley, M. C., Crosier, P. S., Grabher, C., Heath, J. K., Renshaw, S. A., & Lieschke, G. J. (2012). Neutrophil-delivered myeloperoxidase dampens the hydrogen peroxide burst after tissue wounding in zebrafish. *Current Biology, 22*(19), 1818–1824. DOI: 10.1016/j.cub.2012.07.060 PMID: 22940471

Pastar, I., Stojadinovic, O., Krzyzanowska, A., Barrientos, S., Stuelten, C., Zimmerman, K., Blumenberg, M., Brem, H., & Tomic-Canic, M. (2010). Attenuation of the transforming growth factor β-signaling pathway in chronic venous ulcers. *Molecular Medicine (Cambridge, Mass.)*, *16*(3–4), 92–101. DOI: 10.2119/molmed.2009.00149 PMID: 20069132

Pereira, R. V. S., EzEldeen, M., Ugarte-Berzal, E., Martens, E., Malengier-Devlies, B., Vandooren, J., Vranckx, J. J., Matthys, P., & Opdenakker, G. (2023). Physiological fibrin hydrogel modulates immune cells and molecules and accelerates mouse skin wound healing. *Frontiers in Immunology*, *14*, 1170153. Advance online publication. DOI: 10.3389/fimmu.2023.1170153 PMID: 37168862

Pisarik, P. (2016). Choosing tap water vs. sterile saline for wound irrigation. *American Family Physician*, *94*(2), 83–84. https://www.aafp.org/pubs/afp/issues/2016/0715/p83.html PMID: 27419322

Reinke, J. M., & Sorg, H. (2012). Wound repair and regeneration. *European Surgical Research*, *49*(1), 35–43. DOI: 10.1159/000339613 PMID: 22797712

Richardson, R., Slanchev, K., Kraus, C., Knyphausen, P., Eming, S., & Hammerschmidt, M. (2013). Adult zebrafish as a model system for cutaneous wound-healing research. [Note: Original DOI corrected to match article]. *The Journal of Investigative Dermatology*, *133*(6), 1655–1665. DOI: 10.1038/jid.2013.16 PMID: 23325040

Rodrigues, M., Kosaric, N., Bonham, C. A., & Gurtner, G. C. (2019). Wound healing: A cellular perspective. *Physiological Reviews*, *99*(1), 665–706. DOI: 10.1152/physrev.00067.2017 PMID: 30475656

Schultz, G. S., Davidson, J. M., Kirsner, R. S., Bornstein, P., & Herman, I. M. (2011). Dynamic reciprocity in the wound microenvironment. *Wound Repair and Regeneration*, *19*(2), 134–148. DOI: 10.1111/j.1524-475X.2011.00673.x PMID: 21362080

Schultz, G. S., & Wysocki, A. (2009). Interactions between extracellular matrix and growth factors in wound healing. *Wound Repair and Regeneration*, *17*(2), 153–162. DOI: 10.1111/j.1524-475X.2009.00466.x PMID: 19320882

Sen, C. K. (2019). Human wounds and its burden: An updated compendium of estimates. *Advances in Wound Care*, *8*(2), 39–48. DOI: 10.1089/wound.2019.0946 PMID: 30809421

Shi, Y., Shu, B., Yang, R., Xu, Y., Xing, B., & Liu, J.. (2015). Wnt and Notch signaling pathway involved in wound healing by targeting separately c-Myc and Hes1. *Stem Cell Research & Therapy*, *6*, 120. Advance online publication. DOI: 10.1186/s13287-015-0103-4 PMID: 26076648

Sorg, H., & Sorg, C. G. G. (2023). Skin wound healing: Of players, patterns, and paradigms. *European Surgical Research, 64*(2), 141–157. DOI: 10.1159/000528271 PMID: 36417847

Sorrentino, S., Studt, J. D., Medalia, O., & Tanuj Sapra, K. (2015). Roll, adhere, spread and contract: Structural mechanics of platelet function. *European Journal of Cell Biology, 94*(3–4), 129–138. DOI: 10.1016/j.ejcb.2015.01.001 PMID: 25655000

Strbo, N., Yin, N., & Stojadinovic, O. (2014). Innate and adaptive immune responses in wound epithelialization. *Advances in Wound Care, 3*(7), 492–501. DOI: 10.1089/wound.2012.0435 PMID: 25032069

Sun, B. K., Siprashvili, Z., & Khavari, P. A. (2014). Advances in skin grafting and treatment of cutaneous wounds. *Science, 346*(6212), 941–945. DOI: 10.1126/science.1253836 PMID: 25414301

Swann, G. (2010). The skin is the body's largest organ. *Journal of Visual Communication in Medicine, 33*(4), 148–149. DOI: 10.3109/17453054.2010.525439 PMID: 21087182

Szpaderska, A. M., Egozi, E. I., Gamelli, R. L., & DiPietro, L. A. (2003). The effect of thrombocytopenia on dermal wound healing. *The Journal of Investigative Dermatology, 120*(6), 1130–1137. DOI: 10.1016/S0022-202X(18)32320-0 PMID: 12787144

Takeuchi, O., & Akira, S. (2010). Pattern recognition receptors and inflammation. *Cell, 140*(6), 805–820. DOI: 10.1016/j.cell.2010.01.022 PMID: 20303872

Tan, C., Aziz, M., & Wang, P. (2021). The vitals of NETs. *Journal of Leukocyte Biology, 110*(4), 797–808. DOI: 10.1002/JLB.3RU0620-375R PMID: 33378572

Teller, P., & White, T. K. (2009). The physiology of wound healing: Injury through maturation. *The Surgical Clinics of North America, 89*(3), 599–610. DOI: 10.1016/j.suc.2009.03.006 PMID: 19465199

Tomasek, J. J., Gabbiani, G., Hinz, B., Chaponnier, C., & Brown, R. A. (2002). Myofibroblasts and mechano-regulation of connective tissue remodelling. *Nature Reviews. Molecular Cell Biology, 3*(5), 349–363. DOI: 10.1038/nrm809 PMID: 11988769

Tomasek, J. J., Haaksma, C. J., Schwartz, R. J., & Howard, E. W. (2013). Whole animal knockout of smooth muscle alpha-actin does not alter excisional wound healing or the fibroblast-to-myofibroblast transition. *Wound Repair and Regeneration, 21*(1), 166–176. DOI: 10.1111/wrr.12001 PMID: 23253249

Tome-Garcia, J., Li, D., Ghazaryan, S., Shu, L., & Wu, L. (2014). ERBB2 increases metastatic potentials specifically in androgen-insensitive prostate cancer cells. *PLoS One*, *9*(6), e99525. Advance online publication. DOI: 10.1371/journal.pone.0099525 PMID: 24937171

Velnar, T., Bailey, T., & Smrkolj, V. (2009). The wound healing process: An overview of the cellular and molecular mechanisms. *The Journal of International Medical Research*, *37*(5), 1528–1542. DOI: 10.1177/147323000903700531 PMID: 19930861

Vestita, M., Tedeschi, P., & Bonamonte, D. (2022). Anatomy and physiology of the skin. In Maruccia, M., & Giudice, G. (Eds.), *Textbook of plastic and reconstructive surgery*. Springer., DOI: 10.1007/978-3-030-82335-1_1

Vestweber, D. (2015). How leukocytes cross the vascular endothelium. *Nature Reviews. Immunology*, *15*(11), 692–704. DOI: 10.1038/nri3908 PMID: 26471775

Wallace, H. A., Basehore, B. M., & Zito, P. M. (2023). Wound healing phases. In *StatPearls*. StatPearls Publishing., https://www.ncbi.nlm.nih.gov/books/NBK470443/

Wang, P. H., Huang, B. S., Horng, H. C., Yeh, C. C., & Chen, Y. J. (2018). Wound healing. *Journal of the Chinese Medical Association*, *81*(2), 94–101. DOI: 10.1016/j.jcma.2017.11.002 PMID: 29169897

Wilgus, T. A., Roy, S., & McDaniel, J. C. (2013). Neutrophils and wound repair: Positive actions and negative reactions. *Advances in Wound Care*, *2*(7), 379–388. DOI: 10.1089/wound.2012.0383 PMID: 24527354

Wilkinson, H. N., & Hardman, M. J. (2020). Wound healing: Cellular mechanisms and pathological outcomes. *Open Biology*, *10*(9), 200223. Advance online publication. DOI: 10.1098/rsob.200223 PMID: 32993416

Woo, Y. C., Park, S. S., Subieta, A. R., & Brennan, T. J. (2004). Changes in tissue pH and temperature after incision indicate acidosis may contribute to postoperative pain. *Anesthesiology*, *101*(2), 468–475. DOI: 10.1097/00000542-200408000-00029 PMID: 15277931

Wu, W.-S., Chen, L.-R., & Chen, K.-H. (2025). Platelet-rich plasma (PRP): Molecular mechanisms, actions and clinical applications in human body. *International Journal of Molecular Sciences*, *26*(21), 10804. Advance online publication. DOI: 10.3390/ijms262110804 PMID: 41226837

Zheng, S. Y., Wan, X. X., Kambey, P. A., Luo, Y., Hu, X. M., Liu, Y. F., Shan, J. Q., Chen, Y. W., & Xiong, K. (2023). Therapeutic role of growth factors in treating diabetic wound. *World Journal of Diabetes*, *14*(4), 364–395. DOI: 10.4239/wjd.v14.i4.364 PMID: 37122434

Chapter 3
Wound Healing Demystified:
Cellular and Molecular Insights

Zahid Ahmad Thoker
http://orcid.org/0009-0002-6897-5835
Lovely Professional University, Phagwara, India

Lovleen Marwaha
http://orcid.org/0000-0002-5175-2590
Lovely Professional University, India

ABSTRACT

Wound healing is a complex and highly regulated biological process involving coordinated cellular activities and molecular signalling events that restore tissue integrity following injury. Disruption of these mechanisms often results in chronic, non-healing wounds that impose significant clinical and economic burdens worldwide. Recent advances have highlighted the therapeutic potential of phytoconstituents, plant-derived bioactive compounds in modulating key phases of wound repair, including inflammation, angiogenesis, oxidative stress regulation, and extracellular matrix remodelling. This chapter elucidates the cellular and molecular foundations of wound healing and critically examines preclinical and emerging clinical evidence supporting phytochemical-based interventions. It further discusses current challenges, translational limitations, and prospects, emphasising the role of systems biology, nanotechnology, and personalised medicine in advancing safe, effective, and sustainable wound-healing strategies.

DOI: 10.4018/979-8-3373-2908-6.ch003

1. INTRODUCTION

Wound healing is a basic biological phenomenon that is involved in restoring tissue continuity and functionality after trauma. It is a complex, strictly regulated cascade, which entails haemostasis, inflammation, proliferation and remodelling and demands a careful temporal and spatial organization of various cell types (platelets, neutrophils, macrophages, fibroblasts, keratinocytes and endothelial cells), the extracellular matrix (ECM) and a sophisticated array of growth factors, cytokines, as well as proteolytic enzymes (Krishnaswamy, Mintz, and Sagi, 2017). The successful closure of wounds with restoration of barrier function and acceptable tissue architecture is achieved when this coordinated programme works properly; it is disturbed by underlying disease, persistent infection, or local biochemical imbalance, leading to the development of chronic, non-healing wounds that are a significant source of morbidity (Sen, 2021).

1.1 Concept and clinical importance of wound healing

A wound, in clinical terms, is any discontinuity of the integrity of skin or underlying tissues due to either trauma, surgery, burns, pressure, ischaemia, or metabolic disease. Wound healing is not a simple mechanical bandaging but a biologically active process restoring physiological barrier functions, immune defence, and, in the case of potential tissue damage, tissue architecture. The non-occurrence of this process leads to considerable suffering in the patient: chronic pain, immobilization, recurrent infections, the threat of sepsis, and, in difficult situations, amputation of a limb (Sen, 2021). Wound care is resource-intensive in terms of health-systems, with its treatment methods often requiring frequent clinic visits, lengthy treatment courses, specialised dressings, and multidisciplinary care teams, and, therefore, is a burning clinical and economic issue (Martinengo et al., 2019; Sen, 2021).

1.2 Global burden of acute and chronic wounds

There is no simple way of quantifying the magnitude of wounds around the world due to heterogeneous definitions, different study designs, and variable reporting methods. However, systematic reviews and population studies reveal that chronic wounds have a non-negligible prevalence among populations in the world. An observational study meta-analysis cited combined prevalence rates of chronic wounds (including leg ulcers) and highlighted the fact that leg ulcers represent a significant proportion of chronic wounds in the healthcare environment (Martinengo et al., 2019). The national and regional databases indicate the extent of the issue in older and comorbid populations: in high-income countries, chronic wounds impact a significant percentage of geriatric patients and cost the healthcare system billions of dollars a year (Sen, 2021). According to the Global Burden of Disease reports and successive reviews, the prevalence of skin and subcutaneous disorders as the broader term that includes many types of wounds has increased significantly over the past decades, which is indicative of demographic change (ageing), increasing non-communicable diseases (especially diabetes and obesity) and unmet preventive care (Martinengo et al., 2019; Sen, 2021). Concisely, acute and chronic wounds represent an increasing threat to human health that has both profound humanistic and economic implications.

1.3 Rationale for a cellular and molecular understanding

Historical clinical approaches to the management of wounds have put a focus on local care (cleaning, debridement, infection control, moisture balance, and dressings) and management of comorbid systemic comorbidities. Nevertheless, molecular and cell biology discoveries have explained why such clinical interventions can produce vastly different effects on different patients and wound types: wound healing is determined by dynamic interactions among growth factors (e.g., TGF- 2, PDGF, VEGF), inflammatory mediators (TNF- 2, IL- 1), redox balance, ECM turnover (matrix metalloproteinases and their inhibitors) and cellular phenotypes (e.g. macrophage polarisation between pro-inflammatory and pro-repair states) (K Chronic wounds generally share ubiquitous molecular pathways - continued inflammation, high levels of protease action, excessive oxidative stress and impaired angiogenesis, that offer mechanistic vulnerability to treatment. In such a way, an interventional approach to the wound microenvironment needs to be designed rationally with the help of a cellular and molecular lens that will help to remove the symptoms and transform the wound microenvironment into a new, regenerative state instead of a chronic and non-productive condition (Krishnaswamy et al., 2017; Liarte et al., 2020).

1.4 Scope and objectives of the chapter

The chapter will be an attempt to integrate the existing cellular and molecular understanding of wound healing with a conscious emphasis on translational application of systems in phytoconstituent-based therapy. The goals are fourfold: (1) to provide a succinct and yet rigorous narrative of the cellular participants and molecular pathways that may coordinate healing; (2) to identify the general molecular malpractices underlying chronic wounds and consequently define therapeutic targets; (3) to study how biologically active substances (so-called phytoconstituents) of plants can regulate such cellular and molecular activities; and (4) to address the translational imperative, emerging technology (such as nano-delivery and standardised dressings) and research priorities to bring evidence-based ph All this will be accompanied by the focus on the evidence based on recent peer-reviewed literature (2015-2025) in an attempt to provide the reader with the latest mechanistic and clinical outlook aligned with the interests of researchers, clinicians, and developers of wound treatments (Trinh et al., 2022; Carvalho et al., 2021; Zulkefli et al., 2023).

2. CLASSIFICATION OF WOUNDS

Wounds are heterogeneous in origin, depth, duration, and biological behaviour. To make clinical decisions, or to interpret experimental or clinical evidence, a clear classification framework is required, since even the underlying cellular and molecular environment varies significantly across wound types. Expansively, wounds can be classified depending on the duration (acute or chronic), aetiology (surgical or traumatic or burn or burn), the presence or absence of the local or systemic factors that hinder normal healing. One of these groups is linked to the specific pathophysiological processes affecting healing patterns and responsiveness to treatment (Martinengo et al., 2019; Sen, 2021).

2.1 Acute vs. chronic wounds

Acute wounds are those that follow the normal progressive stages of healing and normally attain closure within a predicted time span, which is days to weeks, depending on the size and depth. These might be simple surgical incisions and minor traumatic injuries. Inflammation is short-lived, growth factor signalling is intact, angiogenesis is functional, and the extracellular matrix (ECM) remodelling of these wounds is well balanced between matrix metalloproteinases (MMPs) and their inhibitors (Krishnaswamy, Mintz, & Sagi, 2017).

Chronic wounds, on the other hand, never undergo the usual healing phases and they take months and even years to heal. The usual ones are diabetic foot ulcers, venous leg ulcers, and pressure ulcers. On the molecular level, chronic wounds are associated with persistent inflammation, overexpression of protease activity, inhibited growth factor signalling, resident cell impaired responsiveness, and sustained oxidative stress (Lizarte, Bernabé-Garcia, and Nicolas, 2020). These characteristics generate an unfavorable microenvironment where new tissue is constantly damaged, and it cannot be closed permanently. The ability to differentiate between acute and chronic wounds is therefore essential, as chronic wounds necessitate treatments that actively restore molecular malfunctions, but not merely assist the natural healing process (Sen, 2021).

2.2 Surgical, traumatic, and burn wounds

Wounds can also be categorised depending on the cause. Surgical wounds are purposefully made by designing them under controlled conditions, and they are typically clean, having distinct edges. Simple ones tend to heal by secondary intention, and with little tissue loss and scarring. Yet, contamination, inefficient perfusion, or systemic illness may slow down the healing process and predispose to infection (Martinengo et al., 2019).

Mechanical trauma that causes traumatic wounds includes cuts, abrasions, or crush wounds. Such wounds are usually abnormal, can be characterized by large tissue damage, and are often contaminated with foreign bodies or microbes. The inflammatory response in traumatic wounds is more severe, and the recovery results are highly reliant on early debridement, control of infection, and perfusional restoration (Sen, 2021).

Burn injuries are a unique group that is typified by thermal, chemical, or electrical damage. Burns cause massive tissue destruction, disturb vascular capillaries, and cause severe inflammatory and oxidative effects. Deep partial- and full-thickness burns necessitate a long-term healing process or surgical procedure and have high propensities of infection and scarring. Burn wounds have significant changes in cellular cytokine profiles, angiogenesis, and fibroblast activity, which may complicate regeneration and induce fibrosis (Miricescu et al., 2021).

2.3 Infected and non-healing wounds

A crucial predictor of wound outcome and wound classification is microbial burden. The infecting wounds contain pathogenic microorganisms that overwhelm host defences, causing damage to tissues, prolonged inflammatory response, and delayed healing. Despite lacking overt infection, microbial biofilms, which are immune-resistant and thus antibiotic-resistant, can be found in many chronic wounds. Biofilms sustain a chronic inflammatory response and upsurge the protease isolated impact, consequently destroying these growth factors and ECM components of repair (Krishnaswamy et al., 2017).

The non-healing wounds do not only qualify as such due to infection but also due to a lack of response to normal treatment. These wounds tend to have a complex of factors, i.e., ischaemia, neuropathy, recurring mechanical stress, and molecular defects like impaired macrophage polarisation and failed movement of keratinocytes. Realizing the presence of a wound infection, biofilm, or non-responsiveness to therapy will guide clinical care and the choice of advanced or adjunctive therapy, such as biologics or phytoconstituent-based interventions (Sen, 2021; Trinh et al., 2022).

2.4 Factors affecting wound healing outcomes

There is an interaction between local and systemic factors, thus influencing healing outcomes. The local factors involve the size and depth of the wound, perfusion and oxygenation levels of the tissue, mechanical stress, microbial load, moisture balance, and biochemical composition of the wound microenvironment (e.g., levels of proteases and redox conditions). Systemic factors involve age, nutritional condition, metabolic conditions including diabetes, vascular diseases, immune competency, and medication including corticosteroids or cytotoxic drugs (Martinengo et al., 2019; Sen, 2021).

On a molecular level, these factors all intersect on shared pathways: hypoxia and ischaemia inhibit angiogenic signalling; hyperglycaemia prolongs inflammation and oxidative stress; ageing changes the responsiveness of fibroblast and keratinocyte to growth factors (Liarte et al., 2020). This means that it is possible to manage wounds well only when the classification is made and the biological situation in which healing takes place is respected. This knowledge gives a reason as to why focused therapies such as phytoconstituents are capable of regulating inflammatory responses, oxidative stress, microbial load, and ECM remodelling contextually (Carvalho et al., 2021; Zulkefli et al., 2023).

3. PHASES OF WOUND HEALING

Wound healing is a complex biological process consisting of multiple and functionally different stages: haemostasis, inflammation, proliferation, and remodelling or maturation. Each of them is dictated by a particular cellular activity and a molecular signal, but none of them takes place independently; instead, the successful repair process relies on the sequentiality of transition between phases. The failure at any phase can affect the healing effectiveness and cause pathological results in the form of non-healing wounds or excessive scarring (Krishnaswamy, Mintz, and Sagi, 2017; Sen, 2021).

3.1 Haemostasis phase

The haemostasis stage commences as soon as tissue is damaged and is used to stop bleeding and form a temporary wound bed. Vascular constriction restricts the loss of blood, and platelets are aggregated and form a clot. Platelets are found to be central biologically as well as coagulation: once activated, platelets secrete a broad spectrum of growth factors and mediators, including platelet-derived growth factor (PDGF), transforming growth factor- ß (TGF- β), vascular endothelial growth factor

(VEGF), and epidermal growth factor (EGF) (Liarte, Bernabé-García, & Nicolás, 2020).

The resulting fibrin-fibronectin clot in the process of haemostasis also supplies a temporary extracellular matrix (ECM) scaffold that facilitates the migration of the inflammatory cells, fibroblasts, and endothelial cells to the wound site. Such a provisional matrix is also a reservoir of signalling molecules that control later healing stages. Though haemostasis is a rather short process, insufficient clot formation or excessive bleeding may destabilize the wound microenvironment and disrupt downstream inflammatory and proliferative reactions (Krishnaswamy et al., 2017).

3.2 Inflammatory phase

A typical inflammatory period occurs hours later and may take several days. It is defined by the fact that it recruits innate immune cells that eliminate necrotic tissue and limit infection. The first responders are the neutrophils that rush to the site of the wound and phagocytose debris and microorganisms. They secrete reactive oxygen species (ROS) and proteolytic enzymes, which play a role in antimicrobial defence and can also destroy adjacent tissue unless controlled and in excess or sustained (Miricescu et al., 2021).

The monocytes then develop into macrophages, which are regarded as the key controls of wound healing. Macrophages belong to the category of functional plasticity, and they switch between a pro-inflammatory phenotype during the initial stages of inflammation to the reparative one, which facilitates tissue regeneration. They also release cytokines and growth factors, including TGF-b, PDGF, and VEGF, thus connecting the inflammatory stage to the proliferative stage. It is vital that the inflammation is resolved in time; chronic wounds are characterised by sustained inflammatory signalling that is a key impediment to successful healing (Liarte et al., 2020; Sen, 2021).

3.3 Proliferative phase

The proliferative stage is characterized by a fast process of forming and repairing vascular systems. Fibroblasts then migrate into the wounding area and secrete extracellular matrix components, which are mainly type III collagen, to create granulation tissue. At the same time, keratinocytes multiply and move out of the edges of the wound and the surrounding epidermal tissue to reproduce the surface (Martinengo et al., 2019).

A characteristic of this phase is angiogenesis. Endothelial cells react to hypoxia and growth factor gradients, especially to VEGF and fibroblast growth factors, to develop new capillary networks, which deliver oxygen and nutrients to the regenerating tissue. Fibroblasts could also develop into myofibroblasts that produce contractile forces that reduce the size of the wound. Tightly controlled communication between fibroblast activity, keratinocyte migration, and angiogenesis is crucial to strong tissue regeneration (Miricescu et al., 2021).

3.4 Remodelling and maturation stage

This maturation/remodelling can continue for several months or years following wound closure. At this phase, the granulation tissue is replaced over time with a more organised and mechanically resilient matrix. The collagen type III is broken down and substituted with the collagen type I, which is rearranged along tension lines and cross-linked to enhance tension strength (Krishnaswamy et al., 2017).

The remodelling of ECM is tightly controlled by matrix metalloproteinases (MMPs) and endogenous metalloproteinase inhibitors (tissue inhibitors of metalloproteinases, TIMPs). The density of cells in the wound is reduced through the process of apoptosis of fibroblasts and endothelial cells, and the vascular network returns to almost normal values. Regenerated tissue is seldom able to reach the strength of normal skin, but when remodelling is successful, it leads to functional restoration with little scarring. At this phase, dysregulation can cause hypertrophic scars, keloids, or tissue of weak structural strength (Liarte et al., 2020). The various acute tissue repair processes that are induced by tissue injury are then clustered into four time-dependent phases, as illustrated below in Fig.2.

3.5 Dysregulation of healing phases in chronic wounds

Chronic wounds are a pathological condition, when there is an impairment in the regular process of passing through the stages of healing, usually in the inflammatory stage. The continuous inflammation, overactivation of proteases, inability to polarize the macrophages, and maintain the oxidative stress on the tissue all inhibit the switch to effective proliferation and remodelling (Krishnaswamy et al., 2017). Growth factors can be there but are made ineffective by degradation by high levels of MMP or decreased cellular responsiveness.

Systemic pathologies like diabetes, vascular inadequacy, and age also contribute to the aggravated dysregulation by disorderly angiogenesis, fibroblast activity, and keratinocyte relocation (Sen, 2021). Chronic wounds are consequently in a vicious cycle of inflammation with little constructive tissue being produced. The comprehension of these phase-dependent failures offers a mechanistic explanation of a specific intervention, such as the use of phytoconstituents capable of regulating inflammation, oxidative stress, and matrix remodelling to reestablish a more physiological healing response (Carvalho et al., 2021; Trinh et al., 2022).

4. CELLULAR PLAYERS IN WOUND HEALING

Wound healing is a coordinated process of numerous cell populations, which dynamically interact with the extracellular matrix (ECM), as well as a complex network of soluble mediators that regulate wound healing. The functions of each type of cell are different and mutually supportive in each of the various stages of the healing process. The activity, timing, or communication of any of these cellular players may lead to impairment of tissue repair and foster chronicity or pathological scarring (Liarte, Bernabé-García, and Nicolás, 2020; Sen, 2021).

4.1 Platelets and their signalling roles

The initial cellular responders to tissue injury are platelets, which are critical in the haemostasis phase. In addition to their classical role in clot formation, platelets are powerful signalling units that release a great number of bioactive molecules from their alpha-granules: platelet-derived growth factor (PDGF), transforming growth factor-beta (TGF-beta), vascular endothelial growth factor (VEGF), epidermal growth factor (EGF), and insulin-like growth factor-1 (IGF-1) (Miricescu et al., 2021).

These mediators induce and coordinate the following repair processes by calling inflammatory cells, promoting the migration and proliferation of fibroblasts, and inducing angiogenesis. Platelet aggregation also forms a fibrin-fibronectin clot that gives a temporary ECM scaffold within which cells adhere and migrate. Unregulated platelet activation or parallel dysfunction of growth factor release can slow down healing and has been associated with impaired repair in diabetes and old age (Liarte et al., 2020).

4.2 Neutrophils and macrophages

It is dominated by neutrophils and macrophages that are imperative in host defence and wound debridement. Neutrophils are quickly attracted to the wound, where they ingest dead tissue and microorganisms and produce reactive oxygen species (ROS) and proteases. These measures are essential to contain the infection, but an excess of neutrophils or their prolonged activation may lead to collateral tissue damage and the loss of inflammation (Krishnaswamy, Mintz, and Sagi, 2017).

Macrophages can be considered the key regulators of wound healing. Macrophages are derived from circulating monocytes and possess a high degree of functional plasticity. Early in healing, they assume a pro-inflammatory phenotype that enhances immune responses; however, later they switch to a reparative phenotype enhancing fibroblast activation, angiogenesis, and matrix deposition by secreting growth factors such as TGF- β, PDGF, and VEGF (Liarte et al., 2020). This is a hallmark of non-healing wounds that are chronic in nature (Sen, 2021).

4.3 Fibroblasts and myofibroblasts

The key players in wound repair are fibroblasts, which are the main producers of extracellular matrix. In the proliferative phase, they invade the wound bed on the basis of chemotactic influences and make collagen, fibronectin, and proteoglycans that produce granulation tissue. These fibroblasts are influenced by TGF- 2 and mechanical forces to become myofibroblasts that express alpha-smooth muscle actin (Liarte et al., 2020).

The myofibroblasts produce contractile forces that decrease the size of the wound and aid in the remodelling of the matrix. During the healing process, the myofibroblasts undergo usual apoptosis; continued existence of them is related to excessive fibrosis, hypertrophic scarring, and keloid development. On the other hand, a lack of fibroblast activity results in poor matrix deposition and tensile strength (Krishnaswamy et al., 2017).

4.4 Keratinocytes and re-epithelialisation

The major cells of the epidermis are the keratinocytes that play the role of repairing the epithelial barrier via re-epithelialisation. After injury, there is a phenotypic change in keratinocytes at the wound edge, leading to migration, proliferation, and differentiation. Growth factors involved in these processes include EGF and keratinocyte growth factor, and integrin-based interactions with the provisional ECM (Martinengo et al., 2019).

Wound closure and protection against invasion by microbes depend on the process of effective re-epithelialisation. Persistent inflammation, degradation of growth factors, and changes in ECM composition by proteases, and impaired keratinocyte migration and proliferation in chronic wounds usually halt the epithelial repair (Sen, 2021).

4.5 Endothelial cells and angiogenesis

Angiogenesis is the process through which existing blood vessels differentiate into new ones and is the major characteristic of wound healing, and is endothelially-mediated. The angiogenic factors, especially the VEGF, are caused by hypoxia in the wound microenvironment and activate endothelial cell proliferation, migration, and tube formation (Miricescu et al., 2021).

The newly developed microvasculature feeds the regenerating tissue and helps to bring repairing cells by supplying oxygen and nutrients. With the healing process, overgrown vessels shrink back to regain more or less normal vascular density. Angiogenic impairment, which is typical of ischaemic and diabetic wounds, is an additive factor in delayed wound healing and tissue necrosis (Liarte et al., 2020).

4.6 Stem and progenitor cells in tissue repair

Stem and progenitor cells have become significant constituents in wound healing directly by differentiation and indirectly by paracrine signalling. Themes and engulf growth factors, cytokines, and extracellular vesicles, which control inflammation, increase angiogenesis, and activate fibroblast and keratinocytes (Trinh et al., 2022).

Though their direct role in the regeneration of tissue is a research question, the evidence points towards the support of stem and progenitor cells in the regenerative microenvironment, and not complete replacement of damaged cells. Taking advantage of their healing potential, either on their own or with bioactive agents, including phytoconstituents, is one of the promising directions of further wound treatment (Zulkefli et al., 2023).

5. MOLECULAR MECHANISMS UNDERLYING WOUND REPAIR

Wound repair depends on a tightly coordinated network of cell recruitment, proliferation, migration, differentiation, and extracellular matrix (ECM) turnover signals, which are managed by a finely tuned molecular network. These are highly context-specific mechanisms that work in a well-coordinated temporal chain. A combination of pro-inflammatory and pro-repair mediators, moderated protein catabolism of the matrix, and fine regulation of gene expression are all necessary elements of successful healing. These molecular processes are disrupted, which is one of the main characteristics of non-healing and chronic wounds (Krishnaswamy, Mintz, and Sagi, 2017; Liarte, Bernabé-García, and Nicolás, 2020).

5.1 Growth factors and cytokines

Growth factors are influential control systems of cellular behaviour when a wound is healing, and they primarily work by autocrine and paracrine signalling. One of the first signals to be released in the wound site is platelet-derived growth factor (PDGF), which is fundamental in attracting fibroblasts and smooth muscle cells, as well as promoting their growth. Transforming growth factor- 8 (TGF -8) is a versatile cytokine that modulates inflammation, fibroblast activation, collagen production, and differentiation of myofibroblasts. It is very context-dependent in its effects, and the dysregulated TGF-b signalling has been associated with chronic wounds as well as excessive fibrosis (Liarte et al., 2020).

Angiogenesis, identified as well as keratinocytes and epithelial regeneration, are all required by other growth factors, including vascular endothelial growth factor (VEGF), fibroblast growth factors (FGFs), and epidermal growth factor (EGF). Early inflammatory responses are triggered and facilitated by cytokines such as tumour necrosis factor-alpha (TNF- a), interleukin-1 (IL- 1), and interleukin-6 (IL- 6), but are strictly regulated to enable passage to the proliferative phase. Chronic wounds have an imbalance of pro-inflammatory and pro-repair cytokines as a characteristic molecular characteristic (Miricescu et al., 2021; Sen, 2021).

5.2 Chemokines and inflammatory mediators

Chemokines are low-weight chemoattractant cytokines, which direct the immigration of immune and repair cells towards the wound. The CC and CXC group of chemokines control neutrophil, monocyte, and lymphocyte recruitment during the inflammatory response and angiogenesis and fibroblast activity later on. As an example, CXCL8 (IL-8) facilitates neutrophil inflammation, as well as has pro-angiogenic effects (Martinengo et al., 2019).

Vascular permeability, immune cell activation, and redox balance are also further regulated by inflammatory mediators like prostaglandins, leukotrienes, and nitric oxide. Although necessary to the host defence and initial wound stabilisation, sustained production by these mediators leads to tissue injury, activation of proteases, and cell dysresponsiveness. Chronic wounds are, in turn, described by the presence of an ongoing inflammatory environment that does not allow for resolution and moving on to the regeneration of tissues (Krishnaswamy et al., 2017; Sen, 2021).

5.3 Extracellular matrix remodelling

The extracellular matrix is not only a structural scaffold, but it is also an active particle of cell behaviour and molecular signalling. The ECM, through wound healing, is subjected to constant remodelling that commences with a provisional fibrin-rich structure and transitions into a collagen-based structure. The growth factors and integrins bind to fibronectin and proteoglycans, thus affecting cell adhesion, migration, and differentiation (Liarte et al., 2020).

With the progression of the healing, the temporary matrix is substituted by the granulation tissue that contains type III collagen, and that is subsequently replaced by the type I collagen that has a higher tensile strength. ECM remodelling can be controlled to improve tissue integrity, but excessive degeneration or deposition may result in chronic wound conditions or pathological scarring. The equilibrium between matrix synthesis and degradation is thus an important factor that determines the healing outcomes (Krishnaswamy et al., 2017).

5.4 Matrix metalloproteinases and their regulation

This is because matrix metalloproteinases (MMPs) are zinc-dependent endopeptidases that are at the core of the process of ECM degradation and remodelling. They enable migration of cells to angiogenesis, tissue reorganisation by breaking down collagens, gelatin, and other components of the matrix. The activity of MMP is regulated strictly on various levels, such as on the level of gene transcription, activation, and inactivation of pro-enzymes by tissue inhibitors of metalloproteinases (TIMPs) (Krishnaswamy et al., 2017).

In chronic wounds, this control system is often disturbed, causing too much MMP activity and less TIMP expression. The resulting destruction of ECM and growth factors continues the process of inflammation and does not allow stable tissue development. The therapeutic methods focused on restoring the protease balance by inhibiting the excessive activity of MMP or preventing the destruction of growth factors have thus received significant interest in the field of wound healing (Liarte et al., 2020).

5.5 Oxidative stress and redox signalling

Reactive oxygen species (ROS) have a two-fold effect in wound healing. ROS are signalling molecules at physiological levels, which control cell migration, angiogenesis, and antimicrobial defence. In the first phase of inflammation, ROS are produced by neutrophils and macrophages as a part of the innate immune system that helps eliminate the pathogen (Miricescu et al., 2021).

Nevertheless, too much or too long oxidative stress causes cellular membrane damage, cellular protein and nucleic acid damage, growth factor signalling dysfunction, and inflammation. Oxidative stress is commonly high in chronic wounds because of their ongoing inflammation, ischaemia, as well as defective antioxidant defences. Redox restoration with endogenous antioxidant defenses or exogenous antioxidants is thus regarded as one of the treatment goals with regard to enhancing the healing time (Carvalho et al., 2021).

5.6 Epigenetic and transcriptional regulation

In addition to classical signalling pathways, wound healing is being identified as regulated on an epigenetic and transcriptional level. The expression of inflammatory, angiogenic, and matrix remodelling genes is regulated through epigenetic processes that involve DNA methylation, histone regulation, and non-coding RNAs, including microRNAs. These regulatory layers allow cells to quickly respond to the evolving wound microenvironment (Trinh et al., 2022).

Specifically, the microRNAs have been found to fine-tune macrophage polarisation, fibroblast activity, and keratinocyte migration. The impaired recovery and fibrosis have been associated with dysregulation of epigenetic control mechanisms, and the regulation of epigenetics can be suggested as a potential future avenue in the curative therapy of wounds. A combination of epigenetic knowledge and traditional molecular knowledge offers a better way to understand wound pathophysiology and to discover new treatment targets (Zulkefli et al., 2023). In order to enable easy categorisation of the key molecular mediators during wound repair, Table 1 provides concise information on the key growth factors, cytokines, proteases, and redox mediators, and their cellular sources, biological functions, and respective phases in wound healing.

Table 1. Key Molecular Mediators Involved in Wound Healing and Their Biological Functions

Molecular Mediator	Primary Source Cells	Major Functions in Wound Healing	Associated Healing Phase(s)
Platelet-Derived Growth Factor (PDGF)	Platelets, macrophages	Fibroblast recruitment, cell proliferation, and granulation tissue formation	Haemostasis, Proliferation
Transforming Growth Factor-β (TGF-β)	Platelets, macrophages, fibroblasts	Collagen synthesis, myofibroblast differentiation, and immune regulation	Inflammation, Proliferation, Remodelling
Vascular Endothelial Growth Factor (VEGF)	Keratinocytes, macrophages, and endothelial cells	Angiogenesis, vascular permeability	Proliferation
Epidermal Growth Factor (EGF)	Platelets, keratinocytes	Keratinocyte migration, re-epithelialisation	Proliferation
Tumour Necrosis Factor-α (TNF-α)	Macrophages, neutrophils	Inflammatory cell recruitment, immune activation	Inflammation
Matrix Metalloproteinases (MMPs)	Fibroblasts, macrophages	ECM degradation, cell migration	Remodelling

continued on following page

Table 1. Continued

Molecular Mediator	Primary Source Cells	Major Functions in Wound Healing	Associated Healing Phase(s)
Tissue Inhibitors of Metalloproteinases (TIMPs)	Fibroblasts, keratinocytes	Regulation of ECM degradation	Remodelling
Reactive Oxygen Species (ROS)	Neutrophils, macrophages	Antimicrobial defence, redox signalling	Inflammation, Proliferation

6. SIGNALLING PATHWAYS INVOLVED IN WOUND HEALING

During wound healing, cellular processes are coordinated by a small set of highly conserved intracellular signalling pathways that combine extracellular signals into a set of gene expression programmes. The pathways control inflammation, cell migration, proliferation, differentiation, angiogenesis, and extracellular matrix (ECM) remodelling. Although stringent control of signalling has been beneficial in repairing wounds efficiently, its malregulation leads to chronic yet non-healing wounds or fibrosis proliferation. The knowledge of these pathways is a mechanistic rationale behind specific therapeutic interventions, such as the application of bio-active phytoconstituents capable of regulating the activity of the pathways (Liarte, Bernabé-Garcia, and Nicolás, 2020; Trinh et al., 2022).

6.1 TGF-β/Smad signalling pathway

One of the most significant regulators of wound healing is the transforming growth factor- 0 (TGF- 0). The TGF-B ligands act on the type I and II serine/threonine kinase receptors and trigger the phosphorylation of receptor-regulated Smad proteins (Smad2 and Smad3), which form complexes with Smad4 and translocate to the nucleus to regulate the transcription of target genes (Liarte et al., 2020).

In wound repair, the activation of fibroblasts, differentiation of myofibroblasts, collagen production, and immune regulation are regulated by TGF-B/Smad signalling. Although activation of this pathway early in development is required to form granulation tissue and contract the wound, sustained or excessive signalling encourages fibrosis and pathological scarring. The response of TGF-B signalling is altered in chronic wounds, which contributes to loss of keratinocyte migration and malformation of matrix deposition. Therefore, the Smad/TGF-b signal is an illustration of how space and time are required when it comes to wound signalling (Liarte et al., 2020; Sen, 2021).

6.2 Wnt/β-catenin pathway

WNT/ -catenin signaling is a crucial pathway that controls cell fate, tissue regeneration, and proliferation. Wnt ligand binding in the canonical pathway inhibits β-catenin degradation, which permits β-catenin to accumulate in the cytoplasm and translocate to the nucleus, where it regulates transcription of genes involved in cell proliferation and differentiation (Zulkefli et al., 2023).

WNT/ -catenin Signalling in wound healing Wnt/ -catenin signalling is especially significant in the proliferation of keratinocytes, activation and re-epithelialisation of hair follicle stem cells. It has been shown that transient epidermal activation promotes epidermal regeneration, but prolonged or excessive epidermal activation has been linked to fibrotic response and aberrant scarring. The dual regenerative and pathological potential of the Wnt signalling has also been reported in delayed healing and tissue architecture in chronic wounds, thus suggesting that impaired Wnt signalling can be both regenerative and pathological (Trinh et al., 2022).

6.3 PI3K/Akt and MAPK pathways

The signalling pathways of wound healing, which are central in growth factor signalling, are the phosphoinositide 3-kinase (PI3K)/Akt and mitogen-activated protein kinase (MAPK). PI3K/Akt stimulates cell survival, cell migration, and cell proliferation, especially in keratinocytes, fibroblasts, and endothelial cells. This pathway is also protective as it restricts the apoptosis in the hostile wound microenvironment (Miricescu et al., 2021).

Extracellular signal-regulated kinase (ERK), c-Jun N-terminal kinase (JNK), and p38 MAPK are MAPK pathways that mediate minimal inflammation, cell migration, and stress adaptation. The various growth factors, including EGF, PDGF, and VEGF, stimulate MAPK cascades to initiate re-epithelialisation, fibroblast proliferation, and angiogenesis. The dysregulation of these pathways and especially chronic activation of stress-related MAPKs are contributors to chronic inflammation and impaired healing (Krishnaswamy, Mintz, and Sagi, 2017).

6.4 NF-κB pathway in inflammation

NF- kB pathway is the great controller of inflammatory gene expression. NF-kB also reacts to inflammatory stimuli and microbial products or oxidative stress by dissociating from its inhibitory complex and translocating to the nucleus, inducing the transcription of pro-inflammatory cytokines, chemokines, and adhesion molecules (Sen, 2021).

Early host defence and immune cell recruitment during wound healing requires transient NF-KB activation. Chronic activation, however, preserves a chronic inflammatory state, which is characterised by a high level of cytokines, high growth of protease activities, and cell insensitivity. NF-KB signalling is always prolonged in chronic wounds, and thus it is an attractive target of therapy. Several phytoconstituents were found to inhibit the NF-kB activation and induce the inflammatory process to resolve and transit to the proliferative stage (Carvalho et al., 2021; Trinh et al., 2022).

6.5 VEGF-mediated angiogenic signalling

Angiogenesis is vital to the provision of oxygen and nutrition to the regenerative tissue, and the key figure of this is the vascular endothelial growth factor (VEGF). VEGF interacts with its tyrosine kinase receptors on endothelial cells, resulting in the activation of downstream signalling pathways including PI3K/ Akt and MAPK to stimulate endothelial cell proliferation, migration, and tube formation (Miricescu et al., 2021).

One of the key triggers of VEGF expression in the wound microenvironment is hypoxia, and much of the regulation is by hypoxia-inducible factor-1alpha (HIF-1alpha). Good VEGF signalling presence is sufficient to cause timely granulation tissue vascularisation, whereas a compromised VEGF expression or receptor sensitivity is a factor that leads to poor angiogenesis in ischaemic and diabetic wounds. Despite the encouraging results of the therapeutic VEGF delivery, uncontrolled angiogenic signalling can result in leaky or malfunctioning vessels, and it is necessary to control the activation during the healing process (Sen, 2021).

7. ROLE OF PHYTOCONSTITUENTS IN MODULATING CELLULAR AND MOLECULAR EVENTS

Phytoconstituents: bioactive compounds extracted from medicinal plants have been of significant scientific interest because of their ability to achieve several wound healing stages concomitantly. Phytochemicals may also have pleiotropic actions, promoting inflammation, oxidative stress, angiogenesis, extracellular matrix (ECM) deposition, and immune regulation in a coordinated response, unlike single-target synthetic drugs. This multi-target character is especially beneficial in complicated wound settings, including chronic and infected wounds, where a number of molecular pathways are dysregulated simultaneously (Carvalho et al., 2021; Trinh et al., 2022). Recent developments in molecular biology have helped to demystify the interactions of particular classes of phytoconstituents with intracellular signalling cascades and cellular processes that are pertinent to tissue repair.

7.1 Anti-inflammatory phytochemicals

Early wound defence requires inflammation; however, its chronicity is a significant factor contributing to delayed wound healing. Many phytochemicals possess a strong anti-inflammatory effect due to the regulation of major signalling pathways like NF-KB, MAPK, and STAT. Phenolic acids and terpenoids such as flavonoids (e.g., quercetin, kaempferol, luteolin) have been demonstrated to inhibit the expression of pro-inflammatory cytokines such as tumour necrosis factor-alpha (TNF- 6), interleukin-1B (IL-1B), and interleukin-6 (IL-6) (Carvalho et al., 2021).

Mechanistically, such compounds suppress nuclear translocation of NF-KB and nuclear cyclo-oxygenase-2 (COX-2) and inducible nitric oxide synthase (iNOS), hence inhibiting the production of inflammatory mediators. The anti-inflammatory phytochemicals, when used topically or systemically in experimental wound models, shorten the inflammatory response and increase the rate of transition to proliferation. These effects are especially applicable in non-acute wounds, in which the effect of long-term inflammation is the continuation of tissue damage and the activity of proteases (Zulkefli et al., 2023).

7.2 Antioxidant and free radical scavenging compounds

Non-healing wounds are characterized by oxidative stress. The presence of excessive reactive oxygen species (ROS) damages fibroblast and keratinocyte functions, destroys growth factors, and destroys ECM components. A significant portion of phytoconstituents are effective antioxidants, which allows them to stabilize redox balance in the microenvironment of the wound and eliminate free radicals (Miricescu et al., 2021).

Polyphenols, which are flavonoids and tannins, reduce the electrons to neutralise the ROS and also increase the levels of antioxidant enzymes, which are synthesised endogenously, such as superoxide dismutase, catalase, and glutathione peroxidase. Empirical research indicates that plants that contain antioxidant extracts contribute to the inhibition of lipid peroxidation, maintenance of cell integrity, and improvement of wound contraction and epithelialisation. These compounds indirectly promote angiogenesis, collagen synthesis, and tissue regeneration in general by inhibiting oxidative damage (Carvalho et al., 2021; Trinh et al., 2022).

7.3 Phytoconstituents promoting angiogenesis

Angiogenesis plays a fundamental role in the provision of oxygen and nutrients to the regenerating tissue, and neovascularisation is impaired in ischaemic and diabetic wounds. Some phytoconstituents were found to mediate angiogenesis by stimulating vascular endothelial growth factor (VEGF) and activating downstream signalling through PI3K/Akt and ERK/MAPK (Zulkefli et al., 2023).

Medicinal plants have been found to provide pro-angiogenic effects with flavonoids, saponins, and polysaccharides in an in vivo wound model and the in vitro wound model, which resulted in the increase of capillary density and granulation tissue formation. Notably, phytochemicals tend to induce physiological and not excessive angiogenesis, which makes it less likely to have abnormal or leaky vessels. The specified, regulated response to angiogenesis is especially beneficial in chronic wounds, where the vascular signalling is dysregulated, leading to tissue hypoxia and non-healing (Miricescu et al., 2021).

7.4 Plant-derived compounds enhancing collagen synthesis

Restoration of tensile strength and structural integrity of healed tissue requires collagen deposition and remodelling. Plant compounds can target the proliferation and collagen production of fibroblasts by various mechanisms, mainly by altering TGF- B / Smad signalling and expression of ECM-related genes (Liarte et al., 2020).

There is experimental evidence that phytochemicals have the ability to raise hydroxyproline concentration, improve collagen fibre organisation, and promote wound contraction. Moreover, certain compounds inhibit the activity of the matrix metalloproteinases (MMPs), which prevent the overabundant breakdown of collagen and allow the maintenance of a balanced remodelling of the ECM. These roles lead to better strength of the wound, a decrease in re-injury risk, and a constrained excessive fibrosis in cases where signalling is properly balanced (Krishnaswamy, Mintz, and Sagi, 2017; Carvalho et al., 2021).

7.5 Immunomodulatory effects of medicinal plants

In addition to anti-inflammatory effects, a wide variety of medicinal plants have more extensive immunomodulatory effects, which also affect wound healing. The phytoconstituents have the capacity to modulate macrophage polarisation in favour of a switch to reparative phenotypes over pro-inflammatory phenotypes to drive angiogenesis, fibroblast activation, and matrix deposition (Trinh et al., 2022).

Immunomodulation of this nature increases the risk of inflammation without depressing host defence, thus providing a microenvironment in which tissue regeneration can occur. There are also antimicrobial and anti-biofilm properties of some plant compounds, which, although indirectly, contribute to immune activity by decreasing the intensity of the microbial burden. These immunomodulatory effects, combined with each other, make phytoconstituents highly attractive as adjuncts to standard wound management, especially in chronic and infected wounds in which immune dysregulation is intense (Zulkefli et al., 2023). Because plant-derived bioactives are diverse and multi-target, Table 2 summarises some of the major classes of phytoconstituents, typical compounds, key molecular targets, and their main wound-healing activity.

Table 2. Major Classes of Phytoconstituents and Their Mechanistic Roles in Wound Healing

Phytoconstituent Class	Representative Compounds	Primary Molecular Targets	Key Wound-Healing Effects
Flavonoids	Quercetin, Kaempferol, Luteolin	NF-κB, MAPK, PI3K/Akt	Anti-inflammatory, antioxidant, enhanced re-epithelialisation
Phenolic acids	Gallic acid, Caffeic acid	ROS, inflammatory cytokines	Free radical scavenging, reduced oxidative stress
Terpenoids	Betulinic acid, Ursolic acid	TGF-β, collagen pathways	Fibroblast activation, collagen synthesis
Alkaloids	Berberine, Capsaicin	AMPK, inflammatory mediators	Antimicrobial, immunomodulatory
Saponins	Asiaticoside	VEGF, TGF-β/Smad	Angiogenesis, ECM remodelling
Plant polysaccharides	Aloe vera polysaccharides	Macrophage polarisation	Immune regulation, tissue regeneration

8. PHYTOCHEMICAL-BASED THERAPEUTIC STRATEGIES FOR WOUND HEALING

The growing understanding of the cellular and molecular basis of wound repair has stimulated the development of phytochemical-based therapeutic strategies that extend beyond traditional topical remedies. Contemporary approaches aim to improve the stability, bioavailability, targeting, and clinical reproducibility of plant-derived bioactives, thereby enhancing their translational potential. These strategies integrate pharmacological insights with advances in biomaterials and nanotechnology to address the multifactorial nature of acute and chronic wounds (Trinh et al., 2022; Zulkefli et al., 2023).

8.1 Herbal extracts and bioactive fractions

Herbal extracts remain the most widely used form of phytochemical intervention in wound healing. Crude extracts contain a complex mixture of bioactive compounds, including flavonoids, phenolic acids, terpenoids, alkaloids, and polysaccharides, which may act synergistically to modulate inflammation, oxidative stress, angiogenesis, and extracellular matrix (ECM) deposition (Carvalho et al., 2021). This multi-component nature is particularly advantageous in chronic wounds, where multiple molecular pathways are dysregulated simultaneously.

However, variability in plant species, geographical origin, harvesting conditions, and extraction methods can lead to substantial differences in phytochemical composition and biological activity. To address this limitation, increasing emphasis has been placed on the use of standardised extracts and bioactive fractions enriched in defined marker compounds. Fractionation and phytochemical profiling using chromatographic and spectrometric techniques enable improved reproducibility and facilitate mechanistic studies linking specific constituents to biological effects (Trinh et al., 2022). Such approaches represent an important step towards the clinical acceptance of herbal wound therapies.

8.2 Phytoconstituent-loaded dressings

Advanced wound dressings incorporating phytoconstituents have emerged as a promising strategy to enhance local delivery and therapeutic efficacy. Hydrogels, films, foams, and nanofibrous mats can be impregnated with plant extracts or purified compounds to provide a moist wound environment while enabling sustained release of bioactives (Cedillo-Cortezano et al., 2024).

Phytoconstituent-loaded dressings have been shown to reduce microbial burden, attenuate inflammation, and promote granulation tissue formation in experimental wound models. Importantly, localised delivery minimises systemic exposure and potential toxicity while maintaining therapeutically relevant concentrations at the wound site. Such dressings may be particularly beneficial for chronic ulcers, where prolonged topical treatment is required. Nevertheless, optimisation of release kinetics, stability, and shelf life remains a key challenge for widespread clinical adoption (Zulkefli et al., 2023).

8.3 Nano-delivery systems for plant bioactives

Nanotechnology-based delivery systems have been extensively explored to overcome the inherent limitations of many phytochemicals, including poor aqueous solubility, low stability, and limited skin penetration. Nano-carriers such as liposomes, polymeric nanoparticles, solid lipid nanoparticles, and nanoemulsions can encapsulate plant bioactives, protecting them from degradation and enabling controlled and sustained release (Chiangnoon et al., 2022).

In wound healing applications, nano-delivery systems have demonstrated enhanced penetration into wound tissue, improved bioavailability, and superior therapeutic outcomes compared with free phytochemicals. Additionally, surface modification of nanoparticles allows targeted delivery and responsiveness to wound-specific stimuli such as pH, enzymes, or oxidative stress. While preclinical evidence is encouraging, further studies addressing long-term safety, scalability, and regulatory considerations are essential before these technologies can be routinely translated into clinical practice (Cedillo-Cortezano et al., 2024).

8.4 Synergistic effects of phytoconstituents and conventional therapies

An emerging strategy in wound management involves the combination of phytoconstituents with conventional therapies such as antibiotics, growth factors, debridement techniques, and standard dressings. Phytochemicals may enhance the efficacy of conventional treatments by reducing inflammation and oxidative stress, improving tissue responsiveness, and mitigating adverse effects associated with prolonged drug use (Carvalho et al., 2021).

Synergistic approaches are particularly relevant in infected and chronic wounds, where phytoconstituents with antimicrobial and anti-biofilm properties can complement antibiotic therapy and reduce the risk of resistance. Moreover, combining phytochemicals with advanced biomaterials or cell-based therapies may further enhance regenerative outcomes. Despite promising preclinical data, robust clinical trials are required to establish optimal combinations, dosing regimens, and long-term safety profiles (Trinh et al., 2022; Zulkefli et al., 2023).

9. PRECLINICAL AND CLINICAL EVIDENCE OF PHYTOCONSTITUENT-MEDIATED WOUND HEALING

Phytoconstituents have their therapeutic value in wound management, which has been utilized in numerous studies by way of in vitro studies, in vivo animal studies, and, to a lesser degree, clinical trials. All these lines of evidence favor the biological feasibility of using plant-based compounds in the modulation of major wound-healing pathways, as well as suggest important translational and regulatory issues. Strict assessment in the experimental models is necessary to address the gap between the traditional usage and evidence-based clinical implementation (Carvalho et al., 2021; Trinh et al., 2022).

9.1 In vitro cellular studies

In vitro models provide a controlled environment that can be used to study the direct effects of phytoconstituents on particular cell types in wound healing. Some of the most common assays are fibroblast and keratinocyte proliferation assays, scratch (migration) assays, collagen production assays, macrophage polarisation studies, and endothelial tube-formation assays. Such strategies have shown that many phytochemicals do have the ability to stimulate fibroblast growth, increase the motility of keratinocytes, and induce angiogenic behavior of endothelial cells (Zulkefli et al., 2023).

Mechanistic research indicates that plant-based compounds have the potential to regulate intracellular signalling pathways of NF-kB, PI3K/ Akt, and MAPK, leading to the decreased production of inflammatory mediators and the increased expression of growth factors. The antioxidant phytochemicals have also been found to prevent damage to the cultured cells by oxidative stress, which maintains cellular viability and functionality. Although in vitro results offer important mechanistic understanding, they are not accurate reflections of the complexity of the wound microenvironment and should, therefore, be used with much caution when applied to in vivo conditions (Miricescu et al., 2021).

9.2 In vivo animal models

Animal models are an essential transitional phase between cell-based experiments and clinical trials. Rodent excisional, incisional, and burn wound models are most widely used to determine the effectiveness of phytoconstituents in physiologically spreading conditions. Topical or systemic applicability of plant extracts and individual compounds in these models has been linked to faster wound healing, stronger granulation tissue growth, prominence of collagen deposition, and tensile strength (Carvalho et al., 2021).

The results of the animal research commonly indicate a decrease in inflammatory cell infiltration, increased angiogenesis, and a balanced matrix remodelling after treatment with the phytochemicals. In more recent studies, the nano-formulations of phytoconstituents have proven to be better than the free ones as they exhibit better stability and sustained release at the wound site (Chiangnoon et al., 2022). Although such promising results have been achieved, variations in skin architecture, immune response, and healing dynamics between animals and humans only restrict direct translation and indicate that careful interpretation of preclinical results should be implemented (Trinh et al., 2022).

9.3 Clinical trials and translational challenges

There is relatively limited clinical evidence in support of phytoconstituent-based wound therapies as compared to the numerous preclinical studies that have been conducted. Current clinical trials are usually small, heterogeneous in nature, and variable in outcome measures. Other researchers have found an increase in the rate of epithelialisation, a decrease in infection, and healing of chronic wounds in the presence of herbal preparations or homemade dressings made of plants, whereas others report marginal or inconsistent improvements (Sen, 2021).

Translational obstacles encompass inconsistency in phytochemical expression, absence of standardised formulations, differences in dosing systems, and absence of mechanistic endpoints. Also, in human patients, chronic wounds are also affected by the more complex systemic factors, e.g., diabetes, vascular disease, and age, which are hard to recapitulate in experimental models. In order to promote clinical translation, future studies need to take on rigorous randomised controlled designs, use standardised and well-characterised products, and include clinically meaningful endpoints, including time to complete healing, recurrence rates, and patient-reported outcomes (Cedillo-Cortezano et al., 2024).

9.4 Safety, toxicity, and regulatory considerations

Evaluation of safety is an important element of the formation of phytoconstituents in wound care. Even though many medicinal plants were traditionally used over a long period of time, this does not mean that they are safe in the contemporary clinical setting or at greater dosages. The risks that may occur are local irritation, allergic response, contact with heavy metals or pesticides, and interactions with traditional drugs (Cedillo-Cortezano et al., 2024).

In vitro and in vivo toxicological studies typically suggest a good tolerance record of many plant-derived compounds used topically, but systematic toxicity data are few on many formulations. In different jurisdictions, regulation has been different, with herbal wound products being regulated as medicines, medical devices, or cosmetics, based on their composition and the claims. In order to achieve successful clinical translation, clear regulatory pathways, compliance with good manufacturing practices, and extensive safety documentation are required. The management of these issues will be crucial to the inclusion of phytoconstituent-based therapies in the mainstream wound care (Trinh et al., 2022; Zulkefli et al., 2023).

10. CHALLENGES, LIMITATIONS, AND KNOWLEDGE GAPS

Despite substantial experimental and emerging clinical evidence supporting the wound-healing potential of phytoconstituents, several scientific, technical, and translational challenges continue to limit their widespread clinical adoption. These challenges stem from the intrinsic complexity of plant-derived products, gaps in mechanistic understanding, and difficulties in translating preclinical success into reproducible clinical outcomes. Addressing these limitations is essential for advancing phytochemical-based wound therapies from experimental promise to evidence-based clinical practice (Trinh et al., 2022; Cedillo-Cortezano et al., 2024).

10.1 Variability in phytochemical composition

One of the most significant challenges in phytoconstituent-based wound healing research is the inherent variability in phytochemical composition. The concentration and profile of bioactive compounds within a plant are influenced by numerous factors, including species and cultivar, geographical origin, soil composition, climate, harvesting season, and post-harvest processing. As a result, extracts prepared from the same plant species may exhibit substantial differences in biological activity (Carvalho et al., 2021).

This variability complicates reproducibility across studies and undermines the comparability of experimental and clinical findings. In the context of wound healing, where precise modulation of molecular pathways is required, inconsistent phytochemical composition may lead to variable therapeutic outcomes. Addressing this issue requires systematic characterisation of plant materials and careful documentation of sourcing and processing conditions in research and clinical studies (Zulkefli et al., 2023).

10.2 Standardisation and quality control issues

Closely related to compositional variability is the challenge of standardisation and quality control. Many published studies employ crude plant extracts without defining marker compounds or establishing quantitative benchmarks for bioactive constituents. This lack of standardisation hinders dose optimisation, mechanistic interpretation, and regulatory evaluation (Trinh et al., 2022).

Advanced analytical techniques such as high-performance liquid chromatography, mass spectrometry, and nuclear magnetic resonance spectroscopy offer powerful tools for phytochemical profiling and quality assurance. However, their adoption is inconsistent across laboratories and clinical developers. Furthermore, ensuring batch-to-batch consistency at an industrial scale remains challenging. Robust quality control frameworks and adherence to good manufacturing practices are therefore essential to ensure the safety, efficacy, and reproducibility of phytoconstituent-based wound therapies (Cedillo-Cortezano et al., 2024).

10.3 Mechanistic ambiguities

Although numerous studies demonstrate beneficial wound-healing effects of phytoconstituents, the precise molecular mechanisms underlying these effects are often incompletely defined. Many investigations report phenotypic outcomes, such as accelerated wound closure or enhanced collagen deposition, without linking these effects to specific signalling pathways, gene expression changes, or cellular targets (Carvalho et al., 2021).

The pleiotropic nature of phytochemicals further complicates mechanistic analysis, as a single compound may influence multiple pathways simultaneously. While this multi-target activity may be therapeutically advantageous, it poses challenges for mechanistic clarity and rational drug development. Integrative approaches combining transcriptomics, proteomics, and metabolomics are needed to systematically map phytochemical–target interactions and to distinguish primary effects from secondary or compensatory responses (Trinh et al., 2022).

10.4 Translational barriers

Translating phytoconstituent-based therapies from laboratory research to clinical practice faces several additional barriers. Preclinical models often fail to replicate the complexity of human chronic wounds, which are influenced by systemic conditions such as diabetes, vascular disease, and ageing. Consequently, therapeutic effects observed in animal models may not translate directly to human patients (Sen, 2021).

Clinical trials of herbal wound therapies are frequently limited by small sample sizes, heterogeneous formulations, and inconsistent outcome measures, reducing the strength of evidence available for regulatory approval. Regulatory pathways themselves are often unclear, with phytochemical products variably classified as medicines, medical devices, or cosmetic products depending on jurisdiction and claims. Overcoming these translational barriers will require well-designed, adequately powered clinical trials, harmonised regulatory frameworks, and interdisciplinary collaboration among researchers, clinicians, manufacturers, and regulatory agencies (Cedillo-Cortezano et al., 2024; Zulkefli et al., 2023).

11. FUTURE PERSPECTIVES AND EMERGING TRENDS

The new perspectives of wound-healing studies can be characterized by the rapid development of molecular biology, computational sciences, and biomedical engineering. In the case of phytoconstituent-based therapies, the emerging technologies are providing new possibilities to address the current limitations in the areas of mechanistic ambiguity, variability, and translational inefficiency. It is anticipated that by combining these innovations with conventional pharmacognostic wisdom, the creation of efficient, individualised, and sustainable wound-healing techniques will become significantly quicker (Trinh et al., 2022; Zulkefli et al., 2023).

11.1 Systems biology and omics approaches

Systems biology offers an overall picture of wound healing as a living complex network and not a series of actions. Omics technologies such as genomics, transcriptomics, proteomics, and metabolomics make it possible to profile the wound microenvironment in all aspects of cellular state and binding interactions. These methods have indicated complicated regulatory systems that control inflammation, angiogenesis, extracellular matrix remodelling, and immune resolution (Liarte, Bernabé-Garcia, & Nicolas, 2020).

Omics-based analysis can determine molecular targets and signaling by plant-derived compounds in the context of phytoconstituents, explaining how bioactives work and revealing synergistic interactions between bioactives. A combination of multi-omics data and network modelling should also be used to differentiate between the beneficial adaptive response and maladaptive signalling of chronic wounds. These insights should also inform rational design of the formulation and optimise the timing and dosing of the therapy (Trinh et al., 2022).

11.2 AI-assisted phytoconstituent screening

As part of drug discovery and development, artificial intelligence (AI) and machine learning are being used more often to screen phytoconstituents as wound-healing agents. To predict bioactivity, toxicity, and pathway modulation, computational models are able to analyse large amounts of chemical structures, biological activities, and molecular targets. The tools can save a considerable amount of time and money used on experimental screening (Zulkefli et al., 2023).

Artificial intelligence-based methods can also assist in the discovery of synergistic phytochemical blends and the maximisation of formulations towards particular wound-healing goals, e.g., anti-inflammatory or pro-angiogenic activity. AI-assisted screening paired with experimental validation can revolutionize the research on phytoconstituents that is mostly guided by experimentation into a predictive and data-driven science (Cedillo-Cortezano et al., 2024).

11.3 Personalised wound-healing therapies

Wound management is being more and more oriented towards personalised medicine. Chronic wounds are extremely heterogeneous in their causes, molecular phenotypes, and therapeutic reactions. The combination of molecular diagnostics and progress in the field of biomarker discovery makes the stratification of wounds according to the inflammatory state, protease activity, the number of microbes, and the individual characteristics of the patient, including metabolic control and immune competence, possible (Sen, 2021).

Individualised methods can permit the choice of individual phytoconstituents or preparations to individual wound phenotypes. Indicatively, antioxidant-containing compositions can be selected when the wound has a high oxidative stress, and immunomodulatory phytochemicals can be used when persistent inflammation is the predominant wound condition. The precision strategies may increase the efficacy of the therapeutic process and reduce the number of needless interventions (Trinh et al., 2022).

11.4 Sustainable and green therapeutic development

The idea of sustainability is getting more relevant in the context of developing wound-healing therapies in the future. Phytoconstituents provide natural merits in this respect, because they are obtained as a result of renewable biological resources and are biodegradable in many cases. Nevertheless, mass use of medicinal plants brings up issues connected to the loss of biodiversity, ecology, and the chain of supply and demand (Cedillo-Cortezano et al., 2024).

Future studies are thus likely to be directed towards practices of sustainable cultivation, green extraction methods, and exploiting plant cell cultures or synthetic biology in producing bioactive compounds that have fewer adverse effects on the environment. The concept of sustainability in developing phytochemicals is coherent with the international principles of environmentally friendly healthcare and contributes to the improved sustainability of wound care using plants (Zulkefli et al., 2023).

12. CONCLUSION

Wound healing is a very complicated and highly coordinated biological process that incorporates cellular reactions, molecular signal transduction, and dynamic extracellular matrix bioremodelling to reestablish tissue integrity after injury. The disturbance of this sensitively balanced system, be it due to chronic inflammation, oxidative stress, blocked angiogenesis, or protease imbalance, is the basis of the formation of non-healing wounds as a significant clinical and socio-economic concern across the globe (Sen, 2021).

This chapter has given me a wide understanding of the cellular and molecular basis of wound healing and the increased scientific support of phytoconstituent-based therapeutic interventions. In vitro evidence, animal models, and new clinical studies have proven that plant-derived bioactives have the capacity to positively affect several wound biology processes, such as the resolution of inflammation, redox homeostasis, angiogenesis, collagen synthesis, and immune regulation. Phytoconstituents are especially suitable due to the pleiotropy of their nature, which is particularly suitable in the context of multifactorial pathophysiology of chronic wounds (Carvalho et al., 2021; Trinh et al., 2022).

Despite this promise, there are still substantial challenges insofar as standardisation, mechanistic clarity, translational consistency, and regulatory acceptance are concerned. These problems will be challenging to solve without the assistance of interdisciplinary methods comprising systems biology, sophisticated analytical methods, nanotechnology, and powerful clinical trial design. In the future, personalised and sustainable phytochemical-based wound treatments with artificial intelligence and omics-informed insights can become the change that will turn the mostly supportive wound management into a more targeted mechanism-driven therapeutic approach. Phytoconstituents potentially continue to be central to the future generation of safe, efficacious, and affordable wound-healing biologics with further scientific rigour and translational focus.

REFERENCES

Carvalho, M. T. B., Araújo-Filho, H. G., Barreto, A. S., Quintans-Júnior, L. J., Quintans, J. S., & Barreto, R. S. S. (2021). Wound healing properties of flavonoids: A systematic review highlighting the mechanisms of action. *Phytomedicine*, *90*, 153636. DOI: 10.1016/j.phymed.2021.153636 PMID: 34333340

Cedillo-Cortezano, M., Martínez-Cuevas, L. R., López, J. A. M., Barrera López, I. L., Escutia-Pérez, S., & Petricevich, V. L. (2024). Use of medicinal plants in the process of wound healing: A literature review. *Pharmaceuticals (Basel, Switzerland)*, *17*(3), 303. DOI: 10.3390/ph17030303 PMID: 38543089

Chiangnoon, R., Samee, W., Uttayarat, P., Jittachai, W., Ruksiriwanich, W., Sommano, S. R., Athikomkulchai, S., & Chittasupho, C. (2022). Phytochemical analysis, antioxidant, and wound healing activity of *Pluchea indica* branch extract nanoparticles. *Molecules (Basel, Switzerland)*, *27*(3), 635. DOI: 10.3390/molecules27030635 PMID: 35163900

Krishnaswamy, V. R., Mintz, D., & Sagi, I. (2017). Matrix metalloproteinases: The sculptors of chronic cutaneous wounds. *Biochimica et Biophysica Acta (BBA) –. Biochimica et Biophysica Acta. Molecular Cell Research*, *1864*(11, 11 Pt B), 2220–2227. DOI: 10.1016/j.bbamcr.2017.08.003 PMID: 28797647

Liarte, S., Bernabé-García, A., & Nicolás, F. J. (2020). Role of TGF-β in skin chronic wounds: A keratinocyte perspective. *Cells*, *9*(2), 306. DOI: 10.3390/cells9020306 PMID: 32012802

Martinengo, L., Olsson, M., Bajpai, R., Soljak, M., Upton, Z., Schmidtchen, A., Car, J., & Järbrink, K. (2019). Prevalence of chronic wounds in the general population: Systematic review and meta-analysis of observational studies. *Annals of Epidemiology*, *29*, 8–15. DOI: 10.1016/j.annepidem.2018.10.005 PMID: 30497932

Miricescu, D., Totan, A., & Stanescu, S.. (2021). Growth factors, reactive oxygen species, and wound healing. *International Journal of Molecular Sciences*, *22*(17), 9512. DOI: 10.3390/ijms22179512 PMID: 34502429

Sen, C. K. (2021). Human wound and its burden: Updated compendium of estimates. *Advances in Wound Care*, *10*(5), 281–292. DOI: 10.1089/wound.2021.0026 PMID: 33733885

Trinh, X.-T., & Nguyen-Van Long, L. T. V. A. (2022). A comprehensive review of natural compounds for wound healing: Targeting bioactivity perspective. *International Journal of Molecular Sciences*, *23*(17), 9573. DOI: 10.3390/ijms23179573 PMID: 36076971

Zulkefli, N., Che Zahari, C. N. M., Sayuti, N. H., Kamarudin, A. A., Saad, N., Hamezah, H. S., Bunawan, H., Baharum, S. N., Mediani, A., Ahmed, Q. U., Ismail, A. F. H., & Sarian, M. N. (2023). Flavonoids as potential wound-healing molecules: Emphasis on the pathways perspective. *International Journal of Molecular Sciences*, *24*(5), 4607. DOI: 10.3390/ijms24054607 PMID: 36902038

Section 2

Phytoconstituents in Wound Management

Chapter 4
Phytoconstituents: Nature's Arsenal for Healing

Abraham Olufemi Asuku
http://orcid.org/0000-0003-0071-0531
Bioresources Development Centre, National Biotechnology Research and Development Agency, Nigeria

Ekundayo Temitope Ola
Bioresources Development Centre, National Biotechnology Research and Development Agency, Nigeria

ABSTRACT

In both traditional and modern medicine, the bioactive chemicals found in plants, also known as phytoconstituents, have long been recognized for their medicinal properties. These comprise substances with a wide range of pharmacological characteristics, like anti-inflammatory, antibacterial, antioxidant, and anticancer effects, such as alkaloids, terpenoids, glycosides, flavonoids, and polyphenols. With an emphasis on their mechanisms of action, therapeutic value, and potential for drug development, this chapter explores the various roles that medicinal plants and their phytoconstituents play in wound prevention and care. The chapter also covers advancements in phytochemical research, including bioavailability enhancements and pharmaceutical inclusion. Notwithstanding their enormous therapeutic potential, restrictions including safety, standardization, and regulatory issues are still necessary for their widespread use. By understanding and applying nature's abundance of medicinal compounds, phytoconstituents offer a practical and effective way to improve global health.

DOI: 10.4018/979-8-3373-2908-6.ch004

INTRODUCTION

Wounds occur when the skin is injured physically, chemically, thermally, infectiously, or immunologically, disrupting the continuity of cells. It is described as a disturbance of a living tissue's cellular, anatomical, and functional continuity. Worldwide, wounds provide a serious challenge to patients and medical personnel. In addition to having a substantial financial impact on patients, wounds have an impact on the physical and mental well-being of millions of people. One of the main causes of physical limitations is wounds. The restoration of tissue integrity is the hallmark of successful wound healing. The stimulation and infiltration of inflammatory cells, macrophages, and neutrophils which generate pro-inflammatory cytokines such as interleukin-1 beta (IL-1 beta) and tumor necrosis factor alpha (TNF-α)) is necessary for the healing of wounds (Portou et al., 2015).

Growth factors like transforming growth factor (TGF) and other fibroblast growth factors are activated by these inflammatory cytokines, which leads to the activated fibroblasts proliferating and infiltrating to the wound site (Portou et al., 2015).

However, obesity, aging, and endocrine disorders like diabetes mellitus can all hinder these natural healing mechanisms (Guo & Dipietro, 2010). Although the right amount of each cytokine is necessary for healing, excessive fibroblast proliferation brought on by inflammatory cytokine levels can lead to hypertrophic scarring, which can seriously deform the skin. Both industrialized and developing nations have recently shown an increasing interest in phytoconstituents, which are products from plants and traditional remedies.

Promoting the benefits of all treatments, natural or synthetic is crucial, but the growing need for treatments based on phytochemicals calls for further research into these products. More data have recently emerged supporting the potential uses of topical phytochemicals to improve the healing of acute and chronic wounds.

WOUND HEALING PROCESS

During the stages of homeostasis, inflammation, proliferation, and remodeling as part of the wound healing process, numerous cell types with diverse roles move in unison (figure 1). This process begins with hemostasis; when injury occurs, there is a rapid contraction of the affected blood vessels and the formation of a blood clot in order to halt exsanguination brought on by damage to the vessels (Rodrigues et al., 2019; Wilkinson & Hardman, 2020). Cytokines and growth factors, like platelet-derived, insulin-like, epidermal, and transforming growth factors, are abundant in platelets. These substances operate as repair agents in the healing of wounds by stimulating neutrophils, which bring on endothelial cells, macrophages, and fibroblasts.

Then comes the inflammatory phase, beginning with the early inflammatory reaction, which starts in the late coagulation stage and concludes soon after. In the wound area, this phase causes neutrophils to aggregate, where their main job is to clean the wound. Within 24 to 36 hours of damage, products from platelets and bacteria release a variety of chemotactic substances that attract neutrophils to the wound site for phagocytosis, which eliminates the bacteria and prevents infection. Apoptosis eliminates neutrophils after all contaminating bacteria have been destroyed; wound macrophages then carry on with the phagocytosis process 48–72 hours later. Macrophages are vital regulatory cells that provide a plenty supply of fibroblasts, endothelial cells, and strong tissue growth factors. They also possess a greater lifespan than neutrophils. Accordingly, cytokines, growth factors, proteases, coagulation factors, and reactive oxygen species are produced when neutrophils and macrophages eliminate dead tissue and foreign substances (AbdJalil et al., 2017).

As the wound moves closer to complete healing by reconstructing the extracellular matrix (ECM) basement membrane, fibroblasts and keratinocytes are essential for the development of the granulation tissue that seals the wound and offers a structural platform for cell adhesion, growth, migration, and differentiation (Baron et al., 2020). Fibroblasts are the main cell type that remodels the extracellular matrix of the wound by producing mature collagen fibrils (Rippa et al., 2019). Foreign bodies can cause infection, ischemia, edema, or increased pressure, among other problems that hinder the healing process of wounds. Therefore, proper diagnosis and interventions are essential to promote the healing of wounds and avoid further issues (Haalboom, 2018).

Figure 1. Stages of Wound Healing

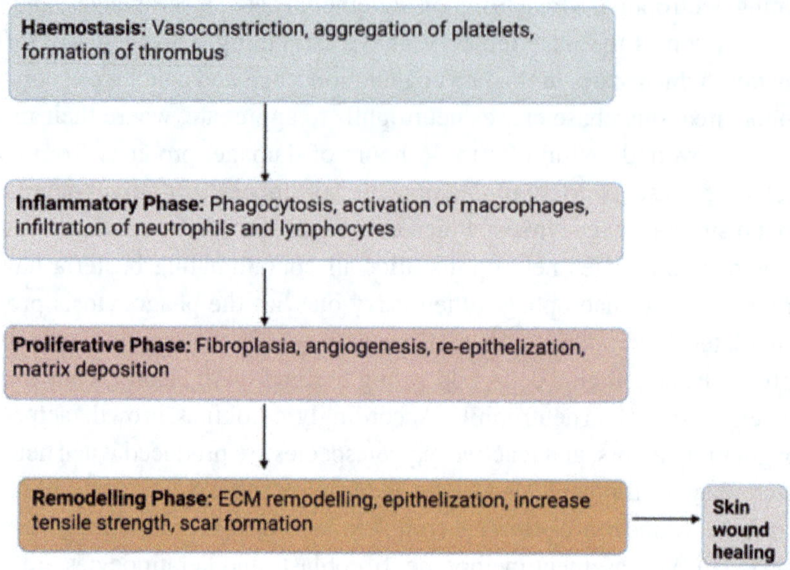

PHYTOCHEMICALS AND THEIR CLASSIFICATION

The term "phytochemicals" refers to a large category of naturally occurring bioactive compounds, such as polyphenols, flavonoids, steroids, saponins, and vitamins, which have a significant physiological effect on the human body.

Alkaloids

Alkaloids are basic compounds having heterocyclic nitrogen atoms that combine to generate salts with acids. The taste of most alkaloids is bitter. The kind of heterocyclic ring system that alkaloids contain determines their classification.

Flavonoids

Vascular plants contain a range of polyphenolic compounds termed flavonoids, such as aglycones, glucosides, and methylated derivatives. The six-membered ring that condenses with the benzene ring is either pyrone (flavonols and flavones) or its dihydro derivative (flavan-3-ols and flavanones). The placement of the benzenoid substituent distinguishes the two types of flavonoids, flavone and isoflavone. The majority of flavonoids can be categorized as monoglycosidic, diglycosidic, etc. within a specific class and are found in nature in conjugated form with sugar.

Phenolic acids

One carboxylic acid functional group is present in phenolic acids. The two unique carbon frameworks found in naturally occurring phenolic acids are the hydroxycinnamic and hydroxybenzoic structures. Simple esters of glucose and hydroxycarboxylic acids are used to create hydroxycinnamic acid molecules. Phenolic molecules biogenetically move through two metabolic pathways: the acetic acid pathway, which primarily produces simple phenol, and the shikimic acid pathway, which primarily produces phenylpropanoids. The phenylpropanoid pathway is used to produce the majority of phenolic chemicals found in plants. Flavonoids, the most abundant class of phenolic chemicals found in nature, are produced when both processes come together.

Saponins

A class of secondary metabolites from plants called saponins is known to produce stable foam in aqueous solutions like soap. Steroid alkaloids, triterpenoids, and glycosylated steroids are examples of saponins. Steroid aglycones come in two primary varieties: derivatives of furostan and spirostan. Sugar moieties (one or more) containing glucose, galactose, xylose, rhamnose, arabinose, or glucuronic acid glycosidically bound to a sapogenin which make up the carbohydrate portion.

Tannins

A diverse class of high molecular weight polyphenolic chemicals, tannins can combine with nucleic acids, polysaccharides, proteins, alkaloids, minerals, and other substances to produce both reversible and irreversible complexes.

Terpenoids

Five-carbon isoprene units were the source of the terpenoids. The multicyclic structures of the majority of terpenoids are distinguished from one another by their fundamental carbon skeletons and functional groups. Terpenes are found throughout nature, mostly in plants where they constitute components of essential oils. The hydrocarbon isoprene, $CH_2 = C(CH_3)\ CH{=}CH_2$, is their fundamental unit.

NATURAL MEDICINAL PRODUCTS AND THEIR PHYTOCONSTITUENTS FOR WOUND HEALING

Natural compounds that have antibacterial properties, help manage inflammatory conditions, and may promote skin tissue regeneration include extracts, phytochemicals, and products from plants (Alherz et al., 2022). They may reduce inflammation and eliminate oxidative stress (Shah & Amini-Nik 2017). In animal models of wound healing, the capacity of several plant extracts and their bioactives to heal wounds was validated. In animal models with and without diabetes, these plants enhanced angiogenesis, collagen deposition, and epithelial cell proliferation (Binsuwaidan et al., 2022).

Aloe vera

Aloe vera L. is a perennial herb. Its applications in the production of mucilaginous gel in pharmacology and cosmetics are numerous (Long, 2016). Vitamins A, C, E, and B12; enzymes; minerals; carbohydrates; polyphenols; sterols; and hormones are among its more than 75 distinct constituents (Sánchez-Machado et al., 2017). This plant has been used to treat various skin issues, such as wound healing, rejuvenation, and other dermatological diseases like inflammation and burns. In fact, scientific research on its physiological role in wound healing has just lately been carried out, despite the fact that it is widely used as a folk treatment (Sánchez-Machado et al., 2017).

The leaves of A. vera are made up of mucilaginous tissue used to make a variety of commercially available gel-based cosmetics and medications. But according to Sánchez-Machado et al. (2017), the gel has been connected to a number of positive outcomes for inflammation of the skin. When it is administered to animals having acute wounds, certain glycoproteins are in charge of the positive effects, and individual gel ingredients may encourage healing of wounds in animal models. In rat cutaneous wound models, topical use of A. vera has been demonstrated to dramatically accelerate the epithelialization, contraction, and maturation rate of wounds. Additionally, in comparison to the control lesions, it decreases inflammation, shrinks the amount of scar tissue, and improves the organization and alignment of regenerated scar tissue while raising the glycosaminoglycans and collagen concentrations (Oryan et al., 2016). By promoting deposition of collagen, proliferation of fibroblasts, and angiogenesis while preventing the excessive formation and buildup of ECM proteins that result in hypertrophic scarring, A. vera has positive effects (Hormozi et al., 2017). In human skin fibroblasts and keratinocytes in in vitro models (Oliveira et al., 2016), by moderately promoting cell migration and considerably promoting fibroblast and keratinocyte proliferation, A. vera sped up wound healing.

There is no evidence of reducing concentrations of microorganisms or improving maturation of scars in burn wounds, no significance in fibroblast migration dynamics, and no difference between sample groups in terms of wound repair, according to an investigation that reported no remarkable differences in wound healing between the control group and the group administered A. vera (Oliveira et al., 2016). This is most likely the result of an issue with extract concentration standardization. It is possible that some of this research mixed aloe extracts with other inappropriate alternative medicines or employed extracts with insufficient amounts of active chemicals known to have anti-inflammatory and proliferative properties.

Crocus sativus

A spice made from Crocus sativus L. is generally referred to as saffron. Contemporary pharmacological studies have demonstrated that extracts or bioactive compounds of saffron, such as phytosterols, monoterpenoids, apocarotenoids, phenolic acids, and flavonoids, possess numerous therapeutic effects (Abu-Izneid et al., 2020). The primary carotenoids of saffron extracts (SE) are crocetin, crocin, safranal, and picrocrocin (Alonso et al., 2012). The carotenoids that are water-soluble are also very limited, and crocin, a digentiobiosyl ester of crocetin, is one of them (Alonso et al., 2012). Such substances have been shown to have many biological activities, including anti-inflammatory and antioxidant (Abu-Izneid et al., 2020). Recent studies have suggested that crocin, crocetin, and safranal reduce nitric oxide and malondialdehyde and increase glutathione, antioxidant enzymes, and thiol under the influence of oxidative stress (Nanda and Madan, 2021). Few authors have reported the impact of saffron and its carotenoids on wound healing, even though most studies have reported high anti-inflammatory and antioxidant properties (Zeka et al., 2018). An in vitro study was conducted on human epidermal keratinocytes and human dermal fibroblasts to determine the capacity of crocin to fight the inflammatory process and to eliminate ROS (Fagot et al., 2018). Crocin reduced the release of inflammatory cytokines, regulated the expression of genes associated with NF-kB and glycosylation, and reduced dose-dependently peroxidation caused by UVA on cell membrane arachidonic acid and squalene (Fagot et al., 2018). This indicates that crocin may be regarded as a chemical with potentially favorable improvement effects on skin aging, given the effect of UV exposure on the skin quality (Krutmann et al., 2017). Stamens and sepals of C. sativus have polyphenols and anthocyanins that have been used as an anti-inflammatory and antioxidant compounds in skin complications (Gigliobianco et al., 2021).

Zeka et al. (2018) developed a novel hydrogel supplemented with antioxidants of saffron flowers. When fibroblast cell cultures of baby mice were incubated with hydrogel, which was enriched with crocin or kaempferol, cell expansion was larger than the control (Zeka et al., 2018). The antioxidant power of chemicals extracted from saffron could be ventured in the healing of the wound, but further research is needed to have an entire understanding of its pathways and mechanisms of action of wound healing.

Curcumin

Curcumin, which is a chemical constituent of turmeric or Curcuma longa, is among the most studied phytochemicals in wound healing. The anti-inflammatory properties of curcumin are the most important mechanism through which this compound influences wound healing. Curcumin is hence an excellent phytochemical compound in the management of hyper-inflammatory lesions, such as burns and chronic diabetic lesions. Curcumin has a low dermal absorption because it is hydrophobic (El-Refaie et al., 2015). In order to improve the topical use of curcumin, many formulations have been developed, including gels, collagen films, polymetric bandages, and alginate foams. Furthermore, curcumin's skin distribution is much improved by lipid-core nanocapsules (LCN) (Friedrich et al., 2015).

For topical applications, LCN-based curcumin delivery methods thus exhibit great promise. Future cutaneous applications of curcumin appear promising, as with the utilization of hydrogels (Du et al., 2016) and chitosan nanoparticles loaded with curcumin (Karri et al., 2016). Chronic wounds have significant levels of inflammation and proteolysis. Therefore, it is essential to manage this dysregulated inflammation in order to guarantee proper wound healing. In the treatment of persistent wounds that fail to heal, topical curcumin shows promise. Interestingly, topical curcumin treatment on diabetic rats' wounds also resulted in increased angiogenesis, as evidenced by a notable elevation of VEGF (Kant et al., 2015).

High concentrations of radical oxygen species, including superoxide and hydrogen peroxide, are also found in chronic wounds, and they can seriously hinder the healing process. In cutaneous wounds, curcumin exhibits strong antioxidant properties. Transdermal administration of curcumin decreased hydrogen peroxide-induced fibroblast and keratinocyte damage in rat excisional wound models (Gadekar et al., 2012).

Glycyrrhiza glabra L

The root of the herbaceous perennial legume *Glycyrrhiza glabra* yields licorice, a sweet and fragrant flavor. Its antioxidant, antibacterial, antifungal, and antiviral capabilities are among the potential healing qualities that have been documented in numerous articles (Pastorino et al., 2018). Phytoconstituents like glycyrrhizin, saponins, triterpene chalcones, isoflavonoids, and flavonoids are the primary biologically active ingredients in licorice and are in charge of the actions that have been reported (Komes et al., 2016) (Figure 2). Its ability to heal full-thickness cutaneous wounds has been documented in studies (Zangeneh et al., 2019; Siriwattanasatorn et al., 2020).

Prior studies have demonstrated that G. glabra's anti-inflammatory and antioxidant components, including flavonoids and triterpenes, decrease pus and free radicals in the wound region, promoting the healing of the wound (Geethalakshmi et al., 2013). In comparison to the control and basal ointment groups, Zangeneh et al. (2019) found that aqueous extract ointment of G. glabra enhances the parameters of wound healing in rats by increasing wound contracture, fibrocyte, hydroxyproline, and hexuronic acid levels while decreasing the area of the wound, total cells, macrophage, neutrophil, and lymphocyte levels. Recently, Hanafy et al. (2022) conducted an investigation in a guinea pig model to investigate the effect of the extract of this plant on wound healing. In this study, the *G. glabra* extract-bearing cream applied topically accelerated wound healing. The finding of flavonoids could have also helped to achieve the outcome of this discovery (Hanafy et al., 2022).

Figure 2. Phytoconstituents and wound healing mechanisms of action of Glycyrrhiza glabra.

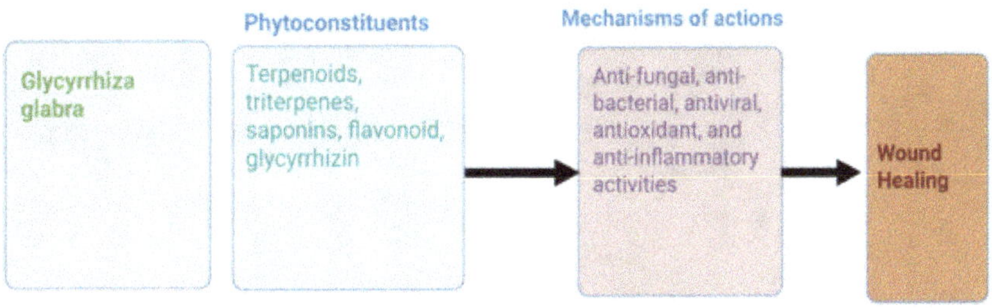

Salvia officinalis

Salvia officinalis L. is an evergreen perennial subshrub. Its leaves are grayish, its stems are woody, and the flower is purplish or blue. These species have over 120 phytocomponents, like glycoside saponins, carbohydrates, fatty acids, alkaloids, phenolic acids, terpenes, polyacetylenes, and flavonoids, according to an investigation of essential oils extracted from plant aerial parts (Badiee et al., 2012). Anti-inflammatory, antimicrobial, and antioxidant properties are among the biological effects of oil and aqueous extracts (Badiee et al., 2012) (Figure 3). The effect of S.

officinalis on the healing of wounds has been demonstrated by numerous studies (de Macedo et al., 2020; Scrima et al., 2020; Farahpour et al., 2020).

Compared to the control group, Farahpour et al. (2020) demonstrated that topical ointments consisting of 2% and 4% sage oil (w/w) had an extraordinary effect on the reduction of pro-inflammatory cytokines and overall bacterial count on full-thickness mouse cutaneous lesions. Based on the outcomes of the study, the treatment promoted the expression of cyclin-D1 that further stimulated the proliferation of fibroblasts and inhibited the pro-inflammatory effect through a reduction in the gene expression of IL-6, TNF-alpha, and IL-1 beta (Farahpour et al., 2020). In vivo research using a rat model of wound used to examine the effects of topical application of different doses of the hydroethanolic extract of this leaf significantly increased the percentage of breaking strength ratio, re-epithelialization period, percentage of wound contraction, and the level of hydroxyproline relative to the control group (Farahpour et al., 2020). Moreover, S. officinalis significantly increased the fibroblast dispersal and the new vessel formation (Karimzadeh & Farahpour, 2017).

Figure 3. Phytoconstituents and wound healing mechanisms of action of Salvia officinalis

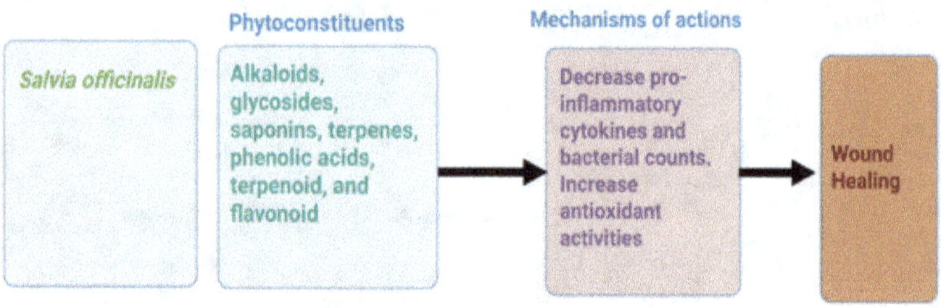

Jojoba

The jojoba plant is now cultivated in territories where the lack of water is acute, as well as in those areas where the profitability of traditional agricultural activities was previously not an option due to its adjustment to a warm and dry climate. It is believed to be resistant to drought, and it is able to withstand high temperatures, and requires minimal irrigation or soil fertility (Al-Obaidi et al., 2017).

Jojoba oil is light yellow, naturally produced oil that may be applied topically and has the ability to repair skin barriers and heal wounds. Along with 97% linear long-chain esters, jojoba oil contains beneficial compounds such as polyphenols, flavonoids, and alkaloids.

Acute lung injury (ALI) animals showed decreased hemorrhage and inflammatory cell infiltrations when dry nanoemulsion powders produced from Jojoba oil were administered, highlighting its effectiveness as a free radical scavenger and anti-inflammatory natural oil utilized for ALI treatment (Zhang et al., 2021). Additionally, a prior study demonstrated that jojoba liquid wax (JLW) remarkably speeds up the wound closure process for both keratinocytes and fibroblasts. It was also found that JLW stimulates fibroblasts to produce more collagen I. Because of its actions on skin cells, JLW may be used in therapeutic settings to heal wounds, according to its pharmacological characterization (Ranzato et al., 2011).

Green Tea

Over two-thirds of people in the world drink green tea (Dou, 2019). Research has demonstrated the health benefits of green tea, including protection against infections, heart disease, cancer, obesity, diabetes, and neurological issues (Hayakawa et al., 2020). Green tea's main constituents are polyphenols, mostly flavonoids. Catechins make up six to sixteen percent of the dried green tea leaves. Epigallocatechin (EGCG), which makes up around 59% of all catechins, is one of the four primary catechins. According to Hayat et al. (2015), it contains roughly 13.6% epicatechin-3-gallate (ECG), 6.4% epicatechin (EC), and 19% epigallocatechin (EGC).

The health advantages of green tea and EGCG in particular have been the subject of numerous research; it is reasonable to assume that the tea's polyphenols play a major role in defining these benefits (Chu et al., 2017). EGCG, which is thought to contain the main bioactivities, is the most prevalent ingredient in green tea leaves. Angiogenic effects, antibacterial, anti-inflammatory, and free radical scavenging properties are some of these bioactivities that lead to the proper healing of the wound and inhibit the onset of infections (Xu et al., 2021).

Panax ginseng

One of the most popular medicinal herbs for treating depression, anxiety, and chronic fatigue syndrome is panax ginseng (Arring et al., 2018). According to reports, this plant contains anti-inflammatory, blood lipid-regulating, vasodilating, and antioxidant properties (Figure 4). These are many bioactive compounds in Panax ginseng, where the most effective active constituents include a family of saponins (Cho et al., 2010). The ginseng plant is made up of diverse bioactive compounds in each of its parts. These are nitrogen, minerals, amino acids, vitamins, ginsenosides, phenolics, alkaloids, carbohydrates, ginseng oils, phytosterols, polypeptides, and selected enzymes (Asuku et al., 2024). It is established that Panax ginseng root extracts protect the skin of C57BL mice against acute UVB exposure and significantly shorten the therapeutic response of excisional wound and laser burn injury (Kim et al., 2013). The studies indicate that Panax ginseng extracts stimulate the production of collagen by the fibroblast of the human skin in vitro (Chen et al., 2017) and stimulate the migration of keratinocytes (Kim et al., 2013). In the same way, at the optimal dosage of 10 ng/ml, P. ginseng promoted the growth of human dermal fibroblast and collagen production (Lee et al., 2016). This could be due to the anti-inflammatory and antioxidative properties of the constituents of this plant.

Figure 4. Phytoconstituents and wound healing mechanism of actions of Panax ginseng

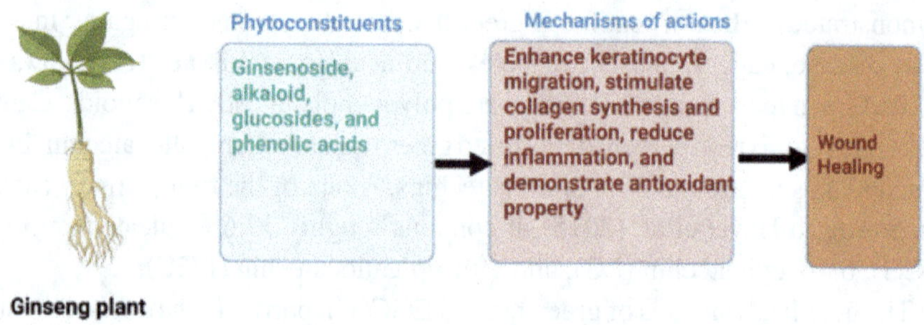

Honey

Honey primarily consists of glucose and fructose but also includes vitamins, minerals, amino acids, enzymes, organic acids, and other compounds. Seasonal variations and the locations where the bees collected the nectar affect the composition of honey, which becomes more concentrated and viscous as the moisture content of the deposited nectar mixture decreases and dries out (Khan et al., 2017; Alotibi et al., 2018). As a viscous fluid, natural honey's jelly-like consistency forms a surface layer over the wound that prevents bacteria from entering and keeps the wound from drying out. Natural honey is made up of carbohydrates, water, proteins, antioxidants, phytochemicals, and minerals. It has been shown that a small number of the ingredients that determine the medical and biological potential of natural honey are likely to vary among the various types of honey. According to reports, honey possesses antioxidant properties (Akintoye et al., 2021).

Trauma, diabetic ulcers, leprosy, burns, cervical varicose ulcers, cancer, boils, scratches, leg ulcers, gastric ulcers, fistulas, amputation, burst abdominal wounds, septic and surgical wounds, cracked nipples, and abdominal wall wounds are just a few of the wound types that honey can effectively treat (Eteraf-Oskouei & Najafi, 2013). Its high sugar content provides an extra supply of glucose for the cellular components in the injured area to flourish by creating a larger osmotic gradient that draws fluid up through the subdermal tissue.

In addition to causing fluid flow that flushes slough, debris, and necrotic tissue as well as germs out of the wound, honey's low water activity inhibits and regulates the growth of bacteria on the wound surface (Sundoro et al., 2012). In addition, honey's low water activity facilitates the movement of nutrients and oxygen from the deep tissue into the wound region. Furthermore, honey's low pH promotes tissue oxygenation, while flavonoids and aromatic acids eliminate free radicals, which cause tissue damage (Molan & White, 2009). Honey has a well-established antimicrobial effect (Vandamme et al., 2013; Al-Nahari et al., 2015). Depending on the kind, honey can have either bactericidal or bacteriostatic activity (Vandamme et al., 2013).

Notably, vitamins, flavonoids, organic acids, and phenolic substances all have antioxidant properties and strengthen honey's antibacterial properties. Flavonoids neutralize the free radicals produced by hydrogen peroxide (Molan and Rhodes, 2015; Minden-Birkenmaier and Bowlin, 2018). Moreover, the immunomodulatory effects of honey aid in wound healing, and its anti-inflammatory and antioxidant properties are explained by the fact that it consists of numerous different components (Majtan, 2014). Also, the elevated nutritional concentration promotes angiogenesis and epithelialization.

The research conducted by Yilmaz and Aygin (2020) is a randomized controlled study that includes a systematic review to determine the effectiveness of honey in the wound-treatment procedure. In their study, honey increased wound contraction and epithelialization in wound healing, reduced pain and inflammation and debridement, guaranteed infection control, accelerated wound healing, and was cost-effective.

Ginger and Gingerols

Ginger has bioactive chemicals that have been demonstrated to possess anti-inflammatory, antioxidant, and anti-cancer effects. It is used to relieve motion nausea, as well as hypoglycemic and hypolipidemic (Friedstat et al., 2017; Fentahun et al., 2021).

As an alternative anti-inflammatory agent, ginger was prescribed in order to prevent the formation of cyclooxygenase (COX)-2 and nitric oxide (NO) (Al-Radabi et al., 2022; Daniels et al., 2022; Yeshi et al., 2022). Its bioactive components comprise terpenes and several phenolic ones, such as paradols, shogaols, and gingerols (Fentahun et al., 2021).

In fact, fresh ginger rhizomes are a rich source of 6-, 8-, and 10- gingerols, which are active compounds with established therapeutic action (Baliga et al. 2013).

6-Gingerol has been reported to exhibit antioxidant and anti-inflammatory properties (Asuku et al., 2024; Ayinla & Asuku, 2025). According to reports on 6-gingerol, it can restrict the production of cytokines and macrophages and promote the epithelization process, which helps to reduce the inflammatory response (Tripathi et al., 2007). This anti-inflammatory effect is beneficial because it promotes wound healing and vascularization. Also, used together with vitamin D, it enhances wound healing among diabetic patients by increasing collagen, hemopexin-like domains (HPX), and fibrillogenesis (FN) expression levels, accelerating the healing process as well as decreasing the time required to epithelialize (Al-Rawaf et al., 2019). Even though the primary constituent of ginger, 6-gingerol, has been observed to possess anti-inflammatory and wound-healing effects, the pharmacokinetic properties of 6-gingerol are poor because of its poor bioavailability on oral administration as well as its low dissolvability in aqueous solutions (Wei et al., 2018). The study of the anti-inflammatory, antiproliferative, and wound-healing effects of 6-gingerol particles loaded into PEGylated nanophytosomes was conducted on lung, pancreatic, and breast cancer cell lines (Al-Samydai et al., 2022). 6-gingerol nanophytosomes were found to have a significant downregulation of inflammatory markers and cytokines, selective antiproliferative activity, and improved wound-healing processes. The 6-gingerol microneedle promotes wound healing in diabetic patients by regulating the macrophage polarization (Yuan et al., 2025).

In this study, 6-G@MN demonstrated significant promise in controlling macrophage polarization to hasten wound healing in diabetic patients. Furthermore, the 6-G@MN hydrogel system demonstrated its safety for internal use. By altering macrophage polarization, its multifunctional qualities improved cell migration, proliferation, and anti-inflammatory responses. Re-epithelialization, collagen remodeling, and enhanced granulation tissue development were all impacted.

Table 1 below shows the natural medicinal plants and their phytoconstituents, their healing mechanisms, and effects

Table 1. **Phytoconstituents, healing mechanisms, and effects of the natural medicinal plants on wound**

Natural Plants	Bioactive Compounds and Phytoconstituents	Mechanisms of Healing	Effects
Ginger	Gingerols, shogaols, paradols, essential oils	Anti-inflammatory, antioxidant, antiemetic, analgesic	Inhibits COX and LOX pathways, reduces oxidative stress, alleviates nausea, modulates inflammatory cytokines
Aloe vera	Polysaccharides, anthraquinones, vitamins, enzymes	Anti-inflammatory, wound healing, antioxidant, immunomodulatory	Stimulates fibroblast proliferation, enhances collagen synthesis, reduces oxidative stress, accelerates wound closure
Honey	Flavonoids, phenolic acids, enzymes, sugars	Antimicrobial, wound healing, anti-inflammatory, antioxidant	Provides antibacterial environment, promotes tissue regeneration, reduces inflammation, scavenges free radicals
Crocus sativa (Saffron)	Crocin, crocetin, safranal, picrocrocin	Antidepressant, antioxidant, neuroprotective, anti-inflammatory	Scavenges free radicals, modulates neurotransmitters (serotonin), reduces neuroinflammation, protects neuronal cells
Panax ginseng	Ginsenosides, polysaccharides, peptides	Immunomodulatory, neuroprotective, adaptogenic, anti-fatigue	Enhances cellular immunity, reduces oxidative stress, improves cognitive function, regulates stress response
Curcumin	Curcuminoids, volatile oils	Anti-inflammatory, antioxidant, anticancer, antimicrobial	Inhibits NF-κB pathway, reduces pro-inflammatory cytokines, neutralizes ROS, modulates apoptosis
Green tea	Polyphenols (EGCG, catechins), caffeine, theanine	Antioxidant, anti-inflammatory, cardioprotective, anticancer	Scavenges free radicals, inhibits tumor cell proliferation, improves endothelial function, modulates immune response
Glycyrrhiza glabra L. (Licorice)	Glycyrrhizin, flavonoids, saponins, coumarins	Anti-inflammatory, antiviral, hepatoprotective, immunomodulatory	Modulates cortisol metabolism, inhibits viral replication, protects liver cells from toxins, reduces inflammation
Jojoba	Wax esters, sterols, tocopherols, flavonoids	Skin healing, anti-inflammatory, moisturizing	Mimics skin sebum, protects against dryness, reduces inflammation in skin disorders

CONCLUSION AND FUTURE DIRECTIONS

Since ancient times, wound healing has been a challenging clinical problem for wound care. The current chapter explains how medicinal plants and phytochemicals with exceptional wound-healing properties are introduced by nature. We now know more about the molecular mechanisms underpinning herbal medicines' effects on wound healing thanks to scientific research. Because of their natural active ingredients' anti-inflammatory, antibacterial, and antioxidant properties, plants can be used to treat and heal skin wounds. Recent research has demonstrated that a variety of natural chemicals, including flavonoids, polysaccharides, phenolic compounds, and saponins, can function at various phases of the healing process via a variety of mechanisms and are principally in charge of the wound-healing qualities of herbal treatments. By controlling and modifying inflammatory responses, polyphenolic substances have been shown to be effective therapeutic agents in wound healing. By improving the proliferation of fibroblasts and/or the synthesis of collagen, diverse phytochemicals found in medicinal plants have been identified as remarkable regulators of homeostasis, re-epithelialization, and regeneration. Through a number of interconnected pathways, scientific study has validated the potent effects of natural substances and their phytoconstituents in wound treatment. In an effort to create a wound management and treatment system that is affordable, effective, stable, and long-lasting, researchers are experimenting with the combinations of novel dressings, formulas, and phytoconstituents. The advancement in nanotechnology has made wound care more efficient and patient-centered. New medication delivery systems for wound treatment are also being made possible by more recent technology, like 3D printing. Regenerative medicine and tissue engineering may be used in the future to create better wound healing systems. Well-designed preclinical and clinical trials, along with enhanced quality control methods for the identification, screening, and measurement of plant components, will support research in wound care management. Patients would have a "green" alternative and virtually fewer adverse effects if they used a biocompatible formulation comprised of natural plant extracts. To make the healing process quick and to prevent complications, it is recommended that multiple treatment modalities be employed simultaneously in the process of wound management, especially when treating chronic wound lesions.

Moreover, several problems also need to be addressed to make medicinal plants and their phytoconstituents more effective and applicable in wound healing. The safety of the phytoconstituents should also be tested, and the adverse effects should be investigated using a double-blinded controlled clinical trial.

REFERENCES

Abd Jalil, M. A., Kasmuri, A. R., & Hadi, H.AbdJalil. (2017). Stingless bee honey, the natural wound healer: A review. *Skin Pharmacology and Physiology, 30*(2), 66–75. DOI: 10.1159/000458416 PMID: 28291965

Abu-Izneid, T., Rauf, A., Khalil, A. A., Olatunde, A., Khalid, A., Alhumaydhi, F. A., Aljohani, A. S. M., Sahab Uddin, M., Heydari, M., Khayrullin, M., Shariati, M. A., Aremu, A. O., Alafnan, A., & Rengasamy, K. R. R. (2020). Nutritional and Health Beneficial Properties of Saffron (Crocus sativus L): A Comprehensive Review. *Critical Reviews in Food Science and Nutrition, 62*(10), 2683–2706. DOI: 10.1080/10408398.2020.1857682 PMID: 33327732

Akintoye, O. O., Ajibare, A. J., Fabunmi, O. A., Asuku, A. O., & Tajudeen, R. A. (2021). Effect of Honey on Oxidative Stress in Brain Tissue of Sleep Deprived Rats. *Nigerian Journal of Basic and Applied Science., 29*(1), 83–89. DOI: 10.4314/njbas.v29i1.10

Al-Nahari, A. A., Almasaudi, S. B., Abd El-Ghany, E. S. M., Barbour, E., Al Jaouni, S. K., & Harakeh, S. (2015). Antimicrobial activities of Saudi honey against Pseudomonas aeruginosa. *Saudi Journal of Biological Sciences, 22*(5), 521–525. DOI: 10.1016/j.sjbs.2015.04.006 PMID: 26288553

Al-Obaidi, J. R., Halabi, M. F., AlKhalifah, N. S., Asanar, S., Al-Soqeer, A. A., & Attia, M. F. (2017). A review on plant importance, biotechno logical aspects, and cultivation challenges of jojoba plant. *Biological Research, 50*(1), 25. DOI: 10.1186/s40659-017-0131-x PMID: 28838321

Al-Rawaf, H. A., Gabr, S. A., & Alghadir, A. H. (2019). Molecular Changes in Diabetic Wound Healing following Administration of Vitamin D and Ginger Supplements: Biochemical and Molecular Experimental Study. Evid.-. *Evidence-Based Complementary and Alternative Medicine : eCAM, 4352470*, 1–13. Advance online publication. DOI: 10.1155/2019/4352470 PMID: 31428171

Al-Samydai, A., Al Qaraleh, M., Alshaer, W., Al-Halaseh, L. K., Issa, R., & Alshaikh, F. (2022). Preparation, Characterization, Wound Healing, and Cytotoxicity Assay of PEGylated Nanophytosomes Loaded with 6-Gingerol. *Nutrients, 14*(23), 5170. DOI: 10.3390/nu14235170 PMID: 36501201

Alherz, F. A., Negm, W. A., Elekhnawy, E., El-Masry, T. A., Haggag, E. M., Alqahtani, M. J., & Hussein, I. A. (2022). Silver nanoparticles prepared using encephalartos laurentianus de wild leaf extract have inhibitory activity against candida albicans clinical isolates. *Journal of Fungi (Basel, Switzerland)*, *8*(10), 1005. DOI: 10.3390/jof8101005 PMID: 36294570

Alonso, G. L., Zalacain, A., & Carmona, M. (2012). Saffron. In *Handbook of Herbs and Spices* (pp. 469–498). Elsevier. DOI: 10.1533/9780857095671.469

Alotibi, I. A., Harakeh, S. M., Al-Mamary, M., Mariod, A. A., Al-Jaouni, S. K., Al-Masaud, S., Alharbi, M. G., & Al-Hindi, R. R. (2018). Floral markers and biological activity of Saudi honey. *Saudi Journal of Biological Sciences*, *25*(7), 1369–1374. DOI: 10.1016/j.sjbs.2018.05.021 PMID: 30505183

Arring, N. M., Millstine, D., Marks, L. A., & Nail, L. M. (2018). Ginseng as a treatment for fatigue: A systematic review. *Journal of Alternative and Complementary Medicine (New York, N.Y.)*, *24*(7), 624–633. DOI: 10.1089/acm.2017.0361 PMID: 29624410

Asuku, A. O., Ajibare, A. J., Ajagun, E. J., Babatunde, S. S., Olajide, T. S., Oyerinde, T. O., Ogungbangbe, G. O., & Odugbesan, O. A. (2024). Herbal Extracts in Antiaging Products. In Yingngam, B. (Ed.), *Eco-friendly Skin Solutions for Natural Cosmeceuticals*. IGI Publisher., DOI: 10.4018/979-8-3693-5473-5.ch012

Asuku, A. O., Ayinla, M. T., Ajibare, A. J., & Olajide, T. S. (2024). Mercury chloride causes cognitive impairment, oxidative stress and neuroinflammation in male Wistar rats: The potential protective effect of 6-gingerol-rich fraction of *Zingiber officinale* via regulation of antioxidant defense system and reversal of pro-inflammatory markers increase. Brain Research, 148741: 1826, ISSN 0006-8993, https://doi.org/ DOI: 10.1016/j.brainres.2023.148741

Ayinla, M. T., & Asuku, A. O. (2025). The neurotoxic effects of lead acetate and the abrogating actions of 6-gingerol-rich extract of ginger via modulation of antioxidant defence system, pro-inflammatory markers, and apoptotic cascade. *Naunyn-Schmiedeberg's Archives of Pharmacology*, *398*(7), 9341–9356. Advance online publication. DOI: 10.1007/s00210-025-03873-x PMID: 39937255

Baliga, M. S., Latheef, L., Haniadka, R., Fazal, F., Chacko, J., & Arora, R. (2013). Bioactive Food as Dietary Interventions for Arthritis and Related Inflammatory Diseases. Elsevier Inc.; Amsterdam, The Netherlands: 2013. Ginger (Zingiber officinale Roscoe) in the Treatment and Prevention of Arthritis; pp. 529–544.

Baron, J. M., Glatz, M., & Proksch, E. (2020). Optimal Support of Wound Healing: New Insights. *Dermatology (Basel, Switzerland), 236*(6), 593–600. DOI: 10.1159/000505291 PMID: 31955162

Binsuwaidan, R., Elekhnawy, E., Elseady, W. S., Keshk, W. A., Shoeib, N. A., Attallah, N. G., Mokhtar, F. A., Abd El Hadi, S. R., Ahmed, E., Magdeldin, S., & Negm, W. A. (2022). Anti-bacterial activity and wound heal ing potential of Cycas thouarsii R. Br n-butanol fraction in diabetic rats supported with phytochemical profiling. *Biomedicine and Pharmacotherapy, 155*, 113763. DOI: 10.1016/j.biopha.2022.113763 PMID: 36182739

Chen, X., Wang, M., Xu, X., Liu, J., Mei, B., Fu, P., Zhao, D., & Sun, L. (2017). Panax ginseng total protein promotes proliferation and secretion of collagen in NIH/3T3 cells by activating extracellular signal-related kinase path way. *Journal of Ginseng Research, 41*(3), 411–418. DOI: 10.1016/j.jgr.2017.02.001 PMID: 28701885

Cho, J.-G., Lee, M.-K., Lee, J.-W., Park, H.-J., Lee, D.-Y., Lee, Y.-H., Yang, D.-C., & Baek, N.-I. (2010). Physicochemical Characterization and NMR Assignments of Ginsenosides Rb1, Rb2, Rc, and Rd Isolated from Panax ginseng. *Journal of Ginseng Research, 34*(2), 113–121. DOI: 10.5142/jgr.2010.34.2.113

Chu, C., Deng, J., Man, Y., & Qu, Y. (2017). Green Tea Extracts Epigallocatechin-3-Gallate for Different Treatments. *BioMed Research International, 2017*(2017), 5615647. DOI: 10.1155/2017/5615647 PMID: 28884125

Daniels, C., Isaacs, Z., Finelli, R., & Leisegang, K. (2022). The Efficacy of Zingiber officinale on Dyslipidaemia, Blood Pressure, and In-flammation as Cardiovascular Risk Factors: A Systematic Review. *Clinical Nutrition ESPEN, 51*, 72–82. DOI: 10.1016/j.clnesp.2022.08.031 PMID: 36184251

de Macedo, L. M., dos Santos, É. M., Militão, L., Tundisi, L. L., Ataide, J. A., Souto, E. B., & Mazzola, P. G. (2020). Rosemary (Rosmarinus officinalis L., Syn Salvia rosmarinus Spenn.) and Its Topical Applications: A Review. *Plants, 2020*(9), 651. DOI: 10.3390/plants9050651 PMID: 32455585

Dou, Q. P. (2019). Tea in Health and Disease. *Nutrients, 2019*(11), 929. DOI: 10.3390/nu11040929 PMID: 31027187

Du, L., Feng, X., Xiang, X., & Jin, Y. (2016). Wound healing effect of an in situ forming hydrogel loading curcumin-phospholipid complex. *Current Drug Delivery, 13*(1), 76–82. DOI: 10.2174/1567201813666151202195437 PMID: 26634789

El-Refaie, W. M., Elnaggar, Y. S., El-Massik, M. A., & Abdallah, O. Y. (2015). Novel curcumin-loaded gel-core hyaluosomes with promising burn-wound healing potential: Development, in vitro appraisal and in vivo studies. *International Journal of Pharmaceutics*, *486*(1-2), 88–98. DOI: 10.1016/j.ijpharm.2015.03.052 PMID: 25818063

Eteraf-Oskouei, T., & Najafi, M. (2013). Traditional and modern uses of natural honey in human diseases: A review. *Iranian Journal of Basic Medical Sciences.*, *16*(6), 731–742. PMID: 23997898

Fagot, D., Pham, D. M., Laboureau, J., Planel, E., Guerin, L., Nègre, C., Donovan, M., & Bernard, B. A. (2018). Crocin, a Natural Molecule with Potentially Beneficial Effects against Skin Ageing. *International Journal of Cosmetic Science*, *40*(4), 388–400. DOI: 10.1111/ics.12472 PMID: 29893408

Farahpour, M. R., Pirkhezr, E., Ashrafian, A., & Sonboli, A. (2020). Accelerated Healing by Topical Administration of Salvia officinalis Essential Oil on Pseudomonas Aeruginosa and Staphylococcus Aureus Infected Wound Model. *Biomedicine and Pharmacotherapy*, *128*, 110120. DOI: 10.1016/j.biopha.2020.110120 PMID: 32460189

Fentahun, N., Anteneh, Y., & Menber, Y. (2021). Malnutrition in the Outcome of Wound Healing at Public Hospitals in Bahir Dar City, Northwest Ethiopia: A Prospective Cohort Study. *Journal of Nutrition and Metabolism*, *8824951*, 1–7. Advance online publication. DOI: 10.1155/2021/8824951 PMID: 33680509

Friedrich, R. B., Kann, B., Coradini, K., Offerhaus, H. L., Beck, R. C., & Windbergs, M. (2015). Skin penetration behavior of lipid-core nanocapsules for simultaneous delivery of resveratrol and curcumin. *European Journal of Pharmaceutical Sciences*, *78*, 204–213. DOI: 10.1016/j.ejps.2015.07.018 PMID: 26215463

Friedstat, J., Brown, D. A., & Levi, B. (2017). Chemical, Electrical, and Radiation Injuries. *Clinics in Plastic Surgery*, *44*(3), 657–669. DOI: 10.1016/j.cps.2017.02.021 PMID: 28576255

Gadekar, R., Saurabh, M., Thakur, G., & Saurabh, A. (2012). Study of formulation, characterisation and wound healing potential of transdermal patches of curcumin. *Asian Journal of Pharmaceutical and Clinical Research*, *5*, 225–230.

Geethalakshmi, R., Sakravarthi, C., Kritika, T., Arul Kirubakaran, M., & Sarada, D. V. L. (2013). Evaluation of Antioxidant and Wound Healing Potentials of Sphaeranthus Amaranthoides Burm.f. *BioMed Research International*, *2013*, 607109. DOI: 10.1155/2013/607109 PMID: 23509751

Gigliobianco, M. R., Cortese, M., Peregrina, D. V., Villa, C., Lupidi, G., Pruccoli, L., Angeloni, C., Tarozzi, A., Censi, R., & Di Martino, P. (2021). Development of New Extracts of Crocus sativus L. By-Product from Two Different Italian Regions as New Potential Active Ingredient in Cosmetic Formulations. *Cosmetics*, *8*(2), 51. DOI: 10.3390/cosmetics8020051

Guo, S., & Dipietro, L. A. (2010). Factors affecting wound healing. *Journal of Dental Research*, *89*(3), 219–229. DOI: 10.1177/0022034509359125 PMID: 20139336

Haalboom, M. (2018). Chronic Wounds: Innovations in Diagnostics and Therapeutics. *Current Medicinal Chemistry*, *25*(41), 5772–5781. DOI: 10.2174/092986732 4666170710120556 PMID: 28699502

Hayakawa, S., Ohishi, T., Miyoshi, N., Oishi, Y., Nakamura, Y., & Isemura, M. (2020). Anti- Cancer Effects of Green Tea Epigallocatchin-3 Gallate and Coffee Chlorogenic Acid. *Molecules (Basel, Switzerland)*, *25*(19), 4553. DOI: 10.3390/ molecules25194553 PMID: 33027981

Hayat, K., Iqbal, H., Malik, U., Bilal, U., & Mushtaq, S. (2015). Tea and Its Consumption: Benefits and Risks. *Critical Reviews in Food Science and Nutrition*, *55*(7), 939–954. DOI: 10.1080/10408398.2012.678949 PMID: 24915350

Hormozi, M., Assaei, R., & Boroujeni, M. B. (2017). The Effect of Aloe vera on the Expression of Wound Healing Factors (TGF 1 and BFGF) in Mouse Embryonic Fibroblast Cell: In Vitro Study. *Biomedicine and Pharmacotherapy*, *88*, 610–616. DOI: 10.1016/j.biopha.2017.01.095 PMID: 28142117

Kant, V., Gopal, A., Kumar, D., Pathak, N. N., Ram, M., Jangir, B. L., Tandan, S. K., & Kumar, D. (2015). Curcumin-induced angiogenesis hastens wound healing in diabetic rats. *The Journal of Surgical Research*, *193*(2), 978–988. DOI: 10.1016/j. jss.2014.10.019 PMID: 25454972

Karimzadeh, S., & Farahpour, M. R. (2017). Topical Application of Salvia officinalis Hydroethanolic Leaf Extract Improves Wound Healing Process. *Indian Journal of Experimental Biology*, *55*, 98–106. PMID: 30183236

Karri, V. V., Kuppusamy, G., Talluri, S. V., Mannemala, S. S., Kollipara, R., Wadhwani, A. D., Mulukutla, S., Raju, K. R., & Malayandi, R. (2016). Curcumin loaded chitosan nanoparticles impregnated into collagen-alginate scaffolds for diabetic wound healing. *International Journal of Biological Macromolecules*, *93*, 1519–1529. DOI: 10.1016/j.ijbiomac.2016.05.038 PMID: 27180291

Khan, K. A., Ansari, M. J., Al-Ghamdi, A., Nuru, A., Harakeh, S., & Iqbal, J. (2017). Investigation of gut microbial communities associated with indigenous honeybee (Apis mellifera jemenitica) from two different eco-regions of Saudi Arabia. *Saudi Journal of Biological Sciences*, *24*(5), 1061–1068. DOI: 10.1016/j.sjbs.2017.01.055 PMID: 28663705

Kim, W. K., Song, S.-Y., Oh, W. K., Kaewsuwan, S., Tran, T. L., Kim, W.-S., & Sung, J.-H. (2013). Wound-healing effect of ginsenoside Rd from leaves of Panax ginseng via cyclic AMP-dependent protein kinase pathway. *European Journal of Pharmacology*, *702*(1-3), 1–3, 285–293. DOI: 10.1016/j.ejphar.2013.01.048 PMID: 23399764

Komes, D., Belščak-Cvitanović, A., Jurić, S., Bušić, A., Vojvodić, A., & Durgo, K. (2016). Consumer Acceptability of Liquorice Root (Glycyrrhiza glabra L.) as an Alternative Sweetener and Correlation with Its Bioactive Content and Biological Activity. *International Journal of Food Sciences and Nutrition*, *67*(1), 53–66. DOI: 10.3109/09637486.2015.1126563 PMID: 26707111

Krutmann, J., Bouloc, A., Sore, G., Bernard, B. A., & Passeron, T. (2017). The Skin Aging Exposome. *Journal of Dermatological Science*, *85*(3), 152–161. DOI: 10.1016/j.jdermsci.2016.09.015 PMID: 27720464

Lee, G. Y., Park, K. G., Namgoong, S., Han, S. K., Jeong, S. H., Dhong, E. S., & Kim, W. K. (2016). Effects of Panax ginseng extract on human dermal fibroblast proliferation and collagen synthesis. *International Wound Journal*, *1*(S1), 42–46. DOI: 10.1111/iwj.12530 PMID: 26507878

Majtan, J. (2014). Honey: An immunomodulator in wound healing. *Wound Repair and Regeneration*, *22*(2), 187–192. DOI: 10.1111/wrr.12117 PMID: 24612472

Minden-Birkenmaier, B. A., & Bowlin, G. L. (2018). Honey-based templates in wound healing and tissue engineering. *Bioengineering (Basel, Switzerland)*, *5*(2), 46. DOI: 10.3390/bioengineering5020046 PMID: 29903998

Molan, P., & Rhodes, T. (2015). Honey: A biologic wound dressing. *Wounds : a Compendium of Clinical Research and Practice*, *27*(6), 141–151. PMID: 26061489

Molan, P. C., & White, R. (2009). *Honey in modern wound management*. Wounds UK Ltd.

Nanda, S., & Madan, K. (2021). TheRoleofSafranal and Saffron Stigma Extracts in Oxidative Stress, Diseases and Photoaging: A Systematic Review. *Heliyon*, *7*(2), e06117. DOI: 10.1016/j.heliyon.2021.e06117 PMID: 33615006

Oliveira, R.N., Mancini, M.C., de Oliveira, F.C.S., Passos, T.M., Quilty, B., Thiré, R.M.d.S.M., & McGuinness, G.B. (2016). FTIR Analysis and Quantification of Phenols and Flavonoids of Five Commercially Available Plants Extracts Used in Wound Healing. Matér. Rio Jan. 21, 767–779.

Oryan, A., Mohammadalipour, A., Moshiri, A., & Tabandeh, M. R. (2016). Topical Application of Aloe vera Accelerated Wound Healing, Modeling, and Remodeling: An Experimental Study. *Annals of Plastic Surgery*, *77*(1), 37–46. DOI: 10.1097/SAP.0000000000000239 PMID: 25003428

Pastorino, G., Cornara, L., Soares, S., Rodrigues, F., & Oliveira, M. B. P. P. (2018). Liquorice (Glycyrrhiza glabra): A Phytochemical and Pharmacological Review. *Phytotherapy Research*, *32*(12), 2323–2339. DOI: 10.1002/ptr.6178 PMID: 30117204

Portou, M. J., Baker, D., Abraham, D., & Tsui, J. (2015). The innate immune system, toll-like receptors and dermal woundhealing: A review. *Vascular Pharmacology*, *71*, 31–36. DOI: 10.1016/j.vph.2015.02.007 PMID: 25869514

Ranzato, E., Martinotti, S., & Burlando, B. (2011). Wound healing properties of jojoba liquid wax: An in vitro study. *Journal of Ethnopharmacology*, *134*(2), 443–449. DOI: 10.1016/j.jep.2010.12.042 PMID: 21211559

Rippa, A. L., Kalabusheva, E. P., & Vorotelyak, E. A. (2019). Regeneration of Dermis: Scarring and Cells Involved. *Cells*, *8*(6), 607. DOI: 10.3390/cells8060607 PMID: 31216669

Rodrigues, M., Kosaric, N., Bonham, C. A., & Gurtner, G. C. (2019). Wound Healing: A Cellular Perspective. *Physiological Reviews*, *99*(1), 665–706. DOI: 10.1152/physrev.00067.2017 PMID: 30475656

Sánchez-Machado, D. I., López-Cervantes, J., Sendón, R., & Sanches-Silva, A. (2017). Aloe vera: Ancient Knowledge with New Frontiers. *Trends in Food Science & Technology*, *61*, 94–102. DOI: 10.1016/j.tifs.2016.12.005

Scrima, M., Melito, C., Merola, F., Iorio, A., Vito, N., Giori, A. M., & Ferravante, A. (2020). Evaluation of Wound Healing Activity of Salvia Haenkei Hydroalcoholic Aerial Part Extract on in Vitro and in vivo Experimental Models. *Clinical, Cosmetic and Investigational Dermatology*, *13*, 627–637. DOI: 10.2147/CCID.S224418 PMID: 32922060

Shah, A., & Amini-Nik, S. (2017). The role of phytochemicals in the inflammatory phase of wound healing. *International Journal of Molecular Sciences*, *18*(5), 1068. DOI: 10.3390/ijms18051068 PMID: 28509885

Siriwattanasatorn, M., Itharat, A., Thongdeeying, P., & Ooraikul, B. (2020). In Vitro Wound Healing Activities of Three Most Commonly Used Thai Medicinal Plants and Their Three Markers. *Evidence-Based Complementary and Alternative Medicine : eCAM*, *2020*(1), e6795383. DOI: 10.1155/2020/6795383 PMID: 32714411

Sundoro, A., Nadia, K., Nur, A., Sudjatmiko, G., & Tedjo, A. (2012). Comparison of physical–chemical characteristic and antibacterial effect between Manuka honey and local honey. *J Plastik Rekonstruksi.*, *2012*(1), 3.

Tripathi, S., Maier, K. G., Bruch, D., & Kittur, D. S. (2007). Effect of 6-Gingerol on Pro- Inflammatory Cytokine Production and Costim-ulatory Molecule Expression in Murine Peritoneal Macrophages. *The Journal of Surgical Research*, *138*(2), 209–213. DOI: 10.1016/j.jss.2006.07.051 PMID: 17291534

Vandamme, L., Heyneman, A., Hoeksema, H., Verbelen, J., & Monstrey, S. (2013). Honey in modern wound care: A systematic review. *Burns*, *39*(8), 1514–1525. DOI: 10.1016/j.burns.2013.06.014 PMID: 23896128

Wei, Q., Yang, Q., Wang, Q., Sun, C., Zhu, Y., Niu, Y., Yu, J., & Xu, X. (2018). Formulation, Characterization, and Pharmacokinetic Studies of 6-Gingerol-Loaded Nanostructured Lipid Carriers. *AAPS PharmSciTech*, *19*(8), 3661–3669. DOI: 10.1208/s12249-018-1165-2 PMID: 30324361

Wilkinson, H. N., & Hardman, M. J. (2020). Wound healing: Cellular mechanisms and pathological outcomes. *Open Biology*, *10*(9), 200223. DOI: 10.1098/rsob.200223 PMID: 32993416

Xu, F. W., Lv, Y. L., Zhong, Y. F., Xue, Y. N., Wang, Y., Zhang, L. Y., Hu, X., & Tan, W. Q. (2021). Beneficial effects of green tea EGCG on skin woundhealing: A comprehensive review. *Molecules (Basel, Switzerland)*, *26*(20), 6123. DOI: 10.3390/molecules26206123 PMID: 34684703

Yeshi, K., Turpin, G., Jamtsho, T., & Wangchuk, P. (2022). Indigenous Uses, Phytochemical Analysis, and Anti-Inflammatory Properties of Australian Tropical Medicinal Plants. *Molecules (Basel, Switzerland)*, *27*(12), 3849. DOI: 10.3390/molecules27123849 PMID: 35744969

Yilmaz, A. C., & Aygin, D. (2020). Honey dressing in wound treatment: A systematic review. *Complementary Therapies in Medicine*, *51*, 102388. DOI: 10.1016/j.ctim.2020.102388 PMID: 32507418

Yuan, L., He, Q., Zhang, Y., Luo, H., Xiang, W., Deng, C., Li, C., Li, X., Yao, L. et al. (2025). 6-Gingerol microneedle promotes diabetic wound healing by regulating macrophage polarization. International Immunopharmacology, 151: 4, 114288

Zangeneh, A., Pooyanmehr, M., Zangeneh, M. M., Moradi, R., Rasad, R., & Kazemi, N. (2019). Therapeutic Effects of Glycyrrhiza glabra Aqueous Extract Ointment on Cutaneous Wound Healing in Sprague Dawley Male Rats. *Comparative Clinical Pathology*, *28*(5), 1507–1514. DOI: 10.1007/s00580-019-03007-9

Zeka, K., Ruparelia, K. C., Sansone, C., Macchiarelli, G., Continenza, M. A., & Arroo, R. R. J. (2018). New Hydrogels Enriched with Antioxidants from Saffron Crocus Can Find Applications in Wound Treatment and/or Beautification. *Skin Pharmacology and Physiology*, *31*(2), 95–98. DOI: 10.1159/000486135 PMID: 29393263

Zhang, G., Xie, F., Sun, Y., Yu, X., Xiao, Z., Fang, R., Li, J., Li, Q., Du, L., & Jin, Y. (2021). Inhalable jojoba oil dry nanoemulsion powders for the treatment of lipopolysaccharide-or H2O2-induced acute lung injury. *Pharmaceutics*, *13*(4), 486. DOI: 10.3390/pharmaceutics13040486 PMID: 33918471

Chapter 5
Polyphenols and Flavanoids:
Antioxidants at the Forefront

Muthukumaran Thulasingam
http://orcid.org/0000-0002-4660-3603
Chettinad School of Pharmaceutical Sciences, Chettinad Academy of Research and Education, Chennai, India

Chitra Vellapandian
SRM College of Pharmacy, SRM Institute of Science and Technology, Kattankulathur, India

Satheeshkumar Sellamuthu
http://orcid.org/0000-0001-5155-4618
Karpagam College of Pharmacy, Coimbatore, India

Mythreyi R. Rajendran
Chettinad School of Pharmaceutical Sciences, Chettinad Hospital and Research Institute, Kelambakkam, India

Priya Dharshini Shanmugam
http://orcid.org/0009-0003-8031-2388
Periyar College of Pharmaceutical Sciences, Tiruchirappalli, India

ABSTRACT

Diabetes and obesity are the main causes of wounds which are regarded as a major issue that impacts the healthcare industry in many nations. Unhealthy lifestyle choices and behaviors exacerbate wounds. After an injury, wound healing is a complex physiological process that is necessary to restore the epithelial barrier. Because of their well-known anti-inflammatory, angiogenesis, re-epithelialization, and antioxidant

DOI: 10.4018/979-8-3373-2908-6.ch005

actions, polyphenols and flavonoids have been shown in numerous studies to have wound-healing capabilities. They have potent anti-inflammatory, antibacterial, and antioxidant qualities and are mostly found in fruits, vegetables, tea, and wine. The complex molecular processes by which polyphenols and flavonoids influence wound healing include immune response modulation, collagen synthesis enhancement, oxidative stress reduction, and angiogenesis promotion. Additionally, the NF-κB and MAPK pathways, two important signaling pathways involved in tissue repair and inflammatory regulation, are influenced by these chemicals. According to research, flavonoids and polyphenols may also support angiogenesis, the process of forming new blood vessels, and extracellular matrix remodeling, both of which are essential for proper wound closure and healing. This chapter provides a comprehensive review of the most current studies on the role of flavonoids and polyphenols in wound healing, with a focus on the potential therapeutic applications of these substances to enhance wound care and prevent chronic wounds.To determine the usefulness of polyphenols and flavonoids in therapeutic settings, more clinical research is necessary to examine their pharmacological characteristics and safety profiles.

OVERVIEW OF POLYPHENOLS AND FLAVONOIDS

Definition and Classification:

Plant secondary metabolites known as polyphenols can be found in a variety of foods, including fruits, vegetables, coffee beans, tea leaves, chocolates, cereals, legumes, and drinks. There are more than 8000 polyphenols in various plants, and their primary function is as antioxidants. Phenolic chemicals are also significant components that affect the nutritional value, flavor, and color of fruits. Fresh fruits, such as grapes, apples, pears, cherries, and so on, contain an average of 300 mg of polyphenols per 100 g, and polyphenols are thought to be responsible for the stress-relieving effects of coffee and tea. Numerous meta-analyses and epidemiological studies have shown that polyphenols and flavonoids have both preventive and therapeutic effects against chronic diseases, such as diabetes mellitus, tumors, cardiovascular, cerebrovascular, and neurodegenerative disorders.

Classification

According to research, polyphenols are made up of a variety of substances, each with a unique complicated structure.Polyphenols divided into flavonoids and non- flavonoids, each with distinct subclasses and health consequences, is based on their structural complexity, which is characterized by phenolic rings and hydroxyl substituents. This group, which includes flavonoids, phenolic acids, stilbenes, and lignans, exemplifies the variety of polyphenols (Durazzo et al., 2019; Mittal et al., 2025; Garg et al., 2019).

The two phenol rings that flavonoids have and the one ring that non-flavonoids have are the main structural differences between the two groups. In flavonoids, the two phenol rings are also joined by a middle pyran ring that contains oxygen. The six subclasses of flavonoids are flavonols, flavones, isoflavones, flavanones, anthocyanidins, and flavanols (Durazzo et al., 2019; Shalaby et al., 2025).

Phenolic acids: They are classified into two classes, hydroxybenzoic acids and hydroxycinnamic acids, and are mostly found in blueberries, coffee, tea, cinnamon, plum, apple, and cherries.

Flavonoids: Fruits, vegetables, legumes, red wine, and green tea are common sources of flavonoids.

Lignans: Flax, linseed, legumes, cereals, grains, fruits, algae, and some vegetables are common sources of these seeds.

Stilbenes: Resveratrol is the most well-known of these, which are made from red wine and peanuts.

Figure 1. Flavonoid classes and their fundamental chemical structures

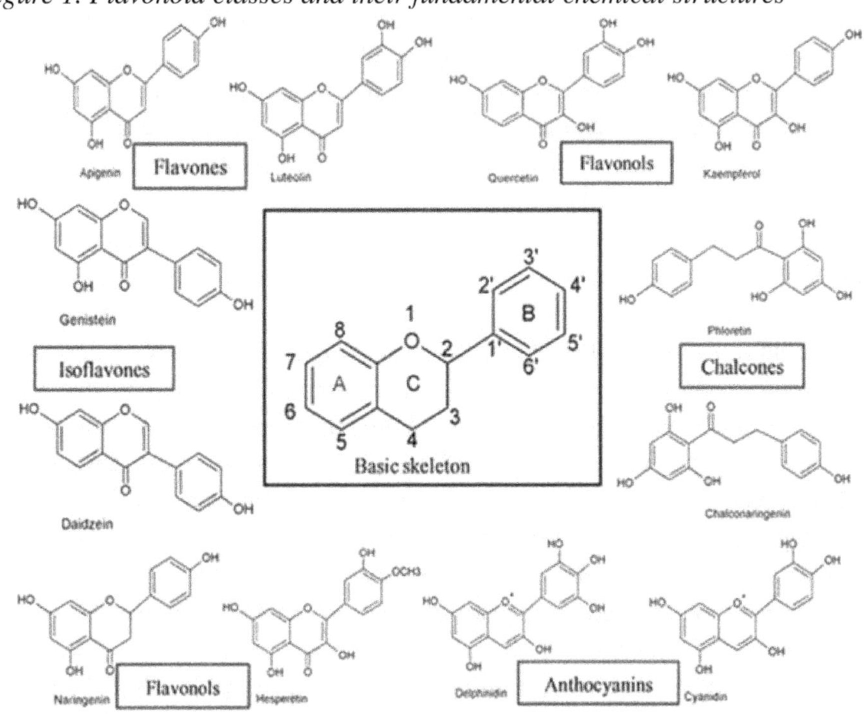

Phenolic Acids

Two distinct classes of phenolic acids are hydroxybenzoic acid and hydroxycinnamic acid. Red fruits like strawberries, raspberries, and blackberries are common sources of hydroxybenzoic acids, which are composed of hydrolyzable tannins such gallotannins and ellagitannins (Zhuang et al., 2023).

With a few notable exceptions, such onions and black radish, which have concentrations in the tens of milligrams per kilogram weight, edible plants typically contain very little hydroxybenzoic acid (Eseberri et al., 2022). Tea leaves frequently contain gallic acid, which has a concentration of about 4.5 g/kg fresh weight (Melrose, 2023).

The nutraceutical potential of plants that contain hydroxybenzoic acid has not been thoroughly investigated because these plants are less frequently used in human meals. The primary components of hydroxycinnamic acids, a more common type of phenolic acid than hydroxybenzoic acids, include compounds like p-coumaric, ferulic, sinapic, and caffeic acids.Green coffee beans contain very high amounts of chlorogenic acid (CGA), a valuable and substantial polyphenol and an inexpensive source of dietary phenol (Zhuang et al., 2023). It is composed of caffeic, ferulic, coumaric, quinic and 3,4 dimethoxycinnamic acid (Criollo-Mendoza et al., 2023). Hydroxycinnamic acids are abundant in the outer layers of ripe fruits. These acids are frequently found in plants in bound forms, which are the esters or glycosylated derivatives of tartaric acid, quinic acid, and shikimic acid.

Blueberries, kiwis, plums, cherries, apples, and other fruits contain between 0.5 to 2 g/kg fresh weight of hydroxycinnamic acid. Cereal grains are rich in ferulic acid; on a dry weight basis, wheat grains contain 0.8–2 g/kg of ferulic acid, which accounts for up to 90% of all polyphenols (Scialò et al., 2017).

Flavonoids

Flavonoids are chemically representative of the family of low molecular weight phenolic compounds that are widely distributed across the kingdom Plantae. In angiosperms, many of these serve as floral pigments. One important family of secondary metabolite phytochemicals having a polyphenolic structure is flavonoids. A wide range of foods, such as fruits, vegetables, tea, chocolate, and wine, contain them. They are called "dietary flavonoids" because they make up a substantial portion of the average human diet. Flavonoids are further classified into numerous subgroups based on stereochemistry and chemical structure, including the location of the B ring's attachment to the C ring, the degree of oxidation of the C ring, and unsaturation. Flavonoids having a B ring joined to position 3 of the C ring are known as isoflavones. However, if the B ring is joined to the C ring at position 4, they are classified as neoflavonoids (Fig. 3). Those in which the B ring is connected at position 2 of the C ring can be further divided into several subgroups based on the oxidation and unsaturation of the C ring3. Catechins, anthocyanins, chalcones, flavones, flavonols, flavanones, and flavanonols are some of these subgroups. The subclasses of flavonoids and the foods that contain them are listed in Table 1.

Table 1. The presence of flavonoid subclasses in food (Teng & Chen, 2019; Middleton, 1998; Shahidi & Naczk, 1995; Belay & Gholap, 2009; Caponio et al., 2022)

S.No	Flavonoid Subclass	Source	Active Compounds	Role in wound healing
(Jomova et al., 2025)	Flavonol	Onion, red wine, kale, olive oil, broccoli apples, cherries, berries, and grapefruit and tea	Kaempferol, quercetin, myricetin and tamarixeti	Act as strong antioxidants and reduce oxidative damage at wound site
(Durazzo et al., 2019)	Flavanol	Apple, tea, cocoa, apple	Catechin, epicatechin, epigallocatechin, glausan-3-epicatechin, proanthocyanidins	Improve blood flow and reduce inflammation
(Mittal et al., 2025)	Flavones	Fruit skins, red wine, buckwheat, red pepper, tomato skin, Parsley, Thyme	Chrysin, apigenin Rutin, luteolin, and glucosidestangeretin	Inhibit inflammatory mediators and support tissue repair
(Garg et al., 2019)	Flavonones	Citrus fruits, grapefruits, lemons, and oranges	Naringin, naringenin, taxifolin, and hesperidin	Promote epithelial regeneration
(Hollman, 2004)	Isoflavones	Soya beans, Legumes	Genistein, daidzein	Enhance collagen synthesis via hormonal modulation
(Shalaby et al., 2025)	Anthocyanidins	Grapes,Cherries, strawberry, rasberry and	Apigenidin, cyaniding, delphinidin, pelargonidin, malvidin	Strengthen capillaries and accelerate wound contraction

Lignans

Two phenylpropane units make up lignans. Linseed, which also contains matairesinol and secoisolariciresinol, is the highest dietary source of lignans. Linseed has a 1000-fold higher concentration of lignans than other cereals, fruits, grains, and some vegetables (Mittal et al., 2025). The gut microbiota biotransforms lignans into the molecules enterolactone and enterodiol. Human and animal seminal fluids, bile, and urine are the main excretion sites for these two moieties (Zhou et al., 2025). Because they function as phytoestrogens, these moieties have antioestrogenic properties.

Stilbenes

Red wine contains trace levels of resveratrol, an antioxidant and antineoplastic compound (Zhou et al., 2025), is the greatest example of stilbenes. Since these are often present in extremely low amounts, dietary consumption of these compounds cannot be linked to antineoplastic effects (Durazzo et al., 2019).

Bioavailability and Absorption:

There is no direct correlation between the amount of polyphenols in diet and the body's ability to absorb or utilize them. When consumed orally, the intestinal mucosa allows polyphenols to enter the bloodstream, where they are subsequently transported to the targeted tissues.

Numerous research confirm that interest in nutraceuticals is rising [4]. Dietary practices have a significant influence on how various metabolic processes are regulated. Food contains bioactive compounds that assist health, such as vitamins, fiber, antioxidants, and polyunsaturated fatty acids, in addition to nourishing the body's metabolic processes (Rodriguez et al., 2008). Eating a balanced diet and all of its components can thereby improve a person's overall health, reduce their chance of developing certain diseases, and increase their quality of life.

Understanding how polyphenols are absorbed and distributed throughout the body is necessary to evaluate their potential impacts on biological activity in live organisms and their overall significance in preventing a variety of illnesses related with oxidative stress (Singer, 2022).

Absorption study is challenging due to the complex chemical composition of meals high in polyphenols as well as other factors including the degree of polymerization and conjugation with other compounds and phenols. The majority of polyphenols are found in food as polymers, glycosides, or esters, and these forms are not absorbed (Garg et al., 2019). In comparison to micro- and macronutrients, polyphenols have relatively low bioavailability because they are identified as xenobiotics after ingestion (Singer, 2022).

According to recent research on polyphenol bioabsorption, when ingested in relatively large doses, their bioavailability is constrained (Gao et al., 2025). The primary barrier to their pharmacological use is their low bioavailability, which is associated with interactions between polyphenols at various stages of digestion, absorption, and distribution that change their molecular structure—specifically, their interactions with food, digestive enzymes, intestinal transporters, and blood proteins (Garg et al., 2019). Very little of the polyphenols that are consumed are absorbed by the small intestine. Enterocytes and hepatocytes can hydrolyze and biotransform simpler polyphenolic substances after absorption. Hydrophilic conjugated metabolites include derivatives of methyl, glucuronide, and sulfate, which are thereafter quickly released into the bloodstream, transported to other organs, or eliminated in urine (Rodriguez et al., 2008). Tracing specific chemicals in the body is made more difficult by the possibility of additional compounds forming during the metabolism of each phenolic molecule (Rodriguez et al., 2008). Usually, phenolic antioxidants undergo metabolic processes that result in completely different molecules, making it practically hard to identify the parent phenol (Zhou et al., 2025). Functional groups like hydroxyl groups are introduced or exposed during oxidation, reduction, or hydrolysis activities in enterocytes during phase I metabolism (Gao et al., 2025).

After being absorbed in the small intestine and entering the systemic circulation, polyphenol metabolites are carried to the liver via the portal vein.The main metabolites found in the portal vein (Zhou et al., 2025) are glucuronides and dimethyl glucuronides. It has been discovered that these polar conjugates can pass through the hepatocyte membrane and undergo additional changes in other cell types, even though the kidneys eliminate the majority of metabolites (Melrose, 2023). The vast majority of polyphenols that are consumed ultimately end up in the colon, including those that are not absorbed from the small intestine and those that are absorbed and digested in the liver or small intestine. They are returned to the colon lumen either by bile or membrane transporters. During the second stage of metabolism, the colon's extensive microbiota breaks down polyphenols and produces a variety of new metabolites (Zhou et al., 2025).

Among the many different processes that bacterial enzymes can catalyze is quick deconjugation. In the colon, flavonoids are broken down by enzymes, producing a variety of new metabolites (Rodriguez et al., 2008). The flavonoid scaffold is broken down by the enzymes of the gut microbiota into phenolic acids and other less complex substances. It is currently unknown how prevalent flavonoid metabolites are in the intestines. Thus, it's critical to comprehend how gut microorganisms impact food's flavonoid distribution and bioactivity (Gao et al., 2025).

To improve the absorption and health benefits of dietary polyphenols, future research and reviews must look into different methods to optimize processing conditions based on specific food compositions (Melrose, 2023). A contemporary solution to the low bioavailability of polyphenolic chemicals is the development and application of food-grade bio-based nanocarriers for their encapsulation, storage, and distribution. By combining nano-delivery methods with the beneficial qualities of polyphenols, the application of bio-based nano-complexes extracted from polyphenol food-grade sources in nutraceuticals offers an intriguing opportunity that could change the development and use of nutraceuticals (Criollo-Mendoza et al., 2023).

Figure 2. Diagram of metabolic fate of polyphenols.

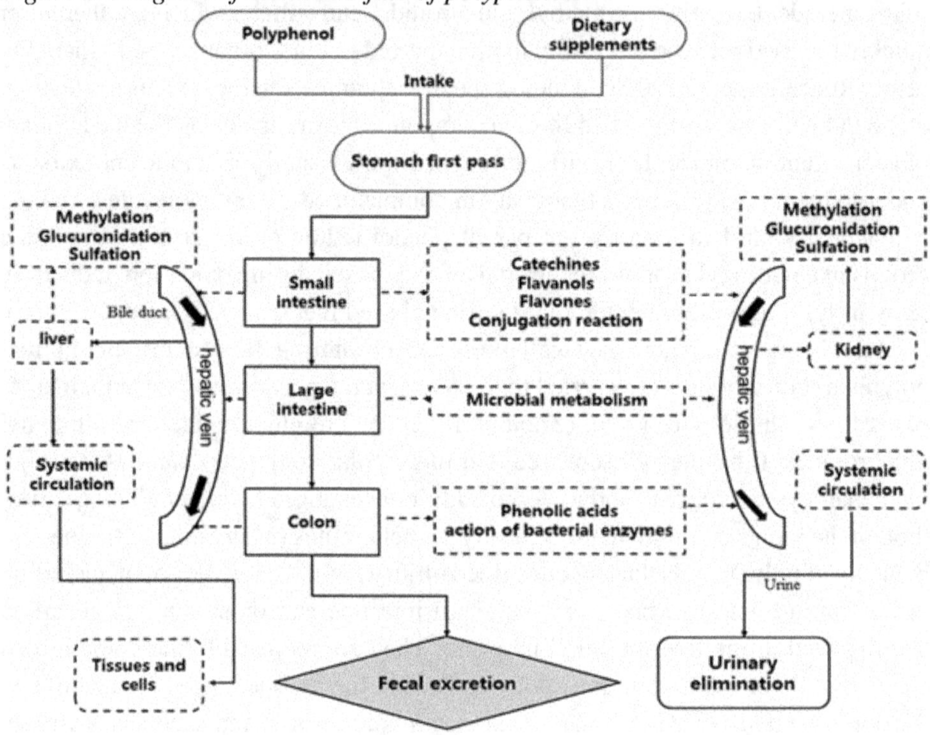

Mechanisms of Action

The main mechanisms behind natural products' potential in wound healing therapy are their anti-inflammatory and antioxidant qualities.

Antioxidant properties

Reactive oxygen species (ROS), which are small molecules produced from oxygen that act as oxidizing agents and cause cell damage, play a major role in the normal wound healing process (Rodero & Khosrotehrani, 2010; Zulkefli et al., 2023). By promoting cell survival signals (Kondo & Ishida, 2010), ROS can effectively promote wound healing and shield tissues from infection at low concentrations.

Excess ROS that causes oxidative stress and cellular damage. Polyphenol act as antioxidants by neutralizing ROS protecting biomolecules and maintaining balanced ROS levels that support wound healing.

In simple terms, polyphenols that promote wound healing by neutralizing excessive reactive oxygen species, protecting cells from oxidative stress and maintaining a favourable environment for tissue repair.

Anti-Inflammatory

The inflammatory phase is an essential part of the healing process, even if excessive inflammation and its prolongation in the early stages of the wound healing process can cause scarring, fibrosis, and a delay in the healing process (Sun et al., 2014). Furthermore, inflammation is necessary to eliminate dead cells and cellular waste as well as to stop microbial infection. Proinflammatory mediators, including chemokines, TNF-α, interleukins, and inducible nitric oxide synthase, are produced during this phase (Pandith et al., 2013). Macrophages with the proinflammatory M1 phenotype are the first to react during the repair phases. In contrast, the pro-repair, anti-inflammatory M2 phenotype is more common in the later stages of wound healing (Rodero & Khosrotehrani, 2010).

Polyphenols support wound healing by reducing excess inflammation, inflammatory mediators regulators and tissue protection from prolonged inflammatory damage.

ROLE OF POLYPHENOLS AND FLAVONOIDS IN WOUND HEALING

Wnt/β-catenin, Hippo, TGF-β, Hedgehog, JNK, (Nrf2/ARE, NF-κB, MAPK/ERK, Ras/Raf/MEK/ERK, PI3K/Akt, NO, etc.) have all been linked to flavonoids, which are well-known for their ability to heal wounds. These pathways include details and a thorough explanation of flavonoids and their underlying mechanisms in treatments for wound healing (Kurahashi & Fujii, 2015). The main mechanisms behind the promise of natural chemicals in wound healing treatments are their antibacterial, anti-inflammatory, and antioxidant qualities (Rodero & Khosrotehrani, 2010). The four continuous, coordinated phases of wound healing are hemostasis, inflammation, proliferation, and remodeling.(**Fig 3**)

Figure 3. The process of wound healing. (1) hemostasis (coagulation); (2) inflammation (early and late); (3) proliferation (migration, proliferation, granulation, and epithelialization); and (4) remodeling (maturation and repair).

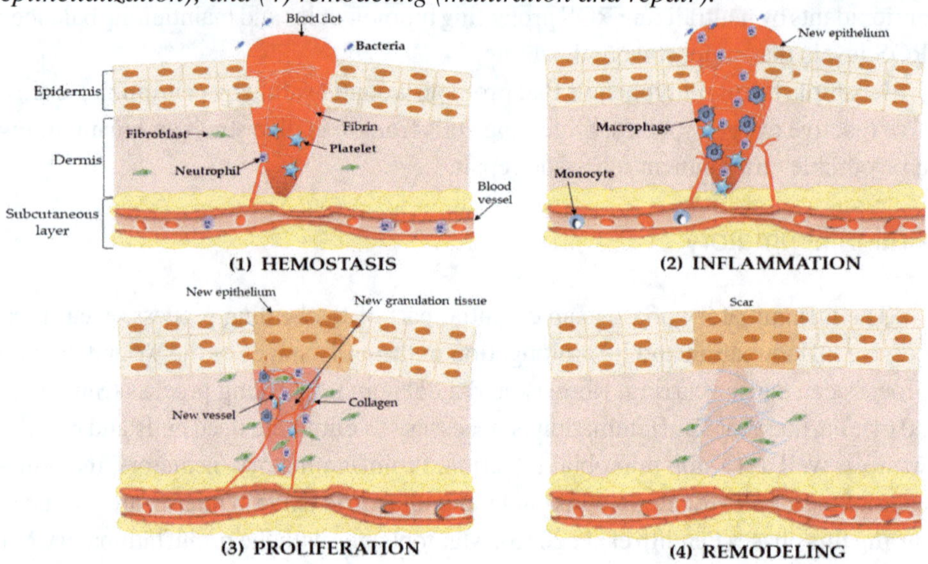

Hemostasis

Within a few minutes of the damage occurring, the first phase, hemostasis, starts, changing the tissue's integrity. Fibronectin, fibrin, vitronectin, and thrombospondin make up blood clots, which are formed when platelets are activated and signal chemicals to form a plug. Additionally, growth factors such as transforming growth factor-β, platelet-derived growth factor, epidermal growth, and chemokines are produced to initiate the repair process (Liu et al., 2020; Salehi et al., 2021). During this stage, the enzyme thromboxane synthase transforms prostaglandin H2 into thromboxane A2 (TXA2). TXA2 is a powerful vasoconstrictor and platelet activator that also aids in the release of neutrophils, endothelial cells, and macrophages—all of which are necessary for the healing of wounds (Al-Roujayee, 2017).

Inflammatory Phase

Following hemostasis, an immunological barrier against microbes (Rudrapal et al., 2024) is formed, the inflammatory phase begins, and certain immune cells start to lessen the damage. During this stage, leukocytes, especially neutrophils, are attracted to the injury site to eliminate bacteria and foreign objects; as part of an inflammatory reaction, neutrophils release proinflammatory cytokines that promote the synthesis of adhesion molecules (Kurahashi & Fujii, 2015). Monocytes enter the wound site at the same time and differentiate into macrophages, which might draw in additional monocytes and increase response[21]. Neutrophils, monocytes, and macrophages are crucial during this phase because they release soluble mediators such as proinflammatory cytokines (interleukins IL-1, IL-6, IL-8, and tumor necrosis factor α (TNF-α)) and growth factors (platelets derived growth factor (PDGF), transforming growth alpha (TGF-α), transforming growth factor beta (TGF-β), fibroblast growth factor (FGF), and insulin-like growth factor-1 (IGF-1). These mediators aid in the activation of the fibroblasts and epithelial cells necessary for phase (Rudrapal et al., 2024).

Proliferative Phase

The proliferative phase, which lasts a few days to weeks, aims to reduce the lesioned tissue area. This phase is characterized by angiogenesis, epidermal regeneration, granulation tissue formation, and collagen deposition (Salehi et al., 2021; Rudrapal et al., 2024). When the activated fibroblast first migrates to the wound, it creates extracellular matrix proteins (fibronectin, hyaluronan, and proteoglycan) to replace fibrin clots and provide the tissue strength. The process by which new extracellular matrix tissue is created within the network of blood vessels using this connective tissue or granulation tissue of which type 3 collagens are the most prevalent is known as angiogenesis, and it is mediated by endothelial cells (Al-Roujayee, 2017).

In order to form a new epidermal barrier, cytokines also induce re-epithelialization, in which keratinocytes, or epithelial cells, proliferate and migrate to the wound (Salehi et al., 2021). The differentiation of fibroblasts into myofibroblasts during this phase also results in decreased proliferation and increased collagen synthesis, which causes the wound to compress and shrink (Al-Roujayee, 2017).

Remodeling Phase

During the last stage (maturation and healing), the extracellular matrix progressively transforms into a mature scar (Rodero & Khosrotehrani, 2010), which may take up to two yearsIn this instance, the collagen is reorganized, the extracellular matrix's collagen type III is converted to collagen type I (this happens when the wound closes), blood flow and new blood vessels are decreased, a cellular environment and mature vascular tissue are formed, and the skin eventually only regains up to 80% of its tensile strength (Al-Roujayee, 2017; Rudrapal et al., 2024). A chronic wound develops when the immune system is dysregulated during the healing process, whereas the previously mentioned wound healing process occurs in a normal or acute wound.

This indicates that there isn't a systematic healing process, and the wound can take more than three months to heal completely (Salehi et al., 2021). Venous ulcers, high blood pressure, obesity, diabetes, and aging are all associated with chronic wounds. Additionally, they have microbial films and ongoing illnesses (Rodero & Khosrotehrani, 2010).

An increase in neutrophils (which release metalloproteinases, or MMPs) alters the inflammatory phase of a chronic wound, resulting in an overabundance of reactive oxygen species (peroxide anion, hydroxyl ion, and superoxide anion) and damaging enzymes (proteases and elastases). Similarly, the breakdown of extracellular matrix elements (collagen, fibronectin, and vitronectin), the decrease in growth factor availability, and the increase in inflammatory cytokines (due to an excess of proinflammatory macrophages) continue this phase. In a similar vein, angiogenesis is hindered, re-epithelialization is difficult, and keratinocytes are hyperproliferative (Salehi et al., 2021).

CLINICAL IMPLICATIONS OF POLYPHENOLS AND FLAVONOIDS IN WOUND HEALING

Therapeutic Potential of Polyphenols and Flavonoids in Chronic Wounds

An increase in neutrophils (which release metalloproteinases, or MMPs) alters the inflammatory phase of a chronic wound, resulting in an overabundance of reactive oxygen species (peroxide anion, hydroxyl ion, and superoxide anion) and damaging enzymes (proteases and elastases). Similarly, the breakdown of extracellular matrix elements (collagen, fibronectin, and vitronectin), the decrease in growth factor availability, and the increase in inflammatory cytokines (due to an excess of proinflammatory macrophages) continue this phase. In a similar vein, angiogenesis is hindered, re-epithelialization is difficult, and keratinocytes are hyperproliferative (Shevelev et al., 2020). Naringenin was embedded with adhesive on the injury site at 1%, 10%, and 20% doses in a full-thickness excisional wound model in male Wistar rats. The results showed that the 20% naringenin group had the best wound closure of all the groups (Choudhary et al., 2020). Naringenin was also found to be helpful in first-degree burns with superficial skin injury by reducing IL-6, IL-1β, caspase-3, TNF-α, PGE2, LTB4, NO, and NF-κB. GPx, GST, CAT, and SOD levels increased in addition to the restoration of GSH and TBARS levels. The capacity of dihydromyricetin and dihydroquercetin to cure wounds was examined in a purulent infection model utilizing male Wistar rats. They were shown to be viable medication candidates for topical use in the management of trophic ulcers and burns (Ajmal et al., 2019).

Topical treatment of quercetin at concentrations of 0.03%, 0.1%, and 0.3% enhanced wound healing in a full thickness wound experiment using Wistar rats by modifying cytokines and growth factors by lowering TNF-α (Ahmed et al., 2018) and raising VEGF, TGF-β1, and IL-10. Another study discovered that applying quercetin topically at dosages of 10, 20, and 40 mg/ml to male diabetic wistar rats with full-thickness wounds reduced inflammatory responses and sped up the healing of diabetic wounds. It was proposed that quercetin reduced IL-6, M1 macrophages, and TNF-α (Rithichai et al., 2024) and increased VEGF-α, M2 macrophages, CD31, and IL10. The utility of ciprofloxacin HCl and quercetin-functionalized nanofiber as a potential wound dressing material was confirmed when CAT and SOD levels were restored in a full-thickness wound model in male wistar rats (Li et al., 2016). When given orally by gavage at a dose of 25 mg/kg38, it improved cytokines involved in inflammation and the antioxidant defense system in addition to promoting wound healing in a study utilizing a full-thickness excisional wound model in male diabetic wistar rats. Goumaet al. (2016) proposed quercetin as a possible alternative treatment for burn wounds (Castangia et al., 2014) based on the findings of their investigation utilizing a full-thickness burn model in female wistar rats. It has also been demonstrated to halt the growth of skin lesions, reversing the many metabolic processes that cause epithelium loss and skin damage (Bagher et al., 2020).

Male wistar rats with full-thickness excisional wounds showed a higher rate of wound closure when hesperidin was administered topically at dosages of 1 and 10% (Haddadi et al., 2018). Furthermore, when given orally at a dose of 100 mg/kg, hesperidin has expedited wound healing, particularly radiation-induced skin damage, in the Cobalt 60 irradiation of rat skin model. Wound healing may be caused by overexpression of VEGF (Li et al., 2018). Furthermore, hesperidin has been demonstrated in a study including oral administration of 25, 50, and 100 mg/kg to SD rats (Abbas et al., 2019) to improve wound healing and accelerate vasculogenesis and angiogenesis by raising the expression of Ang1/Tie-2, SMAD-2/3 mRNA, TGF-β, and VEGF-c.

Curcumin used topically at dosages of 10, 20, and 30 mg/ml has accelerated the healing process and enhanced the rate of wound closure in male rabbits with superficial second-degree burns (Özay et al., 2019). In both diabetic and non-diabetic wistar rats, kaempferol has shown anti-inflammatory benefits on incisional and excisional wounds when administered topically at doses of 0.5 and 1% (Mohammadi et al., 2017). In wistar rats with full-thickness excisional wounds, chrysin therapy at concentrations of 5, 10, and 15% has shown anti-inflammatory effects at several stages of the wound-healing process. It was proposed to increase MMP-2, TIMP-1, and IL-6 and reduce TIMP-2 of iNOS (Kar et al., 2019).

Epigallocatechin gallate is the most common catechin found in tea.Epigallocatechin gallate treatment increased wound shrinking rate, expedited healing, and had anti-inflammatory effects in a full-thickness wound model in wistar rats. It was claimed to increase VEGF and KI67, but there was no discernible change in the levels of AST, ALT, creatinine, or urea. Both TNF-α47 and IL-6 decreased. On the other hand, intraperitoneal injection of epigallocatechin gallate (10 mg/kg) has expedited skin wound healing in male wistar rats with full-thickness wounds. This is most likely due to their combined effects, which reduce TNF-α, IL-1β, and IL-6 (Castangia et al., 2014) levels to control chronic inflammation. Oral treatment of epigallocatechin gallate increased the percentage of wound closure and strengthened the incision wound in male Sherman rats with full-thickness wounds (Kim et al., 2008). By accelerating reepithelialization and angiogenesis through increased immunoreactivity of CD31, Ki-67, and a-smooth muscle actin49, topical application of epigallocatechin gallate in a full-thickness wound model in male mice with genetic diabetes and obese phenotype has improved wound healing in diabetic mice. In a full-thickness wound model in male wistar rats, puerarin has been shown to increase SOD and GPx while decreasing MDA (Parveen et al., 2025) when administered topically at dosages of 0.5 and 10 mg/ml.In male wistar rats with open excision-type wounds, icariin administered topically at doses of 0.004, 0.02, 0.1, and 0.5% has been shown to accelerate the healing of cutaneous wounds. TNF-α and NF-κB51 may be downregulated and IL-10 may be upregulated as a result of increased CD-31 expression and decreased MMP-2 and MMP-9 levels. Apigenin may be applied topically as an antibacterial agent to fight S. aureus and encourage wound healing, according to research. Apigenin (0.1 g) reduced the levels of IFN-γ, IL-1β, and TNF-α when administered topically to female pathogen-free BALB/c mice in a skin injury model employing autoclave tape (Cheng et al., 2018). The improved control of GSH, CAT, and SOD (Shukla et al., 2016) has revealed apigenin's antioxidant properties.Using a full-thickness wound model, topical application of 0.5% and 1% luteolin improved the skin tissue's capacity to repair wounds in male wistar rats with and without diabetes54. Icariin (30 μM) injected into the surrounding tissue of the wound accelerated the migration and proliferation of keratinocytes and accelerated the healing process of skin wounds by activating ERK and AKT, reducing TNF-α and IL-6, and promoting IL-10 (Özay et al., 2018). In male diabetic wister rats with full-thickness open excision wound models, topical administration of morin at 0.5% and 1% concentrations has reduced the reepithelialization time and accelerated the pace of wound contraction by accelerating collagen synthesis (Mi et al., 2018). In a full-thickness excisional wound model in male wistar rats, topical use of naringin at doses of 1.5%, 3%, and 6% has accelerated collagen synthesis, increasing the rate of wound closure (Ponrasu et al., 2018). By increasing growth factor and reducing inflammatory and apoptotic expression, naringin therapy of full-thickness excisional and incisional wounds in male wistar rats demonstrated promise for wound healing. While levels of MDA, MPO, GSH, SOD, and NO were recovered, those of IL-1β, IL-6, IL-8, NF-κB, SMAD-7, Bax, and TNF-α were dropped (Salehi et al., 2018). Topical use of naringin at dosages of 1.5%, 3%, and 6% has accelerated collagen synthesis and increased the pace of wound closure in a full-thickness excisional wound model in male wistar rats (Ponrasu et al., 2018). Naringin treatment of full-thickness excisional and incisional wounds in male wistar rats showed potential for wound healing by boosting growth factor and decreasing inflammatory and apoptotic expression. MDA, MPO, GSH, SOD, and NO levels were restored, whereas IL-1β, IL-6, IL-8, NF-κB, SMAD-7, Bax, and TNF-α levels decreased (Seo et al., 2017). COX2, NFκB, TNF-α, and iNOS decreased in female ICR mice used as models for acute diabetes when genistein was added to the diet at 0.025% and 0.1%. The anti-oxidant defense system (HO-1, GPx, CAT, and Nrf2) either normalizes or increases. returned caspase-1, ASC, and NLRP3 to their initial levels. By reducing oxidative stress and inflammation during the inflammatory stages, it may be useful in both preventing and treating delayed wound healing (Asfour et al., 2017). Subcutaneous treatment inhibited oxidative stress, iNOS activity, and FoxO1 in a full-thickness wound model in male C57BL/6 mice, which improved wound angiogenesis and sped up the healing process (Eo et al., 2016). Genistein supplementation in the diet reduced Zn-SOD, Mn-SOD, and Cu-SOD expression in female ICR mice with full-thickness wounds, but CAT expression did not change. It altered the expression of pro-inflammatory cytokines and reduced oxidative damage in the initial phases of wound healing (Tie et al., 2013). When administered topically, the proanthocyanidin cinnamtannin B-1 promoted mesenchymal stem cell migration in vivo (Park et al., 2011) and improved wound healing in a full-thickness wound model in female db/db mice.

In a full-thickness wound model, epicatechin gallate given intradermally to male neonatal wistar rats increased total NOS and iNOS activity, promoted incisional wound healing, and reduced scar formation in the type-2 diabetes mellitus rat model (Fujita et al., 2015). In a full-thickness wound model in male SD rats, intradermal administration of epicatechin gallate can significantly improve the effectiveness of wound healing and scar formation by up-regulating iNOS, VEGF, and COX67. When given topically to male and female wister rats with full-thickness cutaneous abnormalities, silymarin has been demonstrated to enhance cellular and tissue maturity by regulating inflammatory processes and total cellularity (Kapoor et al., 2004). Oral Daidzein treatment at 70–75 mg/day has reduced the wound area's growth during the inflammatory stages in a full-thickness wound model in female mice69 and promoted the process of re-epithelialization. In a chemical burn (acetic acid) paradigm, male wistar rats treated topically with taxifolin showed reduced inflammation and increased epidermal proliferation (Yamaguchi et al., 2012). In a full-thickness wound model in male wistar rats, rutin has been demonstrated to improve skin wound healing by increasing CAT (Shubina & Shatalin, 2012).

In a clinical research involving 56 individuals with diabetes mellitus (28 men and 28 women), a nano-hydrogel containing quercetin and oleic acid was applied to the foot skin lesions for eight months.

The application of quercetin increased keratinocyte proliferation activity. For comparison, hyaluronic acid served as a positive measure. Quercetin's promising potential as a flavonoid type in the treatment of wound healing is demonstrated by the combination therapy's significant superiority over hyaluronic acid's efficacy. According to the study, quercetin improved tissue viscoelasticity, which decreased skin lesions and might have simplified the treatment of chronic wounds.It was suggested that modifying cytokines, growth factors, and protease would aid in the treatment of chronic wounds and the healing of skin lesions caused by diabetes (Almeida et al., 2012). This ability was also linked to quercetin's remarkable antibacterial action. Additionally, the experiment confirmed that patients reported reduced discomfort and no new infections. There was also no indication of a local recurrence.

Likewise, wound healing ability was ascribed to quercetin's remarkable antibacterial properties. The trial also confirmed that patients reported no new infections and less discomfort. Furthermore, there was no indication of a local recurrence. Micronized diosmin and flavonoid fraction make up MPFF (micronized purified flavonoid fraction). Its flavonoid fraction can be hesperidin, diosmetin, or linarin. Diosmin and hesperidin together were found to be a superior wound management technique for chronic venous leg ulcers (Table 1) (Kant et al., 2021). The clinical study includes the use of a stocking or bandage for compression therapy in addition to the administration of medication or another treatment with Micronised purified flavonoid fractionnamed MPFF. About 723 people with venous leg ulcers got healed significantly faster when compression and oral treatment with MPFF were combined (Table 1) (Coleridge-Smith, 2009). MPFF can easily surpass standard therapy by 32% in terms of healing rates, according to clinical tests conducted over a six-month period. This result implies that MPFF can protect microcirculation from damage brought on by elevated vein ambulatory pressure (Coleridge-Smith et al., 2005). There have been notable outcomes when using flavonoids alone to treat wound healing. For instance, forty men agreed to take part in a clinical investigation in which plant-based quercetin cream was applied to oral sores and mucosa. The study found that the cream was able to heal wounds in two to four days in 35% of cases (Lyseng-Williamson & Perry, 2003). Another clinical study focused on propolis, the common name for a material produced by honeybees and a well-known source of flavonoids. Thirty-three patients (aged 18 to 45) who had straightforward sacrococcygeal pilonidal sinus wound surgery were evaluated for propolis's ability to speed up wound healing (Table 1) (Hamdy & Ibrahem, 2010). These patients had marsupialization surgery, and their wounds were treated with a 15% water solution after each dressing change. The wound healed much more quickly after 14 days of treatment. Necrosis, allergic reactions, or consequences from local infections were not reported during the investigation (Kubat et al., 2021). Another clinical investigation on flavonoid-rich Latin American medicinal plants examined the efficacy of flavonoids in the treatment of chronic leg ulcers. Five medicinal plants from Latin America were compared for their capacity to heal wounds. Mimosa tenuiflora demonstrated the best wound-healing effectiveness among the plants under investigation in a clinical research involving less than 50 patients with persistent venous leg ulcers over a one-year monitoring period, reducing ulcer size by 93% in the eighth week. (Table 1) (Miguel & Antunes, 2011).

Table 1. Clinical outcome of Flavonoids in Wound Healing

Flavonoids	Clinical Trial Design	Outcome	Ref.
Quercetin + oleic acid	56 patients (28 men and 28 women) who administered the nano-hydrogel containing quercetin and oleic acid to the lesion and studied for 8 months	Quercetin enhanced tissue viscoelasticity, decreased skin lesions, and may make treating chronic wounds easier.	(Salazar-Gómez & Alonso-Castro, 2022)
Micronized purified flavonoid fraction + compression treatment	723 patients consumed the MPFF orally and were monitored for six months	MPFF and the compression technique accelerated the healing of venous leg ulcers.	(Kant et al., 2021)
Quercetin	Forty male volunteers received quercetin-containing cream to apply to their oral ulcers, and they were monitored for ten days.	Within two to four days, quercetin cream induced full healing and relieved pain.	(Lyseng-Williamson & Perry, 2003)
Anthocyanin	Anthocyanin gel was applied to the oral wounds of 68 orthodontic patients, who were then monitored for seven days.	In acute oral wounds, the anthocyanin-containing mucoadhesive gel facilitated wound closure.	(Gallelli et al., 2020)
Anthocyanin	60 volunteers participated in a double-blind clinical trial in which anthocyanin gel was applied to oral wounds and monitored for seven days.	By speeding up wound closure, the anthocyanin niosome gel improved patients' quality of life by reducing oral wound pain.	(Limsitthichaikoon et al., 2018)

continued on following page

Table 1. Continued

Flavonoids	Clinical Trial Design	Outcome	Ref.
Flavonoids (class not mentioned)	33 patients underwent simple surgery to repair wounds in the sacrococcygeal pilonidal sinuses, and they were monitored for 28 days.	Wound recovery was aided by the 15% propolis water solution treatment, which demonstrated a notable rate of healing from the initial week.	(Hamdy & Ibrahem, 2010)
Flavonoids (class not mentioned)	The clinical study was divided into three groups: single-blinded, double-blinded and randomized. 50 individuals with chronic venous leg ulcers were monitored for eight weeks.	The study examined 305 plant species that have the ability to heal wounds. Twenty-five compounds were identified from the plants that had significant flavonoid content. By decreasing the ulcer size to 93% in the eighth week, the extract from Mimosa tenuiflora exhibited the strongest wound-healing action.	(Miguel & Antunes, 2011)
Flavonoids (class not mentioned)	140 uniparous women were used as participant groups in a double-blind clinical trial to examine the wound-healing action of epiosomy incisions, and they were monitored for 14 days.	Edema, redness, and pain can all be lessened by the treatment. There was no discernible difference between the treatment and control groups in terms of lowering the risk of wound secretion and dehiscence. According to the study, it's critical to use the right dosage during treatment.	(Damrongrungruang et al., 2021)

By lowering inflammation, encouraging tissue regeneration, and increasing the rates of wound closes, polyphenol and flavonoid-based therapies can generally improve wound healing results, according to current clinical investigations and randomized controlled trials. These results demonstrate their great potential for practical application in wound care.

Formulations and Delivery Systems:

Polyphenols' limited stability and bioavailability limit their medicinal efficacy. By enhancing polyphenol stability and enabling a controlled release at the wound site, the various micro-and nanoencapsulation technologies covered here have shown promise as a means of getting over these restrictions.Rats with and without diabetes were used to test lutein ointment for incision and excision wound models. Glycol stearate, propylene glycol, and liquid paraffin were combined in a 3:6:1 ratio to create an ointment base that increased the bioavailability of luteolin, which has a low water solubility. The ointment base made the luteolin more soluble and contributed to the creation of a homogenous combination with better penetration.

Both diabetic and non-diabetic patients benefited from luteolin ointments' enhanced skin tissue healing capabilities. On day 14, macroscopic results showed that the excision and incision wounds treated with 0.5% (w/w) luteolin ointment showed the best signs of wound healing (96.1% and 97.6%, respectively) (Khan et al., 2025). LysGH15-Apigenin-Aquaphor (LAA) ointment showed bactericidal activity towards S. aureus, in addition to preventing hemolysis. The mean bacterial colony count in mice with an MRSA-infected skin wound treated with LAA dropped to about 10^2 CFU/mg 18 hours after therapy, while it was about 10^2 CFU/mg of tissue in mice not treated. Additionally, in the mouse model, LAA ointment decreased pro-inflammatory cytokine levels (TNF-alpha, IL-1beta, and IFN-gamma) and quickened wound healing. These results suggest that apigenin is a promising therapeutic option for the management of inflammation. Aquaphor, the base of the ointment, acts as a hydrating agent for the LAA ointment and has no influence on the hemolysis inhibition of Apigenin or the lytic activity of LysGH15. Apigenin and LysGH15 may therefore be able to effectively penetrate skin wounds and eliminate MRSA thanks to this foundation (Benedek et al., 2007).

In vitro studies further proved the sufficient thermodynamic and physical stability of curcumin nanoemulsion. Both in vivo and ex vivo tests were conducted on the ideal formulation, which contained 40% Tween 80 as surfactants, PEG 400 as co-surfactants in a 2:1 ratio, and Triacetin: 15% Labrafac PG (1:1) as the oil phase. Curcumin was better able to enter the wound through the skin layers because of the nanoemulsion's ability to break down the outermost layers of skin lipids. For a drug that is poorly soluble, they also provide a better reservoir because they can improve solubilization. The depot effect of curcumin in the skin, which adds to the medication's longer-lasting presence, is another benefit of nanoemulsion (Seema et al., 2023). New tea polyphenol nanospheres (TPN) were encapsulated in a PVA/alginate hydrogel and administered to diabetic rats to promote wound healing and regulate the immune response.Research on animal models and molecular mechanisms revealed that hydrogel could improve wound healing in diabetic rats by controlling the PI3K/AKT signaling pathway. Additionally, it was demonstrated that TPN-hydrogel outperformed ordinary hydrogel in speeding up wound healing (Cheng et al., 2018). QSEG, or sericin-stabilized quercetin emulgel, is designed especially to help diabetics recover their wounds. Shorter intercellular distances were found in in vitro scratch tests. Significant improvements in wound closure were observed in in-vivo wound healing experiments, demonstrating QSEG's powerful wound healing properties. Sericin's inherent stabilizing qualities increase the chance of diabetic wound healing when combined with QSEG. To sum up, QSEG shows promise as a therapeutically beneficial substitute for traditional diabetic wound care methods (Thomas et al., 2017).

For the management of skin injury, Turkish gall extract (TGE), which contains a higher concentration of polyphenols and flavonoids, was successfully created as a sprayable hydrogel. Polyphenols can undergo cross-linking via a number of mechanisms, including as Michael addition/Schiff base reactions, π-π interactions, and hydrogen bonding, to generate polyphenol-based hydrogels because they include a large number of catechol and pyrogallol structures. A natural carboxymethyl chitosan (CMCS) has remarkable qualities like its water-soluble nature, biodegradability, biocompatibility, and a high moisture retention capacity. Moreover, CMCS can bind different biomolecules to produce homogenous composite scaffolds. Turkish gall extract (TGE) and CMCS were combined and sprayed using a dual syringe to create TGE-CMCS hydrogel (TC gel) in a single step. Antioxidant, antibacterial, hemostatic, anti-inflammatory, sprayable, quick gelation, and wound-healing qualities were all displayed by the TC gel. Several rheological studies revealed that TC gel has injectable, thixotropic, and self-healing qualities, indicating that it might be used as a wound dressing.

In particular, TC gel showed good biocompatibility while preserving the biological activity of the assembly component TGE. The TC gel demonstrated exceptional hemostatic properties in both the mouse liver damage model and the mouse tail amputation experiment. TC gel effectively promoted full-thickness skin defect wounds in rats. All of the findings demonstrated the potential importance of wound care's clinical advancement and environmental sustainability (Chen et al., 2020). Tobacco stem polyphenol fraction-AgNP gel has been shown to improve stability and accelerate wound contraction.The re-epithelization of skin wounds has been confirmed by histopathological and collagen content studies. MD simulation showed stability with PI3K and AKT. AgNP gel can be utilized as a nano-drug delivery system to stop the development of chronic wounds, and tobacco stem polyphenol-loaded nanoparticles show potent wound-healing qualities (Burud et al., 2024).

By overcoming the limitations and controlling their release under specific physiological conditions, a variety of nano-formulations are being created to efficiently deliver and transport polyphenols across the wound site. Solid-lipid nanoparticles are designed to help transport polyphenols (Xue et al., 2024) because of their composition, which is similar to biological membranes, and their strong affinity to hydrophobic materials. Polymeric nanoparticles are easily manufactured and can have their surfaces changed to include polyphenols due to their increased chemical and physical stability in physiological situations (Sharma et al., 2024). Comparably, the ability to load of polyphenols can be adjusted by the high surface area, shape, and size of both metallic and non-metallic nanoparticles (Amini & Akbari, 2019). Nano-emulsions, nanogels, and nanofibers have shown enhanced polyphenol encapsulation efficiency, making them effective carriers of polyphenol and flavonoids in addition to being biocompatible and biodegradable (Mohammadi et al., 2017; Amanat et al., 2020). These novel polyphenol delivery techniques enhance the polyphenols' bioavailability, biodegradation, water solubility, absorption, and stability, all of which boost their therapeutic efficacy. These polyphenols encapsulated in nanoparticles may potentially replace overused medications and enhance wound healing at different phases. (Table 2).

Table 2. Different Polyphenol Formulations studied in Healing of wounds

Polyphenol	Formulation	Properties	Type of Wound	Ref.
Curcumin	Nanoemulsion gel	Antioxidant	Incision wound	(Seema et al., 2023)
Quercetin	Emulgel	Antioxidant & Antimicrobial	Diabetic wounds	(Thomas et al., 2017)
Luteolin	Ointments	Antioxidant	Excision and Incision Diabetic wounds	(Khan et al., 2025)
Apigenin	Hydrogels	Antioxidant	Diabetic wounds	(Shukla et al., 2016)
Kaempferol	Ointment	Antioxidant	Diabetic wounds	(Mohammadi et al., 2017)
Hyaluronic acid	Hydrogel	Antioxidant	Diabetic wounds	(Amanat et al., 2020)
Curcumin	Hydrogel	Antioxidant	Diabetic wounds	(Madhyastha et al., 2020)
Thymol	Film	Antioxidant & Antimicrobial	Burn wounds	(Xu et al., 2022)
Carcavol	Film	Antioxidant & Antimicrobial	Chronic wounds	(Krausz et al., 2015)
Tannic acid	Hydrogel	Antioxidant & Antimicrobial	Not specific	(Jiji et al., 2019)
Ferulic acid	Ointment	Antioxidant	Diabetic wounds	(Tenci et al., 2017)
Rutin	Pickering Emulsion	Antioxidant	Full-thickness wounds	(Seo et al., 2017)
Morin	Hydrogel	Antioxidant	Diabetic wounds	(Mi et al., 2018)
Naringin	Ointment	Antioxidant & Antimicrobial	Excision and Incision wounds	(Salehi et al., 2018)
Hesperidin	Hydrogel	Antioxidant & Antimicrobial	Full-thickness wounds	(Haddadi et al., 2018)

Safety Profile and Side Effects:

In general, polyphenols are safe to eat. They are beneficial natural compounds that help prevent and treat a variety of ailments or offer protection against various forms of stress. As a result, we should consume polyphenols, especially those found in fruits, vegetables, nuts, tea, spices, and juices. However, taking large amounts of polyphenols as food supplements without a doctor's supervision increases the risk of negative side effects. High concentrations of polyphenols, which are impossible to obtain through ordinary food consumption, are often found in such supplements. Instead of glycosides, which are present in "natural" foods, these polyphenols are usually found as chemically purified aglycones. Individuals undergoing treatment for a range of chronic conditions who are required to take daily prescription medications are also more vulnerable to the harmful effects of polyphenols when they eat a diet rich in polyphenols or take supplements that contain polyphenols (Xu et al., 2015).

Oral administration of grape seed proanthocyanidin extract at 0.5 or 2 g/kg body weight did not result in any harm in rats or mice.

Additionally, no toxicity was observed after a 60 g/kg diet of punicalagin, a ellagitannin found in pomegranate juice, was administered (Ghaisas et al., 2014). However, rats that received large amounts of quercetin (2% or 4%) experienced chronic nephropathy (Duda-Chodak & Tarko, 2023). When mice were fed quercetin (0.1%), their lifespan was significantly reduced (Cerdá et al., 2003). In high quantities or dosages, certain polyphenols may have genotoxic or carcinogenic consequences (Dunnick & Hailey, 1992). For instance, rats and mice developed kidney and stomach cancers when allowed to consume 2% of Caffeic acid through their food (Hagiwara et al., 1991). Quercetin raises kidney levels of 2-hydroxyestrodiol and 4-hydroxyestrodiol by 60–80% by blocking the O-methylation of catecholestrogens, which may contribute to estradiol-induced carcinogenesis. The genotoxic effects shown in vitro106 could be explained by polyphenols changing into prooxidants at high concentrations. Quercetin's prooxidative metabolic activities have been demonstrated in cellular in vitro investigations by the formation of glutathionyl quercetin adducts in tyrosinase-rich B16F-10 melanoma cells and the myeloperoxidase-rich human HL-60 cell line (Sakihama et al., 2002). This suggests that tissues with elevated oxidative enzyme levels may be particularly vulnerable to the prooxidant toxicity of quercetin. Finally, green tea catechins at 1% or 0.1% of the diet have been found to stimulate the formation of tumors in the colon of male F344 rats. Additionally, quercetin has been found to increase cancer cell proliferation at low concentrations (1–5 µmol/L) (Awad et al., 2002), although it may prevent it at high concentrations. Numerous flavonoids can interfere with the production of thyroid hormones and inhibit thyroid peroxidase (Hirose et al., 2001). Rats administered vitexin, a C-glycosylflavone abundant in millet (Ferreira et al., 2002), had increased thyroid weight and decreased thyroid hormone levels.

Additionally, rats given a genistein-supplemented diet showed decreased thyroid peroxidase activity (Doerge & Chang, 2002). In cases of inadequate iodine intake, these effects of genistein on the functioning of the thyroid are more apparent (Munro et al., 2003).

Both covalent and nonspecific forces can bind flavonoids to proteins to form complexes (Kumar & Pandey, 2013). Polyphenols and proteins can combine to generate soluble or insoluble complexes that change the proteins' structure, solubility, thermal stability, hydrophobicity, isoelectric point, and susceptibility to digestive enzymes. The activity of digestive enzymes like lipases, amylases, and proteases—all of which are proteins—can be impacted by binding with polyphenols. Reduced activity and disturbances in the biochemical reactions or processes that the enzyme catalyzes can result from structural changes in the enzymes (Brudzynski & Maldonado-Alvarez, 2015; Sęczyk et al., 2019) (Table 3).

Table 3. List of polyphenols that inhibited the activity of various digestive enzymes

Digestive Enzyme	Polyphenols	Ref
Trypsin	7,8-Dihydroxyflavone, Quercetin, Luteolin, Kaempferol, Apigenin, Myricetin, Morin, Galangin, Isorhamnetin, Fisetin, Kaempferol, Acacetin, Rutin Apigenin, Baicalein, Naringenin, Biochanin A, Hesperetin	(Li et al., 2014; Maliar et al., 2004; Martinez-Gonzalez et al., 2021)
α-Amylase	Apigenin, Scutellarein, Hispidulin, Nepetin, Hesperetin Luteolin, Quercetin, Catechin, Rutin, Young Apple Polyphenols, Tea Polyphenols, Catechins, Theaflavin, Sorghum Procyanidins Tetramer, Caffeic & Tartaric acid derivates, Ellagitannins, and Chlorogenic acid	(Sun et al., 2016; Sysak et al., 2023; Zhao et al., 2018; Sun et al., 2018; Bai et al., 2020; Li et al., 2007; Zheng et al., 2020)
α-Glucosidase & α-Amylase	Gallocatechin Gallate, Epicatechin Gallate, Quercetin, Kaempferol, Guaijaverin, Avicularin, Myricetin, Hyperin, Apigenin isolated from guava leaves, extracts of raw and heat-processed African pears (Dacryodes edulis), herbal extracts containing purified and extracted Rosmarinic acid, Tea Polyphenols (TP) and various teas, extracted and enriched flavonoids from Rubus Corchorifolius, and ferulic acid.	(Dev et al., 2023; Wang et al., 2022; Wu et al., 2018; Wu et al., 2019; Wang et al., 2010; Oboh et al., 2015; Yang & Kong, 2016; Tian et al., 2021)
Lactase, Maltase, Sucrase, and α-Amylase	5-Caffeoylqunic acid, proanthocyanidins, gallate esters, flavonols, and theaflavins	(Stevens Barrón et al., 2023)
Maltase and Sucrase are α-glucosidases.	5-Caffeoylquinic acid, Epicatechin (4β,8), and Epigallocatechin-3-gallate Polyphenol-rich green tea extract-Epicatechin gallate, Epicatechingallate, and Epicatechin	(Zheng et al., 2020; Williamson, 2013)
Lactase	Epigallocatechin-3-gallate	(Pyner et al., 2017)
Pepsin	Baicalein, Apigenin, Luteolin, Keampferol, Quercetin, Morin, Liquiritigenin, Naringenin, Daidzein, Genistein, and caffeic acid	(Li et al., 2019; Naz et al., 2011)
Pancreatic Lipase Phospholipase A2 & Trypsin	Catechins, Epigallocatechin-3-gallate, Theaflavin-3, 3'-Digallate, Theaflavin-3'-gallate, Theaflavin-3-gallate, Theaflavin (TF)	(Hashemi-Shahraki et al., 2023)

continued on following page

Table 3. Continued

Digestive Enzyme	Polyphenols	Ref
Pancreatic lipase	Methanolic extract of the leaves of Eremochloaophiuroides	(Zeng et al., 2015)
Lipase, α-Amylase, and α-Glucosidase	Black tea, green tea, blueberry, blackberry, red cabbage, broccoli, black turtle bean, and black soybean extracts that have been semi-purified	(Glisan et al., 2017)

Emerging Research in Polyphenol-based Wound Healing Treatments:

Healing of wounds has been the concern of medical science since time immemorial, especially with reference to its effects on chronic illnesses, surgical procedures, and trauma management. Recent studies have paved way for the possibilities offered by natural molecules in healing processes, among which polyphenols have gained primary importance. Many fruits, vegetables, tea, and herbs contain polyphenols, a broad class of phytochemicals. They have established antibacterial, anti-inflammatory, and antioxidant properties. Interest in using these chemicals as therapeutic treatments is growing as we seek to better understand the complex molecular pathways driving wound healing (Lee et al., 2010). Future research elucidates the intricate functions of polyphenols in wound healing. They promote cellular activities essential to tissue healing, such as proliferation, migration, and angiogenesis, in addition to controlling oxidative stress, or the imbalance of free radicals vs antioxidants. Because of the intricate connection between these functions, polyphenols are a great target for innovative therapies, particularly when dealing with issues like infections and wounds that don't heal (Tan & Chang, 2017).

Alternative remedies are becoming increasingly important due to the growing problem of antibiotic resistance and the negative consequences of manufactured pharmaceuticals. Polyphenol-based therapies represent a natural treatment that matches with the increasing popularity of integrative and holistic health methods. Studies are more focussed on knowing how certain polyphenols like flavonoids, tannins, and phenolic acids can be used in topical remedies or systemic treatments to improve the outcomes of wound healing (Lee et al., 2020). The synergistic activity of these chemicals with other healing agents may lead to more effective treatment regimens, as is evident when we examine the specifics of polyphenol-based wound healing therapies.

This section intends to discuss the recent developments in this area, providing information regarding the mechanisms of action, future prospects, and possible applications and future directions of polyphenol research in wound healing. By combining existing studies and emphasizing innovative uses, this chapter will facilitate greater insight into how natural products can be harnessed in contemporary medical practice to enhance patient care and healing.

Advances in nanotechnology and biotechnology to improve the delivery and efficacy of polyphenol and flavonoid-based therapies

Fruits, vegetables, tea, and wine are the main sources of polyphenols and flavonoids, which are bioactive components with antioxidant, anti-inflammatory, and anticancer properties.Despite these qualities, they have drawbacks such rapid metabolism, low bioavailability, and instability, which results in inadequate ingredient delivery. New developments in biotechnology and nanotechnology present opportunities to maximize the effectiveness and delivery of these substances. Additionally, it enables the improvement of weakly water-soluble substances' bioavailability (Sharma et al., 2021).

Challenges in Formulation and Advanced Delivery Strategies

Polyphenols and flavonoids have considerable formulation problems, including low water solubility, poor stability, fast metabolism, and restricted bioavailability, despite their promising medicinal promise. Their consistent therapeutic effectiveness and successful clinical translation are hampered by these limitations. To get around these obstacles, sophisticated delivery techniques—especially those based on nanotechnology—are being developed. These techniques improve localized distribution, increase stability, and allow for controlled and sustained release at the wound site (Sharma et al., 2021).

Nanotechnology in Polyphenol and Flavonoid Delivery

The delivery of polyphenols and flavonoids plays an important role in nanotechnology. This can be explained due to the application of nanoparticle carriers, nanostructured materials, targeted delivery systems, and improved formulation techniques.

Nanoparticle Carriers

Figure 4. Nanoparticle Lipid carriers

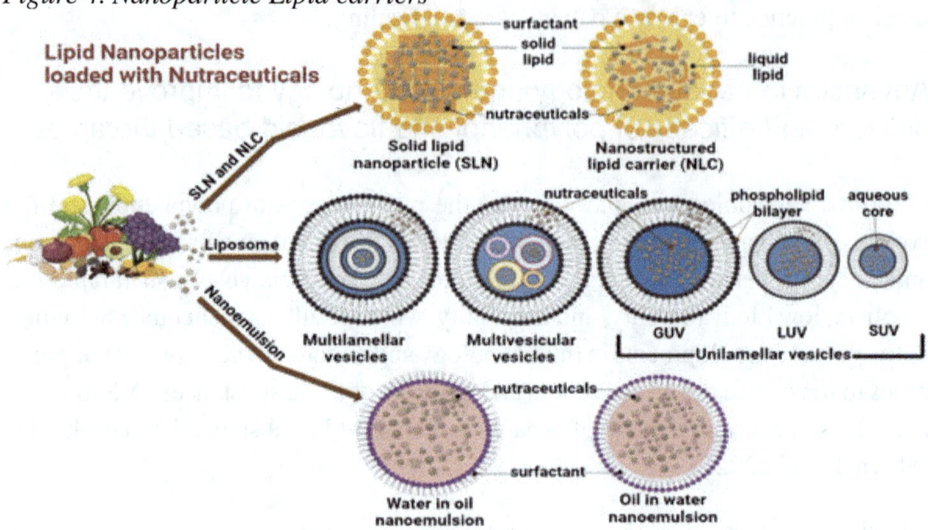

Liposomes:

Polyphenols and flavonoids have considerable formulation problems, including low water solubility, poor stability, fast metabolism, and restricted bioavailability, despite their promising medicinal promise. Their consistent therapeutic effectiveness and successful clinical translation are hampered by these limitations. To get around these obstacles, sophisticated delivery techniques—especially those based on nanotechnology are being developed. These techniques improve localized distribution, increase stability, and allow for controlled and sustained release at the wound site. Liposomes facilitate improved absorption in the gastrointestinal tract, that is essential for maximizing the bioavailability of these compounds. It is reported that when polyphenols are encapsulated within liposomes, they can achieve significantly higher plasma concentrations compared to their free counterparts, leading to enhanced therapeutic outcomes (Sharma et al., 2021).

Solid Lipid Nanoparticles (SLNs):

Figure 5. Solid Lipid Nanoparticles

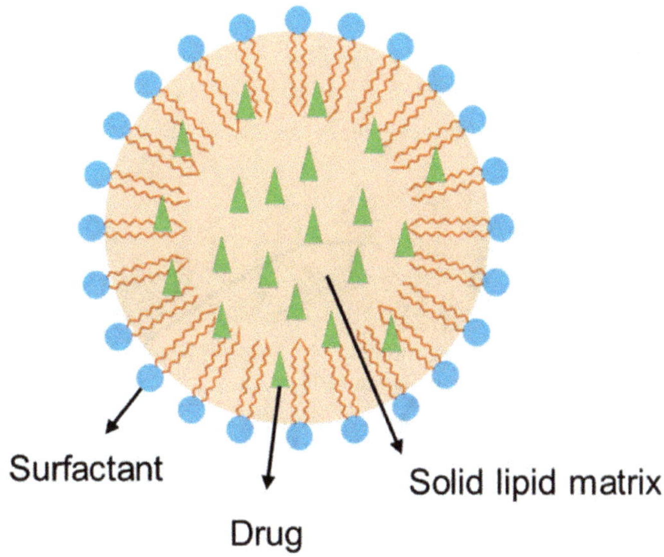

A novel approach to the delivery of polyphenolic chemicals, which are well-known for their many health advantages, is the use of solid lipid nanoparticles (SLNs). The lipid matrices that make up these nanoparticles stay solid at normal temperature, forming a stable microenvironment that shields and maintains the active substances they contain.

The capacity of SLNs to offer controlled release kinetics is one of their main benefits.This means that they can gradually release the encapsulated polyphenols over an extended period, which can enhance their bioavailability and therapeutic efficacy. The controlled release mechanism not only ensures a more sustained delivery of these compounds but also helps maintain their effective concentration in the desired target area for a longer duration (Liu et al., 2022).

Moreover, SLNs significantly improve the stability of polyphenolic compounds during both storage and administration. These delicate substances are protected from oxidative stress and degradation brought on by environmental elements like light and temperature changes by the protective lipid matrix. This protective activity is essential for preserving the polyphenols' efficacy and integrity, guaranteeing that they continue to be active over the whole delivery process. Solid lipid nanoparticles have two advantages when used as a delivery mechanism for polyphenolic compounds: they allow for the controlled and progressive release of these advantageous substances and increase their stability, which maximizes their potential health benefits.

Liposomes are spherical vesicles made of lipid bilayers that contain bioactive substances like flavonoids and polyphenols. Liposomes help the gastrointestinal system absorb these substances better, which is crucial for optimizing their bioavailability.

Polymeric Nanoparticles:

Figure 6. Nanoparticles made of polymers

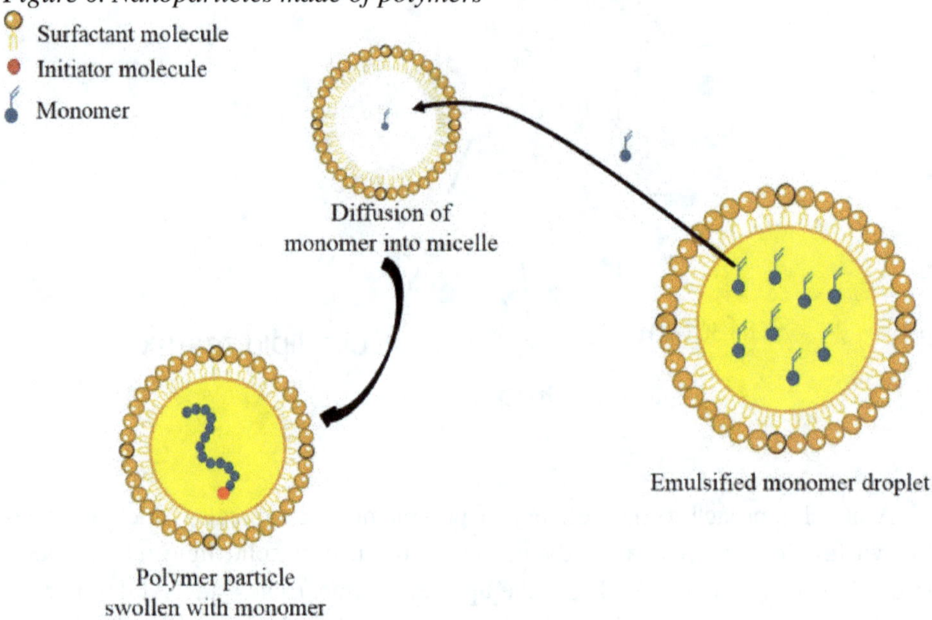

Biodegradable polymers are used to create polymeric nanoparticles, which allow for the targeted administration of medicinal substances like polyphenols.

These nanoparticles can be engineered to achieve prolonged release profiles, thereby enhancing the therapeutic effects and allowing for sustained release over time. Common materials used for polymeric nanoparticles include chitosan, a naturally occurring biopolymer derived from crustacean shells, and PLGA (polylactic-co-glycolic acid)), a synthetic polymer recognized for its biocompatibility and biodegradability. The versatility of these polymers allows for tailoring the release rates and target specific tissues or cells, optimizing the delivery of polyphenols for various therapeutic applications (Nascimento et al., 2020).

All things considered, the creation of nanoparticle carriers is a major breakthrough in the delivery of bioactive substances, enhancing their stability, absorption, and therapeutic efficacy.

Nanostructured Materials

Figure 7.

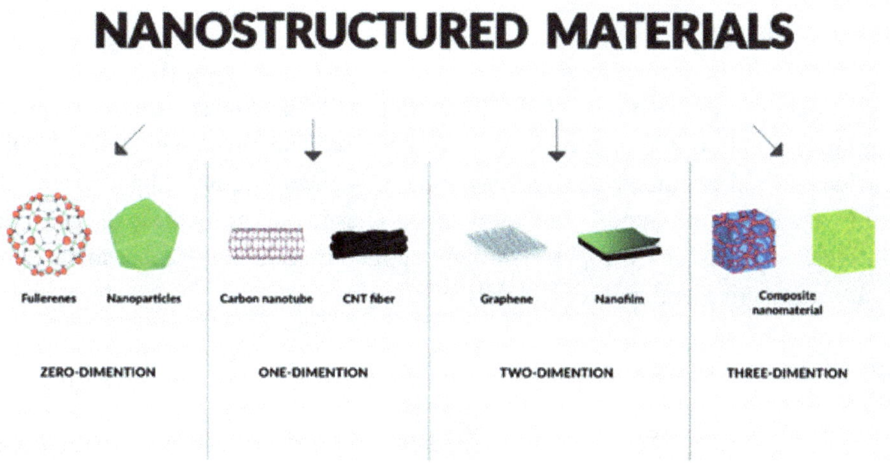

Materials with structural characteristics on the nanoscale, usually between 1 and 100 nanometers, are referred to as nanostructured materials. Because of their small size and high surface area-to-volume ratio, these materials have distinct physical, chemical, and biological characteristics when compared to their bulk counterparts.

In the context of polyphenol and flavonoid delivery, nanostructured materials can include various forms such as nanoparticles, nanofibers, and nanosheets.

Nanofibers

Nanofibers are extremely small fibers, usually ranging in diameter from 1 to 100 nanometers. Nanofibers have special qualities that make them useful in a variety of applications because of their high surface area-to-volume ratio. Using methods like electrospinning or phase separation, they can be made from a variety of materials, such as metals, ceramics, and polymers. Sodium Alginate Nanofibers: These nanofibers are used to encapsulate different flavonoids and provide a sustained release profile in addition to protecting the active ingredients.

Applications of Nanofibers:

1.Biomedical Uses: Nanofibers are utilized in tissue engineering as scaffolds because their structure can mimic the extracellular matrix, which promotes cell adhesion and growth. 2.Filtration: Due to their fine structure, nanofibers are highly effective in filtering out particles, including bacteria and viruses. This makes them valuable in air and water purification systems. 3.Textiles: Nanofibers are incorporated into fabrics to enhance durability, breathability, and antimicrobial properties.

Nanoemulsion

Nanoemulsions are mixtures of oil and water stabilized by surfactants, typically ranging from 20 to 200 nanometers in droplet size. The small size of the droplets provides several advantages, including increased surface area and enhanced bioavailability of active compounds. Nanoemulsions: These systems increase the solubility and bioavailability of hydrophobic polyphenolic compounds. Nanoemulsions can facilitate effective oral and topical delivery.

Applications of Nanoemulsions:

Pharmaceuticals: They can improve the solubility and absorption of poorly soluble drugs, increasing their effectiveness. Cosmetics: Used in skincare products for better penetration and improved hydration. Food Industry: Nanoemulsions can enhance flavor release and stability in food products, improving texture and shelf-life.

Nanosheets

Two-dimensional materials called nanosheets are usually made up of one or more atomic layers. They can be created from a variety of materials, such as graphene, transition metal dichalcogenides, and other nanostructured materials, and have a thickness on the order of nanometers (Zhang et al., 2019).

Applications of Nanosheets:

Electronics: They are good candidates for use in transistors, sensors, and other electronic devices due to their mechanical and electrical conductivity. Energy Storage: Because of its large surface area, which allows for quicker charge and discharge rates, nanosheets can enhance the performance of batteries and supercapacitors. Biomedical Applications: Because of their large surface area and capacity to alter their surfaces for targeted therapies, they are employed in drug delivery systems and imaging.

Each of these nanostructures nanofibers, nanoemulsions, and nanosheets plays a unique role in advancing technology across various fields, from medicine to materials science (Vignesh et al., 2020).

The major attributes and advantages of nanostructured materials include:

1. Increased Stability: Nanostructured materials can preserve bioactive components from the impacts of the surroundings, such as light, oxygen, and humidity, resulting in increased stability.
2. Increased Bioavailability: With the aid of nanostructures, the bioavailability of less soluble compounds, such as certain polyphenols, can significantly be enhanced. This makes their absorption within the body more effective.
3. Controlled Release: Nanostructured materials can be formulated to give controlled release of the encapsulated material. Sustained release can sustain therapeutic concentrations of the active ingredients over an extended period.
4. Targeted Delivery: These materials can be developed to target a specific cell or tissue, and thus more effectively and locally treat the condition with reduced side effects.
5. Flexibility: Nanostructured materials may be designed for a range of applications other than wound healing, such as drug delivery systems, biosensors, and diagnostic agents. With the combination of nanostructured materials with polyphenols and flavonoids, scientists are developing new approaches to improve the therapeutic effects of these bioactive compounds in different medical applications.

Targeted Delivery Systems

Nanoparticles Functionalized with Ligands:

A significant advancement in targeted drug delivery methods is ligand-functionalized nanoparticles. Ligand-functionalized nanoparticles represent a major breakthrough in targeted drug delivery techniques.Such a strategy makes therapy more effective by directing the treatment exactly where it is required, avoiding off-target effects and lowering possible side effects. For instance, in cancer therapy, conjugating monoclonal antibodies or peptide ligands with a high receptor affinity, which are overexpressed on cancer cell surfaces, can allow nanoparticles to target tumors with high specificity (McClements & Gumus, 2016).

This approach provides for the targeted delivery of therapeutics, i.e., flavonoids or chemotherapeutic drugs, into the cancerous cells directly, thereby optimizing their pharmacological action with minimal exposure to normal tissues (Wendlocha et al., 2023). Furthermore, the cellular uptake can be achieved through receptor-mediated endocytosis by employing ligand-functionalized nanoparticles, which also enhances the bioavailability and efficacy of delivered therapeutic agents (Zhang et al., 2015; Raghavendra & Mohan, 2018). Overall, this targeted delivery technique has immense potential for even more effective treatment protocols in oncology and other therapeutic fields, which demonstrates the revolutionary potential of nanotechnology in healthcare (Davis et al., 2008).

Advanced Formulation Methods

Advanced formulation processes are important for enhancing the bioavailability, solubility, and stability of bioactive ingredients like flavonoids and polyphenols that are vital for human health.

Co-crystallization and Complexation

Co-crystallization and complexation techniques involve the complexation with carriers like cyclodextrins. These carriers can significantly enhance the solubility and stability of flavonoids and polyphenols and make them more bioavailable in most applications, particularly in pharmaceuticals and nutraceuticals (Allen & Cullis, 2013). The inclusion of cyclodextrins can help encapsulate hydrophobic compounds, thereby increasing their water solubility and thermal stability (Sriwatanametanon et al., 2015).

Nano-encapsulation

Nano-encapsulation is a potential technique that protects sensitive molecules from degradation during digestion, thereby improving bioavailability. Nano-encapsulation can be utilized to avoid degradation of molecules in harsh conditions of the gastrointestinal tract by encapsulating polyphenols within nanocarriers, thereby improving their therapeutic potential (Rodriguez et al., 2008). The technique can offer controlled release profiles so that the bioactive molecules can be effectively delivered at the target site in the body (Fathi et al., 2019).

Biotechnology Applications

1. Biotransformation

Biotransformation is microbial fermentation or enzymatic processing that can convert polyphenols into more bioavailable forms. Such conversion makes these molecules more bioactive and promotes better absorption in the gastrointestinal tract, causing more therapeutic effects (Zhang et al., 2020).

Various microorganisms can be employed in such actions, thus converting natural polyphenols into bioactive metabolites (Possemiers et al., 2013).

2. Plant Genetic Engineering

Plant genetic engineering offers a platform for increasing the concentration of polyphenols or the biosynthesis of specific flavonoids, which results in their increased presence in health-promoting compounds. Biotechnology provides the possibility to produce transgenic plants overexpressing polyphenol biosynthetic genes, which yields a more affluent source of these useful compounds (Hwang et al., 2018).

3. Tailored Delivery Systems

Molecular Farming: This comprises culturing polyphenolic compounds at elevated concentrations in targeted plants using biotechnological advancements. This method allows for more effective harvesting and extraction processes so that high-quality plant-derived compounds can be easily accessed (Ghosh et al., 2018).

Intelligent Delivery Systems: Conjoining nanotechnology with responsive stimuli materials provides a means for sophisticated delivery systems to be synthesized. Such delivery systems can leverage temperature-sensitive or pH-sensitive carriers, which allow them to discharge polyphenols on the basis of certain physiological situations. By incorporating this type of targeted discharge method, there may be heightened curative properties displayed by the polyphenols when they respond appropriately (Ghosh et al., 2020).

Obstacles and Limitations

1.Safety and Toxicity: The design of new nanocarriers requires a comprehensive analysis of their safety profiles and cytotoxicity. Knowledge of nanomaterials' interactions with biological systems is crucial to ascertain consumer safety (Paredes et al., 2021).

2.Regulatory Hurdles: Regulatory issues surrounding nanomedicine are complicated, and compliance with these regulations while finding a way around the framework constitutes a major problem in the generation and market introduction of new therapy (Ribeiro et al., 2021).

3.Scalability: For advanced formulations and biotechnological processes to be commercially successful, the production of nanomaterials and other active ingredients needs to be scalable. For these cutting-edge technologies to be successfully implemented, they must be developed at a reasonable cost and made widely accessible (Duran et al., 2019).

Co-crystallization and Complexation: Polyphenols and flavonoids can be made more soluble and stable by employing techniques that create complexes with cyclodextrins or other carriers (Ranjha et al., 2022).

Nano-encapsulation: It safeguards vulnerable compounds from digestion, resulting in increased bioavailability.

Biotechnology Applications

1. Biotransformation: Microbial fermentation or enzymatic reactions can transform polyphenols into more bioavailable forms, making them more bioactive and enabling them to be absorbed better in the gastrointestinal tract (Lohray et al., 2006).
2. Plant Genetic Engineering: Genetic alteration of plants for increasing their content of polyphenols or generating certain flavonoids can translate into greater supply of these beneficial compounds. Transgenic plants could be engineered, for instance, to overproduce genes responsible for polyphenol biosynthesis (Kim et al., 2019).
3. Tailor-made Delivery Systems:

Molecular Farming: Engineered production of polyphenolic compounds in high concentration in chosen plants using biotechnology makes for more effective harvest and extraction.

Smart Delivery Systems: The integration of nanotechnology with stimuli-responsive materials, like pH-sensitive or temperature-sensitive carriers, has the potential to develop delivery systems that release polyphenols based on certain physiological conditions (Kaur et al., 2021).

4. Safety and Toxicity: The safety profiles of new nanocarriers and the possible toxicity of nanomaterials need to be carefully assessed.
5. Regulatory Barriers: It may be difficult to navigate the regulatory environment for nanomedicine, and achieving compliance while taking new therapies to the market is a serious challenge.
6. Scalability: Scalability of the production of nanomaterials and biotech processes is needed to make them available and affordable (Trottier et al., 2018).

Exploration of combination therapies with polyphenols, other bioactive compounds, and traditional wound healing treatments.

WOUND HEALING: A MULTIFACETED BIOLOGICAL PROCESS

Wound healing is a sophisticated and intricate biological process that encompasses several stages, including hemostasis, inflammation, proliferation (tissue formation), and remodeling. Traditional wound management approaches primarily employ topical agents, such as antiseptics and dressings, to foster healing (Fu et al., 2020). However, recent advancements suggest that incorporating polyphenols and other bioactive compounds into wound healing therapies could significantly enhance recovery outcomes. This contemporary integrative method takes advantage of the distinct characteristics of these compounds particularly, antioxidative, anti-inflammatory, and antibacterial properties creating a synergy of effects that foster faster and more efficient wound healing (Kelikian et al., 2021).

Polyphenols in Wound Healing

Flavonoids, phenolic acids, and tannins are some groups of polyphenols, which are a diverse group of naturally occurring compounds found in many plants (Ahmed et al., 2018). It has been demonstrated that polyphenols are of importance because they may contribute to wound healing through a number of ways, including:

1. **Antioxidant Activity:** Polyphenols are recognized to successfully scavenge free radicals, unstable molecules which are capable of creating oxidative stress on the site of the wound. By eliminating oxidative stress, polyphenols avoid further damage to the tissues and create a better environment for the healing process.
2. **Anti-inflammatory Action:**These compounds have the ability to control the expression and function of inflammatory cytokines, which are important signaling molecules that cause inflammation.By reducing inflammation, polyphenols promote the restoration to the proliferative stage of wound healing more rapidly, mitigating the pain and discomfort of the acute wound.
3. **Angiogenesis Promotion:** The development of new blood vessels, or angiogenesis, is crucial for supplying the wound site with the nutrition and oxygen it needs. Angiogenesis, which is crucial for effective wound healing and tissue regeneration, has been shown to be significantly enhanced by substances like quercetin, a flavonoid that has been thoroughly studied.
4. **Collagen Synthesis:** Collagen is one of the main structure proteins of wound healing, contributing tensile strength and integrity to healed tissue. Some polyphenols can enhance the activation of fibroblasts, the cells that produce collagen, and subsequently enhance the deposition of collagen and the quality of the healed tissue (Basak et al., 2021).

Combination with Other Bioactive Compounds

Figure 8.

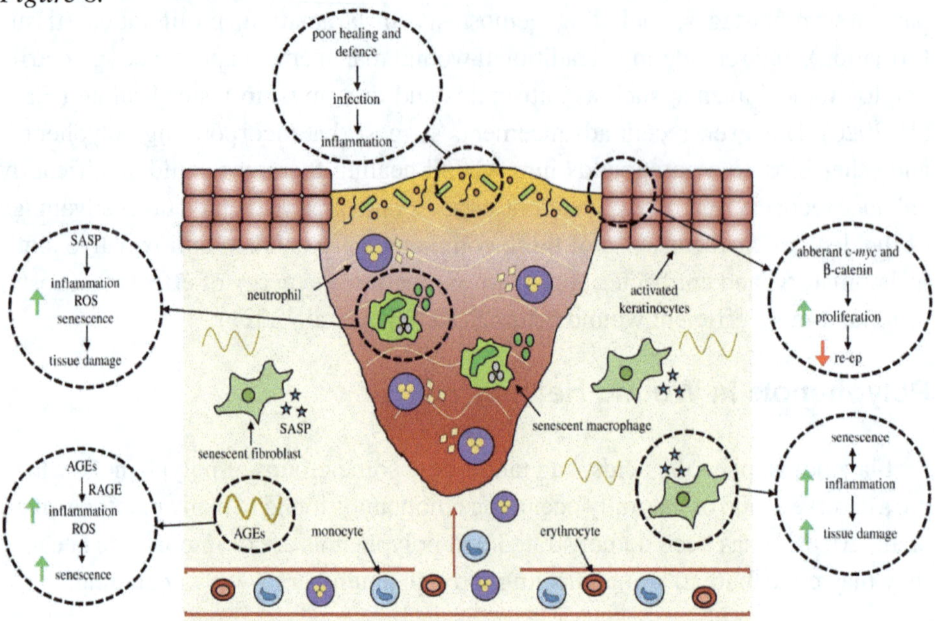

The therapeutic effects of polyphenols can be significantly amplified when combined with other bioactive compounds. This synergy can enhance wound healing potentials in several ways:

Essential Oils: Natural oils such as lavender and tea tree oil are renowned for their antibacterial properties. When these essential oils are used in conjunction with polyphenols, they not only augment antimicrobial activity but also offer soothing benefits that can alleviate discomfort associated with wounds (Li et al., 2016).

Probiotics: Probiotics help balance the microbiome, which is important for preventing infections in wounds. The integration of probiotics with polyphenols could work synergistically to enhance tissue repair and further reduce inflammation by promoting a healthy inflammatory response (Zhuang et al., 2020).

Integration with Traditional Treatments

Figure 9.

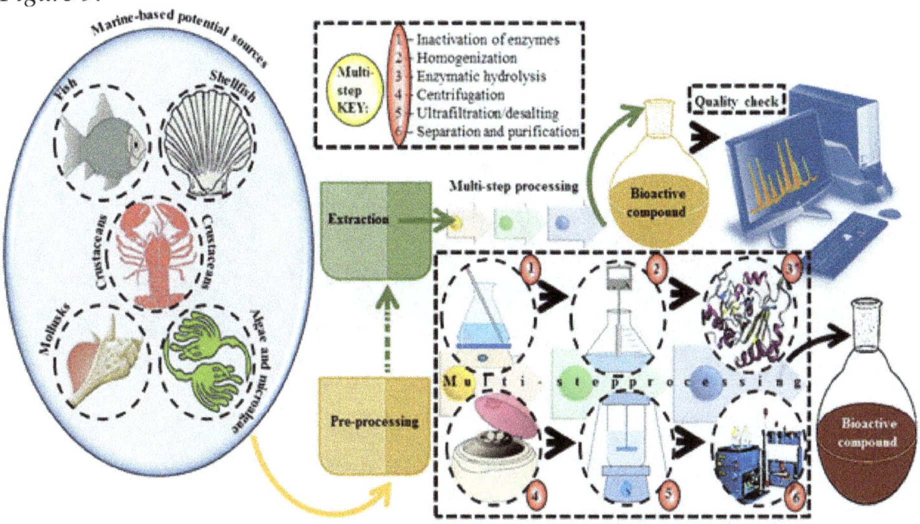

Figure 10.

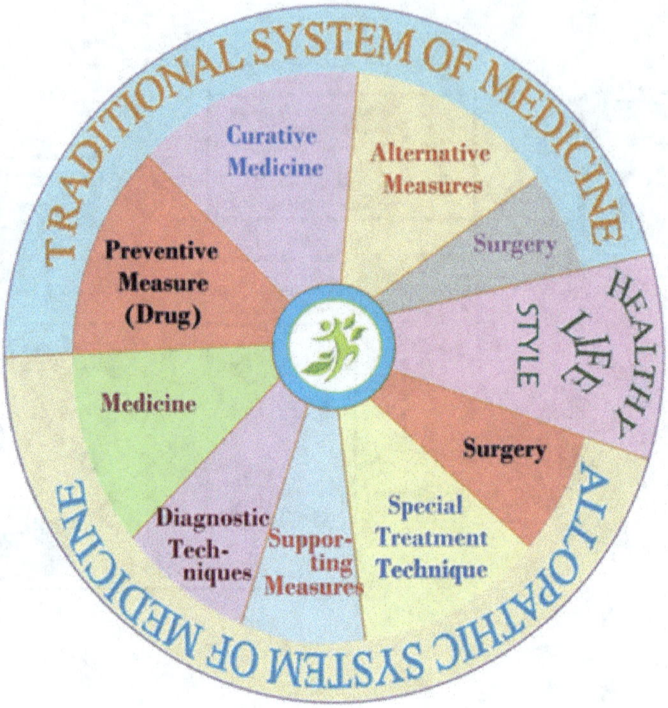

The interaction of polyphenols with other bioactive compounds with known wound care therapies can result in significant enhancements in clinical outcomes. Some exciting uses include (Gang et al., 2024).

1. **Dressings and Hydrogels:** The delivery of polyphenols into sophisticated hydrogel dressings has the potential to provide a moist wound environment while, at the same time, releasing bioactive agents that enhance healing. Additionally, dressings incorporating silver with polyphenolic extracts provide antimicrobial protection while accelerating the healing process.
2. **Topical Uses**: Polyphenol and essential oil-rich creams, ointments, or gels may be applied directly to wounds. These preparations are formulated to promote wound healing at the same time as they ease pain and discomfort, offering a two-way benefit.
3. **Nutrition:** The body's systemic wound healing mechanisms may be aided by consuming foods high in polyphenols, such as fruits, vegetables, tea, and spices. This dietary strategy supports holistic wound care approaches, offering a multifaceted approach to complete recovery (Hwang & Lee, 2021).

Challenges and Considerations

While the use of polyphenols in wound healing procedures is encouraging, there are several obstacles that must be addressed to optimize their efficacy:

1. **Formulation Stability:** Stability and bioavailability of polyphenols and other bioactive molecules in topical formulations are essential. Instability can undermine their therapeutic potential, requiring careful formulation and testing.
2. **Individual Heterogeneity:** The individual may react in a varying manner toward combination therapies based on individual biological and physiological differences. It emphasizes the necessity to adopt customized forms of treatment to achieve maximum therapeutic benefit from wound management.
3. **Regulatory Approvals:** Approving the safety, efficacy, and clinical benefit of such combinations involves a large amount of testing with strict regulatory compliance, a time-consuming and complicated task.

Potential for Personalized Wound Care:

Personalized Polyphenol-Based Wound Care

Treatments can now be tailored to an individual's genetic composition thanks to advancements in nutrigenomics and personalized medicine (Mohammadi et al., 2017). This is particularly significant when it comes to polyphenol-based therapy, as genetic polymorphisms are critical to the metabolism, bioavailability, and therapeutic activity of these compounds (Cavanagh & Miller, 2019).

1. Genetic Variation and Polyphenol Metabolism

The human body undergoes extensive metabolism of polyphenols through phase I and phase II enzymes as well as gut microbiota-mediated responses. Cytochrome P450s (CYPs), UDP-glucuronosyltransferases (UGTs), sulfotransferases (SULTs), and catechol-O-methyltransferase (COMT) are a few important enzymes. The activity of these enzymes and, hence, the biological availability of active polyphenol metabolites may be influenced by genetic variation in these genes Patients with low-activity COMT genotypes, for example, will have higher systemic levels of certain flavonoids, which will result in more potent biological effects including anti-inflammatory and antioxidant properties that are crucial for wound healing (Dargère et al., 2020). Similarly, the effectiveness of polyphenol breakdown and conversion into bioactive metabolites can be impacted by variations in gut microbiota makeup, which is partially influenced by host genetics.

2. Implications for Wound Healing

As wound healing is a delicately balanced interplay of oxidative stress regulation, inflammation, angiogenesis, and tissue remodeling, the efficacy with which polyphenols affect these processes could be quite different in different individuals. It's possible that not every patient will benefit from a "one-size-fits-all" approach to polyphenol-based wound treatment.In genetically predisposed individuals with increased oxidative stress or compromised inflammatory responses, polyphenols might yield greater benefits. On the other hand, poor metabolizers might need dose-down or alternative delivery systems (e.g., nanoformulations) to attain therapeutic levels at the wound site (Smith & Kaur, 2022).

3. Personalized Treatment Strategies

Personalized wound care is a concept that entails integrating a patient's genetic history, metabolic status, and lifestyle considerations into an overall treatment regimen. This might involve:

Pharmacogenetic testing to detect relevant polymorphisms influencing polyphenol metabolism. Application of targeted delivery systems, e.g., nanocarriers functionalized to specific genetic or inflammatory markers. Incorporation of dietary polyphenol guidelines according to individual absorption and metabolic potential. Modulation of the microbiome (e.g., through prebiotics or probiotics) to increase polyphenol biotransformation in the gut.

Regulatory Considerations and Future Clinical Trials:

1. Regulatory Approval Pathway

A strict regulatory framework established by agencies such as the U.S. Food and Drug Administration (FDA), the European Medicines Agency (EMA), and India's Central Drugs Standard Control Organization (CDSCO) is necessary for the development of polyphenol and flavonoid-based wound healing products from discovery to market. These goods can be classified as topical medicines, medical devices, or nutraceuticals, each of which has distinct regulatory requirements (Zahra et al., 2024).

Steps in the Regulatory Pathway:

Preclinical Research: To determine the safety profile, understand the mechanism of action, and assess the effectiveness of the wound healing formulations, this first stage involves both in vitro (in the lab) and in vivo (on animals) testing. To ensure that the substances are safe and function as intended, this research is crucial.

Formulation Development: Here, scientists concentrate on making the formulation stable and bioavailable for delivery of active ingredients to the injury site (Rubeena et al., 2021; Bhattacharya & Hossain, 2021). This is achieved by using proper excipients and creating a delivery system with optimum therapeutic potential.

Investigational New Drug (IND) Application: Following positive preclinical results, an IND application has to be filed, detailing the design, methodology, and ethical issues of the proposed clinical study.

Clinical Trial can be classified as follows.

Phase I Clinical Trial: Analyzing the safety profile and determining the ideal dosage range for the polyphenol preparation are the main goals of the Phase I trial, which is conducted on healthy volunteers. During this stage, pharmacokinetics may also be assessed to determine the substances' absorption, distribution, metabolization, and excretion.

Phase II Clinical Trial: A larger cohort of patients with the target disease is treated in this phase, with an eye toward further evaluating effectiveness and tracking any side effects. Information collected during this trial offers useful insight into therapeutic outcomes and informs dose adjustments needed.

Phase III Clinical Trial: Comprises large and multicenter randomized controlled trials (RCTs) with hundreds to thousands of patients. Phase III is designed to establish the product's efficacy and safety in a variety of populations and different demographics to obtain key data for regulatory approval.

New Drug Application (NDA) or Device Application: A final NDA or device application brings together all the data and evidence gathered in earlier phases to seek approval for marketing the product.

Post-Marketing Surveillance (Phase IV): Following approval, Phase IV studies are carried out to track long-term safety and effectiveness in the general population. This phase is important for detecting rare side effects and ensures continued compliance with regulatory requirements.

For polyphenol products, there is a specific focus on:

Standardization: Maintaining consistent quality and concentration of active ingredients in various batches. Source Consistency: Using dependable sourcing techniques to ensure uniformity in the geographic and botanical origin of the bioactive compounds (Rudrapal et al., 2024). Active Ingredient Quantification: Using sophisticated analytical techniques to measure accurately the concentration of the active components, which may vary widely due to extraction methods (Zhang et al., 2022).

2. Requirement for Strong Clinical Trials

1. Numerous polyphenols, including resveratrol, quercetin, and curcumin, exhibit great promise in preclinical models of wound healing; however, there are few well-conducted human clinical trials that support these findings.
2. Effectiveness in Varying Populations: The therapeutic action of polyphenols may be affected by a number of factors such as genetic heterogeneity, dietary intake, and environmental exposures. For instance, variations in gut microbiota between individuals can significantly influence the bioavailability of these compounds, affecting their efficacy (Zhang et al., 2022).
3. Safety & Dosing: While polyphenols are generally considered safe, it is important to perform systematic studies of possible dose-dependent toxicity and interactions with other drugs. Having a clear safety profile will assist in guiding dosage recommendations for clinical use. Formulation Evaluation: Various formulations, nanoformulations, hydrogels, and dressings infused with somethingneed strict validation to determine their ability for dermal absorption, long-term stability, and patient compliance in actual practice (Lee et al., 2023).

Clinical Trial Pyramid for Polyphenol Therapies:

The pyramid of clinical evidence required for polyphenol therapy validation is as follows:

1. Phase IV: Post-market surveillance involves real-world data, side effect incidence, and assessment of rare risks related to the therapy.
2. Phase III: Multi-center RCTs with a large patient cohort that yield sound evidence of the product's safety and efficacy.
3. Phase II: Proof of concept trials in affected patient populations to identify further therapeutic benefit.
4. Phase I: First safety and pharmacokinetic evaluations in healthy volunteers.
5. Preclinical: Data from cell lines and animal models used to underpin early-stage development.

Each phase of clinical trial must include: Standardized Extracts/Doses: Reproducibility of results depends on consistent formulations (O'Brien et al., 2022). Validated Biomarkers of Wound Healing: These may include quantifiable measures such as collagen deposition and angiogenesis that show tangible evidence of effectiveness (Zhang et al., 2020). Patient-Reported Outcomes: Including participants' feedback on pain management and quality of life increases understanding of the treatment effect.

3. Special Considerations

Regulatory Category: The designation of polyphenolic products will depend—classified as drugs, cosmeceuticals, or medical devicesupon their intended application and claims regarding their therapeutic effect. The category may have a major impact on the timelines to approval and individual test requirements.

Good Manufacturing Practice (GMP): Compliance with rigid GMP requirements at every step of the process of production is imperative to warrant the quality and reproducibility of the extraction of natural products, which are paramount for ensuring regulatory compliance.

Ethical Concerns: It is important to highlight informed consent in clinical trials so that the participants are clearly aware of the nature of the study and risk involved. Importantly, inclusivity and diversity in clinical trials are also very important, especially for under-represented populations like diabetic, elderly, or immunocompromised patients who can have differing responses to wound healing treatments (Wu et al., 2021; Miller et al., 2022).

In summary, navigating the regulatory world for polyphenol-derived wound therapies is a multifaceted strategy that includes strict scientific scrutiny and compliance with ethical, safety, and quality guidelines.

To guarantee the effective clinical adoption of polyphenol-based wound healing therapies, substantial, well-structured clinical trials and clearly defined regulatory standards are required, even though preclinical and early clinical research indicate encouraging outcomes.

5. CONCLUSION

The exploration and integration of combination therapies involving polyphenols, other bioactive compounds, and traditional wound healing treatments represent a promising avenue for enhancing wound management strategies. These multifaceted approaches exploit the synergistic potential of multiple compounds, leading to quicker and more efficient healing processes. Continued research and clinical trials are essential for validating these strategies, determining optimal applications, and ultimately improving patient outcomes in wound care. In conclusion, managing the regulatory environment for polyphenol-based wound treatments requires a multimodal strategy that includes strict adherence to safety, ethical, and quality criteria as well as rigorous scientific examination.

REFERENCES

Abbas, M., Hussain, T., Arshad, M., Ansari, A. R., Irshad, A., Nisar, J., Hussain, F., Masood, N., Nazir, A., & Iqbal, M. (2019). Wound healing potential of curcumin cross-linked chitosan/polyvinyl alcohol. *International Journal of Biological Macromolecules, 140*, 871–876. DOI: 10.1016/j.ijbiomac.2019.08.153 PMID: 31437503

Ahmed, O. M., Mohamed, T., Moustafa, H., Hamdy, H., Ahmed, R. R., & Aboud, E. (2018). Quercetin and low level laser therapy promote wound healing process in diabetic rats via structural reorganization and modulatory effects on inflammation and oxidative stress. *Biomedicine and Pharmacotherapy, 101*, 58–73. DOI: 10.1016/j.biopha.2018.02.040 PMID: 29477473

Ajmal, G., Bonde, G. V., Thokala, S., Mittal, P., Khan, G., Singh, J., Pandey, V. K., & Mishra, B. (2019). Ciprofloxacin HCl and quercetin functionalized electrospun nanofiber membrane: Fabrication and its evaluation in full thickness wound healing. *Artificial Cells, Nanomedicine, and Biotechnology*, *47*(1), 228–240. DOI: 10.1080/21691401.2018.1548475 PMID: 30688107

Al-Roujayee, A. S. (2017). Naringenin improves the healing process of thermally-induced skin damage in rats. *The Journal of International Medical Research*, *45*(2), 570–582. DOI: 10.1177/0300060517692483 PMID: 28415935

Allen, T. M., & Cullis, P. R. (2013). Liposomal drug delivery systems: From concept to clinical applications. *Advanced Drug Delivery Reviews*, *65*(1), 36–48. DOI: 10.1016/j.addr.2012.09.037 PMID: 23036225

Almeida, J., Benvegnú, D., Boufleur, N., Reckziegel, P., Barcelos, R., Coradini, K., de Carvalho, L. M., Bürger, M. E., & Beck, R. C. R. (2012). Hydrogels containing rutin intended for cutaneous administration: Efficacy in wound healing in rats. *Drug Development and Industrial Pharmacy*, *38*(7), 792–799. DOI: 10.3109/03639045.2011.628676 PMID: 22066462

Amanat, S., Taymouri, S., Varshosaz, J., Minaiyan, M., & Talebi, A. (2020). Carboxymethyl cellulose-based wafer enriched with resveratrol-loaded nanoparticles for enhanced wound healing. *Drug Delivery and Translational Research*, *10*(5), 1241–1254. DOI: 10.1007/s13346-020-00711-w PMID: 31981141

Amini, S. M., & Akbari, A. (2019). Metal nanoparticles synthesis through natural phenolic acids. *IET Nanobiotechnology / IET*, *13*(8), 771–777. DOI: 10.1049/iet-nbt.2018.5386 PMID: 31625516

Asfour, M. H., Elmotasem, H., Mostafa, D. M., & Salama, A. A. (2017). Chitosan based Pickering emulsion as a promising approach for topical application of rutin in a solubilized form intended for wound healing: In vitro and in vivo study. *International Journal of Pharmaceutics, 534*(1-2), 325–338. DOI: 10.1016/j.ijpharm.2017.10.044 PMID: 29074391

Awad, H. M., Boersma, M. G., Boeren, S., van der Woude, H., van Zanden, J., van Bladeren, P. J., Vervoort, J., & Rietjens, I. M. C. M. (2002). Identification of o-quinone/quinone methide metabolites of quercetin in a cellular in vitro system. *FEBS Letters, 520*(1-3), 30–34. DOI: 10.1016/S0014-5793(02)02754-0 PMID: 12044865

Bagher, Z., Ehterami, A., Safdel, M. H., Khastar, H., Semiari, H., Asefnejad, A., Davachi, S. M., Mirzaii, M., & Salehi, M. (2020). Wound healing with alginate/chitosan hydrogel containing hesperidin in rat model. *Journal of Drug Delivery Science and Technology, 55*, 101379. DOI: 10.1016/j.jddst.2019.101379

Bai, F., Wang, Y., Zhang, S., Wang, Y., Zhang, J., Cao, J., & Sun, L. (2020). Caffeoyl substitution changes the inhibition mode of tartaric acid against α-amylase: Analysis of the enzyme inhibition by four caffeic and tartaric acid derivates. *Lebensmittel-Wissenschaft + Technologie, 133*, 109942. DOI: 10.1016/j.lwt.2020.109942

Basak, S., Ghosh, N., & Saha, S. (2021). Anti-inflammatory mechanisms of polyphenols and their application in medicinal chemistry: A review. *Medicinal Chemistry (Shariqah, United Arab Emirates), 17*(1), 20–35. DOI: 10.2174/15734064156662 10715110612

Belay, A., & Gholap, A. V. (2009). Characterization and determination of chlorogenic acids (CGA) in coffee beans by UV-VIS spectroscopy. *African Journal of Pure and Applied Chemistry, 3*(11), 234–240.

Benedek, B., Kopp, B., & Melzig, M. F. (2007). Achillea millefolium L. s.l.–Is the anti-inflammatory activity mediated by protease inhibition? *Journal of Ethnopharmacology, 113*(2), 312–317. DOI: 10.1016/j.jep.2007.06.014 PMID: 17689902

Bhattacharya, A., & Hossain, S. (2021). Nutritional support in wound healing: The role of polyphenolic-rich diets. *Nutrition Research Reviews, 34*(2), 146–158. DOI: 10.1017/S0954422421000014

Brudzynski, K., & Maldonado-Alvarez, L. (2015). Polyphenol-protein complexes and their consequences for the redox activity, structure and function of honey. A current view and new hypothesis—A review. *Polish Journal of Food and Nutrition Sciences, 65*(2), 71–80. DOI: 10.1515/pjfns-2015-0030

Burud, A., Galatage, S., Manjappa, A., Salawi, A., Nadaf, S., Holam, M., Harale, S., Kumbhar, R., Peram, M. R., & Suryadevara, V. (2024). Sericin stabilized emulgel for improving therapeutic efficacy of quercetin in treatment of diabetic wound healing. *Journal of Dispersion Science and Technology*, 1–21. Advance online publication. DOI: 10.1080/01932691.2024.2444980

Caponio, G. R., Lippolis, T., Tutino, V., Gigante, I., De Nunzio, V., Milella, R. A., Gasparro, M., & Notarnicola, M. (2022). Nutraceuticals: Focus on anti-inflammatory, anti-cancer, antioxidant properties in gastrointestinal tract. *Antioxidants*, *11*(7), 1274. DOI: 10.3390/antiox11071274 PMID: 35883765

Castangia, I., Nácher, A., Caddeo, C., Valenti, D., Fadda, A. M., Díez-Sales, O., Ruiz-Saurí, A., & Manconi, M. (2014). Fabrication of quercetin and curcumin bionanovesicles for the prevention and rapid regeneration of full-thickness skin defects on mice. *Acta Biomaterialia*, *10*(3), 1292–1300. DOI: 10.1016/j.actbio.2013.11.005 PMID: 24239901

Cavanagh, H. M. A., & Miller, J. J. (2019). Essential oils: A new perspective on antimicrobial activity. *International Journal of Aromatherapy*, *19*(1-2), 1–7.

Cerdá, B., Cerón, J. J., Tomás-Barberán, F. A., & Espín, J. C. (2003). Repeated oral administration of high doses of the pomegranate ellagitannin punicalagin to rats for 37 days is not toxic. *Journal of Agricultural and Food Chemistry*, *51*(11), 3493–3501. DOI: 10.1021/jf020842c PMID: 12744688

Chen, G., He, L., Zhang, P., Zhang, J., Mei, X., Wang, D., Zhang, Y., Ren, X., & Chen, Z. (2020). Encapsulation of green tea polyphenol nanospheres in PVA/alginate hydrogel for promoting wound healing of diabetic rats by regulating PI3K/AKT pathway. *Materials Science and Engineering C*, *110*, 110686. DOI: 10.1016/j.msec.2020.110686 PMID: 32204114

Cheng, M., Zhang, L., Zhang, H., Li, X., Wang, Y., Xia, F., Wang, B., Cai, R., Guo, Z., Zhang, Y., Ji, Y., Sun, C., Feng, X., Lei, L., Yang, Y., Han, W., & Gu, J. (2018). An ointment consisting of the phage lysin LysGH15 and apigenin for decolonization of methicillin-resistant Staphylococcus aureus from skin wounds. *Viruses*, *10*(5), 244. DOI: 10.3390/v10050244 PMID: 29734776

Cheng, M., Zhang, L., Zhang, H., Li, X., Wang, Y., Xia, F., Wang, B., Cai, R., Guo, Z., Zhang, Y., Ji, Y., Sun, C., Feng, X., Lei, L., Yang, Y., Han, W., & Gu, J. (2018). An ointment consisting of the phage lysin LysGH15 and apigenin for decolonization of methicillin-resistant Staphylococcus aureus from skin wounds. *Viruses*, *10*(5), 244. DOI: 10.3390/v10050244 PMID: 29734776

Choudhary, A., Kant, V., Jangir, B. L., & Joshi, V. (2020). Quercetin loaded chitosan tripolyphosphate nanoparticles accelerated cutaneous wound healing in Wistar rats. *European Journal of Pharmacology*, *880*, 173172. DOI: 10.1016/j.ejphar.2020.173172 PMID: 32407724

Coleridge-Smith, P., Lok, C., & Ramelet, A.-A. (2005). Venous leg ulcer: A meta-analysis of adjunctive therapy with micronized purified flavonoid fraction. *European Journal of Vascular and Endovascular Surgery*, *30*(2), 198–208. DOI: 10.1016/j.ejvs.2005.04.017 PMID: 15936227

Coleridge-Smith, P. D. (2009). Leg ulcer treatment. *Journal of Vascular Surgery*, *49*(3), 804–808. DOI: 10.1016/j.jvs.2009.01.003 PMID: 19268785

Criollo-Mendoza, M. S., Contreras-Angulo, L. A., Leyva-López, N., Gutiérrez-Grijalva, E. P., Jiménez-Ortega, L. A., & Heredia, J. B. (2023). Wound healing properties of natural products: Mechanisms of action. *Molecules (Basel, Switzerland)*, *28*(2), 598. Advance online publication. DOI: 10.3390/molecules28020598 PMID: 36677659

Damrongrungruang, T., Paphangkorakit, J., Limsitthichaikoon, S., Khampaenjiraroch, B., Davies, M. J., Sungthong, B., & Priprem, A. (2021). Anthocyanin complex niosome gel accelerates oral wound healing: In vitro and clinical studies. *Nanomedicine; Nanotechnology, Biology, and Medicine*, *37*, 102423. DOI: 10.1016/j.nano.2021.102423 PMID: 34214683

Dargère, S., Dube, M., & Gervais, M. (2020). Probiotic treatment improves wound healing in diabetic mice. *Microorganisms*, *8*(6), 835. Advance online publication. DOI: 10.3390/microorganisms8060835 PMID: 32498344

Davis, M. E., Chen, Z. G., & Golomb, H. M. (2008). Policy forum: Nanotechnology. *Science*, *319*(5869), 1673–1675.

Dev, S. S., Farghadani, R., Abidin, S. A. Z., Othman, I., & Naidu, R. (2023). Flavonoids as receptor tyrosine kinase inhibitors in lung cancer. *Journal of Functional Foods*, *110*, 105845. Advance online publication. DOI: 10.1016/j.jff.2023.105845

Doerge, D. R., & Chang, H. C. (2002). Inactivation of thyroid peroxidase by soy isoflavones, in vitro and in vivo. *Journal of Chromatography. B, Analytical Technologies in the Biomedical and Life Sciences*, *777*(1-2), 269–279. DOI: 10.1016/S1570-0232(02)00214-3 PMID: 12270219

Duda-Chodak, A., & Tarko, T. (2023). Possible side effects of polyphenols and their interactions with medicines. *Molecules (Basel, Switzerland)*, *28*(6), 2536. DOI: 10.3390/molecules28062536 PMID: 36985507

Dunnick, J. K., & Hailey, J. R. (1992). Toxicity and carcinogenicity studies of quercetin, a natural component of foods. *Toxicological Sciences*, *19*(3), 423–431. DOI: 10.1093/toxsci/19.3.423 PMID: 1459373

Duran, N.. (2019). Regulatory challenges for nanomedicine: The demand for clarity in the regulatory landscape. *Nanomedicine (London)*, *14*(4), 400–402.

Durazzo, A., Lucarini, M., Souto, E. B., Cicala, C., Caiazzo, E., Izzo, A. A., & Santini, A. (2019). Polyphenols: A concise overview on the chemistry, occurrence, and human health. *Phytotherapy Research*, *33*(9), 2221–2243. DOI: 10.1002/ptr.6419 PMID: 31359516

Eo, H., Lee, H.-J., & Lim, Y. (2016). Ameliorative effect of dietary genistein on diabetes induced hyper-inflammation and oxidative stress during early stage of wound healing in alloxan induced diabetic mice. *Biochemical and Biophysical Research Communications*, *478*(3), 1021–1027. DOI: 10.1016/j.bbrc.2016.07.039 PMID: 27431618

Eseberri, I., Trepiana, J., Léniz, A., Gómez-García, I., Carr-Ugarte, H., González, M., & Portillo, M. P. (2022). Variability in the beneficial effects of phenolic compounds: A review. *Nutrients*, *14*(9), 1925. DOI: 10.3390/nu14091925 PMID: 35565892

Fathi, M.. (2019). Nano-encapsulation of bioactive compounds improves bioavailability and efficacy. *Food Chemistry*, *276*, 817–828.

Ferreira, A., Lisboa, P., Oliveira, K., Lima, L., Barros, I., & Carvalho, D. (2002). Inhibition of thyroid type 1 deiodinase activity by flavonoids. *Food and Chemical Toxicology*, *40*(7), 913–917. DOI: 10.1016/S0278-6915(02)00064-9 PMID: 12065212

Fu, J., Huang, J., Lin, M., Xie, T., & You, T. (2020). Quercetin promotes diabetic wound healing via switching macrophages from M1 to M2 polarization. *The Journal of Surgical Research*, *246*, 213–223. DOI: 10.1016/j.jss.2019.09.011 PMID: 31606511

Fujita, K., Kuge, K., Ozawa, N., Sahara, S., Zaiki, K., Nakaoji, K., Hamada, K., Takenaka, Y., Tanahashi, T., Tamai, K., Kaneda, Y., & Maeda, A. (2015). Cinnamtannin B-1 promotes migration of mesenchymal stem cells and accelerates wound healing in mice. *PLoS One*, *10*(12), e0144166. DOI: 10.1371/journal.pone.0144166 PMID: 26657737

Gallelli, G., Cione, E., Serra, R., Leo, A., Citraro, R., Matricardi, P., Di Meo, C., Bisceglia, F., Caroleo, M. C., Basile, S., & Gallelli, L. (2020). Nano-hydrogel embedded with quercetin and oleic acid as a new formulation in the treatment of diabetic foot ulcer: A pilot study. *International Wound Journal*, *17*(2), 485–490. DOI: 10.1111/iwj.13299 PMID: 31876118

Gang, G., Gao, R., Zhao, H., Xu, Y., Xing, Y., Jin, X., Hong, L., Yan, S., & Shi, B. (2024). Effects of water extracts of Artemisia annua L. on rumen immune and antioxidative indexes, fermentation parameters and microbials diversity in lambs. *Frontiers in Microbiology, 15*, 1485882. Advance online publication. DOI: 10.3389/fmicb.2024.1485882 PMID: 39493850

Gao, H., Pei, X., Song, X., Wang, S., Yang, Z., Zhu, J., Lin, Q., Zhu, Q., & Yang, X. (2025). Application and development of CRISPR technology in the secondary metabolic pathway of the active ingredients of phytopharmaceuticals. *Frontiers in Plant Science, 15*, 1477894. Advance online publication. DOI: 10.3389/fpls.2024.1477894 PMID: 39850214

Garg, S. K., Shukla, A., & Choudhury, S. (2019). Polyphenols and flavonoids. In Gupta, R., Srivastava, A., & Lall, R. (Eds.), *Nutraceuticals in veterinary medicine* (pp. 171–184). Springer., DOI: 10.1007/978-3-030-04624-8_13

Ghaisas, M. M., Kshirsagar, S. B., & Sahane, R. S. (2014). Evaluation of wound healing activity of ferulic acid in diabetic rats. *International Wound Journal, 11*(5), 523–532. DOI: 10.1111/j.1742-481X.2012.01119.x PMID: 23236955

Ghosh, S.. (2018). Genetic engineering of plants for enhanced polyphenol production. *Plant Biotechnology Journal, 16*(6), 1035–1047.

Ghosh, S.. (2020). Molecular farming in plants for therapeutic protein production. *Biomolecules, 10*(9), 1–14.

Glisan, S. L., Grove, K. A., Yennawar, N. H., & Lambert, J. D. (2017). Inhibition of pancreatic lipase by black tea theaflavins: Comparative enzymology and in silico modeling studies. *Food Chemistry, 216*, 296–300. DOI: 10.1016/j.foodchem.2016.08.052 PMID: 27596423

Haddadi, G., Abbaszadeh, A., Mosleh-Shirazi, M. A., Okhovat, M. A., Salajeghe, A., & Ghorbani, Z. (2018). Evaluation of the effect of hesperidin on vascular endothelial growth factor gene expression in rat skin animal models following cobalt-60 gamma irradiation. *Journal of Cancer Research and Therapeutics, 14*(12, Suppl. 5), S1098–S1104. DOI: 10.4103/0973-1482.202892 PMID: 30539852

Hagiwara, A., Hirose, M., Takahashi, S., Ogawa, K., Shirai, T., & Ito, N. (1991). Forestomach and kidney carcinogenicity of caffeic acid in F344 rats and C57BL/6N × C3H/HeN F1 mice. *Cancer Research, 51*(20), 5655–5660. PMID: 1913684

Hamdy, A., & Ibrahem, M. (2010). Management of aphthous ulceration with topical quercetin: A randomized clinical trial. *The Journal of Contemporary Dental Practice, 11*(4), E009–E016. DOI: 10.5005/jcdp-11-4-9 PMID: 20953559

Hashemi-Shahraki, F., Shareghi, B., Farhadian, S., & Yadollahi, E. (2023). A comprehensive insight into the effects of caffeic acid (CA) on pepsin: Multi-spectroscopy and MD simulations methods. *Spectrochimica Acta. Part A: Molecular and Biomolecular Spectroscopy*, *289*, 122240. DOI: 10.1016/j.saa.2022.122240 PMID: 36527971

Hirose, M., Hoshiya, T., Mizoguchi, Y., Nakamura, A., Akagi, K., & Shirai, T. (2001). Green tea catechins enhance tumor development in the colon without effects in the lung or thyroid after pretreatment with 1,2-dimethylhydrazine or 2,2′-dihydroxy-di-n-propylnitrosamine in male F344 rats. *Cancer Letters*, *168*(1), 23–29. DOI: 10.1016/S0304-3835(01)00502-X PMID: 11368873

. Hollman, P. C. H. (2004). Absorption, bioavailability, and metabolism of flavonoids. Pharmaceutical Biology, 42(sup1), 74–83. DOI: 10.3109/13880200490893492

Hwang, J. S.. (2018). Microbial transformations of polyphenolic compounds. *Biotechnology Letters*, *40*(4), 481–490.

Hwang, S. H., & Lee, H. J. (2021). Stimulation of fibroblast growth and collagen synthesis by flavonoids in skin wound healing. *Journal of Dermatological Science*, *103*(2), 114–121. DOI: 10.1016/j.jdermsci.2021.06.007

Jiji, S., Udhayakumar, S., Rose, C., Muralidharan, C., & Kadirvelu, K. (2019). Thymol enriched bacterial cellulose hydrogel as effective material for third degree burn wound repair. *International Journal of Biological Macromolecules*, *122*, 452–460. DOI: 10.1016/j.ijbiomac.2018.10.192 PMID: 30385344

Jomova, K., Alomar, S. Y., Valko, R., Liska, J., Nepovimova, E., Kuca, K., & Valko, M. (2025). Flavonoids and their role in oxidative stress, inflammation, and human diseases. *Chemico-Biological Interactions*, *111489*, 111489. https://www.sciencedirect.com/science/article/pii/S000927972500119X. DOI: 10.1016/j.cbi.2025.111489 PMID: 40147618

Kant, V., Jangir, B. L., Sharma, M., Kumar, V., & Joshi, V. G. (2021). Topical application of quercetin improves wound repair and regeneration in diabetic rats. *Immunopharmacology and Immunotoxicology*, *43*(5), 536–553. DOI: 10.1080/08923973.2021.1950758 PMID: 34278923

Kapoor, M., Howard, R., Hall, I., & Appleton, I. (2004). Effects of epicatechin gallate on wound healing and scar formation in a full thickness incisional wound healing model in rats. *American Journal of Pathology*, *165*(1), 299–307. DOI: 10.1016/S0002-9440(10)63297-X PMID: 15215184

Kar, A. K., Singh, A., Dhiman, N., Purohit, M. P., Jagdale, P., Kamthan, M., Singh, D., Kumar, M., Ghosh, D., & Patnaik, S. (2019). Polymer-assisted in situ synthesis of silver nanoparticles with epigallocatechin gallate (EGCG) impregnated wound patch potentiate controlled inflammatory responses for brisk wound healing. *International Journal of Nanomedicine, 14*, 9837–9854. DOI: 10.2147/IJN.S228462 PMID: 31849472

Kaur, B.. (2021). The role of microbial fermentation in the transformation of polyphenols. *Critical Reviews in Food Science and Nutrition, 61*(10), 1682–1709.

Kelikian, A., Silva, F., & Viana, G. (2021). Antioxidant role of polyphenols in wound healing: A review. *Antioxidants, 10*(2), 188. Advance online publication. DOI: 10.3390/antiox10020188 PMID: 33525629

. Khan, I., et al. (2025). Recent trends in natural antioxidants: A decade of progress in phenols, flavonoids and artemisinin research (2015–2025).

Kim, H., Kawazoe, T., Han, D. W., Matsumara, K., Suzuki, S., Tsutsumi, S., & Hyon, S.-H. (2008). Enhanced wound healing by an epigallocatechin gallate-incorporated collagen sponge in diabetic mice. *Wound Repair and Regeneration, 16*(5), 714–720. DOI: 10.1111/j.1524-475X.2008.00422.x PMID: 19128267

Kim, K.. (2019). Advances in nanotechnology for drug delivery. *Current Drug Targets, 20*(4), 389–400.

Kondo, T., & Ishida, Y. (2010). Molecular pathology of wound healing. *Forensic Science International, 203*(1-3), 93–98. DOI: 10.1016/j.forsciint.2010.07.004 PMID: 20739128

Krausz, A. E., Adler, B. L., Cabral, V., Navati, M., Doerner, J., Charafeddine, R. A., Chandra, D., Liang, H., Gunther, L., Clendaniel, A., Harper, S., Friedman, J. M., Nosanchuk, J. D., & Friedman, A. J. (2015). Curcumin-encapsulated nanoparticles as innovative antimicrobial and wound healing agent. *Nanomedicine; Nanotechnology, Biology, and Medicine, 11*(1), 195–206. DOI: 10.1016/j.nano.2014.09.004 PMID: 25240595

Kubat, M., Karabulut, Z., & Şengül, S. (2021). Effect of propolis on wound healing in sacrococcygeal pilonidal disease: A randomized controlled clinical trial. *Pakistan Journal of Pharmaceutical Sciences, 34*(1). PMID: 34602433

Kumar, S., & Pandey, A. K. (2013). Chemistry and biological activities of flavonoids: An overview. *TheScientificWorldJournal, 2013*(1), 162750. DOI: 10.1155/2013/162750 PMID: 24470791

Kurahashi, T., & Fujii, J. (2015). Roles of antioxidative enzymes in wound healing. *Journal of Developmental Biology*, *3*(2), 57–70. DOI: 10.3390/jdb3020057

Lee, E. M., Lee, S. S., Chung, B. Y., Cho, J.-Y., Lee, I. C., Ahn, S. R., Jang, S. J., & Kim, T. H. (2010). Pancreatic lipase inhibition by C-glycosidic flavones isolated from Eremochloa ophiuroides. *Molecules (Basel, Switzerland)*, *15*(11), 8251–8259. DOI: 10.3390/molecules15118251 PMID: 21081855

Lee, J. H., Lee, J. H., & Lee, Y. C. (2020). Polyphenols and their therapeutic role in wound healing: A review. *Antioxidants*, *9*(9), 697.

Lee, Y., Kim, S., & An, S. (2023). Patient variability in response to wound healing therapies. *International Wound Journal*, *20*(5), 1610–1622. DOI: 10.1111/iwj.13986

Li, H., Tanaka, T., Zhang, Y.-J., Yang, C.-R., & Kouno, I. (2007). Rubusuaviins A–F, monomeric and oligomeric ellagitannins from Chinese sweet tea and their α-amylase inhibitory activity. *Chemical & Pharmaceutical Bulletin*, *55*(9), 1325–1331. DOI: 10.1248/cpb.55.1325 PMID: 17827756

Li, M., Xu, J., Shi, T., Yu, H., Bi, J., & Chen, G. (2016). Epigallocatechin-3-gallate augments therapeutic effects of mesenchymal stem cells in skin wound healing. *Clinical and Experimental Pharmacology & Physiology*, *43*(11), 1115–1124. DOI: 10.1111/1440-1681.12652 PMID: 27557877

Li, Q., Wei, Q., Yuan, E., Yang, J., & Ning, Z. (2014). Interaction between four flavonoids and trypsin: Effect on the characteristics of trypsin and antioxidant activity of flavonoids. *International Journal of Food Science & Technology*, *49*(4), 1063–1069. DOI: 10.1111/ijfs.12401

Li, W., Kandhare, A. D., Mukherjee, A. A., & Bodhankar, S. L. (2018). Hesperidin, a plant flavonoid accelerated the cutaneous wound healing in streptozotocin-induced diabetic rats: Role of TGF-β/Smads and Ang-1/Tie-2 signaling pathways. *EXCLI Journal*, *17*, 399–415. PMID: 29805347

Li, W.-T., Chuang, Y.-H., & Hsieh, J.-F. (2019). Characterization of maltase and sucrase inhibitory constituents from Rhodiola crenulata. *Foods*, *8*(11), 540. DOI: 10.3390/foods8110540 PMID: 31684079

Limsitthichaikoon, S., Khampaenjiraroch, B., Damrongrungruang, T., Limphirat, W., Thapphasaraphong, S., & Priprem, A. (2018). Topical oral wound healing potential of anthocyanin complex: Animal and clinical studies. *Therapeutic Delivery*, *9*(5), 359–374. DOI: 10.4155/tde-2017-0123 PMID: 29681233

Liu, J., Zhang, Y., & Zhang, M.. (2022). The roles of polyphenols in wound healing: A comprehensive review. *Frontiers in Nutrition*, *9*, 891575.

Liu, Y.-Y., Wu, J.-Q., Fan, R.-Y., He, Z.-H., Li, C.-Y., & He, M.-F. (2020). Isoliquiritin promote angiogenesis by recruiting macrophages to improve the healing of zebrafish wounds. *Fish & Shellfish Immunology, 100*, 238–245. DOI: 10.1016/j.fsi.2020.02.071 PMID: 32135341

Lohray, B. B.. (2006). Pharmaceutical applications of cyclodextrins. *Biopharmaceutics & Drug Disposition, 27*(4), 145–156.

Lyseng-Williamson, K. A., & Perry, C. M. (2003). Micronised purified flavonoid fraction: A review of its use in chronic venous insufficiency, venous ulcers and haemorrhoids. *Drugs, 63*(1), 71–100. DOI: 10.2165/00003495-200363010-00005 PMID: 12487623

Madhyastha, H., Halder, S., Queen Intan, N., Madhyastha, R., Mohanapriya, A., & Sudhakaran, R.. (2020). Surface refined Au quercetin nanoconjugate stimulates dermal cell migration: Possible implication in wound healing. *RSC Advances, 10*(62), 37683–37694. DOI: 10.1039/D0RA06690G PMID: 35515178

Maliar, T., Jedinák, A., Kadrabová, J., & Šturdík, E. (2004). Structural aspects of flavonoids as trypsin inhibitors. *European Journal of Medicinal Chemistry, 39*(3), 241–248. DOI: 10.1016/j.ejmech.2003.12.003 PMID: 15051172

Martinez-Gonzalez, A. I., Díaz-Sánchez, Á. G., de la Rosa, L. A., Bustos-Jaimes, I., & Alvarez-Parrilla, E. (2021). A novel approach to trypsin inhibition by flavonoids. *Journal of Food Bioactives: An Official Scientific Publication of the International Society of Nutraceuticals and Functional Foods (ISNFF)*, •••, 14. DOI: 10.31665/JFB.2021.14272

McClements, D. J., & Gumus, C. E. (2016). Nanoemulsions: Formation, properties, and applications. *Comprehensive Reviews in Food Science and Food Safety, 15*(4), 630–646.

Melrose, J. (2023). The potential of flavonoids and flavonoid metabolites in the treatment of neurodegenerative pathology in disorders of cognitive decline. *Antioxidants, 12*(3), 663. DOI: 10.3390/antiox12030663 PMID: 36978911

Mi, B., Liu, J., Liu, G., Zhou, W., Liu, Y., Hu, L., Xiong, L., Ye, S., & Wu, Y. (2018). Icariin promotes wound healing by enhancing the migration and proliferation of keratinocytes via the AKT and ERK signaling pathway. *International Journal of Molecular Medicine, 42*(2), 831–838. DOI: 10.3892/ijmm.2018.3676 PMID: 29767263

Middleton, E. J.Jr. (1998). Effect of plant flavonoids on immune and inflammatory cell function. *Advances in Experimental Medicine and Biology, 439*, 175–182. DOI: 10.1007/978-1-4615-5335-9_13 PMID: 9781303

Miguel, M. G., & Antunes, M. D. (2011). Is propolis safe as an alternative medicine? *Journal of Pharmacy & Bioallied Sciences*, *3*(4), 479–495. DOI: 10.4103/0975-7406.90101 PMID: 22219581

Miller, A., Barbera, R., & Adriani, W.. (2022). Personalized approaches to wound healing: Evaluating the role of polyphenol delivery in chronic wound management. *Journal of Wound Care*, *31*(4), 192–201.

Mittal, G., Prashanth, A., Dhali, A., Prasad, R., Yogesh, S., Nurani, K. M., & Găman, M. A. (2025). Plant extracts with antioxidant and hepatoprotective benefits for liver health: A bibliometric analysis of drug delivery systems. *World Journal of Gastroenterology*, *31*(18), 105836. https://pmc.ncbi.nlm.nih.gov/articles/PMC12146923. DOI: 10.3748/wjg.v31.i18.105836 PMID: 40496363

Mohammadi, Z., Sharif Zak, M., Majdi, H., Seidi, K., Barati, M., Akbarzadeh, A., & Latifi, A. M. (2017). The effect of chrysin-loaded nanofiber on wound healing process in male rat. *Chemical Biology & Drug Design*, *90*(6), 1106–1114. DOI: 10.1111/cbdd.12996 PMID: 28388004

Munro, I. C., Harwood, M., Hlywka, J. J., Stephen, A. M., Doull, J., Flamm, W. G., & Adlercreutz, H. (2003). Soy isoflavones: A safety review. *Nutrition Reviews*, *61*(1), 1–33. DOI: 10.1301/nr.2003.janr.1-33 PMID: 12638461

Nascimento, R., Lima, J., & Oliveira, J.. (2020). Nanotechnology as a tool for enhancing the bioavailability and effectiveness of polyphenols and flavonoids. *Nutrients*, *12*(8), 2450. PMID: 32824061

Naz, S., Siddiqi, R., Dew, T. P., & Williamson, G. (2011). Epigallocatechin-3-gallate inhibits lactase but is alleviated by salivary proline-rich proteins. *Journal of Agricultural and Food Chemistry*, *59*(6), 2734–2738. DOI: 10.1021/jf103072z PMID: 21348516

O'Brien, J., Wang, A., & Miller, D. (2022). Regulatory framework for bioactive compounds in wound care: Challenges and perspectives. *Regulatory Toxicology and Pharmacology*, *129*, 105056. Advance online publication. DOI: 10.1016/j.yrtph.2022.105056

Oboh, G., Ademosun, A., Olasehinde, T., Oyeleye, S., & Ehiakhamen, E. (2015). Effect of processing methods on the antioxidant properties and inhibition of α-amylase and α-glucosidase by African pear (Dacryodes edulis) fruit. *NUTRAfoods : International Journal of Science and Marketing for Nutraceutical Actives, Raw Materials, Finish Products*, *14*(1), 19–26. DOI: 10.1007/s13749-014-0065-7

Özay, Y., Guzel, S., Erdoğdu, İ., Yıldırım, Z., Pehlivanoğlu, B., Turk, B., & Darcan, S. (2018). Evaluation of the wound healing properties of luteolin ointments on excision and incision wound models in diabetic and non-diabetic rats. *Records of Natural Products, 12*(4), 350–366. DOI: 10.25135/rnp.38.17.08.135

Özay, Y., Güzel, S., Yumrutaş, Ö., Pehlivanoğlu, B., Erdoğdu, İ. H., Yildirim, Z., Türk, B. A., & Darcan, S. (2019). Wound healing effect of kaempferol in diabetic and nondiabetic rats. *The Journal of Surgical Research, 233*, 284–296. DOI: 10.1016/j.jss.2018.08.009 PMID: 30502261

Pandith, H., Zhang, X., Liggett, J., Min, K. W., Gritsanapan, W., & Baek, S. J. (2013). Hemostatic and wound healing properties of Chromolaena odorata leaf extract. *ISRN Dermatology, 2013*, 168269. DOI: 10.1155/2013/168269 PMID: 23984087

Paredes, A. J.. (2021). Smart polymeric systems for drug delivery of bioactive compounds. *Journal of Drug Targeting, 29*(3), 224–239.

Park, E., Lee, S. M., Jung, I.-K., Lim, Y., & Kim, J.-H. (2011). Effects of genistein on early-stage cutaneous wound healing. *Biochemical and Biophysical Research Communications, 410*(3), 514–519. DOI: 10.1016/j.bbrc.2011.06.013 PMID: 21679688

Parveen, B., Rajinikanth, V., & Narayanan, M. (2025). Natural plant antioxidants for food preservation and emerging trends in nutraceutical applications. *Discover Applied Sciences, 7*(8), 845. Advance online publication. DOI: 10.1007/s42452-025-07464-6

Ponrasu, T., Veerasubramanian, P. K., Kannan, R., Gopika, S., Suguna, L., & Muthuvijayan, V. (2018). Morin incorporated polysaccharide–protein (psyllium–keratin) hydrogel scaffolds accelerate diabetic wound healing in Wistar rats. *RSC Advances, 8*(5), 2305–2314. DOI: 10.1039/C7RA10334D PMID: 35541447

Possemiers, S.. (2013). The role of biotransformation in the bioavailability of dietary polyphenols. *Food Research International, 54*(1), 213–229.

Pyner, A., Nyambe-Silavwe, H., & Williamson, G. (2017). Inhibition of human and rat sucrase and maltase activities to assess antiglycemic potential: Optimization of the assay using acarbose and polyphenols. *Journal of Agricultural and Food Chemistry, 65*(39), 8643–8651. DOI: 10.1021/acs.jafc.7b03678 PMID: 28914528

Raghavendra, G., & Mohan, B. (2018). Nanostructured materials for drug delivery. *Nanomedicine; Nanotechnology, Biology, and Medicine, 14*(1), 215–230.

Ranjha, M.. (2022). Scalability of nanomaterial production: Issues and strategies. *Journal of Nanomedicine & Nanotechnology, 13*(2), e1234.

Ribeiro, A.. (2021). Safety and toxicological assessment of nanomaterials in drug delivery systems. *Nano Today*, *36*, 100969.

Rithichai, P., Jirakiattikul, Y., Nambuddee, R., & Itharat, A. (2024). Effect of salicylic acid foliar application on bioactive compounds and antioxidant activity in holy basil (Ocimum sanctum L.). *International Journal of Agronomy*, *2024*, 8159886. Advance online publication. DOI: 10.1155/2024/8159886

Rodero, M. P., & Khosrotehrani, K. (2010). Skin wound healing modulation by macrophages. *International Journal of Clinical and Experimental Pathology*, *3*(7), 643–653. PMID: 20830235

Rodriguez, P. G., Felix, F. N., Woodley, D. T., & Shim, E. K. (2008). The role of oxygen in wound healing: A review of the literature. *Dermatologic Surgery*, *34*(9), 1159–1169. DOI: 10.1097/00042728-200809000-00001 PMID: 18513296

Rubeena, P., Kumari, G., & Siqun, M. L. (2021). Polyphenolic ointment for wound healing: An innovative approach. *Journal of Pharmacy & Bioallied Sciences*, *13*(1), 67–73. DOI: 10.4103/jpbs.JPBS_84_20

Rudrapal, M., Rakshit, G., Singh, R. P., Garse, S., Khan, J., & Chakraborty, S. (2024). Dietary polyphenols: Review on chemistry/sources, bioavailability/metabolism, antioxidant effects, and their role in disease management. *Antioxidants*, *13*(4), 429. https://www.mdpi.com/2076-3921/13/4/429. DOI: 10.3390/antiox13040429 PMID: 38671877

Rudrapal, M., Rakshit, G., Singh, R. P., Garse, S., Khan, J., & Chakraborty, S. (2024). Dietary polyphenols: Review on chemistry/sources, bioavailability/metabolism, antioxidant effects, and their role in disease management. *Antioxidants*, *13*(4), 429. DOI: 10.3390/antiox13040429 PMID: 38671877

Sakihama, Y., Cohen, M. F., Grace, S. C., & Yamasaki, H. (2002). Plant phenolic antioxidant and prooxidant activities: Phenolics-induced oxidative damage mediated by metals in plants. *Toxicology*, *177*(1), 67–80. DOI: 10.1016/S0300-483X(02)00196-8 PMID: 12126796

Salazar-Gómez, A., & Alonso-Castro, A. J. (2022). Medicinal plants from Latin America with wound healing activity: Ethnomedicine, phytochemistry, preclinical and clinical studies—A review. *Pharmaceuticals (Basel, Switzerland)*, *15*(9), 1095. DOI: 10.3390/ph15091095 PMID: 36145316

Salehi, M., Ehterami, A., Farzamfar, S., Vaez, A., & Ebrahimi-Barough, S. (2021). Accelerating healing of excisional wound with alginate hydrogel containing naringenin in rat model. *Drug Delivery and Translational Research, 11*(1), 142–153. DOI: 10.1007/s13346-020-00731-6 PMID: 32086788

Salehi, M., Vaez, A., Naseri-Nosar, M., Farzamfar, S., Ai, A., Ai, J., Tavakol, S., Khakbiz, M., & Ebrahimi-Barough, S. (2018). Naringin-loaded poly(ε-caprolactone)/gelatin electrospun mat as a potential wound dressing: In vitro and in vivo evaluation. *Fibers and Polymers, 19*(1), 125–134. DOI: 10.1007/s12221-018-7528-6

Scialò, F., Fernández-Ayala, D. J., & Sanz, A. (2017). Role of mitochondrial reverse electron transport in ROS signaling: Potential roles in health and disease. *Frontiers in Physiology, 8*, 428. DOI: 10.3389/fphys.2017.00428 PMID: 28701960

Sęczyk, Ł., Świeca, M., Kapusta, I., & Gawlik-Dziki, U. (2019). Protein–phenolic interactions as a factor affecting the physicochemical properties of white bean proteins. *Molecules (Basel, Switzerland), 24*(3), 408. DOI: 10.3390/molecules24030408 PMID: 30678067

Seema, B. R., Jyothi, Y., Bhovi, C. N., Aradhya, M. V., Lekhak, M., & Mane, S. R. (2023). Advanced up and down methodology for acute toxicity assessment with reliable LD50 verified by aqueous extract of curly kale using Wistar rats. *Research Journal of Pharmacy and Technology, 16*(10), 4519–4524. https://www.indianjournals.com/article/rjpt-16-10-006

Seo, G. Y., Lim, Y., Koh, D., Huh, J. S., Hyun, C., Kim, Y. M., & Cho, M. (2017). TMF and glycitin act synergistically on keratinocytes and fibroblasts to promote wound healing and anti-scarring activity. *Experimental & Molecular Medicine, 49*(3), e302. DOI: 10.1038/emm.2016.167 PMID: 28303029

Shahidi, F., & Naczk, M. (1995). *Food phenolics: Sources, chemistry, effects, applications*. Technomic Publishing.

Shalaby, E. A., Aboul-Enein, A., El-Shemy, H., & Nassrallah, A. (2025). Active ingredients of some Egyptian plants species with anticancer activity and its mode of action. *Arab Journal of Biotechnology, 24*(1), 18–40. https://journals.ekb.eg/article_432494_09dce9501746bb2f9536fa511639e7fc.pdf. DOI: 10.21608/arjb.2025.365025.1000

Sharma, V., Moshage, H., & Haznadar, M. (2021). Polyphenols: Potential role in cutaneous wound healing processes. *Foods, 10*(5), 992. PMID: 34062860

Sharma, Y., Bhardwaj, R., Kaur, A., Singh, G., Kulkarni, S., & Bhatia, A.. (2024). Nano-gel formulation of polyphenolic fraction of tobacco stem for wound healing and its inhibitory efficacies against the receptors of chronic wound development. *Journal of Biomolecular Structure & Dynamics*. Advance online publication. PMID: 38433407

Shevelev, A. B., La Porta, N., Isakova, E. P., Martens, S., Biryukova, Y. K., Belous, A. S., Sivokhin, D. A., Trubnikova, E. V., Zylkova, M. V., Belyakova, A. V., Smirnova, M. S., & Deryabina, Y. I. (2020). In vivo antimicrobial and wound-healing activity of resveratrol, dihydroquercetin, and dihydromyricetin against Staphylococcus aureus, Pseudomonas aeruginosa, and Candida albicans. *Pathogens (Basel, Switzerland)*, *9*(4), 296. DOI: 10.3390/pathogens9040296 PMID: 32316572

Shubina, V., & Shatalin, Y. V. (2012). Skin regeneration after chemical burn under the effect of taxifolin-based preparations. *Bulletin of Experimental Biology and Medicine*, *154*(2), 152–157. DOI: 10.1007/s10517-012-1897-z PMID: 23330113

Shukla, R., Kashaw, S. K., Jain, A. P., & Lodhi, S. (2016). Fabrication of apigenin loaded gellan gum–chitosan hydrogels (GGCH-HGs) for effective diabetic wound healing. *International Journal of Biological Macromolecules*, *91*, 1110–1119. DOI: 10.1016/j.ijbiomac.2016.06.075 PMID: 27344952

Singer, A. J. (2022). Healing mechanisms in cutaneous wounds: Tipping the balance. *Tissue Engineering. Part B, Reviews*, *28*(6), 1151–1167. DOI: 10.1089/ten.teb.2021.0114 PMID: 34915757

Smith, J., & Kaur, J. (2022). Advanced dressings infused with bioactive compounds for enhanced wound healing. *Wound Repair and Regeneration*, *30*(2), 174–185. DOI: 10.1111/wrr.12952

Sriwatanametanon, S.. (2015). Nanoparticle-assisted pharmacotherapy for cancer. *Asian Pacific Journal of Cancer Prevention : APJCP*, *16*(4), 1477–1485.

Stevens Barrón, J. C., Chapa González, C., Álvarez Parrilla, E., & De la Rosa, L. A. (2023). Nanoparticle-mediated delivery of flavonoids: Impact on proinflammatory cytokine production: A systematic review. *Biomolecules*, *13*(7), 1158. Advance online publication. DOI: 10.3390/biom13071158 PMID: 37509193

Sun, B. K., Siprashvili, Z., & Khavari, P. A. (2014). Advances in skin grafting and treatment of cutaneous wounds. *Science*, *346*(6212), 941–945. DOI: 10.1126/science.1253836 PMID: 25414301

Sun, L., Gidley, M. J., & Warren, F. J. (2018). Tea polyphenols enhance binding of porcine pancreatic α-amylase with starch granules but reduce catalytic activity. *Food Chemistry, 258*, 164–173. DOI: 10.1016/j.foodchem.2018.03.017 PMID: 29655719

Sun, L., Warren, F. J., Netzel, G., & Gidley, M. J. (2016). 3 or 3′-galloyl substitution plays an important role in association of catechins and theaflavins with porcine pancreatic α-amylase: The kinetics of inhibition of α-amylase by tea polyphenols. *Journal of Functional Foods, 26*, 144–156. DOI: 10.1016/j.jff.2016.07.012

Sysak, S., Czarczynska-Goslinska, B., Szyk, P., Koczorowski, T., Mlynarczyk, D. T., Szczolko, W., Lesyk, R., & Goslinski, T. (2023). Metal nanoparticle-flavonoid connections: Synthesis, physicochemical and biological properties, as well as potential applications in medicine. *Nanomaterials (Basel, Switzerland), 13*(9), 1531. Advance online publication. DOI: 10.3390/nano13091531 PMID: 37177076

Tan, Y., & Chang, S. K. (2017). Digestive enzyme inhibition activity of the phenolic substances in selected fruits, vegetables and tea as compared to black legumes. *Journal of Functional Foods, 38*, 644–655. DOI: 10.1016/j.jff.2017.04.005

Tenci, M., Rossi, S., Aguzzi, C., Carazo, E., Sandri, G., Bonferoni, M., Grisoli, P., Viseras, C., Caramella, C. M., & Ferrari, F. (2017). Carvacrol/clay hybrids loaded into in situ gelling films. *International Journal of Pharmaceutics, 531*(2), 676–688. DOI: 10.1016/j.ijpharm.2017.06.024 PMID: 28619454

Teng, H., & Chen, L. (2019). Polyphenols and bioavailability: An update. *Critical Reviews in Food Science and Nutrition, 59*(13), 2040–2051. DOI: 10.1080/10408398.2018.1437023 PMID: 29405736

Thomas, L., Zakir, F., Mirza, M. A., Anwer, M. K., Ahmad, F. J., & Iqbal, Z. (2017). Development of curcumin loaded chitosan polymer based nanoemulsion gel: In vitro, ex vivo evaluation and in vivo wound healing studies. *International Journal of Biological Macromolecules, 101*, 569–579. DOI: 10.1016/j.ijbiomac.2017.03.066 PMID: 28322948

Tian, J.-L., Si, X., Wang, Y.-H., Gong, E.-S., Xie, X., Zhang, Y., Li, B., & Shu, C. (2021). Bioactive flavonoids from Rubus corchorifolius inhibit α-glucosidase and α-amylase to improve postprandial hyperglycemia. *Food Chemistry, 341*, 128149. DOI: 10.1016/j.foodchem.2020.128149 PMID: 33039745

Tie, L., An, Y., Han, J., Xiao, Y., Xiaokaiti, Y., Fan, S., Liu, S., Chen, A. F., & Li, X. (2013). Genistein accelerates refractory wound healing by suppressing superoxide and FoxO1/iNOS pathway in type 1 diabetes. *The Journal of Nutritional Biochemistry, 24*(1), 88–96. DOI: 10.1016/j.jnutbio.2012.02.011 PMID: 22819564

Trottier, L.. (2018). Regulatory challenges for nanomedicine: Perspectives and future directions. *Nanomedicine (London), 14*(8), 1129–1138.

Vignesh, K., Thangavelu, A., & Suresh, S. (2020). Nanofibers: Properties and applications. *Nanotechnology, 31*(30), 305603.

Wang, H., Du, Y.-J., & Song, H.-C. (2010). α-Glucosidase and α-amylase inhibitory activities of guava leaves. *Food Chemistry, 123*(1), 6–13. DOI: 10.1016/j.foodchem.2010.03.088

Wang, X., Yang, J., Li, H., Shi, S., & Peng, X. (2022). Mechanistic study and synergistic effect on inhibition of α-amylase by structurally similar flavonoids. *Journal of Molecular Liquids, 360*, 119485. DOI: 10.1016/j.molliq.2022.119485

Wendlocha, D., Krzykawski, K., Mielczarek-Palacz, A., & Kubina, R. (2023). Selected flavonols in breast and gynecological cancer: A systematic review. *Nutrients, 15*(13), 2938. Advance online publication. DOI: 10.3390/nu15132938 PMID: 37447264

Williamson, G. (2013). Possible effects of dietary polyphenols on sugar absorption and digestion. *Molecular Nutrition & Food Research, 57*(1), 48–57. DOI: 10.1002/mnfr.201200511 PMID: 23180627

Wu, H., Estevez-Santiago, R., & Angelino, D.. (2021). The influence of gut microbiota on polyphenol metabolism and the resulting bioactivity. *Frontiers in Nutrition, 8*, 36.

Wu, X., Ding, H., Hu, X., Pan, J., Liao, Y., Gong, D., & Zhang, G. (2018). Exploring inhibitory mechanism of gallocatechin gallate on α-amylase and α-glucosidase relevant to postprandial hyperglycemia. *Journal of Functional Foods, 48*, 200–209. DOI: 10.1016/j.jff.2018.07.022

Wu, X., Hu, M., Hu, X., Ding, H., Gong, D., & Zhang, G. (2019). Inhibitory mechanism of epicatechin gallate on α-amylase and α-glucosidase and its combinational effect with acarbose or epigallocatechin gallate. *Journal of Molecular Liquids, 290*, 111202. DOI: 10.1016/j.molliq.2019.111202

Xu, F., Weng, B., Gilkerson, R., Materon, L. A., & Lozano, K. (2015). Development of tannic acid/chitosan/pullulan composite nanofibers from aqueous solution for potential applications as wound dressing. *Carbohydrate Polymers, 115*, 16–24. DOI: 10.1016/j.carbpol.2014.08.081 PMID: 25439862

Xu, Z., Liu, G., Liu, P., Hu, Y., Chen, Y., Fang, Y., Sun, G., Huang, H., & Wu, J. (2022). Hyaluronic acid-based glucose-responsive antioxidant hydrogel platform for enhanced diabetic wound repair. *Acta Biomaterialia, 147*, 147–157. DOI: 10.1016/j.actbio.2022.05.047 PMID: 35649507

Xue, R., He, L., Wu, J., Kong, X., Wang, Q., Chi, Y., Liu, J., Wang, Z., Zeng, K., Chen, W., Ren, H., & Han, B. (2024). Multifunctional sprayable carboxymethyl chitosan/polyphenol hydrogel for wound healing. *International Journal of Biological Macromolecules, 275*, 133303. DOI: 10.1016/j.ijbiomac.2024.133303 PMID: 38917923

Yamaguchi, K., Urai, T., Okuwa, M., & Nakatani, T. (2012). The effect of isoflavone oral medication on cutaneous wound healing in ovariectomized female mice. *Wound Repair and Regeneration, 20*(1), A5.

Yang, X., & Kong, F. (2016). Effects of tea polyphenols and different teas on pancreatic α-amylase activity in vitro. *Lebensmittel-Wissenschaft + Technologie, 66*, 232–238. DOI: 10.1016/j.lwt.2015.10.035

Zahra, M., Abrahamse, H., & George, B. P. (2024). Green nanotech paradigm for enhancing sesquiterpene lactone therapeutics in cancer. *Biomedicine and Pharmacotherapy, 173*, 116426. Advance online publication. DOI: 10.1016/j.biopha.2024.116426 PMID: 38471274

Zahra, M., Abrahamse, H., & George, B. P. (2024). Flavonoids: Antioxidant powerhouses and their role in nanomedicine. *Antioxidants, 13*(8), 922. DOI: 10.3390/antiox13080922 PMID: 39199168

Zeng, H.-J., Yang, R., Liang, H., & Qu, L.-B. (2015). Molecular interactions of flavonoids to pepsin: Insights from spectroscopic and molecular docking studies. *Spectrochimica Acta. Part A: Molecular and Biomolecular Spectroscopy, 151*, 576–590. DOI: 10.1016/j.saa.2015.06.059 PMID: 26162346

Zhang, S., Ou, Q., Xin, P., Yuan, Q., Wang, Y., & Wu, J. (2019). Polydopamine/puerarin nanoparticle-incorporated hybrid hydrogels for enhanced wound healing. *Biomaterials Science, 7*(10), 4230–4236. DOI: 10.1039/C9BM00991D PMID: 31393463

Zhang, S., Yang, K., & Zhang, M. (2022). Stability of polyphenols in various topical formulations: A review. *Pharmaceutics, 14*(9), 1818. Advance online publication. DOI: 10.3390/pharmaceutics14091818 PMID: 36145565

Zhang, Y.. (2019). Polymeric nanoparticles for targeted delivery of therapeutics. *Journal of Controlled Release, 301*, 56–69.

Zhang, Y.. (2020). Nanocarriers for the delivery of bioactive compounds: A review. *Trends in Food Science & Technology, 99*, 61–70.

Zhang, Y., Liu, Y., & Zhang, W. (2015). Nanosheets in electronics. *Advanced Materials, 27*(44), 6833–6848. PMID: 26449732

Zhang, Y., Xie, C., Wang, L., Huang, L., & Wei, Y. (2020). The role of catechol-O-methyltransferase (COMT) polymorphisms in the metabolism and bioavailability of flavonoids: Implications for their protective effects on health. *Phytotherapy Research, 34*(6), 1303–1312. PMID: 31840883

Zhao, L., Wang, F., Lu, Q., Liu, R., Tian, J., & Huang, Y. (2018). Study on interaction between human salivary α-amylase and sorghum procyanidin tetramer: Binding characteristics and structural analysis. *International Journal of Biological Macromolecules, 118*, 1136–1141. DOI: 10.1016/j.ijbiomac.2018.06.189 PMID: 30001600

Zheng, Y., Tian, J., Yang, W., Chen, S., Liu, D., Fang, H., Zhang, H., & Ye, X. (2020). Inhibition mechanism of ferulic acid against α-amylase and α-glucosidase. *Food Chemistry, 317*, 126346. DOI: 10.1016/j.foodchem.2020.126346 PMID: 32070843

Zheng, Y., Yang, W., Sun, W., Chen, S., Liu, D., Kong, X., Tian, J., & Ye, X. (2020). Inhibition of porcine pancreatic α-amylase activity by chlorogenic acid. *Journal of Functional Foods, 64*, 103587. DOI: 10.1016/j.jff.2019.103587

Zhou, F., Li, G., Tan, R., Wu, G., & Deng, C. (2025). Artemisinins in autoimmune diseases: Effects and mechanisms in systemic lupus erythematosus and rheumatoid arthritis. *British Journal of Pharmacology, 182*(15), 3411–3427. Advance online publication. DOI: 10.1111/bph.70086 PMID: 40474495

Zhuang, W. B., Li, Y. H., Shu, X. C., Pu, Y. T., Wang, X. J., Wang, T., & Wang, Z. (2023). The classification, molecular structure and biological biosynthesis of flavonoids, and their roles in biotic and abiotic stresses. *Molecules (Basel, Switzerland), 28*(8), 3599. DOI: 10.3390/molecules28083599 PMID: 37110833

Zhuang, Y., Hu, K., & Wei, Y.. (2020). Quercetin promotes angiogenesis via upregulation of VEGF and bFGF expression in the wound healing process. *Molecules (Basel, Switzerland), 25*(12), 2903. Advance online publication. DOI: 10.3390/molecules25122903 PMID: 32599753

Zulkefli, N., Che Zahari, C. N. M., Sayuti, N. H., Kamarudin, A. A., Saad, N., Hamezah, H. S., & Sarian, M. N. (2023). Flavonoids as potential wound-healing molecules: Emphasis on pathways perspective. *International Journal of Molecular Sciences, 24*(5), 4607. DOI: 10.3390/ijms24054607 PMID: 36902038

Chapter 6
Terpenoids and Essential Oils: Aromatic Agents in Tissue Repair

Manju Jhakar
School of Pharmacy and Emerging Sciences, Baddi University of Emerging Sciences and Technology, Baddi, India

Nidhi Rani
http://orcid.org/0000-0003-2549-3481
Chitkara College of Pharmacy, Chitkara University, India

Ashish Kashyap
Bhojia Dental College and Hospital, Baddi, India

Thakur Gurjeet Singh
Chitkara College of Pharmacy, Chitkara University, India

ABSTRACT

Plant-derived bioactive compounds such as terpenoids and essential oils (EOs) are increasingly recognised for their therapeutic value in wound healing and tissue regeneration. There are many biological activities, e.g., anti-inflammatory, antioxidant, antibacterial, and wound-healing activity, which are exhibited by these volatile oils. Current science validates their roles in current medical practices, which have traditionally been the cornerstone of herbal medicine. This chapter explores the multifaceted functions of terpenoids and essential oils in healing tissue, with a specific interest in their mode of action in a range of therapeutic settings. Their role in repairing and healing joints is highlighted, particularly by antioxidant and anti-inflammatory mechanisms that mitigate oxidative stress and enhance tissue regeneration. Neuroprotective activity, especially in the treatment of neurodegener-

DOI: 10.4018/979-8-3373-2908-6.ch006

ative diseases, is also examined. Additionally, the chapter explores their application in stimulating tissue development and collagen synthesis. Lavender, chamomile, rosemary, and thyme are essential oils that have also shown potential in bone and wound healing, which can be used along with scaffolds. Advances in extraction methods such as hydrodistillation and supercritical fluid extraction have enhanced the therapeutic properties of these compounds. Importantly, although the preclinical data are compelling for many agents of interest, there is a significant translational gap: with few exceptions, such therapies have not been adopted in routine clinical practice due to unmet needs related to standardization, bioavailability of delivery engineering, and regulatory validation. This chapter therefore reframes the discussion from cataloguing preclinical promise to critically analysing the translational barriers facing terpenoids and EOs, and proposes a structured four-stage research strategy to guide their validation as clinically approved wound-healing therapeutics.

1. INTRODUCTION TO TERPENOIDS AND ESSENTIAL OILS

Terpenoids and essential oils (EOs) are plant-origin compounds of natural origin that have medicinal and therapeutic applications (Nguyen et al., 2021). Terpenoids (or isoprenoids) are a broad and diversified class of naturally occurring organic substances formed from terpenes (C_5H_8), which are chemical compounds consisting of repeated isoprene (C_5H_8) units ((Nguyen et al., 2021; Upadhyay & Bano, 2023). Plants synthesise these chemicals through two pathways: mevalonate (MVA) and methylerythritol phosphate (MEP). The classifications are based on the number of isoprene units (monoterpenoids: C_{10}, sesquiterpenoids: C_{15}, diterpenoids: C_{20}, triterpenoids: C_{30}, etc.). Terpenoids are highly effective in pharmacology due to their lipid-philic properties, allowing for easy interaction with biological membranes (Ashour et al., 2010; Tetali, 2019).

In contrast, essential oils are strong, fragrant substances that are obtained from different plant parts, such as flowers, leaves, stems, bark, and roots. They are manufactured by techniques such as steam distillation, cold pressing, or solvent extraction and are predominantly made up of monoterpenes and sesquiterpenes with other bioactive compounds such as phenols, aldehydes, and esters. Essential oils are essential ecophysiological players in plants, being involved in the attraction of pollinators, repellence of herbivores and protection against microorganisms. Lavender, eucalyptus, tea tree and peppermint, for example which are common essential oils used have been reported to possess diverse biological activities (Noriega, 2020; Matera et al., 2023).

Such inherently powerful agents, both terpenoids and essential oils, are integrated into traditional-medicine practices used by cultures around the world, including

Ayurvedic and Chinese medicine (TCM), as well as being used in Middle Eastern & European Colleges of Medicine since at least 1AD. They can be used for healing wounds, pain reduction, and as an antimicrobial and stress reducer (Dash et al., 2021). Recent pharmacological investigations have provided extensive evidence, based on the anti-inflammatory, antioxidant, antimicrobial, and neuroprotective properties of plants, supporting their use in today's systems of medicine. Considering their role in cell reparative, oxidative stress and immune processes, they were also investigated in other therapeutic applications, including wound dressing, tissue regeneration and tumor therapy (Pereira et al., 2022). Moreover, these naturally derived materials may be further employed in single-use biological applications such as drug delivery devices and regenerative medicine as the research progresses (Teixeira et al., 2020). In this study, distinct terpenoids and essential oils along with their chemical nature, plant resources and biodiversity have been summarised in Table 1.

Table 1. Types of Terpenoids and Essential Oils with Their Abundance in Plants

Type of Compound	Chemical Structure	Common Plant Sources	Amount Present (%)	Ref.
Monoterpenoids	$C_{10}H_{16}$ (2 isoprene units)	Lavender, eucalyptus, peppermint	0.5% – 3% in essential oils	(Zielińska-Błajet & Feder-Kubis, 2020)
Sesquiterpenoids	$C_{15}H_{24}$ (3 isoprene units)	Chamomile, ginger, cloves	0.1% – 5% in plant extracts	(Stephane, 2020)
Diterpenoids	$C_{20}H_{32}$ (4 isoprene units)	Rosemary, ginseng, sage	0.01% – 0.5% in leaves and roots	(El-Saber Batiha et al., 2020)
Triterpenoids	$C_{30}H_{48}$ (6 isoprene units)	Olive, neem, liquorice	0.1% – 1% in plant extracts	(Cox-Georgian & Basu, 2019)
Tetraterpenoids (Carotenoids)	$C_{40}H_{56}$ (8 isoprene units)	Carrot, tomato, paprika	0.01% – 0.3% in fruits and flowers	(Muchlinski et al., 2020)
Polyterpenoids	$(C_5H_8)n$ (many isoprene units)	Rubber tree (Hevea brasiliensis)	30% – 40% in latex	(Sahu et al., 2020)

1.1. Traditional vs. Modern Uses of Terpenoids and Essential Oils

Terpenoids and volatile oils are major components of these extracts that have been widely applied in traditional medicine for their therapeutic benefits, from bringing a wound to easing anxiety. Presumed hard arcs. Since science has progressed to a better knowledge of their mechanics, and, therefore, they can be included in contemporary medical therapy. These compounds are now widely employed in a variety.

therapeutic applications, including antibacterial agents, analgesics, and anticancer drugs (Jagwani & Krishna, 2021). Table 2 Ancient and modern use of terpenoids and essential oils, showing their changing significance in medicine.

Table 2. Traditional and Modern Uses of Terpenoids and Essential Oils

Category	Traditional Uses	Modern Applications	Examples	Ref.
Wound Healing	Used in herbal pastes and oils for cuts, burns, and ulcers.	Stimulates fibroblast activity enhances collagen synthesis and reduces scar formation.	Lavender oil, Chamomile oil, Tea tree oil	(Samuelson et al., 2020)
Pain Relief	Applied as balms or infusions for muscle pain and arthritis.	Acts as an analgesic and anti-inflammatory agent via COX-2 inhibition.	Eugenol (Clove oil), Menthol (Peppermint oil)	(Przekora, 2020)
Antimicrobial	Used for disinfecting wounds and treating infections.	Broad-spectrum activity against bacteria, fungi, and viruses.	Tea Tree oil, Oregano oil, Eucalyptus oil	(Peter et al., 2021)
Respiratory Health	Inhaled for colds, congestion, and cough relief.	Acts as an expectorant, bronchodilator, and anti-inflammatory agent.	Eucalyptus oil, Camphor oil, Peppermint oil	(Horváth & Ács, 2015)
Digestive Health	Used in herbal infusions for indigestion and bloating.	Modulates gut microbiota and reduces inflammation in the gastrointestinal tract.	Ginger oil, Fennel oil, Caraway oil	(Tanveer et al., 2020)
Neuroprotection & Stress Relief	Used in aromatherapy for calming effects and cognitive enhancement.	Supports neuroprotection, reduces anxiety, and enhances cognitive function.	Linalool (Lavender oil), Borneol (Rosemary oil)	(Alam et al., 2023)
Cancer Treatment	Traditional herbal extracts are used for tumors.	Some terpenoids induce apoptosis and inhibit tumor growth in cancer cells.	Taxol (from Yew tree), Curcumin (Turmeric)	(Safarzadeh et al., 2014)

2. PHYSIOLOGICAL ACTIVITIES CONTRIBUTING TO TISSUE REPAIR

The biological activities of terpenoids and their essential oils (EOs) have an importance at the levels of tissue repair and regeneration. However, instead of listing these activities generically, a more clinically meaningful approach may be to relate each activity directly to the specific pathophysiological targets that characterise chronic wound failure especially in the case of diabetic foot ulcers (DFUs). DFU pathophysiology is marked by four interrelated defects: persistent microbial biofilm, aberrant chronic inflammation, deficient angiogenesis, and damaged extracellular matrix (ECM) reorganisation. In the subsequent sequences, we address potential therapeutic target clinical areas that each of these classes of compound may have therapeutically useful activity in (as demonstrated by preclinical findings) , along with a similar (for terpenoids only), OV-LV assessment 36 for each drug candidate.

2.1. Therapeutic Target: Anti-Biofilm and Antimicrobial Defence

Biofilms are one of the key factors behind why chronic wounds, such as leg and foot ulcers (which affect primarily those with diabetes), don't heal even after treatment with antibiotics. Developed biofilms in DFUs, meanwhile from Staphylococcus aureus and Pseudomonas aeruginosa for the most part, constitute a physical barrier that filters bacteria off immune cells and increases MICs malevolently. Several terpenes have exhibited striking anti-biofilm activity: thymol and carvacrol (the latter from thyme and oregano oils) disrupt bacterial membrane integrity and quorum sensing signals required for biofilm maintenance, while the primarily active terpinen-4-ol in tea tree oil has been demonstrated to act against pre-established S. aureus biofilms at sub-MIC concentrations, and eugenol (clove oil) prevents adhesion by interference with surface protein expression. However, the translational caveat here is large; by far the majority of anti-biofilm data originate from in vitro static biofilms (e.g., microtiter plate) with poor relation to the three-dimensional nutrient-sparse polymicrobial biofilms encountered in vivo. Penetration of EO bactericidal actives through mature biofilm matrix at concentrations attainable in wound tissue without cytotoxicity is infrequent. This is a well-defined engineering problem that has to be solved in order for anti-biofilm efficacy statements to be clinically substantiated (Huang et al., 2019).

2.2.Target of Therapy: Termination of Chronic Inflammation

In contrast to acute inflammation, which resolves and advances into proliferation, excessive pro-inflammatory cytokine (TNF-alpha, IL-1beta, IL-6) and also matrix metalloproteinase (MMPs) expression is maintained in chronic wound inflammation, degrading newly synthesised extracellular matrix components and inhibiting reepithelialisation (Nguyen et al., 2021). Several terpenoids have demonstrated mechanistic potential in addressing this phenotype: beta-caryophyllene (black pepper, copaiba oil) is a selective CB2 receptor agonist that downregulates NF-kappaB signalling, decreasing TNF-alpha and IL-6 secretion in macrophage models; limonene suppresses NLRP3 inflammasome activation; and rosmarinic acid (rosemary) inhibits COX-2 and 5-LOX pathways. This inflammatory status is further enhanced by the overproduction of ROS. The antioxidant potential of eugenol, carvacrol, and rosmarinic acid scavenges free radicals and reduces lipid peroxidation, thus preventing oxidative damage to collagen progenitors. The translational disconnect in this population relates to the lack of a pharmacokinetic (PK) endpoint: optic evidence that topically delivered EO actives actually achieve therapeutic concentrations within dermal tissue at the wound bed, specifically in patients with limited microcirculation associated with DFU. Randomised studies assessing the penetration depth of EO and changes in tissue-level cytokine levels are needed in order to translate this into the clinic (Samanta et al., 2016; Pérez-Recalde et al., 2018).

2.3. Therapeutic Target: Pro-Angiogenic Activity

Deficient angiogenesis is a key feature of DFU pathophysiology: impaired neovascularisation leads to hypoxic wound beds, which are not able to support the metabolic needs of healing.

Restoration of blood vessel formation - primarily through upregulation of vascular endothelial growth factor (VEGF) and promotion of endothelial cell migration - is therefore a critical therapeutic target. Preliminary evidence suggests several terpenoids may support angiogenesis: borneol (rosemary) has been shown to promote VEGF expression and endothelial tube formation in in vitro models; linalool (lavender oil) enhances fibroblast and endothelial cell proliferation; and alpha-pinene promotes macrophage-mediated angiogenic signalling. However, the translational gap in this target domain is among the most severe. Pro-angiogenic claims are almost entirely based on cell-culture experiments using non-diabetic cell lines, which do not replicate the hyperglycaemic, oxidatively stressed endothelial dysfunction characteristic of DFU. No clinical studies to date have measured VEGF upregulation or vessel density changes in human chronic wounds in response to EO treatment. This represents a critical unmet research need before pro-angiogenic claims can be substantiated (Adukwu et al., 2016; Caputo et al., 2020).

2.4. Therapeutic Target: ECM Remodelling and Re-epithelialization

Effective wound closure requires both the deposition of organised collagen matrix by fibroblasts and the coordinated migration of keratinocytes across the wound surface. Dysregulated MMP activity in chronic wounds degrades collagen faster than it is synthesised, preventing stable ECM formation. Several terpenoids have demonstrated ECM-supportive activity: thymol stimulates osteoblast differentiation and collagen deposition in bone repair models; rosmarinic acid promotes fibroblast proliferation and suppresses MMP-1 and MMP-3 expression; lavender and chamomile EOs have been shown to accelerate re-epithelialization and wound contraction in rodent excision models; and limonene supports myelination in neural repair contexts, suggesting potential for the neuropathic component of DFU. The neuroprotective properties of borneol and limonene - including oxidative stress reduction, pro-inflammatory cytokine suppression, and modulation of neurotransmitter activity - are also relevant to the neurological deficits that define DFU. The translational limitation here is methodological: existing data derive predominantly from non-diabetic animal excision models, which heal fundamentally differently from the ischaemic, neuropathic, and biofilm-colonised environment of a clinical DFU. Future studies must use streptozotocin-induced diabetic wound models as a minimum standard to generate translationally meaningful efficacy data (Xu et al., 2022; Li et al., 2021; González-Fuentes et al., 2018).

3. APPLICATIONS IN TISSUE REPAIR

Terpenoids and essential oils (eos) play an important role in tissue repair through cell proliferation stimulation and antimicrobial defence mechanisms. Terpenoids and EOs are important in collagen genesis. It plays an important role in wound healing through bioactive compounds that stimulate fibroblast proliferation, producing glycosaminoglycans and collagen, while facilitating the deposition of an extracellular matrix, and through the release of bioactive factors that modulate inflammatory processes to regenerate tissue (Chircov et al., 2021). Figure 1.

Figure 1. Terpenoids and essential oils in tissue repair

Characteristic	Terpenoids & Essential Oils
Cell Proliferation	Stimulates cell growth
Antimicrobial Defence	Provides defence mechanisms
Collagenogenesis	Important role in collagen creation
Wound Healing	Stimulates fibroblast proliferation
Bone Regeneration	Promotes osteoblast differentiation
Biomaterial Engineering	Enables controlled release of compounds

3.1. Wound Healing

Lavender and chamomile essential oils stimulate wound healing through enhanced collagen deposition, wound closure, and re-epithelialization and are therefore effective in the treatment of skin wounds, burns, and ulcers (Bahadur & Fatima, 2024). Their antibacterial properties prevent infection and maintain the cleanliness of the wound. Lavender oil has linalool, which activates fibroblast activity and stimulates renewal of the skin. Chamomile has α-bisabolol, which reduces inflammation and stimulates healing of the epithelium. Tea tree oil has a high amount of terpinen-4-ol, a broad-spectrum antibacterial agent that is known to prevent infections of wounds (Albahri et al., 2023).

3.2. Bone Regeneration

Apart from the healing of soft tissues, bone reconstruction is helped by essential oils. Thymol, which is from thyme oil, and Rosmarinic acid, which is from rosemary oil, boost osteoblast growth and mineralisation of the bones, thereby being helpful for fracture repair, integration of a bone transplant, and prevention against osteoporosis. The osteoblastic activity and calcification are promoted by rosemary oil, while thyme oil, which possesses thymol and carvacrol, contributes to bone recovery (Ivanova et al., 2023; Elbahnasawy & Rakhimov, 2019).

3.3. Advanced Applications

New developments in terpenoid and essential oil applications in biomaterials engineering have focused on scaffolds, hydrogels, and coatings of implants. Through these developments, controlled release of bioactive chemicals is achieved, which sustains ongoing anti-inflammatory and antibacterial functions and enhances cell adhesion, survival, function, and integration with tissue. Chitosan scaffolds that are biodegradable and are impregnated with tea tree oil, for instance, enhance antibacterial protection and wound healing. Eucalyptus oil-based hydrogels help to maintain the wound moist and reduce microbial infection. In addition, essential oil-coated implants like rosemary and thyme oils prevent post-surgical infections and enhance biocompatibility (Unalan & Boccaccini, 2021; Chelu, 2024).

4. MECHANISMS OF ACTION

Terpenoids and EOs exert their therapeutic effects in tissue repair through a variety of biochemical and molecular mechanisms. As a result, their power to modulate inflammation, oxidative stress, and cellular repair is key in speeding up wound healing, bone regeneration, and neural tissue recovery (Sharifi-Rad, 2017; Sheppard et al., 2022).

4.1. The Biochemical Mechanisms Involved in Tissue Repair:

Terpenoids and essential oils modulate tissue regeneration mainly by the regulation of gene expression, enzyme activity, and receptor interactions. This leads to more cell proliferation, extracellular matrix changes, and subsequent tissue healing. Linalool and β-caryophyllene are the terpenoids that induce wound healing by accelerating fibroblast migration and collagen synthesis. Both the TTO and EEO oils are widely recognised for their anti-bacterial activity, cleaving bacterial membranes and preventing biofilm formation, thus avoiding chronic wound infections. The angiogenesis, the formation of new vessels necessary to bring oxygen and nutrients to a tissue in need of repair, is also regulated by compounds such as rosmarinic acid and thymol, which have been shown they influence this important event (Ivanova et al., 2023; Sharifi-Rad et al., 2017; Raza et al., 2024; Koyama et al., 2019; Ngeow et al., 2022).

4.2. Effect on Inflammation:

Inflammation is the first phase of tissue regeneration; however, abundant inflammation results in chronic wounds and fibrosis. By inhibiting proinflammatory cytokines (such as TNF-α, IL-6, and IL-1β) and controlling important signalling pathways, terpenoids and essential oils reduce inflammation. By preventing the NF-κB pathway, reducing COX-2 activity, and managing macrophage activity, these measures reduce inflammation. Successful tissue repair depends on the induction of a shift from the pro-inflammatory M1 to the anti-inflammatory M2 phenotype. Clove oil and rosemary oil both possess β-caryophyllene, a cannabinoid receptor agonist that lowers inflammation and heals wounds (Abbas et al., 2017; Zhao et al., 2016; Quintans et al., 2019; Baradaran Rahimi & Askari, 2022).

4.3. Effect on Oxidative Stress:

Oxidative stress, which is caused by an excess of reactive oxygen species (ROS), weakens cellular structures and delays the healing process. High concentrations of natural antioxidants found in most essential oils neutralise ROS and stop damage from free radicals. Clove oil contains eugenol, a free radical scavenger that guards against DNA damage and lipid peroxidation. Carvacrol, which is included in oregano oil, prevents oxidative stress in cells by boosting the activity of antioxidant enzymes such as catalase (CAT) and superoxide dismutase (SOD). Rosmarinic acid in rosemary oil reduces mitochondrial dysfunction and enhances energy production in regenerated tissues (Dunnill et al., 2017; Gülçin, 2011; Madkour et al., 2024).

4.4. Effect on Cellular Repair and Regeneration:

Terpenoids and essential oils with tissue remodelling properties induce cell proliferation, migration, and matrix deposition. Limonene, found in citrus oils, induces nerve formation and myelination, with neuroprotective effects. Thymol, which is derived from thyme oil, induces bone formation through the regulation of osteoblast differentiation. Lavender oil, particularly its linalool constituent, enhances keratinocyte proliferation and epidermal renewal (da Silva Carneiro et al., 2025; Aziz et al., 2018). Fig. 2.

Figure 2. Mechanisms of Action of Terpenoids and Essential Oils in Tissue Repair

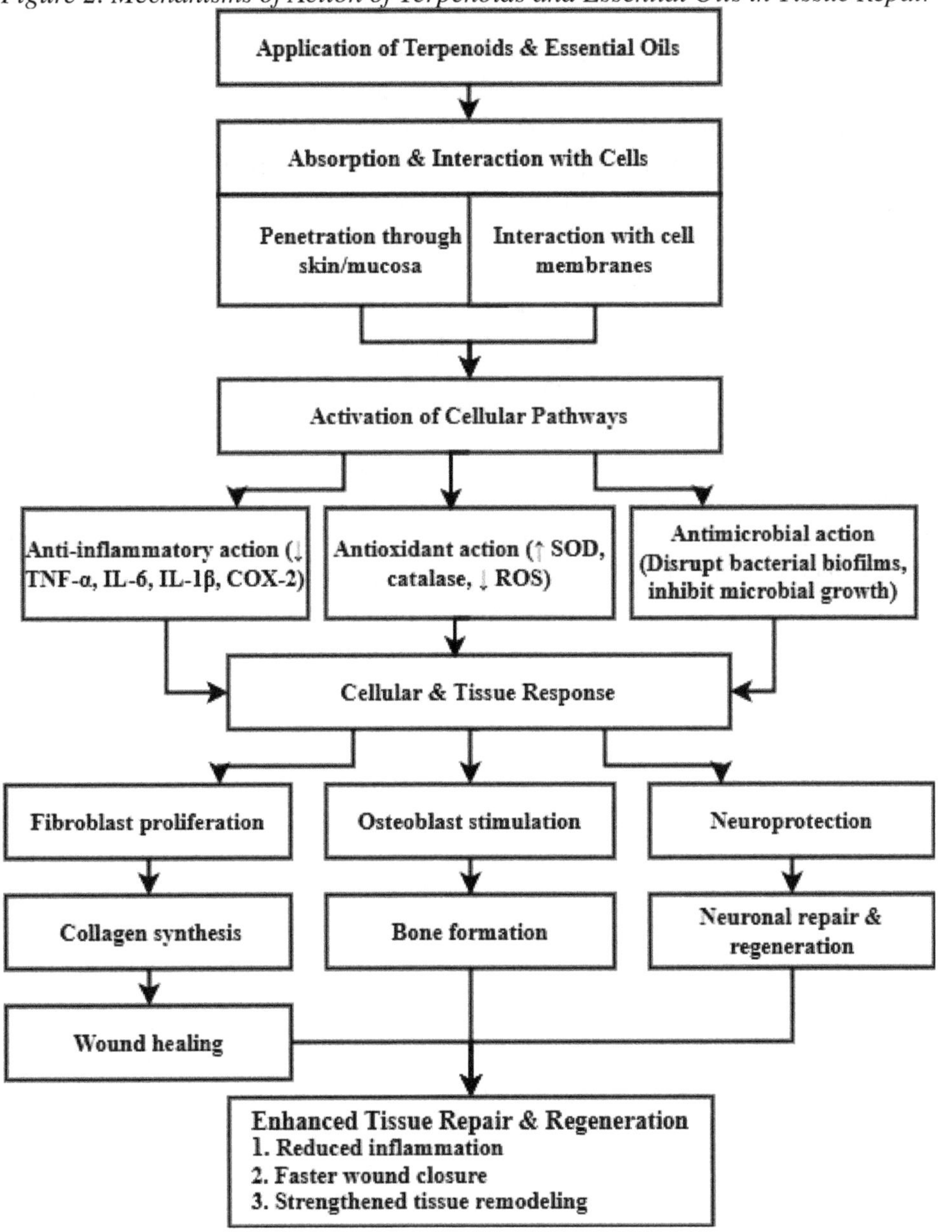

5. EXTRACTION METHODS AND ADVANCES

The quality, content, and production of essential oils, all of which ultimately impact their therapeutic effect, are highly dependent on the method of extraction. For example, the plant material must be boiled to facilitate hydrodistillation in order to release the volatile chemicals. But in this process, the substances are frequently subjected to heat, which might degrade heat-sensitive elements and reduce the efficacy of the oil.

On the other hand, the preferred technique is now supercritical fluid extraction (SFE), particularly using supercritical carbon dioxide (CO_2). This approach takes advantage of the unique solvent properties of supercritical fluids, meaning a substance that is intermediate between gas and liquid, in contrast to typical dense liquids or gases, it may behave as a G909Idense fluid having variable solubility, density and gas densities. It operates at temperatures and pressures above the critical point of CO_2. This is what allows penetration of the plant-matrix and selective extraction. SFE is also advantageous due to the fact that it can be performed at low temperature, which avoids decomposition of thermolabile compounds, and recovery of extracts with superior yield and purity. Save time and money for the survey, as benefits of SFE over traditional techniques are fully proved (Sethunga et al., 2022; Reyes-Jurado et al., 2015; Rajput et al., 2023). Based on a similar study of SFE (subcritical water extraction) from kinnow mandarin peels, it was found that SFE had the highest essential oil yield (1.57%) with antioxidant activity and total phenolic content (Mukhopadhyay, 2009). This approach showed that the method raised advantages and generated rich, good-quality bioactive-loaded essential oils. Fig 3.

Figure 3. Extraction Methods and Their Impact on Essential Oils

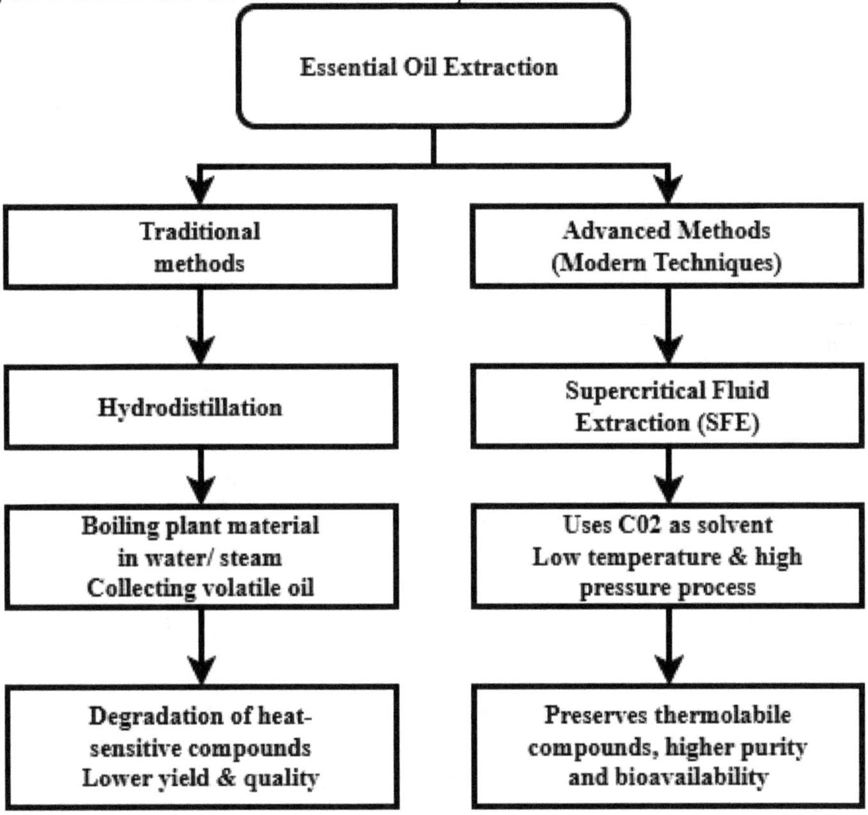

Moreover, the SFE technique is quite flexible and an appropriate selection of conditions (e.g., temperature, pressure) can be applied to lead the extraction of selected components in oils that have been identified as having therapeutic properties. This precision is paramount in applications requiring high levels of both purity and concentration, for example, in pharmaceuticals, or cosmetics (Yousefi et al., 2019).

Table 3. Traditional vs. Advanced Extraction Method

Aspect	Hydrodistillation (Traditional)	Supercritical Fluid Extraction (SFE)	Ref.
Principle	Boiling plant material to release essential oils	CO_2 in a supercritical state extract oils	(Dashtian et al., 2024)
Temperature	High (risk of degradation)	Low (preserves sensitive compounds)	(Costa et al., 2012)
Yield	Lower yield and less potent extracts	Higher yield with purer extracts	(Perović et al., 2024)
Purity	Contains impurities	High purity and stability	(Augoye et al., 2024)
Bioactive Compound Loss	Significant due to heat exposure	Minimal due to lower temperatures	(Ali et al., 2023)
Cost & Accessibility	Low-cost, widely accessible	Expensive, requires specialised equipment	(Aćimović et al., 2020)
Applications	General use, traditional medicine	Pharmaceuticals, cosmetics, and advanced therapy	(Machado et al., 2013)

5.1. Extraction methods: Hydro distillation and SFE

It is the process of removing essential oils and terpenoids from the herbal material, a key step in exploiting their bioactive as well as medicinal activities for various biomedical applications. Among the different extraction techniques, hydro distillation and supercritical fluid extraction are two of the most commonly used techniques with their own advantages and uses.

Hydro distillation is one of the traditional techniques for obtaining essential oils. This is accomplished by submerging the plant material in water and boiling it or suspending it over boiling water. Vaporous material of steam from the plant material is removed by water vapour during boiling. This steam is then condensed, and the oil, being heavier than water, rises to the top. Hydro distillation is a cheap process and satisfactory for all aromatics. However, the operations can be at high temperatures for long periods of time that break down heat-sensitive compounds and reduce total oil yield and quality (Pourmortazavi & Hajimirsadeghi, 2007; Souiy, 2023).

5.2. Supercritical fluid extraction (SFE):

is a technique that uses carbon dioxide (CO_2) as the solvent, followed by high pressure and temperature conditions, allowing study of kinetics and thermodynamics, taken as standard recent practice to traditional methods. In SFE, the CO_2 is operated above its critical temperature and critical pressure and behaves as both liquid and gas. Supercritical CO_2 easily collects terpenoids, and essential oil is isolated with high precision. The method is performed at cold conditions, thereby not exposing thermolabile compounds. CO_2 is a non-toxic, eco-friendly, and solvent-free post-treatment, which is why it is being used in the pharmaceutical, cosmetic and food industries (Selvaraj et al., 2022; Babbar et al., 2015).

5.3. The importance of extraction procedures in the effectiveness of essential oils.

In assessing quality, strength, purity, and general utility of essential oils, extraction methods come into play. The bioactive molecules responsible for the medicinal value of essential oils are highly dependent on temperature, pressure, and chemical conditions. Therefore, to maintain their natural integrity as well as elevate their biological action, an optimal extraction method is employed (Zio et al., 2020; Varzakas, 2016)

- **Preservation of Bioactive Substances**: The chemical composition of essential oils is determined by various extraction methods. The volatility of elements like terpenes, phenolics, and aldehydes is impacted differently by steam distillation, cold pressing, solvent extraction, and supercritical fluid extraction (Mira et al., 1999). Poor extraction leads to the loss or degradation of such constituents, which diminishes the therapeutic potential of the oil (Belwal et al., 2020).
- **Purity and Safety:** Residual solvents or contaminants added during poor or aggressive extraction methods can threaten the safety of the essential oil (Nafiu et al., 2017). Solvent extraction, for instance, can leave dangerous residues if not properly refined, making both cutaneous and internal application of the oil unsafe. Low-contamination and high-purity procedures are thus favoured for aromatherapy, medicinal, and cosmetic uses (Alam et al., 2024).
- **Yield & Concentration:** Efficiency of the extraction method impacts the amount of essential oil extracted from a volume of plant material (Da Silva et al., 2022). Supercritical CO_2 extraction can be used to enhance yields while preserving a broader spectrum of volatile and non-volatile chemicals. It makes the extract complete and more concentrated, accordingly more effective (Aljaafari et al., 2021).
- **Standardisation and consistency:** Industrial and clinical applications require consistent quality oils. Processings at controlled parameters (e.g., pressure, temperature and time) result in a reproducible product, which is essential for standardizing quality of oils that is essential for pharmacological conformation and legal compliance (Rani et al., 2023; Vora et al., 2024).
- **Environmental friendliness and Sustainability:** Extraction methods today also take into account the environmental impact. CO_2 extraction is effi-

cient and green because it uses non-toxic solvents and produces no waste. Sustainable harvesting practices help in maintaining the ecological balance and also promising for future supply of medicinal plants (Baptista-Silva et al., 2020; Da Silva et al., 2022).

6. CHALLENGES AND LIMITATIONS

This chapter describes the hindrances and barriers to the clinical use of essential oils and terpenoids for medical purposes-. One of the main issues is inconsistency in efficacy, which depends on factors such as plant origin, extraction process and variations in dose. This can vary considerably between species, environments and pest-type infestations, and time to harvesting (Aljaafari et al., 2021; Gurjar et al., 2012). Furthermore, the content of bioactive compounds is also substantially different in terms of quality, purity and concentration because the two extraction methods involved (hydrodistillation vs supercritical fluid extraction). Also, the therapeutic potential of the oils obtained may highly depend upon the extraction process, so it is challenging to retain uniform clinical efficacies across different protocols. The absence of standard protocols may make the dosage uniform, and they are not readily repeatable, and therefore, when used in clinics or therapy performance parameters for all time (Rani et al., 2023; Vora et al., 2024; Baptista-Silva et al., 2020).

One major obstacle is the lack of standardized ways to treat in the clinical environment. Despite the high number of studies which point to the therapeutic effects of essential oils, a consensus on dosage, formulation and application does not yet exist. Non-standardisation has made the reproducibility of clinical results, safety and efficacy in different patients difficult. Moreover, the quality of EOs is poor because different extraction methods are used, and several plants are considered to produce pharmaceutical-grade products with defined therapeutic activities. The only solution for these problems is clinical validation and actionable therapeutic strategies. This includes establishing standardised methods for evaluating the quality of essential oil, improving processes for their extraction and pursuing large clinical trials to demonstrate their safety and efficacy in medical practice. A critical but underexamined dimension of the challenges facing terpenoids and EOs is the structural mismatch between the conditions under which preclinical efficacy is demonstrated and the biological reality of clinical chronic wounds. This chasm spans several layers and comprises a series of well-defined engineering, pharmacological, and regulatory issues.

Pharmacokinetic and delivery issues: Compounds in EOs are mostly hydrophobic, volatile constituents. Topically, they are found to rapidly evaporate, be poorly soluble in aqueous solutions and fail to penetrate wound eschar deeply, resulting in few effective levels reached at the site of intended action – within the wound bed. In contrast to drug actives developed for strippable transdermal delivery, all studies of EO in the literature report on unrefined oils applied directly and do not provide data on actual tissue concentrations attained. In the absence of confirmed pharmacokinetics (Cmax, Tmax, AUC at the wound site), no efficacy statement could be deemed clinically relevant. Nanoparticle encapsulation, hydrogel matrices and cyclodextrin inclusion complexes are potential engineering solutions; no such carrier system has ever been combined with an antiseptic terpenoid proven by a controlled fistula healing trial in humans. Model-to-clinic disconnect: The classic excision wound model in healthy animals the workhorse of EO wound healing research -- heals by contraction, with this process contributing ~90% to closure in rodents but <20% in humans. It does not model biofilm colonisation, diabetic microangiopathy or neuropathy, nor the chronically inflamed phenotype of a clinical DFU. Positive results in these bow and arrow models clearly carry diminished translational weight and need to be acknowledged as hypothesis-generating more than efficacy-confirming.

Lack of regulatory pathway: Most jurisdictions do not have a dedicated regulation system for phytochemical wound therapeutics as with pharmaceuticals. Depending upon its formulation and geographical jurisdiction, EOs could be categorised as cosmetics, phytopharmaceuticals or medical devices with varied evidentiary requirements. This muddy regulatory environment discourages pharma investment and means even good compounds lack a clean pathway to approval. This needs to be addressed through proactive dialogue with the regulatory agencies (FDA, EMA) to establish a data package for phytochemical wound therapeutic approval (Khedkar, 2025; Chelu, 2024; Pavela & Benelli, 2016; Wang et al., 2023).

Table 4. Challenges and Limitations of Essential Oils in Clinical Applications

Challenge	Description	Impact on Clinical Use	Possible Solution	Ref.
Efficacy Variability	Differences in plant source, extraction method, and dosage lead to inconsistent chemical composition and therapeutic effects.	Reduces reproducibility and reliability of clinical outcomes.	Standardized cultivation and extraction protocols.	(Rani et al., 2023; Vora et al., 2024; Baptista-Silva et al., 2020; Khedkar, 2025; Chelu, 2024; Pavela & Benelli, 2016; Wang et al., 2023)
Extraction Procedures	Hydrodistillation and SFE yield varying purity and bioactivity levels.	Inconsistent potency and effectiveness.	Use of advanced methods like SFE for higher purity.	
Dosage Inconsistencies	Lack of uniform dosing standards for clinical applications.	Difficulty in determining safe and effective doses.	Establish dosage guidelines through clinical trials.	
Lack of Standardization	No universal framework for formulation, dosing, or application in medical settings.	Challenges in regulatory approval and clinical reproducibility.	Develop standardized clinical protocols.	
Stability Issues	Some essential oils degrade over time due to oxidation or improper storage.	Reduced therapeutic efficacy.	Use of protective packaging and antioxidants.	
Limited Clinical Validation	Insufficient large-scale clinical trials to verify efficacy and safety.	Slower acceptance in mainstream medicine.	More rigorous and large-scale clinical studies.	

7. A FOUR-STAGE RESEARCH STRATEGY FOR TRANSLATIONAL VALIDATION OF TERPENOIDS AND ESSENTIAL OILS IN CHRONIC WOUND HEALING

7.1 Phytochemical Standardisation Protocol

Stage 1: No compound can progress through a translational pathway without a reproducible, characterised starting material. Stage 1 mandates rigorous phytochemical standardisation of any candidate terpenoid or EO before it advances to mechanistic or in vivo testing. This must include:

- GC-MS fingerprinting to define the full chemical profile of each batch, with minimum thresholds set for primary bioactive constituents (e.g., thymol > 45% for thyme oil designated for anti-biofilm studies; linalool > 25% for lavender EO designated for anti-inflammatory studies) (Handa et al., 2008; Figueiredo et al., 2008).
- Batch-to-batch reproducibility testing across a minimum of five independent extraction batches from standardised plant material (Groeber et al., 2016).
- Stability profiling under relevant storage and formulation conditions (temperature, light exposure, pH) (European Medicines Agency. (2011). Guideline on specifications for herbal substances and preparations., 2011)
- Cytotoxicity screening on primary human dermal fibroblasts and keratinocytes to establish a therapeutic index before any efficacy testing proceeds. Only compounds achieving a defined standardisation specification should advance to Stage 2. This forms a compound-specific chemical passport and ensures that any positive efficacy readout is reproducible and attributable not to a crude extract but to a defined chemical entity (Kunle et al., 2012; Azeredo et al., 2017).

Standardisation should be defined in a Compound Development Dossier (CDD), like a drug DMF, that then goes with the candidate throughout all further development steps. However, this widely used strategy in the development of pharmaceuticals is virtually unexplored in EO research: it represents the low benchmark for any future regulatory filing and useful pooling of findings across studies (Handa et al., 2008; Figueiredo et al., 2008; Groeber et al., 2016; European Medicines Agency. (2011). Guideline on specifications for herbal substances and preparations., 2011; Kunle et al., 2012; Azeredo et al., 2017).

7.2. Validation with disease-relevant in vitro model and mechanistic assay on dregearene effect

Stage 2: progresses from early activity screening to mechanism validation in cell systems reflecting the biology of the target indication. For DFU-targeted candidates, this would include Efficacy against biofilms testing using 3D flow-based biofilm models (e.g., CDC Biofilm Reactor), with clinical isolates of S. aureus and P. aeruginosa from DFUs. Viable cells within the biofilm mass should also be described as endpoints with MBECs determined and compared to cytotoxic threshold levels in Phase 1 (Olson et al., 2002; Han et al., 2024). Anti-inflammatory mechanistic validation in TNF-alpha-stimulated macrophages (THP-1 cells or primary human macrophages) model upon hyperglycemic conditions (25 mM glucose) that mimic the DFU inflammatory microenvironment. Issues to be addressed are multiplex cytokine determinations (TNF-alpha, IL-1beta and IL-6, IL-10), NF-kappaB nuclear translocation assays and MMP- 1/MMP-3 activities in co-cultured fibroblast models (Veith et al., 2020). Pro-angiogenic activity was determined upon analysis of tube formation and scratch-wound migration on human dermal microvascular endothelial cells (HMVECs) under hyperglycemic conditions; VEGF-A secretion in the supernatant was assessed by ELISA as readout. Evaluation of ECM remodelling by collagen gel contraction assays (to assess collagen synthesis MMP/TIMP ratios, and keratinocyte migration (Falanga et al., 2022). A positive control (e.g. silver sulfadiazine for an anti-biofilm agent or dexamethasone for an anti-inflammatory) and a vehicle control formulary matched to the anticipated delivery vehicle should be incorporated into all Phase 2 studies.

7.3 Translational In Vivo Validation Using Diabetic Wound Models

Stage 3: Consideration of animal models designed specifically to mimic the pathophysiology of a diabetic foot ulcer. The standard rodent excision model in the noncompromised host is inadequate and cannot be considered as prime evidence of graft efficacy for wound healing. The minimal model would be the STZ-induced type 1 diabetic mouse or rat where healing is demonstrably deficient (Galiano et al., 2004). For a stricter translation, the db/db (leptin receptor-deficient) mouse genetic diabetic model, accompanied by spontaneous obesity and insulin resistance, is preferentially resistant (Wilkinson & Hardman, 2020). All in vivo studies are to include:

- Standardised wound creation with 8mm biopsy punch full-thickness excision and splinting of the wounds to prevent contraction-dominant healing,
- Bacterial inoculation with a known polymicrobial biofilm (S. aureus P. aeruginosa) was established 48 hours before treatment.
- Quantification of wound area changes by digital planimetry at defined time points(g., Days0,3,7,10, and 14) (Galiano et al., 2004; Wilkinson & Hardman, 2020; Moura et al., 2013)
- Histological analysis at Day 14, including measurements for collagen density (Masson's trichrome staining), vascularity (CD31 immunostaining), inflammatory infiltrate immunohistochemistry stains of CD68 macrophage staining), and re-epithelialization length.
- Tissue-level availability levels quantifications using the candidate compounds via LC MSSMS from Day 7 wound tissue samples. Confirmation of in vivo results of the third stage should, therefore, be a prerequisite for advancing to early clinical trials (Moura et al., 2013).

7.4 Stage 4: Design of Phase I/II Clinical Trials

Stage 4 converts successful Stage 3 agents into first-in-human trials. Recommended design features include:

- **Formulation**: candidates should be delivered as an engineered formulation that has been designed for penetration into a wound bed, and not as a crude oil. Preferred compositions are based on chitosan nanoparticle systems, sodium alginate hydrogel wound dressings or cyclodextrin inclusion complexes, for the specific actives as validated in a Stage 3 pharmacokinetic sub-study (Everett & Mathioudakis, 2018; Armstrong et al., 2017).
- **Patient population:** Adults, with a verified clinical diagnosis of T2DM and CLoW >4 weeks, Wagner Grade 1-2 when the evidence for biofilm is confirmed. Energy exclusion criteria should be active osteomyelitis, advanced peripheral arterial disease (ABI < 0.5), and present systemic antibiotic therapy (Shoham et al., 2024; Zhang et al., 2017).
- **Phase I primary endpoint**: Safety and tolerability, including the frequency of adverse skin reactions, allergic sensitisation and systemic absorption (plasma pharmacokinetics): phase I primary endpoint (Shoham et al., 2024).
- **Phase II key endpoint:** per cent reduction of wound area at Week 4 based on digital planimetry. Secondary outcomes are: the proportion of wounds with 50% reduction in wound surface area; validated wound scores (PUSH Tool, BWAT); time to 50% closure; quantitative culture for wound biofilm burden; and health-related quality of life (EQ-5D) (Zhang et al., 2017; Guo & DiPietro, 2010; Ngeow et al., 2022).
- **Design:** a multicentre, double-blinded randomised controlled trial with standard-of-care comparator arm. To obtain 80% power for the detection of a clinically relevant 20% between group difference in wound area reduction, at least 120 patients per arm is advised (Zhang et al., 2017).
- **Regulatory interaction**: pre-submission consultation with the relevant national medicines authority (CDSCO in India; FDA or EMA for international studies) is advised to align on product classification and evidentiary requirements for a marketing authorisation (Guo & DiPietro, 2010).

8. CONCLUSION

Terpenoids and essential oils represent a scientifically compelling class of phytochemicals with demonstrated preclinical activity across four clinically critical domains of chronic wound pathophysiology: anti-biofilm defence, resolution of chronic inflammation, pro-angiogenic support, and ECM remodelling. Key candidates, including thymol, carvacrol, beta-caryophyllene, linalool, borneol, rosmarinic acid, and terpinen-4-ol, have mechanistic profiles that map directly onto the therapeutic targets defining diabetic foot ulcer failure. However, their absence from mainstream clinical wound management reflects a series of clearly identifiable translational barriers: lack of phytochemical standardisation, pharmacokinetic data gaps, the use of inadequate animal models, and the absence of regulatory pathways for phytochemical wound therapeutics. These are engineering, scientific, and regulatory challenges amenable to structured solutions, not insurmountable obstacles. This chapter has proposed a structured four-stage translational research strategy to bridge the gap between preclinical promise and clinical validation. Stage 1 mandates phytochemical standardisation and cytotoxicity profiling before any efficacy claim is advanced. Step 2 includes mechanism validation in disease-relevant in vitro models of hyperglycaemia. In Stage 3, STZ-induced diabetic wound models with biofilm inoculation are required as the minimum in vivo criteria. Stage 4 provides a detailed Phase I/II clinical trial design framework, including patient selection criteria, primary and secondary endpoints, and regulatory engagement guidance. Adoption of this framework would transform the evidence base for terpenoids and EOs from a promising but clinically inaccessible collection of preclinical data into a structured pathway toward regulatory approval and patient benefit. The centuries-old use of these aromatic compounds in wound care merits the rigour of modern translational science to validate, standardise, and deliver their therapeutic potential to patients who need it most.

REFERENCES

Abbas, F., Ke, Y., Yu, R., Yue, Y., Amanullah, S., Jahangir, M. M., & Fan, Y. (2017). Volatile terpenoids: Functions, biosynthesis, modulation and genetic manipulation. *Planta, 246*, 803–816. DOI: 10.1007/s00425-017-2749-x PMID: 28803364

Aćimović, M., Tešević, V., Smiljanić, K. T., Cvetković, M., Stanković, J., Kiprovski, B., & Sikora, V. (2020). Hydrolates: Chemical composition and applications. *Advanced Technologies*, *9*(2), 54–70.

Adukwu, E. C., Bowles, M., Edwards-Jones, V., & Bone, H. (2016). Antimicrobial and cytotoxic evaluation of lemongrass essential oil and citral. *Applied Microbiology and Biotechnology*, *100*, 9619–9627. DOI: 10.1007/s00253-016-7807-y PMID: 27562470

Alam, M. M., Greco, A., Rajabimashhadi, Z., & Corcione, C. E. (2024). Green extraction of lignin and applications. Cleaner Materials, 100253.

Alam, W., Hussain, Y., Ahmad, S., Ali, A., & Khan, H. (2023). Neuroprotective effects of essential oils. In *Phytonutrients and neurological disorders* (pp. 305–333). Academic Press. DOI: 10.1016/B978-0-12-824467-8.00006-1

Albahri, G., Badran, A., Hijazi, A., Daou, A., Baydoun, E., Nasser, M., & Merah, O. (2023). Therapeutic wound healing bioactivities of medicinal plants. *Life (Chicago, Ill.)*, *13*(2), 317. DOI: 10.3390/life13020317 PMID: 36836674

Ali, S., Ekbbal, R., Salar, S., Ali, S. A., Jaiswal, A. K., Singh, M., Yadav, D. K., & Kumar, S. (2023). Quality standards and pharmacological interventions of natural oils. *ACS Omega*, *8*(43), 39945–39963. DOI: 10.1021/acsomega.3c05241 PMID: 37953833

Aljaafari, M. N., AlAli, A. O., Baqais, L., Alqubaisy, M., AlAli, M., Molouki, A., Ong-Abdullah, J., Abushelaibi, A., Lai, K. S., & Lim, S. H. (2021). Therapeutic applications of essential oils. *Molecules (Basel, Switzerland)*, *26*(3), 628. DOI: 10.3390/molecules26030628 PMID: 33530290

Aljaafari, M. N., AlAli, A. O., Baqais, L., Alqubaisy, M., AlAli, M., Molouki, A., Ong-Abdullah, J., Abushelaibi, A., Lai, K. S., & Lim, S. H. (2021). Potential therapeutic applications of essential oils. *Molecules (Basel, Switzerland)*, *26*(3), 628. DOI: 10.3390/molecules26030628 PMID: 33530290

Armstrong, D. G., Boulton, A. J. M., & Bus, S. A. (2017). Diabetic foot ulcers and recurrence. *The New England Journal of Medicine*, *376*(24), 2367–2375. DOI: 10.1056/NEJMra1615439 PMID: 28614678

Ashour, M., Wink, M., & Gershenzon, J. (2010). Biochemistry of terpenoids: Monoterpenes, sesquiterpenes and diterpenes. In M. Wink (Ed.), Biochemistry of Plant Secondary Metabolism (Annual Plant Reviews, Vol. 40, pp. 258–303). Wiley.

Augoye, O., Sobowale, A., & Muyiwa-Ajayi, T. P. (2024). Sustainable process design in essential oil extraction.

Azeredo, J., Azevedo, N. F., Briandet, R., Cerca, N., Coenye, T., Costa, A. R., Desvaux, M., Di Bonaventura, G., Hébraud, M., Jaglic, Z., Kačániová, M., Knøchel, S., Lourenço, A., Mergulhão, F., Meyer, R. L., Nychas, G., Simões, M., Tresse, O., & Sternberg, C. (2017). Critical review on biofilm methods. *Critical Reviews in Microbiology, 43*(3), 313–351. DOI: 10.1080/1040841X.2016.1208146 PMID: 27868469

Aziz, Z. A., Ahmad, A., Setapar, S. H., Karakucuk, A., Azim, M. M., & Lokhat, D.. (2018). Essential oils: Extraction techniques and therapeutic potential. *Current Drug Metabolism, 19*(13), 1100–1110. DOI: 10.2174/1389200219666180723144850 PMID: 30039757

Babbar, N., Oberoi, H. S., & Sandhu, S. K. (2015). Nutraceutical potential of bioactives from fruit residues. *Critical Reviews in Food Science and Nutrition, 55*(3), 319–337. DOI: 10.1080/10408398.2011.653734 PMID: 24915390

Bahadur, S., & Fatima, S. (2024). Essential oils of medicinal plants and wound healing activities. *Current Pharmaceutical Biotechnology, 25*(14), 1818–1834. DOI: 10.2174/0113892010282605231218064053 PMID: 38310452

Baptista-Silva, S., Borges, S., Ramos, O. L., Pintado, M., & Sarmento, B. (2020). Essential oils as therapeutic agents. *The Journal of Essential Oil Research, 32*(4), 279–295. DOI: 10.1080/10412905.2020.1746698

Baptista-Silva, S., Borges, S., Ramos, O. L., Pintado, M., & Sarmento, B. (2020). Progress of essential oils as potential therapeutic agents. *The Journal of Essential Oil Research, 32*(4), 279–295. DOI: 10.1080/10412905.2020.1746698

Baradaran Rahimi, V., & Askari, V. R. (2022). Immunomodulatory effects of β-caryophyllene. *BioFactors (Oxford, England), 48*(4), 857–882. DOI: 10.1002/biof.1869 PMID: 35648433

Belwal, T., Chemat, F., Venskutonis, P. R., Cravotto, G., Jaiswal, D. K., Bhatt, I. D., Devkota, H. P., & Luo, Z. (2020). Scaling-up non-conventional extraction techniques. *Trends in Analytical Chemistry, 127*, 115895. DOI: 10.1016/j.trac.2020.115895

Caputo, L., Smeriglio, A., Trombetta, D., Cornara, L., Trevena, G., & Valussi, M.. (2020). Chemical composition and biological activities of Leptospermum petersonii and Eucalyptus gunnii essential oils. *Frontiers in Microbiology*, *11*, 409. DOI: 10.3389/fmicb.2020.00409 PMID: 32351456

Chelu, M. (2024). Hydrogels with essential oils: Recent advances. *Gels (Basel, Switzerland)*, *10*(10), 636. DOI: 10.3390/gels10100636 PMID: 39451288

Chelu, M. (2024). Hydrogels with essential oils: Recent advances in designs and applications. *Gels (Basel, Switzerland)*, *10*(10), 636. DOI: 10.3390/gels10100636 PMID: 39451288

Chircov, C., Miclea, I. I., Grumezescu, V., & Grumezescu, A. M. (2021). Essential oils for bone repair and regeneration—Mechanisms and applications. *Materials (Basel)*, *14*(8), 1867. DOI: 10.3390/ma14081867 PMID: 33918697

Costa, P., Grosso, C., Gonçalves, S., Andrade, P. B., Valentão, P., Bernardo-Gil, M. G., & Romano, A. (2012). Recovery of bioactives from Lavandula viridis by supercritical fluid extraction. *Food Chemistry*, *135*(1), 112–121. DOI: 10.1016/j.foodchem.2012.04.108 PMID: 26434270

Cox-Georgian, D., Ramadoss, N., Dona, C., & Basu, C. (2019). Therapeutic and medicinal uses of terpenes. In *Medicinal plants: From farm to pharmacy* (pp. 333–359). Springer. DOI: 10.1007/978-3-030-31269-5_15

Da Silva, R. F., Carneiro, C. N., de Sousa, C. B., Gomez, F. J., Espino, M., Boiteux, J., Fernández, M. D., Silva, M. F., & Dias, F. D. (2022). Green analytical extraction of medicinal plant bioactives. *Microchemical Journal*, *175*, 107184. DOI: 10.1016/j.microc.2022.107184

Da Silva, R. F., Carneiro, C. N., de Sousa, C. B., Gomez, F. J., Espino, M., Boiteux, J., Fernández, M. D., Silva, M. F., & Dias, F. D. (2022). Sustainable extraction of bioactive compounds from medicinal plants based on green analytical chemistry principles. *Microchemical Journal*, *175*, 107184. DOI: 10.1016/j.microc.2022.107184

da Silva Carneiro, R. E., de Almeida da Costa, M. H., Lima Verde Santos, J. Z., Teixeira Uchoa, V., da Silva Lopes, L. U., & Mendes, A. N. (2025). Therapeutic actions of terpenes in neurodegenerative disorders and ECM remodeling. *Biocell*, *49*(1). Advance online publication. DOI: 10.32604/biocell.2024.058405

Dash, S., Behera, P. M., Mandal, U., Nayak, R., Parida, S., & Mahalik, G. (2021). Potential medicinal plants of Centurion University of Technology and Management and their medicinal uses. *Journal of Medicinal Plants Studies*, *9*(3), 251–258. DOI: 10.22271/plants.2021.v9.i3c.1303

Dashtian, K., Kamalabadi, M., Ghoorchian, A., Ganjali, M. R., & Rahimi-Nasrabadi, M. (2024). Integrated supercritical fluid extraction of essential oils. *Journal of Chromatography. A*, *1723*, 465240. DOI: 10.1016/j.chroma.2024.465240 PMID: 39154494

Dunnill, C., Patton, T., Brennan, J., Barrett, J., Dryden, M., Cooke, J., Leaper, D., & Georgopoulos, N. T. (2017). Reactive oxygen species in wound healing. *International Wound Journal*, *14*(1), 89–96. DOI: 10.1111/iwj.12557 PMID: 26688157

El-Saber Batiha, G., Alkazmi, L. M., Wasef, L. G., Beshbishy, A. M., Nadwa, E. H., & Rashwan, E. K. (2020). Syzygium aromaticum (Myrtaceae): Traditional uses, bioactive constituents, pharmacological and toxicological activities. *Biomolecules*, *10*(2), 202. DOI: 10.3390/biom10020202 PMID: 32019140

Elbahnasawy, A. S., Valeeva, E. R., El-Sayed, E. M., & Rakhimov, I. I. (2019). Impact of thyme and rosemary on osteoporosis prevention in rats. *Journal of Nutrition and Metabolism*, *2019*, 1431384. DOI: 10.1155/2019/1431384 PMID: 31049223

European Medicines Agency. (2011). Guideline on specifications for herbal substances and preparations.

Everett, E., & Mathioudakis, N. (2018). Management of diabetic foot ulcers. *Annals of the New York Academy of Sciences*, *1411*(1), 153–165. DOI: 10.1111/nyas.13569 PMID: 29377202

Falanga, V., Isseroff, R. R., Soulika, A. M., Romanelli, M., Margolis, D., Kapp, S., Granick, M., & Harding, K. (2022). Chronic wounds. *Nature Reviews. Disease Primers*, *8*(1), 50. DOI: 10.1038/s41572-022-00377-3 PMID: 35864102

Figueiredo, A. C., Barroso, J. G., Pedro, L. G., & Scheffer, J. J. (2008). Factors affecting secondary metabolite production in plants. *Flavour and Fragrance Journal*, *23*(4), 213–226. DOI: 10.1002/ffj.1875

Galiano, R. D., Michaels, J. V., Dobryansky, M., Levine, J. P., & Gurtner, G. C. (2004). Murine model of impaired wound healing. *Wound Repair and Regeneration*, *12*(4), 485–492. DOI: 10.1111/j.1067-1927.2004.12404.x PMID: 15260814

González-Fuentes, J., Selva, J., Moya, C., Castro-Vázquez, L., Lozano, M. V., Marcos, P., Plaza-Oliver, M., Rodríguez-Robledo, V., Santander-Ortega, M. J., Villaseca-González, N., & Arroyo-Jimenez, M. M. (2018). Neuroprotective natural molecules from food to brain. *Frontiers in Neuroscience*, *12*, 721. DOI: 10.3389/fnins.2018.00721 PMID: 30405328

Groeber, F., Schober, L., Schmid, F. F., Traube, A., Kolbus-Hernandez, S., Daton, K., Hoffmann, S., Petersohn, D., Schaefer-Korting, M., Walles, H., & Mewes, K. R. (2016). Validation of reconstructed epidermis for skin irritation testing. *Toxicology In Vitro*, *36*, 254–261. DOI: 10.1016/j.tiv.2016.07.008 PMID: 27435616

Gülçin, İ. (2011). Antioxidant activity of eugenol. *Journal of Medicinal Food*, *14*(9), 975–985. DOI: 10.1089/jmf.2010.0197 PMID: 21554120

Guo, S. A., & DiPietro, L. A. (2010). Factors affecting wound healing. *Journal of Dental Research*, *89*(3), 219–229. DOI: 10.1177/0022034509359125 PMID: 20139336

Gurjar, M. S., Ali, S., Akhtar, M., & Singh, K. S. (2012). Efficacy of plant extracts in plant disease management. *Agricultural Reviews (Karnal)*, *33*(2), 125–134.

Han, Y. Z., Du, B. X., Zhu, X. Y., Wang, Y. Z., Zheng, H. J., & Liu, W. J. (2024). Lipid metabolism disorder in diabetic kidney disease. *Frontiers in Endocrinology*, *15*, 1336402. DOI: 10.3389/fendo.2024.1336402 PMID: 38742197

Handa, S. S., Khanuja, S. P. S., Longo, G., & Rakesh, D. D. (2008). *Extraction technologies for medicinal and aromatic plants*. UNIDO.

Horváth, G., & Ács, K. (2015). Essential oils in respiratory tract diseases: Antibacterial and anti-inflammatory actions. *Flavour and Fragrance Journal*, *30*(5), 331–341. DOI: 10.1002/ffj.3252 PMID: 32313366

Huang, Y., Dan, N., Dan, W., & Zhao, W. (2019). Reinforced polycaprolactone/chitosan nanocomposites for controlled curcumin release in wound dressing. *ACS Omega*, *4*(27), 22292–22301. DOI: 10.1021/acsomega.9b02217 PMID: 31909312

Ivanova, E. P., Hasan, J., Webb, H. K., Gervinskas, G., Juodkazis, R., & Truong, V. K.. (2023). Biomaterial-assisted bone regeneration inspired by essential oil bioactives. *Materials (Basel)*, *16*(11), 4012. PMID: 37297146

Jagwani, D., & Krishna, P. H. (2021). Nature's nano-assets: Green synthesis, characterization techniques and applications. *Materials Today: Proceedings*, *46*, 2307–2317. DOI: 10.1016/j.matpr.2021.04.185

Khedkar, D. D. (2025). Plant products and essential oil therapy for emerging viral diseases. In *Fighting multidrug resistance with herbal extracts, essential oils and their components* (pp. 391–419). Academic Press. DOI: 10.1016/B978-0-443-29044-2.00017-5

Koyama, S., Purk, A., Kaur, M., Soini, H. A., Novotny, M. V., Davis, K., Kao, C. C., Matsunami, H., & Mescher, A. (2019). β-Caryophyllene enhances wound healing through multiple routes. *PLoS One*, *14*(12), e0216104. DOI: 10.1371/journal.pone.0216104 PMID: 31841509

Kunle, O. F., Egharevba, H. O., & Ahmadu, P. O. (2012). Standardization of herbal medicines. *International Journal of Biodeversity and Conservation*, *4*(3), 101–112. DOI: 10.5897/IJBC11.163

Li, Y., Ren, M., Wang, J., Ma, R., Chen, H., Xie, Q., Li, H., Li, J., & Wang, J. (2021). Borneol intervention in ischemic stroke: A systematic review. *Frontiers in Pharmacology*, *12*, 606682. DOI: 10.3389/fphar.2021.606682 PMID: 34017247

Machado, B. A., Pereira, C. G., Nunes, S. B., Padilha, F. F., & Umsza-Guez, M. A. (2013). Supercritical CO_2 extraction: Applications and perspectives. *Separation Science and Technology*, *48*(18), 2741–2760. DOI: 10.1080/01496395.2013.811422

Madkour, M., Aboelenin, M. M., Habashy, W. S., Matter, I. A., Shourrap, M., Elolimy, A. A., & Aboelazab, O. (2024). Oregano and rosemary extracts modulate gut and antioxidant responses in heat-stressed broilers. *Poultry Science*, *103*(9), 103996. DOI: 10.1016/j.psj.2024.103996 PMID: 39024691

Matera, R., Lucchi, E., & Valgimigli, L. (2023). Plant essential oils as healthy functional ingredients of nutraceuticals and diet supplements: A review. *Molecules (Basel, Switzerland)*, *28*(2), 901. DOI: 10.3390/molecules28020901 PMID: 36677959

Mira, B., Blasco, M., Berna, A., & Subirats, S. (1999). Supercritical CO_2 extraction of orange peel oil. *The Journal of Supercritical Fluids*, *14*(2), 95–104. DOI: 10.1016/S0896-8446(98)00111-9

Moura, L. I., Dias, A. M., Carvalho, E., & de Sousa, H. C. (2013). Wound dressings for diabetic foot ulcers. *Acta Biomaterialia*, *9*(7), 7093–7114. DOI: 10.1016/j.actbio.2013.03.033 PMID: 23542233

Muchlinski, A., Ibdah, M., Ellison, S., Yahyaa, M., Nawade, B., & Laliberte, S.. (2020). Diversity and function of terpene synthases in carrot aroma and flavor biosynthesis. *Scientific Reports*, *10*(1), 9989. DOI: 10.1038/s41598-020-66866-1 PMID: 32561772

Mukhopadhyay, M. (2009). Extraction and processing with supercritical fluids. *Journal of Chemical Technology and Biotechnology*, *84*(1), 6–12. DOI: 10.1002/jctb.2072

Nafiu, M. O., Hamid, A. A., Muritala, H. F., & Adeyemi, S. B. (2017). Standardization of medicinal plants. In Medicinal spices and vegetables from Africa (pp. 171–204).

Ngeow, W. C., Tan, C. C., Goh, Y. C., Deliberador, T. M., & Cheah, C. W. (2022). Promoting oxygenation and angiogenesis in oral wound healing. *Bioengineering (Basel, Switzerland)*, *9*(11), 636. DOI: 10.3390/bioengineering9110636 PMID: 36354548

Ngeow, W. C., Tan, C. C., Goh, Y. C., Deliberador, T. M., & Cheah, C. W. (2022). Oxygenation and angiogenesis in wound healing. *Bioengineering (Basel, Switzerland)*, *9*(11), 636. DOI: 10.3390/bioengineering9110636 PMID: 36354548

Nguyen, T. T., Le, T. V., Dang, N. N., Nguyen, D. C., Nguyen, P. T., Tran, T. T., Nguyen, Q. V., Bach, L. G., & Pham, D. N. T. (2021). Microencapsulation of essential oils by spray-drying and influencing factors. *Journal of Food Quality*, *2021*, 5525879. DOI: 10.1155/2021/5525879

Noriega, P. (2020). Terpenes in essential oils: Bioactivity and applications. In *Terpenes and terpenoids—Recent advances*. IntechOpen.

Olson, M. E., Ceri, H., Morck, D. W., Buret, A. G., & Read, R. R. (2002). Biofilm bacteria. *Canadian Journal of Veterinary Research = Revue Canadienne de Recherche Vétérinaire*, *66*(2), 86–92. PMID: 11989739

Pavela, R., & Benelli, G. (2016). Essential oils as ecofriendly biopesticides: Challenges and constraints. *Trends in Plant Science*, *21*(12), 1000–1007. DOI: 10.1016/j.tplants.2016.10.005 PMID: 27789158

Pereira, J. A., Casado, N., Porto-Figueira, P., & Câmara, J. S. (2022). The potential of microextraction techniques for the analysis of bioactive compounds in food. *Frontiers in Nutrition*, *9*, 825519. DOI: 10.3389/fnut.2022.825519 PMID: 35257008

Pérez-Recalde, M., Arias, I. E., & Hermida, É. B. (2018). Essential oils in biopolymer wound dressings: A systematic review. *Phytomedicine*, *38*, 57–65. DOI: 10.1016/j.phymed.2017.09.024 PMID: 29425655

Perović, A. B., Karabegović, I. T., Krstić, M. S., Veličković, A. V., Avramović, J. M., Danilović, B. R., & Veljković, V. B. (2024). Lavender essential oil recovery by novel hydrodistillation methods. *Industrial Crops and Products*, *211*, 118244. DOI: 10.1016/j.indcrop.2024.118244

Peter, A. E., Sandeep, B. V., Rao, B. G., & Kalpana, V. L. (2021). Natural immunosuppressants as adjuvants to target cytokine storm in COVID-19. *Frontiers in Pharmacology*, *11*, 583777. DOI: 10.3389/fphar.2020.583777 PMID: 33708109

Pourmortazavi, S. M., & Hajimirsadeghi, S. S. (2007). Supercritical fluid extraction in essential oil analysis. *Journal of Chromatography. A*, *1163*(1–2), 2–4. DOI: 10.1016/j.chroma.2007.06.021 PMID: 17624357

Przekora, A. (2020). Tissue engineered artificial skin grafts for chronic wound treatment: Can functional skin be reconstructed in vitro? *Cells*, *9*(7), 1622. DOI: 10.3390/cells9071622 PMID: 32640572

Quintans, J. S., Shanmugam, S., Heimfarth, L., Araújo, A. A., Almeida, J. R., Picot, L., & Quintans-Júnior, L. J. (2019). Monoterpenes modulating cytokines. *Food and Chemical Toxicology*, *123*, 233–257. DOI: 10.1016/j.fct.2018.10.058 PMID: 30389585

Rajput, S., Kaur, S., Panesar, P. S., & Thakur, A. (2023). Supercritical fluid extraction of essential oils from Citrus reticulata. *Biomass Conversion and Biorefinery*, *13*(16), 14605–14614. DOI: 10.1007/s13399-022-02807-4

Rani, M., Jindal, S., Anand, R., Sharma, N., Ranjan, K. R., Mukherjee, M. D., & Mishra, V. (2023). Essential oils for therapeutic benefits. In Essential oils: Extraction methods and applications (pp. 977–1008).

Rani, M., Jindal, S., Anand, R., Sharma, N., Ranjan, K. R., Mukherjee, M. D., & Mishra, V. (2023). Plant essential oils and their constituents for therapeutic benefits. In Essential oils: Extraction methods and applications (pp. 977–1008). DOI: 10.1002/9781119829614.ch42

Raza, A., Chohan, T. A., Zaidi, S. H., Hai, A., Alzahrani, A. R., Imran, M., & Saleem, H. (2024). Biochemical perspectives on natural products in wound healing: Phytochemicals in tissue repair and scar prevention. *Chemistry & Biodiversity*, *21*(10), e202400615. DOI: 10.1002/cbdv.202400615 PMID: 38958197

Reyes-Jurado, F., Franco-Vega, A., Ramírez-Corona, N., Palou, E., & López-Malo, A. (2015). Essential oils: Antimicrobial activity and extraction modeling. *Food Engineering Reviews*, *7*(3), 275–297. DOI: 10.1007/s12393-014-9099-2

Safarzadeh, E., Shotorbani, S. S., & Baradaran, B. (2014). Herbal medicine as inducers of apoptosis in cancer therapy. *Advanced Pharmaceutical Bulletin*, *4*(Suppl 1), 421. PMID: 25364657

Sahu, P., Bhowmick, A. K., & Kali, G. (2020). Terpene-based elastomers: Synthesis, properties, and applications. *Processes (Basel, Switzerland)*, *8*(5), 553. DOI: 10.3390/pr8050553

Samanta, R., Pattnaik, A. K., Pradhan, K. K., Mehta, B. K., Pattanayak, S. P., & Banerjee, S. (2016). Wound healing activity of silibinin in mice. *Pharmacognosy Research*, *8*(4), 298. DOI: 10.4103/0974-8490.188880 PMID: 27695272

Samuelson, R., Lobl, M., Higgins, S., Clarey, D., & Wysong, A. (2020). Effects of lavender essential oil on wound healing: A review. *Journal of Alternative and Complementary Medicine (New York, N.Y.)*, *26*(8), 680–690. DOI: 10.1089/acm.2019.0286 PMID: 32589447

Selvaraj, K., Theivendren, P., Pavadai, P., Sureshbabu, R. K., Ravishankar, V., & Palanisamy, P.. (2022). Physicochemical parameters affecting extraction of bioactives. *Current Bioactive Compounds*, *18*(4), 11–27.

Sethunga, M., Ranaweera, K. K., Gunathilake, K. D., & Munaweera, I. (2022). Extraction methods of essential oils and oleoresins. *Journal of Food Bioprocess Engineering*, *5*(2), 151–167.

Sharifi-Rad, J., Sureda, A., Tenore, G. C., Daglia, M., Valussi, M., & Tundis, R.. (2017). Biological activities of essential oils. *Molecules (Basel, Switzerland)*, *22*(1), 70. DOI: 10.3390/molecules22010070 PMID: 28045446

Sheppard, A. J., Barfield, A. M., Barton, S., & Dong, Y. (2022). Reactive oxygen species in bone regeneration. *Frontiers in Bioengineering and Biotechnology*, *10*, 836764. DOI: 10.3389/fbioe.2022.836764 PMID: 35198545

Shoham, Y., Snyder, R. J., Levy, Y. K., Zarbiv, K. D., Klinger, E., & Kramer, M.. (2024). Bromelain-based enzymatic debridement. *EClinicalMedicine*, *75*, 102655. PMID: 39763591

Souiy, Z. (2023). Essential oil extraction process. In *Essential oils—Recent advances*. IntechOpen.

Stephane, F. F. (2020). Terpenoids as important bioactives. In *Essential oils: Bioactive compounds, new perspectives and applications*. IntechOpen. DOI: 10.5772/intechopen.91426

Tanveer, M., Wagner, C., ul Haq, M. I., Ribeiro, N. C., Rathinasabapathy, T., Butt, M. S., Shehzad, A., & Komarnytsky, S. (2020). Spicing up gastrointestinal health with dietary essential oils. *Phytochemistry Reviews*, *19*(2), 243–263. DOI: 10.1007/s11101-020-09664-x

Teixeira, M. A., Paiva, M. C., Amorim, M. T., & Felgueiras, H. P. (2020). Electrospun nanocomposites containing cellulose derivatives modified with specialized biomolecules for enhanced wound healing. *Nanomaterials (Basel, Switzerland)*, *10*(3), 557. DOI: 10.3390/nano10030557 PMID: 32204521

Tetali, S. D. (2019). Terpenes and isoprenoids: A wealth of compounds for global use. *Planta*, *249*(1), 1–8. DOI: 10.1007/s00425-018-3056-x PMID: 30467631

Unalan, I., & Boccaccini, A. R. (2021). Essential oils in biomedical applications. *Current Opinion in Biomedical Engineering*, *17*, 100261. DOI: 10.1016/j.cobme.2021.100261

Upadhyay, R., & Bano, S. (2023). A review on terpenoid synthesized nanoparticles and their antimicrobial activity. *Oriental Journal of Chemistry*, *39*(2), 1–10. DOI: 10.13005/ojc/390226

Varzakas, T. (2016). Food additives and contaminants. In Varzakas, T., & Tzia, C. (Eds.), *Handbook of food processing and preservation* (pp. 319–388).

Veith, A. P., Henderson, K., Spencer, A., Sligar, A. D., & Baker, A. B. (2020). Therapeutic strategies for enhancing angiogenesis in wound healing. *Advanced Drug Delivery Reviews*, *146*, 97–125. DOI: 10.1016/j.addr.2018.09.010 PMID: 30267742

Vora, L. K., Gholap, A. D., Hatvate, N. T., Naren, P., Khan, S., Chavda, V. P., Balar, P. C., Gandhi, J., & Khatri, D. K. (2024). Essential oils in clinical aromatherapy. *Journal of Ethnopharmacology*, *318*, 118180. DOI: 10.1016/j.jep.2024.118180 PMID: 38614262

Vora, L. K., Gholap, A. D., Hatvate, N. T., Naren, P., Khan, S., Chavda, V. P., Balar, P. C., Gandhi, J., & Khatri, D. K. (2024). Essential oils for clinical aromatherapy: A comprehensive review. *Journal of Ethnopharmacology*, *318*, 118180. DOI: 10.1016/j.jep.2024.118180 PMID: 38614262

Wang, H., Chen, Y., Wang, L., Liu, Q., Yang, S., & Wang, C. (2023). Advancing herbal medicine through robust quality control practices. *Frontiers in Pharmacology*, *14*, 1265178. DOI: 10.3389/fphar.2023.1265178 PMID: 37818188

Wilkinson, H. N., & Hardman, M. J. (2020). Wound healing. *Open Biology*, *10*(9), 200223. DOI: 10.1098/rsob.200223 PMID: 32993416

Xu, B., Bai, L., Chen, L., Tong, R., Feng, Y., & Shi, J. (2022). Terpenoid natural products exert neuroprotection via PI3K/Akt signaling. *Frontiers in Pharmacology*, *13*, 1036506. DOI: 10.3389/fphar.2022.1036506 PMID: 36313360

Yousefi, M., Rahimi-Nasrabadi, M., Pourmortazavi, S. M., Wysokowski, M., Jesionowski, T., Ehrlich, H., & Mirsadeghi, S. (2019). Supercritical fluid extraction of essential oils. *Trends in Analytical Chemistry*, *118*, 182–193. DOI: 10.1016/j.trac.2019.05.038

Zhang, P., Lu, J., Jing, Y., Tang, S., Zhu, D., & Bi, Y. (2017). Global epidemiology of diabetic foot ulceration. *Annals of Medicine*, *49*(2), 106–116. DOI: 10.1080/07853890.2016.1231932 PMID: 27585063

Zhao, R., Liang, H., Clarke, E., Jackson, C., & Xue, M. (2016). Inflammation in chronic wounds. *International Journal of Molecular Sciences*, *17*(12), 2085. DOI: 10.3390/ijms17122085 PMID: 27973441

Zielińska-Błajet, M., & Feder-Kubis, J. (2020). Monoterpenes and their derivatives—Recent development in biological and medical applications. *International Journal of Molecular Sciences*, *21*(19), 7078. DOI: 10.3390/ijms21197078 PMID: 32992914

Zio, S., Cisse, H., Zongo, O., Guira, F., Tapsoba, F., & Somda, N. S.. (2020). Edible oil refining and contaminants. *Journal of Food Technology Research*, *7*(1), 9–47. DOI: 10.18488/journal.58.2020.71.9.47

Chapter 7
Tannins:
Natural Astringents for Wound Closure

Manish Dhall
College of Pharmacy PGMIS, Pt. B.D. Sharma University of Health Sciences, Rohtak, India

Renu Kadian
Ram Gopal College of Pharmacy, Sultanpur, India

Tarun Kumar
Department of Pharmaceutical Sciences, Central University of Haryana, Mahendergarh, India

Manish Pandey
Department of Pharmaceutical Sciences, Central University of Haryana, Mahendergarh, India

Ashwani Arya
http://orcid.org/0000-0001-9671-4427
Department of Pharmaceutical Education and Research, Bhagat Phool Singh Mahila Vishwavidyalaya, Khanpur Kalan, India

ABSTRACT

Wound closure is a multifaceted biological phenomenon through which hemostasis, inflammatory phase, proliferative or re-epithelialization and maturation are different stages of wound healing. As natural astringents, tannins are available as phytoconstituents in several herbal plants and have become particularly remarkable pharmacological potential for wound closure benefits. Tannins exhibited strong astringent, antimicrobial, anti-inflammatory, and antioxidant characteristics, making

DOI: 10.4018/979-8-3373-2908-6.ch007

them effective in enhancing tissue repair. Recent developments in biomaterials have resulted to the creation of tannin-based wound dressings, hydrogels, and bioactive scaffolds, which allow for a controlled release of tannins, enhancing their stability as well as bioavailability while maintaining moisture balance at the wound site. Tannin-infused nanofibers and hydrocolloid dressings have exhibited promise in supporting faster wound closure by creating an optimal healing environment. With their diverse therapeutic benefits, natural based tannins may provide a valuable herbal alternative in comparison to synthetic agents for wound healing, offering safer and more effective approaches for the management of wound.This chapter highlighted pathophysiology of wound healing, sources and classification of tannins, therapeutic mechanism of tannins, and applications of several naturally derived tannins in wound care and closure as well as the challenges aspects and future perspectives of tannins as natural astringents for the management of wound closure.

1. INTRODUCTION

A wound happens when the integrity of any tissue is disturbed (Almadani et al., 2021). Wound healing is a quite complicated procedure associated with several strictly systematized cellular-level events for the effective recovery of wounded tissue. If there is improper cellular behavior in wound healing due to certain body conditions, healing gets impaired. Thus, it is very important that should be a proper understanding of biological and clinical wound healing mechanisms (Wilkinson and Hardman, 2020). Wound closure is a complex biological process through which, hemostasis, inflammatory phase, proliferative or rebounding and maturation are different stages of wound healing. Haemostasias is the first stage of wound closure and occurs in 3 stages. First is vasoconstriction of the blood vessel wall to stop excessive bleeding, and the second stage is initial phase of the hemostasis, which include the aggregations of platelet as well as formation of plug in response to the collagen exposure in the subendothelial matrix (Furie and Furie, 2008). The secondary hemostasis results in the stimulation of the coagulation pathways, which fibrinogen changed from soluble to insoluble form and leads to formation of fibrin mesh. Furthermore, development of thrombus by the platelet plugs and a fibrin mesh, which consequences to prevent the bleeding and forms a temporary platform for the infiltration of important cells for wound healing (Poo, 1977). Main cellular components involved in this phase are platelets, and the matrix is fibrinogen (Rumbaut, 2010). Inflammatory response is another phase in development of wound healing, which initiate during injury in 24 hours and ended in duration of two weeks in normal conditions and takes longer for severe injury. Neutrophils, mast cells, macrophages and monocytes, are the important cells which play a major role in this phase. Mast cells are mainly responsible for the inflammatory signs, such as swelling, redness, and pain at the sight of injury. They do so by the release of histamines and enzymes. Neutrophils are released in response inflammatory response by bone marrow. The several inflammatory signaling pathways including the chemokines, lipid mediators etc. are generated at site of injured location. Similarly, neutrophils caused oxidative burst and leading the process of phagocytosis of toxins and clearing them (Su and Richmond, 2015; Su et al., 2017). Macrophages, which may initiate the process of wound healing as well as regeneration process of tissue (Hu et al., 2017). Next phase is the proliferative phase, in which, along with wound healing, the new tissue is also formed, new epithelium, neovascularization, and immunomodulation take place (Gurtneret al., 2008). Remodeling, the last phase starts with primary aggregation of fibrin clot and at last development of mature, collagen-rich scar (Li et al., 2007; Schultz and Wysocki, 2009; Young and McNaught, 2011; Han Zhang et al., 2024). Medications are significantly advanced, but they still can't fully overcome wound healing complications. Effective wound healing is not only up to the tissue regeneration, but also the minimization of scar formation. The natural plants and their extracts contain several phytoconstituents such as elastin, fibrin, chitosan etc., which show a notable effect in protection of wound healing (Han Zhang et al., 2024). Tannins may play a major role in the protection of wound healing. It acts as an astringent, boosts the production of fibroblasts and extracellular matrix, and helps in the speedy wound healing by hastening the epithelialization cascade. They also promote angiogenesis, which leads to the new blood vessels, thus providing adequate requirement of nutrients as well as oxygen to infection of injured site. There are many tannin-based wound healing products which enhance tannin release, stability, allow controlled release, increase bioavailability and provide moisture at the wound site. However, there are certain drawbacks which are associated with the wound healing activity of tannins, such as cytotoxicity at high concentrations, stability, bioavailability, etc. (Criollo-Mendoza et al., 2023; Jing et al., 2022). Current formulations require improvement in terms of bioavailability. Tannins available as phytoconstituents in several herbal plants and have become particularly remarkable pharmacological potential for wound healing and closure benefits. Tannins exhibited the strong astringent, antimicrobial, anti-inflammatory, and antioxidant characteristics, making them effective in enhancing tissue repair. As natural astringents, tannins work by binding proteins and forming a protective layer on the wound, which aids in hemostasis process by constricting blood vessels as well as promoting clot formation, ultimately minimizing bleeding effects. Recently, natural bioactive components have been significantly recognized for their pharmacological potential in management of wound care due to their low toxicities as well as the biocompatibility along with versatile properties. With their diverse therapeutic benefits, natural based tannins may provide a valuable herbal alternative in comparison to synthetic agents for wound healing, offering safer as well as more effective and efficient approaches for the management of wound (Almadani et al., 2021; Pizzi, 2021; Jing et al., 2022). Tannins available as natural phytoconstituents from several herbal plants and have become particularly remarkable pharmacological potential for wound healing and closure benefits. Tannins protected the wound healing and wound closure as strong astringents through its multitargeted mechanism such as antimicrobial, anti-inflammatory, and antioxidant characteristics, making them effective in enhancing tissue repair. Tannins contribute to wound healing as well wound closure to accelerating the healing process by serving as antibacterial, anti-inflammatory agents, antioxidants potential as well as promoting angiogenesis, cell proliferation and synthesis of collagen. They help enhance collagen production, stimulate cell growth, and encourage the development of new form of blood vessels, all of which accelerate the healing process. Natural tannins also facilitate wound closure through several mechanisms, such as decreasing the reactive oxygen species (ROS) in order to decrease the oxidative stress and also enhancing fibroblast proliferation, wound contraction as well as increasing the growth of capillary networks (Orlowski et al., 2018; Karuppagounder et al., 2015; Jing et al., 2022). Tannins, may provide a natural and versatile therapeutic approach for the wound healing and closure, utilizing their astringent, anti-microbial potential, anti-inflammatory, as well as regenerative properties. Ongoing research studies into tannin-based biomaterials has the significant potential to transform wound care by offering sustainable as well as an effective healing solutions for a wide range of wound types. More clinical trials need to be done to check effectiveness. Nanotechnology and tissue engineering may help in this regard. Natural tannins from herbs are a safer and effective approach for the management of wounds. The current chapter highlights the natural source of tannins, its mechanism in various stages of wound healing, applications and the challenges associated with the current formulations and how the challenges need to be overcome.

2. SOURCES AND CLASSIFICATION OF TANNINS

Natural tannin sources, such as grape seeds, green tea, oak bark and peel of the pomegranate, have been extensively used for research for their therapeutic potential in terms of wound healing. These natural herbal derived tannins have been incorporated into several pharmaceutical formulations in terms of management of acute as well as chronic wounds. Their ability to promote healing while decreasing scarring makes them a promising available natural therapeutic option for natural wound care managements (Pizzi, 2021)

2.1. Sources of Tannins:

There are two major sources of tannins: natural and artificially formed. The tannin found in the plant has a notable significant influence on the health of humans in terms of positive and negative ways. These are extensivelyspread in so numerous plant species. Plants with higher concentrations of condensed form of tannins are wattle, mangrove, quebracho and hemlock, while hydrolysable tannins are found more in the chestnut and myrobalan species (Pizzi, 2021). The apple comprises various phenolic compounds such as chlorogenic acid, catechins, epicatechins, rutin, flavonoids, phlorizin, condensed tannins, etc. Condensed Tannins are found in several plants as a secondary metabolite and are named as proanthocyanidins. These compounds are consumed daily by us in our daily life derived from fresh fruits like apples and in foods that are processed that are made from them. Other than nutrients, they act as functional food as well (Pizzi, 2021).

2.2. Classification of Tannins:

Plants are consumed for their nourishment, and tannins, which act as plant-based polyphenolic astringents, are present in several portions of herbs (Das et al., 2020). They are mostly found in the gymnosperms and angiosperms, and based on that, they are classified as condensed and hydrolysable tannins (Saad et al., 2014; Navarrete et al., 2013).

2.2.1. Hydrolysable Tannins:

This type of tannin contains glucose or other polyhdric alcohols that are known as gallotannins, as they are gallic acid or hexahydroxy diphenic acid esterified, also known as ellagitannins (Kemppainen et al., 2014). Based on the esterification, these types of tannins are again subdivided into two other types that are gallotannins and ellagitannins (Khanbabaee, and Van Ree. 2001). On the acid hydrolysis of this type of tannin, it will result in gallic acid and ellagic acid or other similar compounds (Hemingway, 1998). In the centre of hydrolysable tannin, there is the presence of the carbohydrate, the OH group in the mainly of the carbohydrate type are completely or half esterified with group congaing the phenolic groups, likely of the gallic acid present in Gallo tannin and the ellagic acid present in ellagotannins. These types of tannins are hydrolyzed by the groups that are weak acids and weak bases, leading to the formation of carbohydrates and phenolic acids. Gallotannins are the simplest hydrolysable tannins that result from the polyol and polyphenolic residues. As most gallotannins are associated to saccharides. While the ellagitannins are the common recognized class of tannins including with above the five hundred herbal compounds, which are composed of the form of gallotannins, further consequences in unit (monomer) with chirality axis, oxidative coupling in two galloyl units, minimum (Arbenz and Avérous, 2015).

2.2.2. Condensed Tannins:

These types of Tannins are also regarded as Proanthocyanidins having 2 to 50 or more polymers of flavonoid units that are combined with carbon-carbon bonds and are not sensitive to separation through hydrolysis. This type of tannin consists of flavolans or polymers of flavan-3,4-diols that are regarded as leucoanthocyanidins, and/or flavan-3-ols are known as catechins (Kemppainen et al., 2014). On the comparison of hydrolysable tannin and condensed tannin, condensed tannin has a huge range of molecular weight (MW), ranging from 500 to almost 20,000, also can generate polymeric materials on binding with the aldehydes. A water-insoluble product named Phlobaphene is produced on the condensed tannins polymerization. Another significant wood-based product named Char, present in higher yield in the condensed tannins in comparison with the hydrolysable tannins because of its condensed structure (Sebestyén et al., 2019). Catechins and epicatechins are two major examples of these types of tannins.

2.2.3. Complex Tannins:

These types of tannins are the other type of tannin that are constitute from the one unit of gallotannin or an ellagitannin, that are bound to the flavan-3-ol which is a catechin monomer through the glycosidic bond (Rajasekaran et al., 2021). The example of this type of tannin is acutissimin, it is composed of the unit named as flavagallonyl that connect D-glucose that connect polyol through the glycosidic bond in carbon one and three esters bonds (Arbenz and Avérous, 2015).

2.2.4. Pseudotannins:

Tannins are generally complex organic compounds that are found in plants. The main challenge to using tannin in biomedical approaches is that these are not stable in physiological conditions as well as the lack of control over the reactivity and purity. So, there is the formation and the identification of the tannin-like polymers with the controlled constitution. And these types of tannins are known as pseudotannins that are formed by the alteration or substitution in the linear dextran chain with the protocatechuic, gallic and other pendant groups to mimic the action of the hydrolysable tannin (Rajasekaran et al., 2012; Cheng et al., 2015).Pseudotannins do not obey the Goldbeater skin test as they have lesser MW compounds. Chlorogenic acid found in coffee and Nux vomica, catechins found in cocoa, and ipecacuanha acid found in the ipecacuanha are the best instances of Pseudotannins. The identification test for the chlorogenic acid is by extracting the drug using water and treating the extract with a solution of ammonia leads to the slow formation of the green color when comes in exposure to air (Pizzi et al., 2021).

3. THERAPEUTIC MECHANISM OF TANNINS IN WOUND CLOSURE

Tannins available as phytoconstituents in several herbal plants and have become particularly remarkable pharmacological potential for wound healing and closure benefits. Tannins protected the wound healing and wound closure as strong astringents through its multitargeted mechanism such as antimicrobial, anti-inflammatory, and antioxidant characteristics, making them effective in enhancing tissue repair. Tannins contribute to wound healing as well wound closure to accelerating the healing process by serving as antibacterial, anti-inflammatory agents, antioxidants potential as well as promoting angiogenesis, cell proliferation and synthesis of collagen. They help enhance collagen production, stimulate cell growth, and encourage the development of new form of blood vessels, all of which accelerate the healing process. Natural tannins also facilitate wound closure through several mechanisms, such as decreasing the reactive oxygen species (ROS) in order to decrease the oxidative stress and also enhancing fibroblast proliferation, wound contraction as well as increasing the growth of capillary networks. Figure-1, represented the multitherapeutic mechanism of tannins towards the management of wound healing.

Figure 1. Multitherapeutic mechanism of tannins towards the prevention of the wound healing

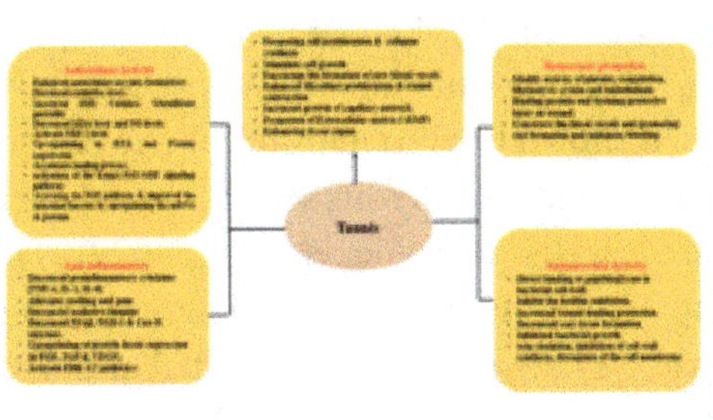

Tannins contribute in protection of wound healing mechanism and help in assisting the closure of the wound to accelerating the healing process by serving as antibacterial, anti-inflammatory agents, antioxidants potential as well as promoting angiogenesis, cell proliferation and the synthesis of collagen. They help enhance collagen production, stimulate cell growth to furthermore, accelerate the healing process. Natural tannins also facilitate wound closure through several mechanisms, such as targeting the ROS mechanism in terms to decrease the oxidative stress and also enhancing fibroblast proliferation, wound contraction, as well as increasing the growth of capillary networks.

3.1. Hemostasis Promotion

Hemostasis is essential for maintaining blood fluidity within intact vessels and for preventing bleeding when vessel injury occurs. This delicate balance is achieved through the simultaneous action of both clot-promoting and clot-inhibiting mechanisms, which involve e platelets, the coagulation cascade, the fibrinolytic system, and the vascular endothelium.

3.1.1. Protein precipitation and vasoconstriction aiding in blood clot formation.

Tannins, naturally occurring compounds found in plants, show promise in reducing the risk of thromboembolic conditions. They contribute significantly to primary hemostasis by promoting the development of platelet plug on the location of vascular damage. Additionally, platelets facilitate secondary hemostasis by presenting phospholipid surfaces that activate coagulation factors. When blood vessels are injured, Willebrand factor, tissue factor along with the released substance form platelets including the epinephrine, adenosine diphosphate and thrombin, become exposed or activated, initiating the hemostatic response (Marcińczyk et al., 2022). Plant-derived extracts, fractions, and individual compounds can influence the hemostatic system through multiple pathways. Tannins, in particular, are known to affect platelet function, blood coagulation, fibrinolysis, and the vascular endothelium. They have also been shown to exert antithrombotic effects. Numerous studies, including those cited in recent reviews highlight the beneficial impact of tannins on the regulation of hemostasis (Marcińczyk et al., 2022). The initial phase of hemostasis begins promptly after tissue injury, typically within minutes, when tissue integrity is compromised. Platelets become activated at the injury site and release signaling molecules that promote clot formation. These clots are primarily made up of thrombospondin, vitronectin, fibrin and fibronectin and they which collectively assists the mechanism of aggregation of platelets and form a protective plug. Additionally, the release of several concerned growth factors including the transforming growth factor- β (TGF- β), epidermal growth factor (EGF) and several chemokines participate a notable involvement in the initiating tissue repair mechanisms. During this phase, prostaglandin (PGG-H2), is enzymatically changed to thromboxane (TX-A2) by thromboxane synthase. TXA2 is a potent stimulator of platelet aggregation and vasoconstriction. It also contributes to the recruitment of the related cell in immune systems, including macrophages, endothelial cells along with neutrophils, may all important for protective effects of wound healingmechanism (Chen et al., 2019; Criollo-Mendoza et al., 2023). Similarly, as natural astringents, tannins work by binding proteins and forming a protective layer around the wound, which aids in hemostasis process by constricting blood vessels as well as promoting clot formation, ultimately minimizing bleeding effects. Wound healing process is a prolonged and intricate biological course which occurs in separate but overlapping phases including the hemostasis phase, inflammatory phase, proliferation, and tissue remodeling. Various plants rich in tannins have demonstrated protective wound healing properties (Yang et al., 2017). Tannic acid (TA), for example, has been revealed to enhance wound recovery in rat models by promoting re-epithelialization and hair follicle regeneration. This effect is achieved through the increased expression of key growth factors including the vascular endothelial growth factor (VEGF), fibroblast growth factor, TGF-β, along with decreasing the several pro-inflammatory responses (IL-6 and IL-1), partly via activation of the signaling ERK1/2 mechanism (Chen et al., 2019). Additionally, pH-responsive composite hydrogels made from tannic acid and carboxylated agarose have exhibited strong anti-inflammatory potential in terms of antagonize the production of nitric oxide (NO) in activated human macrophages in a dose-dependent fashion (Ninan et al., 2016). Studies also show that tannic acid-containing collagen composites (TCCs) significantly enhance closure of wound as compared to unmodified form of collagen scaffolds. In the context of corneal wound repair, growth factors like hepatocyte growth factor (HGF) along with the growth factor of keratinocyte are known to stimulate the ERK 1/2 signaling mechanism, which contributes to process of wound healing (Natarajan et al., 2013; Kung et al., 2008). Tannic acid promotes the secretion of bFGF, which in turn activates the signaling pathways (ERK ½), consequences to improvethe proliferation of cellular level, along with growth of tissue through the process of wound healing (Chen et al., 2019).

3.2. Anti-inflammatory Properties:

3.2.1. Reduction of several pro-inflammatory cytokines.

Moreover, tannins help modulate the inflammatory response by decreasing the several pro-inflammatory markers of IL-6, tumor necrosis factor-α, alleviating swelling, pain, and oxidative damage at the wound site. Tannic acid (TA) mitigates inflammation through a variety of molecular signaling pathways. It interferes with SMAD2-mediated gene expression triggered by TGF-β and also by as well as through controlling the STAT3 and NF-κB signaling cascades (Jing et al., 2022). Additionally, TA helps counteract oxidative stress by influencing the Keap1-Nrf2/ARE pathway, thereby contributing to its anti-inflammatory effects.Prostaglandin E2 (PGE2), a key inflammatory mediator which may associated with the neuroinflammation and pain signaling, particularly by enhancing the sensitivity of neuronal receptors and contributing to central sensitization associated with inflammatory pain. Research has shown that TA possesses anti-nociceptive properties, which include restoring antioxidant defenses, resolving neuroinflammatory responses, and correcting disruptions in neurotransmitter balance in mice exposed to glyphosate (Gly) (Abolarin & Owoyele, 2024). Neuroinflammation is a significant factor in glyphosate-induced hypersensitivity to pain (hyperalgesia), and if left unchecked, it may result in further neuronal damage. Furthermore, TA treatment has been observed to reduce inflammatory marker levels and bring neurotransmitter concentrations, key players in pain modulation, back to normal physiological ranges, thereby supporting its therapeutic potential in managing inflammation and pain (Abolarin and Owoyele, 2024). Tannic acid (TA) has been shown to lower the several inflammatory markers expression including the IL-1, IL-6, TNF- α, COX-II, prostaglandin E2 (PGE2). Research suggests that TA may alleviate symptoms of atopic dermatitis in mouse models by suppressing the activity of NF-κB pathways, cytochrome- C oxidase-III (COX-III) along with inflammatory cytokines, effects that may be mediated through the upregulation of peroxisome proliferator-activated receptor gamma (PPARγ) (Karuppagounder et al., 2015). The therapeutic properties of tannins are largely linked with potential of anti-inflammatory as well as antioxidant capabilities. These effects include increasing the antioxidative enzymes markers of catalase enzymes, SOD as well as glutathione peroxidase, while also reducing the proinflammatory response markers (IL-1β along with TNF-α), and caspase-3 via the inhibition of the NF-κB mechanism (Zhang et al., 2017; Jing et al., 2022).

Furthermore, TA has been found to significantly reduce neurological and behavioral deficits, oxidative stress, and inflammation following traumatic brain injury. This protective role is thought to be linked to the level of signaling pathways (NRF-2 /ARF). TA demonstrated anti-inflammatory effects in terms of decreased oxidative stress via targeting the ROS production and inhibiting the activation ofNF-κB pathway. The suppression of TLR4-mediated NF-κB signaling is considered a key part of its mode of action (Jing et al., 2022). Wound healing is complex process, during which inflammatory cytokines as well as the growth factors play a vital role in tissue repair. According to research by Chen et al., 2019, treatment with TA was found to reduce elevated levels (IL-1, IL-6), which may be caused by skin injuries. This indicates that TA exhibits notable anti-inflammatory properties during the initial inflammatory phase, helping to protect the affected tissue. Inflammatory markers response including the cytokines, growth factors and chemokines is essential in regulating the interactions among different cell types throughout the wound healing process. Findings revealed that TA aids efficiently enhanced the levels of growth factors in comparison to the untreated model's experimental group, while simultaneously lowering IL-1 and IL-6 concentrations. These outcomes suggest that TA not only promotes the release of beneficial growth factors on location of wound exposure, but also helps suppress inflammatory cytokines, supporting tissue regeneration and recovery (Kung et al., 2008).

3.3. Antimicrobial Effects:

3.3.1. Inhibition of growth of the bacterial and fungal.

Antimicrobial properties of tannins, may prevent infections by inhibiting the growth of microorganism of fungi as well as bacteria and decreasing the complications. TA has shown their capability in terms of antagonize intestinal bacterial growth such as Lactobacillus, Enterobacter, Bacteroides, clostridium species. *Clostridium* species, when tested at concentrations ranging from 3 to 200 μM in culture media. It is suggested that TA may suppress bacterial efflux pumps, enhancing the effectiveness of certain antibiotics against resistant strains (Jing et al., 2022). The antibacterial potential of TA may also be linked to its interaction with peptidoglycan in bacterial cell walls, which compromises cell wall integrity. Bacterial biofilm formation, a major contributor to antibiotic resistance and chronic infections, can also be disrupted by TA. Specifically, TA has been found to reduce biofilm development in both methicillin-sensitive microorganism of *S. aureus* (MSSA, ATCC25923) and methicillin-resistant *S. aureus* (MRSA, JP735) strains (Dong et al., 2018). Furthermore, TA has shown effectiveness in inhibiting formation of bio-film by *E. coli* without affecting the bacteria's general viability or metabolic activity (Hancock et al., 2010; Payne et al., 2013). Thus, antimicrobial properties prevent infections by inhibiting the bacterial as well fungal growth, which may decrease the complication's risk. Moreover, tannins help modulate the inflammatory response by decreasing several pro-inflammatory response markers including the interlukin-6, tumor necrosis factor- α, alleviating swelling, pain, and oxidative damage at the wound site.

3.3.2. Protective action against wound infections.

Tannins have been reported to support wound closure and reduce the formation of scar tissue, partly by suppressing the synthesis and facilitating the deletion of reactive oxygen species (ROS). Additionally, to their function in tissue repair, tannins offer several therapeutic benefits, including pain relief, reduction in the risk of secondary infections, prevention of plasma leakage, and support for robust epithelial regeneration. Historically, tannins were used in burn wound management, but their use declined in the mid-20th century due to the emergence of topical antibiotics and reports of liver toxicity in both clinical and experimental settings (Chokotho and Hassel, 2005). Tannins exhibit antimicrobial activity through various mechanisms such as chelating iron, which deprives bacteria of an essential nutrient, disrupting bacterial cell walls and membranes, and interfering with fatty acid synthesis. Additionally, tannins have been identified as potential quorum sensing inhibitors, capable of downregulating bacterial virulence by suppressing the genes expression included in enzyme production, biofilm formation, adhesion, motility, and toxin synthesis (Farha et al., 2020). Their ability to bind and sequester ferric iron from the environment further contributes to their antimicrobial effect by limiting bacterial access to this critical resource. Tannins have demonstrated the capacity to hinder the growth of a diverse variety of microorganisms, including fungi, bacteria and yeasts (Scalbert, 1991). Recent studies also confirm their effectiveness against both Gram-positive and negative bacteria, with the majority of tannins exhibiting bacteriostatic features in contrast to bactericidal properties (Boakye et al., 2016).

3.3.3. Targeted to tissue regeneration and synthesis of collagen

Synthesis of collagen as well as regeneration phase of tissue is the significant component in process of wound healing. Tannins as astringents are associated with boosting fibroblast as well as production of extra-cellular-matrix. They also speed up epithelialization cascade which may further consequences to fastercontraction of wound process and may leads tohealing the tissue in way of strong aspect. Furthermore, tannins may support the angiogenesis and also assist in development of new blood vessels in terms of regeneration process for proper oxygen requirements as well as supply of nutrients (El-Mabood et al., 2014). TA contributes to improved wound healing through both direct and indirect mechanisms. Its direct action includes enzyme inhibition and antibacterial effects, while indirectly, it helps reduce the risk of postoperative wound infections and promotes effective hemostasis due to its astringent effects on mucosal tissues (El-Mabood et al., 2014). In vitro studies reported that natural TA can reduced oxidative and endoplasmic reticulum stress triggered by palmitic acid in alveolar epithelial A549 cells of humans. These resultspropose a potential function for TA in preventing the advancement of pulmonary fibrosis. Transformation of growth factor-beta 1 (TGF-β1) is commonly associated with fibrotic areas in both human cases and experimental models of lung fibrosis. Blocking TGF-β signaling, such as through inhibition of its soluble receptors, has been shown to reduce fibrogenesis. Previous research indicates that TA exhibits anti-fibrotic activity both in human lung cultured fibroblasts and in animal models of bleomycin-stimulated pulmonary fibrosis. This effect is believed to involve a decrease in TGF-β expression and a reduction in Smad2/3 phosphorylation, which are key components of the fibrotic signaling pathway (El-Mabood et al., 2014).

3.3.4. Enhancement of fibroblast activity

Following tissue injury, fibroblasts move to the site of wound, where they experience process of proliferation and differentiation. The sustained presence of such differentiated fibroblasts, termed myofibroblasts, has been concerned with extreme scarring in different tissues. Studies suggest that tannic acid (TA) can suppress differentiation of fibroblast, as evidenced by the down-regulation of key markers such as type I collagen, α-smooth muscle actin, and N-cadherin. Furthermore, TA prevents TGF-β1-stimulated gene expression changes included in extracellular matrix (ECM) remodeling, notably by reducing the expression of matrix metalloproteinases and their inhibitors. TA also suppresses TGF-β1-stimulated proliferation of fibroblast and stimulates G0/G1 phase cell cycle. Additionally, TA exerts its inhibitory responses by reducing the ERK signaling proteins and phosphorylation of Smad. Collectively, these results reveal that TA ensures promise as a therapeutic agent for the prevention of fibrotic tissue remodeling by modulating fibroblast activity and ECM dynamics (Pattarayan et al., 2018). Tannins, particularly tannic acid (TA), are known to enhance fibroblast activity, promoting both proliferation and migration, which is critical for wound closure. This effect is partly attained through the regulation of growth factors namely TGF-β, bFGF, and VEGF, as well as by modulating key signaling pathways such as ERK 1/2. Additionally, tannins play a protective role by preventing the degradation of elastic fibers and supporting elastogenesis, further contributing to tissue repair. Fibrogenesis performs a crucial function in the restoration of tissue structure and function following injury in wound closure. However, when injury persists, it can interfere with normal repair processes, resulting in abnormal accumulation of extracellular matrix (ECM) components and progression to fibrosis. This pathological condition emerges through a cascade of events triggered by persistent tissue damage, including: (1) initial disruption of epithelial or endothelial barriers; (2) secretion of TGF-β1, (3) infiltration of immune cells; (4) generation of ROS; (5) activation of cells responsible for collagen synthesis; (6) differentiation of fibroblasts into myofibroblasts within the ECM; and (7) potential resolution of fibrosis if the injurious stimulus is removed. Although TGF-β1 is essential for normal wound repair, in pathological conditions, it amplifies communication among various cell types, including inflammatory, parenchymal, and collagen-producing cells, ultimately promoting the fibrotic response (Kisseleva and Brenner, 2008). Targeting TGF-β1 has shown promise in preventing fibrosis. For example, blocking soluble TGF-β receptors in experimental and human pulmonary fibrosis models can effectively prevent fibrogenesis. A findingby researchers exhibited that tannic acid exhibits a potent anti-fibrotic action, in both bleomycin-induced in vivo model of pulmonary fibrosis and in vitro human lung fibroblasts (Jing et al, 2022).

3.4. Antioxidant activity

Oxidative stress, initiated by an augmentation in reactive oxygen species (ROS) namelyhydrogen peroxide, superoxide, and hydroxyl radicals, is directly linked to fibrosis. Additionally, the depletion of endogenous antioxidants exacerbates further tissue injury. Acute tissue injury that leads to vascular damage involves several processes, including released of cytokines, blood clot formation and aggregation of platelet as well as the degranulation. In addition to that injury also consequences to enhanced blood vessel permeability (Kisseleva and Brenner, 2008).

3.4.1. Decreasing ROS and enhancing antioxidant enzymes

Tannic acid (TA) has been reported to lessen oxidative stress by stimulating the Keap1-Nrf2/ARE signaling pathway, offering protective effects in vivo (Jing et al., 2022). Tannic acid has achieved significant concentration for its antioxidant features. In vivo findings demonstrated that TA treatment increased glutathione peroxidase (GSH-PX) action and enhanced the manifestation of ZO-1 in the jejunum of weaned piglets. It also decreased serum malondialdehyde (MDA) levels and downregulated Keap1 expression at both the mRNA and protein amounts. These findings indicate that TA improves antioxidant capacity by modifying the cellular redox state and stimulating the Nrf2 pathway. Additionally, TA helped to strengthen the intestinal barrier by up-regulating the expression of tight junction proteins and enhancing transepithelial electrical resistance. Overall, these findings suggest that TA can mitigate oxidative damage and improve intestinal function and antioxidant efficiency through the activation of the Nrf2 signaling pathway, highlighting its potential therapeutic application in reducing oxidative stress in the intestines of weaned piglets (Wang et al., 2022). The Nrf2 signaling pathway plays a critical performance in defending the body against oxidative damage by regulating the expression of various antioxidant genes, thereby enhancing cellular antioxidant ability (Nguyen et al., 2009). Oxidative stress, which results from an excessive accumulation of ROS has a strong association with the development of fibrosis. This condition is further worsened by the depletion of the body's natural antioxidants, leading to increased tissue injury. Acute tissue injury that leads to vascular damage involves several processes, including platelet clumping and release of granules, clot formation, secretion of cytokines, and enhanced blood vessel permeability (Kisseleva and Brenner, 2008).

Tannic acid (TA), a component identified for its potent antioxidant activity, has been found to counteract oxidative stress by the stimulation of the Keap1-Nrf2/ARE signaling pathway, offering protective effects in vivo (Jing et al., 2022). Research has highlighted TA's ability to enhance antioxidant features. For instance, in vivo studies in weaned piglets revealed that TA treatment led to enhanced activity of glutathione peroxidase (GSH-PX) and improved expression of the tight junction protein ZO-1 I n the jejunum. It also lowered serum levels of malondialdehyde (MDA) and suppressed Keap1 gene expression at both the transcriptional and protein levels. These findings indicate that TA enhances redox balance and triggers the Nrf2 pathway to boost the body's antioxidant capacity. Furthermore, TA contributed to improved intestinal barrier integrity by enhancing tight junction protein expression and enhancing transepithelial electrical resistance. These observations underscore TA's ability to lessen oxidative stress and enhance intestinal health through Nrf2 pathway activation, demonstrating its efficacy as a therapeutic moiety for managing oxidative stress in the intestines of weaned piglets (Wang et al., 2022). The Nrf2 pathway is vital in protecting against oxidative damage, as it governs the expression of various antioxidant-related genes, ultimately boosting the cell's antioxidant response (Nguyen et al., 2009). Previous findings have described that TA can mitigate apoptosis, oxidative damage, and inflammation induced through arsenic trioxide by modulation of NF-kappa B/Nrf2 pathway in rodents (Jin et al., 2020). Furthermore, TA treatment has been exhibited to potentially enhance the protein amounts of heme oxygenase-1 and Nrf2, offering protection against distressing injury of brain in rodent models (Salman et al., 2020). TA treatment has also been found to reduce malondialdehyde (MDA) and pro-inflammatory cytokine amounts (such as IL-1β, TNF-α, and IL-6) while enhancing the proportions of anti-inflammatory cytokines (namely IL-4, IL-10, and TGF-β1). Moreover, TA treatment enhanced the actions of antioxidant enzymes namely catalase (CAT), glutathione peroxidase (GPx),superoxide dismutase (SOD) and in Gly-treated mice, suggesting its potent anti-inflammatory and antioxidant responses (Abolarin and Owoyele, 2024). Given these benefits, tannins, particularly tannic acid, present a promising natural therapeutic option for wound closure, utilizing their anti-inflammatory, astringent, regenerative and antimicrobial properties. Ongoing research into tannin-based biomaterials ensuresimmensecapability to reform wound healing by offering sustainable and effective solutions for a diverse variety of wound types.

4. APPLICATIONS OF TANNINS IN WOUND CARE AND CLOSURE

Tannins, particularly tannic acid, have been increasingly investigated for their significant performance in wound management due to their natural bioactivity. A comprehensive exploration of tannins biochemical roles in wound healing is summarized in this section supported by scientific research and potential therapeutic applications which make tannins a potential natural alternative or additive in the development of next-generation wound care materials.

4.1. *Traditional medicinal uses of tannin-rich plants in wound healing*

Several literature findings reported the applications of tannins as natural astringents for management of wound healing and closure. A comprehensive overview of traditional uses of tannin-rich plants and its correlation with scientific research is summarized below (Fraga-Corral et al., 2021).

4.1.1. *Terminalia chebula* (Haritaki)

- **Traditional usage**: It is extensively utilized for treatment of wounds and ulcers in Ayurvedic and Unani medicine in form of pastes and poultices.
- **Scientific research**: The tannin extracts obtained from immature fruits revealed antibacterial action against *Klebsiella pneumoniae* and *Staphylococcus aureus*, and exhibited enhanced cutaneous wound closure in rat models (Li et al., 2011, Suguna et al., 2002).

4.1.2. *Psidium guajava* (Guava)

- **Traditional usage**: It is widely utilized for wound management in various cultures owing to its antimicrobial properties.
- **Scientific research**: The gel has been formulated from tannin-enriched fractions of *Psidium guajava* leaves exhibited potential contraction of wound and promoted healing in diabetic rat models (Jayakumari, 2018).

4.1.3. *Entada phaseoloides* (Stinky bean)

- **Traditional usage**: It is utilized in for its wound healing properties traditional medicine.

- **Scientific research**: The tannins extracted from *Enataphaseoloides* supported wound closure in rats, confirmed through enhanced proliferation of fibroblastsand promoted healing ofwounds (Su et al., 2017).

4.1.4. *Achyranthes aspera* (Prickly chaff flower) and *Ocimum basilicum* (Sweet Basil)

- **Traditional usage**: It is employed for managing inflammation and wounds in traditional medicine.
- **Scientific research**: The tannins extracted from callus cultures in leaves of these plants revealed anti-inflammatory action and supported wound closure in rabbit models (Ambreen & Mirza, 2020; Fikru et al., 2012).

4.1.5. *Sanguisorba officinalis* (Great Burnet)

- **Traditional usage**: It is employed for its hemostatic and wound-healing features in traditional medicine.
- **Scientific research**: The aqueous extracts from the root exhibited enhanced collagen synthesis, contraction of wound, and enhanced angiogenesis in mouse models (Shedoeva et al., 2019, Zhang et al., 2018).

4.1.6 *Hamamelis virginiana* (Witch hazel)

- **Traditional usage**: The decoctions of witch hazel leaves and bark are employed in Native American communities in traditional medicine to manage skin inflammations, insect bites, cuts and minor wounds.
- **Scientific research**: The glycolic witch hazel bark extracts demonstrated modulation of inflammatory responses in skin cells by inhibition of pro-inflammatory cytokines in human keratinocytes (Piazza et al., 2022). The antioxidant action of witch hazel exhibited protective action against oxidative stress in skin cells. The methanolic and aqueous extracts of witch hazel exhibited antimicrobial action by inhibiting skin pathogens namely *Staphylococcus aureus* and *Streptococcus* species in wound infections (Cheesman et al., 2021). The witch hazel extracts have been incorporated with bacterial cellulose membranes for combining the structural advantages of cellulose with therapeutical benefits of witch hazel. These composite materials exhibited a moist wound environment and delivered anti-inflammatory agents directly to the wound area (Molin et al., 2023).

4.1.7. *Quercus* **species(Oak)**

- **Traditional usage**: It is employed for management of wounds, diabetic sores, burns and ulcers in traditional medicine. It forms a protective barrier over wounds and prevents microbial contamination.
- **Scientific research**: An ethanolic extract of *Quercus infectoria* (Aleppo Oak) exhibited enhanced wound healing in vitro in fibroblast cells by reduction of oxidative stress and lessened inflammation. The extract enhanced wound closure by promoting cell viability in vitro (Wunnoo et al., 2024). The *Quercus brantii* (Brant's Oak) extract topically applied in rats on wounds exhibited prospective in augmenting wound healing, owing to its antioxidant and antibacterial features (Haghdoost et al., 2016). The *Quercus glauca* (Ring-Cupped Oak) extract prepared from the seed coat ensured re-epithelialization of wound by supporting migration of fibroblast and lessening dermal inflammation in in vivo and in vitro models (Kim et al., 2024).

4.1.8. Punica granatum (Pomengrante)

- **Traditional usage**: The peels of pomengranate are employed for treating infected wounds traditionally. It is rich in ellagitannins thereby providing antioxidant and antimicrobial effects.
- **Scientific research**: The influence of pomegranate peel polyphenols formulated gel was investigated on wound closure in diabetes induced rats. The formulated gel notably favored infiltration of fibroblast and regeneration of collagen and enhanced wound healing (Yan et al., 2013). The wound closure healing activity of gel prepared from the methanolic extract of pomegranate peels was studied. The 5% formulated gel exhibited potential healing in Wistar rats with enhanced hydroxyproline content and better histopathological properties (Murthy et al., 2004). The pomegranate peel hydro-alcoholic extract based carboxymethyl cellulose (CMC) gel exhibited that wounds treated with this formulated gel healed rapidly than those managed with CMC alone or sterile saline, revealing efficacy of pomegranate peel extract in favoring wound closure (Zekavat et al., 2016).

4.1.9. Camellia sinensis (Tea)

- **Traditional usage**: The black and green tea leaves are employed for washing infected wounds and soothing skin infections in traditional medicine. It is rich in catechins thereby providing anti-inflammatory, antioxidant and antimicrobial benefits.

- **Scientific research**: The influence of ethanolic extract prepared from leaves of green tea was studied on surgical wound closure in rats. The wounds managed with green tea extract exhibited healing faster in comparison to control groups (Asadi et al., 2013). The influence of green tea prepared ointment on wound closure and episiotomy pain in primiparous females was studied. The findings revealed that the green tea ointment group depicted potential reductions in pain and enhanced wound healing in contrast to control (Shahrahmani et al., 2018). The effect of green tea extract on burn induced wounds of second degree in rats was studied. The results demonstrated that green tea extract on topical application enhanced re-epithelialization and diminished burning site, suggesting its efficacy in burn wound closure (Fatemi et al., 2014).

4.1.10. Castanea sativa (European Chestnut) and *Aesculus hippocastanum* (Horse Chestnut)

- **Traditional usage**: The chestnut leaf and bark infusions are employed for managing chronic wounds and varicose ulcers in European traditional medicine. The tannin rich bark infusions lessen inflammation and diminish excessive exudates from wounds.
- **Scientific research**: The effects of horse chestnut seed extract (HCE) on wound closure in rats were studied. The results exhibited that HCE enhanced production of extracellular matrix resulting in better tensile strength of wound (Kovac et al., 2020). The influence of chestnut honey loaded carboxymethyl cellulose based hydrogel on wound healing in diabetic mice was investigated. The findings supported enhanced formation of tissue granulation, and revealed better antibacterial action against *Escherichia coli* and *Staphylococcus aureus* (Park et al., 2017).

4.2. Tannin-based wound dressings and bioactive hydrogels

Tannin-based wound dressings and bioactive hydrogels have acquired potential consideration owing to their multifunctional features, including antimicrobial, hemostatic, antioxidant, and tissue-regenerative responses. The recent research in literature has explored various formulations incorporating tannic acid to support wound closure. An overview of notable research findings is mentioned below:

4.2.1. *Tannin acid–thioctic acid (TATA) network hydrogel types wound dressing*

The hydrogel based wound dressing was prepared by nucleophilic addition reaction of polyphenol sulfur radicals employing thioctic acid and tannic acid. These hydrogels depicted hydrophobic and antibacterial characteristics, better rheological response, antioxidant action, and enhanced blood clotting actions, depicting their efficacy as wound dressings (Hu et al., 2023).

4.2.2. *Self-healing and tissue adhesive hydrogel based dressing with tannic acid and ferric ion complexes*

A hybrid hydrogel wound dressing was synthesized by copolymerization of 3-acrylamido phenylboronic acid, nanoscale tannic acid and ferric ion complexes, acrylamide and chitosan. The hydrogel revealed strong mechanical strength, self-healing ability, ultra stretchability, tissue adhesiveness, antibacterial action, antioxidant features, and supported tissue regeneration in infected wound models (Ke Yang et al., 2022).

4.2.3. Agar polyvinyl alcohol type hydrogel based wound dressing with tannic acid

The hydrogel-based wound dressing employing polyvinyl alcohol and agar along with tannic acid has been designed. These hydrogel dressings revealed significant hemostatic and antibacterial features, better mechanical characteristics, and favoured wound healing in *Staphylococcus aureus*-infected rat models (Cheng et al., 2022).

4.2.4. Tannic acid–maltodextrin–polyvinyl alcohol hydrogel

The hydrogels comprising of maltodextrin, tannic acid and PVA have been prepared through a freeze-thaw method. The hydrogels demonstrated biocompatibility, good mechanical features, antibacterial and antioxidative responses, and potentially enhanced wound closure in full-thickness skin wound models (Zuo et al., 2023).

4.2.5. Gelatin methacrylate/alginate–tannin hydrogels

The gelatin methacrylate, alginate, and tannic acid have been combined by researchers, resulting in hydrogels with stable porous structures, antioxidant and antibacterial features, excellent biocompatibility and improved wound closure in mouse models by modulation of inflammatory action and supporting granulation tissue formation (Wei et al., 2024).

4.3. Synergistic effects of tannins with other wound healing agents

Tannins, particularly tannic acid, have been shown to have synergistic effects when combined with different therapeutic agents in wound closure. These combinations augment features, namely antioxidant action, anti-inflammatory effects, mechanical strength, antimicrobial property and promotion of tissue regeneration. The significant studies highlighting these synergistic interactions are mentioned below:

4.3.1. Zinc oxide nanoparticles and tannic acid in hydrogel

The hybrid hydrogel comprised gelatin, oxidized sodium alginate, zinc oxide-based nanoparticles and tannic acid was studied. The hydrogel favoured regeneration process of skin epithelial tissue in mice. This composite revealed efficient antibacterial action hostile to *Escherichia coli* and *Staphylococcus aureus* and better mechanical features revealing the synergistic responses of zinc oxide-based nanoparticles and tannic acid (Liu et al., 2024).

4.3.2. Silver modified nanoparticles and tannic acid

The silver modified nanoparticles (AgNPs)-tannic acid for wound closure were studied. The tannic acid-modified AgNPs revealed significant antibacterial activity and promoted migration of keratinocytes in vitro. These modified nanoparticles supported better wound healing, epithelialization, tissue formation, and granulation in a mouse wound model in contrast to the control (Orlowski et al., 2018).

4.3.3. Carboxymethyl chitosan and tannic acid-based hydrogels

The antioxidant, antibacterial and biocompatible hydrogels employing fungal resultant carboxymethyl chitosan and TA were developed. The incorporation of TA promoted the adhesive properties of hydrogel and better antibacterial effectiveness hostile to both Gram positive and negative bacteria, thereby supporting its worth as a dressing for infectious wound closure (Rao et al., 2023).

4.3.4. Alginate hydrogels with reduced graphene oxide and tannic acid

The alginate hydrogels were developed by incorporating tannins and reduced graphene oxide (rGO). The modified hydrogels revealed better adhesive properties and mechanical strength, making them potential candidates for significant wound-healing patches (Carrasco et al., 2024).

4.3.5. Oxidized hyaluronic and tannic acid in injectable hydrogels

A self-healing injectable hydrogel was formulated employing oxidised hyaluronic acid and chitosan-tannic acid. This tannin-hyaluronate hydrogel exhibited better antioxidant features, and favoured proliferation and migration of fibroblasts, superior biocompatibility, making it a capable candidate for wound closure benefits (Jin et al., 2024).

5. CHALLENGES AND FUTURE PROSPECTIVES

5.1 Limitations of tannin-based treatments in wound closure

Tannin based treatments have revealed promising potential for wound healing since tannins are natural components possessing anti-inflammatory, antioxidant, antimicrobial, hemostatic and astringent characteristics. However, the recent research has identified several limitations while employing tannins for wound healing:

5.1.1. *Bioavailability and stability issues:*

Tannins specifically tannic acid is prone to oxidation, exhibits low stability and lesser bioavailability thereby resulting in not readily absorption and distribution to the wound portion, further influencing their biological efficacy. Tannins also depict sensitivity to light, thereby show degradation under exposure of light (Guimarare et al., 2023; Earl et al., 2025).

5.1.2. Significant cytotoxicity at higher concentrations:

Although tannins depict valuable characteristics, at higher concentrations specific chemical structure of some tannins can exhibit cytotoxicity, thereby damaging or killing cells, specifically keratinocytes and fibroblasts that are vital for the repairing process and regenerative mechanism of tissues (Fernandez et al., 2023).

5.1.3. *Potential immunogenicity and delayed healing process:*

Though tannins can lessen inflammation, their robust astringent property can excessively dry the wound site by provoking an immune response, influencing properties of mesenchymal cells, damaging the re-epithelialization process and delaying the tissue regeneration and healing process (Abedanzadeh et al., 2024).

5.1.4. *Variability in composition of natural extracts*:

The composition and concentration of tannins from natural sources such as plants, namely oak, myrobalan, witch hazel or chestnut, can vary significantly in their chemical constitution depending on the source of plant, environmental factors and extraction procedures. This inconsistency exhibits challenges in standardizing formulated products and depicting potential therapeutic outcomes (Li et al, 2011).

5.1.5. *Interactions* with other medications:

Tannins can interact with other components in wound-healing formulated products, thereby influencing their potential effectiveness. In some instances, tannins can exhibit complexation with proteins and metal ions, possibly interfering with other topical treatments and affecting the therapeutic activity of other co-administered drugs. This demands careful design of formulations to avoid adverse drug reactions (Son et al., 2018).

5.1.6. *Limited clinical evidence:*

The evidence for tannin wound healing has come from *in vitro* experimental studies and animal models. The research on tannin-based wound therapies is still at the preclinical phase, and large-scale human clinical trials are limited (Tu et al., 2025).

5.2. Advances in tannin-based biomaterials

Tannin-based biomaterials, specifically those integrating tannic acid, have come into view as capable potential candidate for advanced wound closure applications. The multifunctional features of tannin-based biomaterials, including antimicrobial, antioxidant, hemostatic, and tissue-regenerative effects, have been explored for wound management. An overview of recent advancements in this field is reported in literature, supported by relevant research findings in various studies by research scientists (Samavati et al., 2025; Yang et al., 2025; Liu et al., 2024; Suntar et al., 2021; Wang et al., 2024).

5.2.1. **Electrospun polycaprolactone/tannic acid scaffolds**

Electrospinning techniques were utilized to synthesize polycaprolactone (PCL) scaffolds enriched with tannic acid. These scaffolds depicted potential antibacterial action, supporting them as suitable for wound dressing benefits (Chen et al., 2023).

5.2.2. **Tannic acid impregnated with chitosanRGD peptides/ alginate scaffolds**

Scaffolds comprised of chitosan, alginate, and RGD peptides were synthesized for skin regeneration, crosslinked with tannic acid. These scaffolds depicted better mechanical features, favored sustained release of TA, and improved cell viability. These scaffolds enhanced tissue remodeling and diminished scarring in rat wound models in comparison to commercial products (Mndlovu et al., 2023).

5.2.3. **Tannic acid-bridged magnetic responsive multifunctional hydrogels**

Magnetic responsive hydrogels have been created by researchers by bridging tannic acid with cobalt ferrite nanoparticles and polyvinyl alcohol. These hydrogels exhibited better mechanical stimulation-induced early vascularization, depicted antibacterial action, and enhanced wound closure under an influence of external magnetic field (Wang et al., 2022).

5.2.4. **pH-receptive tannic acid carboxylated agarose hydrogels**

The hydrogels combining tannic acid and carboxylated agarose, impregnated with zinc ions have been developed. These hydrogels revealed pH-sensitive release of tannic acid, anti-inflammatory and antibacterial characteristics, and favoured proliferation of fibroblast, supporting them as a promising alternative for wound healing benefits (Ninan et al., 2016).

5.2.5. **Tannic Acid-integrated polyethylene glycol hydrogel sponges**

Hydrogel sponges were prepared by incorporating tannic acid into polyethylene glycol (PEG) matrices. These hydrogel sponges depicted superior antibacterial characteristics hostile to *Staphylococcus aureus* and*Escherichia coli* microbes, mechanicallybetter strength, and favored wound healing, exhibiting their effectiveness for treating infection-prone wounds (Akin Betal, 2024).

5.3 Pharmacokinetics of Tannins

5.3.1 Absorption

Tannins are high-molecular-weight polyphenols (300–20,000 Da), and their absorption from the GIT is poorly efficient owing to size and protein binding. Higher polymerization results in lower absorption (Marcinczyk, 2022). Smaller tannin units or tannin metabolites (e.g., gallic acid) are more readily absorbed than intact polymers. Gallic acid, a hydrolysis product of hydrolysable tannins, depicted an oral bioavailability of about 41.8% in a pharmacokinetic study, indicating that some metabolites can be absorbed systemically (Karas et al., 2017). Polymerized proanthocyanidins remain largely non-resorbable, with minimal levels detected in urine after ingestion. Tannins are large polar molecules that are generally poorly absorbed through the skin, limiting their systemic penetration after topical application (Smeriglio et al., 2017).

5.3.2 Distribution

Due to limited absorption, intact tannins are rarely detected in systemic circulation; instead, distribution is dominated by small phenolic metabolites (e.g., gallic and ellagic acids, urolithins) in presence of plasma (Cosme et al., 2025). In reference to the protein binding, the tannins have a high affinity for proteins, leading to strong complex formation with plasma and tissue proteins, which can influence both distribution and local activity (e.g., binding to tissue enzymes or extracellular proteins) (Zayed et al., 2025).

5.3.3 Metabolism

In reference to the Gut microbiota biotransformation, the microbial enzymes in the colon are crucial mediators of tannin metabolism. Condensed and hydrolysable tannins are metabolized by gut flora into smaller, more bioaccessible compounds (e.g., urolithins, gallic acid derivatives). Urolithins and related microbial metabolites often have higher systemic detectability than the parent tannin structures, and metabolic outcomes depend on individual microbiota composition (Cosme et al., 2025). In the enteric and hepatic transformations, after initial microbial transformation, metabolites undergo phase I/II metabolism (oxidation, conjugation with glucuronide/sulfate), which further modifies their biological action and elimination pathway (Sallam et al., 2021).

5.3.4 Excretion

In the Renal & biliary routes, the conjugated metabolites (e.g., glucuronides/ sulfates of urolithins, gallic acid) are mainly excreted via urine, resulting in trace levels of tannin metabolites detected in urine and plasma after ingestion. Larger un-metabolized tannin polymers are excreted in feces due to poor absorption and strong protein binding in the gut lumen (Cosme et al., 2025).

5.3.5 Bioavailability Challenges for tannins based therapeutic for wound closures

Some of the bioavailability challenges occurring for tannins based therapeutics for wound closures such as faced poor intrinsic bioavailability. In this issue, the large size and high polarity of tannin molecules limit their permeation across enterocytes and skin barriers, resulting in poor systemic bioavailability (Lipinski, 2000). Other challenges such as protein & nutrient interactions may occurs for tannins-based therapeutics, in which the tannins may form complexes with dietary proteins, digestive enzymes, and minerals (e.g., iron), which results in hindrance in absorption and systemic exposure, behaving as anti-nutrients in high doses (Zayed et al., 2025). Other faced issue such as dependence on microbiota, which the bioavailability is highly dependent on the presence and composition of gut microbiota capable of cleaving tannin polymers into absorbable metabolites (Sallam et al., 2021).

5.4. Future research directions in tannin pharmacology for wound healing

However, challenges such as standardizing of tannin extract, ensuring the stability as well as focusing the potential cytotoxicity at highest concentration, required to be overcome in terms of wider potential clinical applications. Moreover, upcoming research should be focus on refining tannin-based formulations in terms of enhancing their bioavailability as well as conducting large-scale clinical studies in terms of confirmation of its clinical effectiveness. Ongoing research studies into tannin-based biomaterials has the notable therapeutic potential in terms of protection of wound care through giving sustainable as well as and effective healing alternatives for a widespread series of wound types. Additionally, tannin-based bio adhesives have been developed as alternatives as compared to synthetic closure materials, which may provide a strong adhesion as well as biocompatibility. Combining tannins with advanced biomaterials, such as nanotechnology as well as tissue engineering, may offers exciting possibilities for future wound care solutions. To recognize the significance of tannin-based treatments in wound healing process, forthcoming research studies should center on the s*tandardization of natural extracts and quality control*, which these standardized extraction procedures and formulation methods are required to be developed for tannin-rich plant extracts to ensure consistency in chemical composition, quality control and stability of tannic acid-based formulations and reproducibility in therapeutic outcomes. The randomized, well-designed controlled trials are necessary to be performed to evaluate the effectiveness and safety of tannic acid-based treatments in diverse patient populations for translation into clinical practice. In addition to that the mechanistic Studies, which may indicate the future study is demanded to clarify the molecular pathways which tannins can support wound healing, including their interactions with cellular pathways, which will aid in optimizing their therapeutic applications in formulations (Quave, 2018). By using the innovative technology such as biomaterial engineering as well as delivery system at nanoscale are the approaches, which may create the new revolution from the natural remedies to reliable approach at clinical level as therapeutic strategies in wound closures.

6. CONCLUSION

Wound closure is a multistep and time-consuming biological phenomenon that includes hemostasis, inflammatory phase, proliferative or re-epithelialization and maturation. Recently, natural bioactive components have been significantly recognized for their pharmacological potential in the management of wound care due to their low toxicity as well as their biocompatibility, along with versatile properties. Wound closure methods have advanced considerably, evolving from basic suturing to adhesive agents, with continuous improvement in techniques. Quick wound closures require external methods which have evolved drastically from suturing to adhesive agents, and the techniques are under continuous improvement. Nowadays, natural phytoconstituents are being considered for the management of wounds to overcome the toxicity, biocompatibility, toxicity issues associated with traditional methods. Tannins, as a bio constituent obtained from several herbal plants, because of their strong astringent, antimicrobial, anti-inflammatory, and antioxidant characteristics, are a potential candidate for the development of formulations linked to wound healing. As natural astringents, tannins work by binding proteins and forming a defensive film on the wound, that aids in the hemostasis process by constricting blood vessels as well as promoting clot formation, ultimately minimizing bleeding effects. There are several formulations which are being used but require more improvement in terms of biocompatibility, toxicity, and controlled delivery, and more clinical trials in this area will be helpful for the preparation of more effective formulations.

REFERENCES

Abedanzadeh, M., Abolmaali, S. S., Heidari, R., Aalaei, E., Kaviani, M., Dara, M., Mohammadi, S., Azarpira, N., & Amaddon, A. M. (2024). Photo-crosslinked hyaluronic acid hydrogels designed for simultaneous delivery of mesenchymal stem cells and tannic acid: Advancing towards scarless wound healing. *International Journal of Biological Macromolecules, 281*, 136394. DOI: 10.1016/j.ijbiomac.2024.136394 PMID: 39406324

Abolarin, P. O., & Owoyele, B. V. (2024). Tannic acid inhibits pain mediators, inflammation and oxidative stress in mice exposed to glyphosate-based herbicide. *Environmental Analysis, Health and Toxicology, 39*(2), e2024019. DOI: 10.5620/eaht.2024019 PMID: 39054833

Akin, B., Akgul, B., Tasdurmazli, S., Abamor, E. S., Ozbek, T., Ozcelik, B., & Ozmen, M. M. (2024). Tannic Acid Incorporated Antibacterial Polyethylene Glycol Based Hydrogel Sponges for Management of Wound Infections. *Macromolecular Bioscience, 24*(8), 2400101. DOI: 10.1002/mabi.202400101 PMID: 38748813

Almadani, Y. H., Vorstenbosch, J., Davison, P. G., & Murphy, A. M. (2021). Wound healing: a comprehensive review. In *Seminars in plastic surgery* (Vol. 35, No. 03, pp. 141-144). Thieme Medical Publishers, Inc. DOI: 10.1055/s-0041-1731791

Ambreen, M., & Mirza, S. A. (2020). Evaluation of anti-inflammatory and wound healing potential of tannins isolated from leaf callus cultures of Achyranthes aspera and Ocimum basilicum. *Pakistan Journal of Pharmaceutical Sciences, 33*(1), 361–36. PMID: 32122869

Arbenz, A., & Avérous, L. (2015). Chemical modification of tannins to elaborate aromatic biobased macromolecular architectures. *Green Chemistry, 17*(5), 2626–2646. DOI: 10.1039/C5GC00282F

Asadi, S. Y., Parsaei, P., Karimi, M., Ezzati, S., Zamiri, A., Mohammadizadeh, F., & Rafieian-Kopaei, M. (2013). Effect of green tea (Camellia sinensis) extract on healing process of surgical wounds in rat. *International Journal of Surgery, 11*(4), 332–337. DOI: 10.1016/j.ijsu.2013.02.014 PMID: 23459184

Boakye, Y. D., Agyare, C., & Hensel, A. (2016). Anti-infective properties and time-kill kinetics of Phyllanthus muellerianus and its major constituent, geraniin. *Medicinal Chemistry (Sunnyvale, Calif.), 6*(2), 2. DOI: 10.4172/2161-0444.1000332

Carrasco, S., González, L., Tapia, M., Urbano, B. F., Aguayo, C., & Fernández, K. (2024). Enhancing alginate hydrogels as possible wound-healing patches: The synergistic impact of reduced graphene oxide and tannins on mechanical and adhesive properties. *Polymers*, *16*(8), 1081. DOI: 10.3390/polym16081081 PMID: 38675000

Cheesman, M. J., Alcorn, S., Verma, V., & Cock, I. E. (2021). An assessment of the growth inhibition profiles of Hamamelis virginiana L. extracts against Streptococcus and Staphylococcus spp. *Journal of Traditional and Complementary Medicine*, *11*(5), 457–465. DOI: 10.1016/j.jtcme.2021.03.002 PMID: 34522640

Chen, X., Zhang, Q., Wang, Y., Meng, J., Wu, M., Xu, H., Du, L., & Yang, X. (2023). Fabrication and characterization of electrospun poly (caprolactone)/tannic acid scaffold as an antibacterial wound dressing. *Polymers*, *15*(3), 593. DOI: 10.3390/polym15030593 PMID: 36771894

Chen, Y., Tian, L., Yang, F., Tong, W., Jia, R., Zou, Y., Yin, L., Li, L., He, C., Liang, X., Ye, G., Lv, C., Song, X., & Yin, Z. (2019). Tannic Acid Accelerates Cutaneous Wound Healing in Rats Via Activation of the ERK 1/2 Signaling Pathways. *Advances in Wound Care*, *8*(7), 341–354. DOI: 10.1089/wound.2018.0853 PMID: 31737421

Cheng, C., Peng, X., Xi, L., Wan, C., Shi, S., Wang, Y., & Yu, X. (2022). An agar–polyvinyl alcohol hydrogel loaded with tannic acid with efficient hemostatic and antibacterial capacity for wound dressing. *Food & Function*, *13*(18), 9622–9634. DOI: 10.1039/D2FO02251F PMID: 36004684

Cheng, H. A., Drinnan, C. T., Pleshko, N., & Fisher, O. Z. (2015). Pseudotannins self-assembled into antioxidant complexes. *Soft Matter*, *11*(39), 7783–7791. DOI: 10.1039/C5SM01224D PMID: 26313262

Cosme, F., Aires, A., Pinto, T., Oliveira, I., Vilela, A., & Gonçalves, B. (2025). A comprehensive review of bioactive tannins in foods and beverages: Functional properties, health benefits, and sensory qualities. *Molecules (Basel, Switzerland)*, *30*(4), 800. DOI: 10.3390/molecules30040800 PMID: 40005115

Criollo-Mendoza, M. S., Contreras-Angulo, L. A., Leyva-López, N., Gutiérrez-Grijalva, E. P., Jiménez-Ortega, L. A., & Heredia, J. B. (2023). Wound healing properties of natural products: Mechanisms of action. *Molecules (Basel, Switzerland)*, *28*(2), 598. DOI: 10.3390/molecules28020598 PMID: 36677659

Das, A. K., Islam, M. N., Faruk, M. O., Ashaduzzaman, M., & Dungani, R. (2020). Review on tannins: Extraction processes, applications and possibilities. *South African Journal of Botany*, *135*, 58–70. DOI: 10.1016/j.sajb.2020.08.008

Dong, G., Liu, H., Yu, X., Zhang, X., Lu, H., Zhou, T., & Cao, J. (2018). Antimicrobial and anti-biofilm activity of tannic acid against Staphylococcus aureus. *Natural Product Research*, *32*(18), 2225–2228. DOI: 10.1080/14786419.2017.1366485 PMID: 28826250

Earl, J. E., Aubert, J. F., Michael, E. O., & Edwards, D. J. (2025). Effects of Soil, Light, and Temperature on Freshwater Tannin Concentrations. *Hydrobiology*, *4*(1), 2. DOI: 10.3390/hydrobiology4010002

El-Mabood, A., El-Sayed, A., Zaher, N. A., & Ali, H. E. (2014). Topical tannic acid application decreases posthemorrhoidectomy pain: A novel idea. *The Egyptian Journal of Surgery*, *33*(4), 213. DOI: 10.4103/1110-1121.147586

Farha, A. K., Yang, Q.-Q., Kim, G., Li, H.-B., Zhu, F., Liu, H.-Y., Gan, R.-Y., & Corke, H. (2020). Tannins as an alternative to antibiotics. *Food Bioscience*, *38*, 100751. DOI: 10.1016/j.fbio.2020.100751

Fatemi, M. J., Nikoomaram, B., Rahimi, A. A. K., Talayi, D., Taghavi, S., & Ghavami, Y. (2014). Effect of green tea on the second degree burn wounds in rats. *Indian Journal of Plastic Surgery : Official Publication of the Association of Plastic Surgeons of India*, *47*(03), 370–374. DOI: 10.4103/0970-0358.146593 PMID: 25593422

Fernández, K., Llanquileo, A., Bustos, M., Aedo, V., Ruiz, I., Carrasco, S., Tapia, M., Pereira, M., Meléndrez, M. F., Aguayo, C., & Atanase, L. I. (2023). Self-Assembled CNF/rGO/Tannin Composite: Study of the Physicochemical and Wound Healing Properties. *Polymers*, *15*(12), 2752. DOI: 10.3390/polym15122752 PMID: 37376399

Fikru, A., Makonnen, E., Eguale, T., Debella, A., & Mekonnen, G. A. (2012). Evaluation of in vivo wound healing activity of methanol extract of Achyranthes aspera L. *Journal of Ethnopharmacology*, *143*(2), 469–474. DOI: 10.1016/j.jep.2012.06.049 PMID: 22771316

Fraga-Corral, M., Otero, P., Cassani, L., Echave, J., Garcia-Oliveira, P., Carpena, M., Chamorro, F., Lourenço-Lopes, C., Prieto, M. A., & Simal-Gandara, J. (2021). Traditional Applications of Tannin Rich Extracts Supported by Scientific Data: Chemical Composition, Bioavailability and Bioaccessibility. *Foods*, *10*(2), 251. DOI: 10.3390/foods10020251 PMID: 33530516

Furie, B., & Furie, B. C. (2008). Mechanisms of thrombus formation. *The New England Journal of Medicine*, *359*(9), 938–949. DOI: 10.1056/NEJMra0801082 PMID: 18753650

Gurtner, G. C., Werner, S., Barrandon, Y., & Longaker, M. T. (2008). Wound repair and regeneration. *Nature*, *453*(7193), 314–321. DOI: 10.1038/nature07039 PMID: 18480812

Haghdoost, F., Baradaran Mahdavi, M. M., Zolfaghari, B., Sanei, M. H., Najafi, S., Zandifar, A., Manouchehri, N., & Javanmard, S. H. (2016). The effect of Quercus brantii gall extract on burn wound healing in rat. *Iranian Journal of Basic Medical Sciences.*, *19*(10), 1144–1150. PMID: 27872712

Hancock, V., Dahl, M., Vejborg, R. M., & Klemm, P. (2010). Dietary plant components ellagic acid and tannic acid inhibit Escherichiacoli biofilm formation. *Journal of Medical Microbiology*, *59*(4), 496–498. DOI: 10.1099/jmm.0.013680-0 PMID: 19959627

Hemingway, R. W. (1998). Practical polyphenolics: From structure to molecular recognition and physiological action, by Edwin Haslam. (Book review). *Journal of Natural Products*, *61*(11), 1454–1455. DOI: 10.1021/np980243t

Hu, M. S., Walmsley, G. G., Barnes, L. A., Weiskopf, K., Rennert, R. C., Duscher, D., Januszyk, M., Maan, Z. N., Hong, W. X., Cheung, A. T., Leavitt, T., Marshall, C. D., Ransom, R. C., Malhotra, S., Moore, A. L., Rajadas, J., Lorenz, H. P., Weissman, I. L., Gurtner, G. C., & Longaker, M. T. (2017). Delivery of monocyte lineage cells in a biomimetic scaffold enhances tissue repair. *JCI Insight*, *2*(19), e96260. DOI: 10.1172/jci.insight.96260 PMID: 28978794

Hu, X., Kong, Q., Li, R., Zhou, C., & Li, Z. (2023). Tannin–thioctic acid network hydrogels with antibacterial and hydrophobic properties. *Fibers and Polymers*, *24*(10), 3373–3384. DOI: 10.1007/s12221-023-00329-y

Jayakumari, S. (2018). Formulation and evaluation of herbal gel from tannin-enriched fraction of Psidium guajava Linn. leaves for diabetic wound healing. [IJGP]. *International Journal of Green Pharmacy*, *12*(03). Advance online publication. DOI: 10.22377/ijgp.v12i03.2009

Jin, W., Xue, Y., Xue, Y., Han, X., Song, Q., Zhang, J., Li, Z., Cheng, J., Guan, S., Sun, S., & Chu, L. (2020). Tannic acid ameliorates arsenic trioxide-induced nephrotoxicity, contribution of NF-κB and Nrf2 pathways. *Biomedicine and Pharmacotherapy*, *126*, 110047. DOI: 10.1016/j.biopha.2020.110047 PMID: 32146384

Jin, X., Wei, C., Li, K., Yin, P., Wu, C., & Zhang, W. (2024). Polyphenol-mediated hyaluronic acid/tannic acid hydrogel with short gelation time and high adhesion strength for accelerating wound healing. *Carbohydrate Polymers*, *342*, 122372. DOI: 10.1016/j.carbpol.2024.122372 PMID: 39048222

Jing, W., Xiaolan, C., Yu, C., Feng, Q., & Haifeng, Y. (2022). Pharmacological effects and mechanisms of tannic acid. *Biomedicine and Pharmacotherapy*, *154*, 113561. DOI: 10.1016/j.biopha.2022.113561 PMID: 36029537

Karas, M., Jakubczyk, A., Szymanowska, U., Złotek, U., & Zielińska, E. (2017). Digestion and bioavailability of bioactive phytochemicals. *International Journal of Food Science & Technology*, *52*(2), 291–305. DOI: 10.1111/ijfs.13323

Karuppagounder, V., Arumugam, S., Thandavarayan, R. A., Pitchaimani, V., Sreedhar, R., Afrin, R., Harima, M., Suzuki, H., Nomoto, M., Miyashita, S., Suzuki, K., Nakamura, M., Ueno, K., & Watanabe, K. (2015). Tannic acid modulates NFκB signaling pathway and skin inflammation in NC/Nga mice through PPARγ expression. *Cytokine*, *76*(2), 206–213. DOI: 10.1016/j.cyto.2015.05.016 PMID: 26049169

Kemppainen, K., Siika-aho, M., Pattathil, S., Giovando, S., & Kruus, K. (2014). Spruce bark as an industrial source of condensed tannins and non-cellulosic sugars. *Industrial Crops and Products*, *52*, 158–168. DOI: 10.1016/j.indcrop.2013.10.009

Khanbabaee, K., & Van Ree, T. (2001). Tannins: Classification and definition. *Natural Product Reports*, *18*(6), 641–649. DOI: 10.1039/b1010611 PMID: 11820762

Kim, S. H., Shin, H. L., Son, T. H., Lim, S. A., Kim, D., Yoon, J. H., Choi, H., Kim, H. G., & Choi, S. W. (2024). Quercus glauca Acorn Seed Coat Extract Promotes Wound Re-Epithelialization by Facilitating Fibroblast Migration and Inhibiting Dermal Inflammation. *Biology (Basel)*, *13*(10), 775. DOI: 10.3390/biology13100775 PMID: 39452084

Kisseleva, T., & Brenner, D. A. (2008). Mechanisms of fibrogenesis. *Experimental Biology and Medicine*, *233*(2), 109–122. DOI: 10.3181/0707-MR-190 PMID: 18222966

Kung, H. N., Yang, M. J., Chang, C. F., Chau, Y. P., & Lu, K. S. (2008). In vitro and in vivo wound healing-promoting activities of β-lapachone. *American Journal of Physiology. Cell Physiology*, *295*(4), C931–C943. DOI: 10.1152/ajpcell.00266.2008 PMID: 18650264

Li, J., Chen, J., & Kirsner, R. (2007). Pathophysiology of acute wound healing. *Clinics in Dermatology*, *25*(1), 9–18. DOI: 10.1016/j.clindermatol.2006.09.007 PMID: 17276196

Li, K., Diao, Y., Zhang, H., Wang, S., Zhang, Z., Yu, B., Huang, S., & Yang, H. (2011). Tannin extracts from immature fruits of Terminalia chebula Fructus Retz. promote cutaneous wound healing in rats. *BMC Complementary and Alternative Medicine*, *11*(1), 86. DOI: 10.1186/1472-6882-11-86 PMID: 21982053

Lipinski, C. A. (2000). Drug-like properties and the causes of poor solubility and poor permeability. *Journal of Pharmacological and Toxicological Methods*, *44*(1), 235–249. DOI: 10.1016/S1056-8719(00)00107-6 PMID: 11274893

Liu, H., Yang, Y., Deng, L., Shen, Z., Huang, Q., Shah, N. G., Chen, W., Zhang, Y., Wang, X., Yu, L., & Chen, Z. (2024). Antibacterial and antioxidative hydrogel dressings based on tannic acid-gelatin/oxidized sodium alginate loaded with zinc oxide nanoparticles for promoting wound healing. *International Journal of Biological Macromolecules*, *279*(Pt 2), 135177. DOI: 10.1016/j.ijbiomac.2024.135177 PMID: 39214222

Marcińczyk, N., Gromotowicz-Popławska, A., Tomczyk, M., & Chabielska, E. (2022). Tannins as hemostasis modulators. *Frontiers in Pharmacology*, *12*, 806891. DOI: 10.3389/fphar.2021.806891 PMID: 35095516

Mndlovu, H., du Toit, L. C., Kumar, P., & Choonara, Y. E. (2023). Tannic acid-loaded chitosan-RGD-alginate scaffolds for wound healing and skin regeneration. *Biomedical Materials (Bristol, England)*, *18*(4), 045009. DOI: 10.1088/1748-605X/acce88 PMID: 37075773

Molin, M. L., Segat, B., Garcia, M. C. F., Pezzin, A. P. T., Silveira, M. L. L., & Schneider, A. L. D. S. (2023). Development and characterization of bacterial cellulose membrane incorporated with Witch hazel extract. Matéria (Rio de Janeiro), 28, e20230008. https://doi.org/ K.N., Reddy, V. K., Veigas, J. M., & Murthy, U. D. (2004). Study on wound healing activity of Punica granatum peel. *Journal of Medicinal Food, 7*(2), 256-259.doi: DOI: 10.1590/1517-7076-RMAT-2023-0008Murthy

Natarajan, V., Krithica, N., Madhan, B., & Sehgal, P. K. (2013). Preparation and properties of tannic acid cross-linked collagen scaffold and its application in wound healing. *Journal of Biomedical Materials Research. Part B, Applied Biomaterials*, *101*(4), 560–567. DOI: 10.1002/jbm.b.32856 PMID: 23255343

Navarrete, P., Pizzi, A., Pasch, H., Rode, K., & Delmotte, L. (2013). Characterization of two maritime pine tannins as wood adhesives. *Journal of Adhesion Science and Technology*, *27*(22), 2462–2479. DOI: 10.1080/01694243.2013.787515

Nguyen, T., Nioi, P., & Pickett, C. B. (2009). The Nrf2-antioxidant response element signaling pathway and its activation by oxidative stress. *The Journal of Biological Chemistry*, *284*(20), 13291–13295. DOI: 10.1074/jbc.R900010200 PMID: 19182219

Ninan, N., Forget, A., Shastri, V. P., Voelcker, N. H., & Blencowe, A. (2016). Antibacterial and anti-inflammatory pH-responsive tannic acid-carboxylated agarose composite hydrogels for wound healing. *ACS Applied Materials & Interfaces*, *8*(42), 28511–28521. DOI: 10.1021/acsami.6b10491 PMID: 27704757

Orlowski, P., Zmigrodzka, M., Tomaszewska, E., Ranoszek-Soliwoda, K., Czupryn, M., Antos-Bielska, M., Szemraj, J., Celichowski, G., Grobelny, J., & Krzyzowska, M. (2018). Tannic acid-modified silver nanoparticles for wound healing: The importance of size. *International Journal of Nanomedicine*, *13*, 991–1007. DOI: 10.2147/IJN.S154797 PMID: 29497293

Park, J. S., An, S. J., Jeong, S. I., Gwon, H. J., Lim, Y. M., & Nho, Y. C. (2017). Chestnut honey impregnated carboxymethyl cellulose hydrogel for diabetic ulcer healing. *Polymers*, *9*(7), 248. DOI: 10.3390/polym9070248 PMID: 30970925

Pattarayan, D., Sivanantham, A., Bethunaickan, R., Palanichamy, R., & Rajasekaran, S. (2018). Tannic acid modulates fibroblast proliferation and differentiation in response to pro-fibrotic stimuli. *Journal of Cellular Biochemistry*, *119*(8), 6732–6742. DOI: 10.1002/jcb.26866 PMID: 29665059

Payne, D. E., Martin, N. R., Parzych, K. R., Rickard, A. H., Underwood, A., & Boles, B. R. (2013). Tannic acid inhibits Staphylococcus aureus surface colonization in an IsaA-dependent manner. *Infection and Immunity*, *81*(2), 496–504. DOI: 10.1128/IAI.00877-12 PMID: 23208606

Piazza, S., Martinelli, G., Vrhovsek, U., Masuero, D., Fumagalli, M., Magnavacca, A., Pozzoli, C., Canilli, L., Terno, M., Angarano, M., Dell'Agli, M., & Sangiovanni, E. (2022). Anti-Inflammatory and Anti-Acne Effects of Hamamelis virginiana Bark in Human Keratinocytes. *Antioxidants*, *11*(6), 1119. DOI: 10.3390/antiox11061119 PMID: 35740016

Pizzi, A. (2021). Tannins medical/pharmacological and related applications: A critical review. *Sustainable Chemistry and Pharmacy*, *22*, 100481. DOI: 10.1016/j.scp.2021.100481

Quave, C. L. (2018). Wound healing with botanicals: A review and future perspectives. *Current Dermatology Reports*, *7*(4), 287–295. DOI: 10.1007/s13671-018-0247-4 PMID: 31106027

Rajasekaran, S., Rajasekar, N., & Sivanantham, A. (2021). Therapeutic potential of plant-derived tannins in non-malignant respiratory diseases. *The Journal of Nutritional Biochemistry*, *94*, 108632. DOI: 10.1016/j.jnutbio.2021.108632 PMID: 33794331

Rao, K. M., Uthappa, U. T., Kim, H. J., & Han, S. S. (2023). Tissue adhesive, biocompatible, antioxidant, and antibacterial hydrogels based on tannic acid and fungal-derived carboxymethyl chitosan for wound-dressing applications. *Gels (Basel, Switzerland)*, *9*(5), 354. DOI: 10.3390/gels9050354 PMID: 37232946

Rumbaut, R. E. (2010). *Platelet-vessel wall interactions in hemostasis and thrombosis*. Morgan & Claypool Life Sciences. DOI: 10.4199/C00007ED1V01Y201002ISP004

Saad, H., Khoukh, A., Ayed, N., Charrier, B., & Charrier-El Bouhtoury, F. (2014). Characterization of Tunisian Aleppo pine tannins for a potential use in wood adhesive formulation. *Industrial Crops and Products*, *61*, 517–525. DOI: 10.1016/j.indcrop.2014.07.035

Sallam, I. E., Abdelwareth, A., Attia, H., Aziz, R. K., Homsi, M. N., von Bergen, M., & Farag, M. A. (2021). Effect of Gut Microbiota Biotransformation on Dietary Tannins and Human Health Implications. *Microorganisms*, *9*(5), 965. DOI: 10.3390/microorganisms9050965 PMID: 33947064

Salman, M., Tabassum, H., & Parvez, S. (2020). Tannic acid provides neuroprotective effects against traumatic brain injury through the PGC-1α/Nrf2/HO-1 pathway. *Molecular Neurobiology*, *57*(6), 2870–2885. DOI: 10.1007/s12035-020-01924-3 PMID: 32399817

Samavati, S. S., Kashanian, S., Derakhshankhah, H., Rabiei, M., Sajadimajd, S., Fakhri, S., & Rashidi, Z. (2025). Accelerated wound healing through tannin-rich Jaft extract: Concentration-dependent efficacy and mechanistic insights from Quercus brantii ointment formulations. *Scientific Reports*, *15*(1), 29004. DOI: 10.1038/s41598-025-13832-4 PMID: 40781116

Schultz, G. S., & Wysocki, A. (2009). Interactions between extracellular matrix and growth factors in wound healing. *Wound Repair and Regeneration*, *17*(2), 153–162. DOI: 10.1111/j.1524-475X.2009.00466.x PMID: 19320882

Sebestyén, Z., Jakab, E., Badea, E., Barta-Rajnai, E., Şendrea, C., & Czégény, Z. S. (2019). Thermal degradation study of vegetable tannins and vegetable tanned leathers. *Journal of Analytical and Applied Pyrolysis*, *138*, 178–187. DOI: 10.1016/j.jaap.2018.12.022

Shahrahmani, H., Kariman, N., Jannesari, S., Rafieian-Kopaei, M., Mirzaei, M., Ghalandari, S., Shahrahmani, N., & Mardani, G. (2018). The effect of green tea ointment on episiotomy pain and wound healing in primiparous women: A randomized, double-blind, placebo-controlled clinical trial. *Phytotherapy Research*, *32*(3), 522–530. DOI: 10.1002/ptr.5999 PMID: 29235159

Shedoeva, A., Leavesley, D., Upton, Z., & Fan, C. (2019). Wound healing and the use of medicinal plants. *Evidence-Based Complementary and Alternative Medicine : eCAM*, *2684108*, 1–30. Advance online publication. DOI: 10.1155/2019/2684108 PMID: 31662773

Smeriglio, A., Barreca, D., Bellocco, E., & Trombetta, D. (2017). Proanthocyanidins and hydrolysable tannins: Occurrence, dietary intake and pharmacological effects. *British Journal of Pharmacology, 174*(11), 1244–1262. DOI: 10.1111/bph.13630 PMID: 27646690

Son, H. Y., Jun, H., Kim, K. R., Hong, C. A., & Nam, Y. S. (2018). Tannin-mediated assembly of gold–titanium oxide hybrid nanoparticles for plasmonic photochemical applications. *Journal of Industrial and Engineering Chemistry, 63*, 420–425. DOI: 10.1016/j.jiec.2018.03.002

Su, X., Liu, X., Wang, S., Li, B., Pan, T., Liu, D., Wang, F., Diao, Y., & Li, K. (2017). Wound-healing promoting effect of total tannins from Entada phaseoloides (L.) Merr. in rats. *Burns, 43*(4), 830–838. DOI: 10.1016/j.burns.2016.10.010 PMID: 28040363

Su, Y., & Richmond, A. (2015). Chemokine regulation of neutrophil infiltration of skin wounds. *Advances in Wound Care, 4*(11), 631–640. DOI: 10.1089/wound.2014.0559 PMID: 26543677

Suguna, L., Singh, S., Sivakumar, P., Sampath, P., & Chandrakasan, G. (2002). Influence of Terminalia chebula on dermal wound healing in rats. *Phytotherapy Research, 16*(3), 227–231. DOI: 10.1002/ptr.827 PMID: 12164266

Suntar, I., Çetinkaya, S., Panieri, E., Saha, S., Buttari, B., Profumo, E., & Saso, L. (2021). Regulatory Role of Nrf2 Signaling Pathway in Wound Healing Process. *Molecules (Basel, Switzerland), 26*(9), 2424. DOI: 10.3390/molecules26092424 PMID: 33919399

Tu, Y., Zheng, W., Ding, Z., Xiang, J., Yang, Q., Liu, Y., Cao, J., Shen, Y., Tang, Z., Lin, S., Fan, L., Xu, Y., & Chen, B. (2025). Exosome-loaded tannic acid-thioctic acid hydrogel enhances wound healing in coagulation disorders. *Materials Today. Bio, 31*, 101496. DOI: 10.1016/j.mtbio.2025.101496 PMID: 39990738

Wang, M., Huang, H., Wang, L., Yin, L., Yang, H., Chen, C., Zheng, Q., & He, S. (2022). Tannic acid attenuates intestinal oxidative damage by improving antioxidant capacity and intestinal barrier in weaned piglets and IPEC-J2 cells. *Frontiers in Nutrition, 9*, 1012207. DOI: 10.3389/fnut.2022.1012207 PMID: 36407512

Wang, T., Zheng, Y., Zhang, J., & Wu, Z. (2024). Targeting ferroptosis promotes diabetic wound healing via Nrf2 activation. *Heliyon, 10*(19), e37477. DOI: 10.1016/j.heliyon.2024.e37477 PMID: 39421383

Wei, Z., Zuo, Y., Wu, E., Huang, L., Qian, Y., Wang, J., & Chen, Z. (2024). Highly biocompatible, antioxidant and antibacterial gelatin methacrylate/alginate-tannin hydrogels for wound healing. *International Journal of Biological Macromolecules, 279*, 135417. DOI: 10.1016/j.ijbiomac.2024.135417 PMID: 39265900

Wilkinson, H. N., & Hardman, M. J. (2020). Wound healing: Cellular mechanisms and pathological outcomes. *Open Biology, 10*(9), 200223. DOI: 10.1098/rsob.200223 PMID: 32993416

Wunnoo, S., Sermwittayawong, D., Praparatana, R., Voravuthikunchai, S. P., & Jakkawanpitak, C. (2024). Quercus infectoria Gall Ethanolic Extract Accelerates Wound Healing through Attenuating Inflammation and Oxidative Injuries in Skin Fibroblasts. *Antioxidants, 13*(9), 1094. DOI: 10.3390/antiox13091094 PMID: 39334753

Yang, F., Fang, Z., Yang, M., Cheng, Z., Tian, Y., Liang, J., & Li, T. (2025). Research hotspots in natural products for wound healing: A bibliometric analysis and literature review. *Clinical, Cosmetic and Investigational Dermatology, 18*, 15. Advance online publication. DOI: 10.2147/CCID.S544826 PMID: 40873990

Yang, H. L., Tsai, Y. C., Korivi, M., Chang, C. T., & Hseu, Y. C. (2017). Lucidone promotes the cutaneous wound healing process via activation of the PI3K/AKT, Wnt/β-catenin and NF-κB signaling pathways. *Biochimica et Biophysica Acta (BBA)-. Molecular Cell Research, 1864*(1), 151–168. PMID: 27816443

Yang, K., Zhou, X., Li, Z., Wang, Z., Luo, Y., Deng, L., & He, D. (2022). Ultrastretchable, self-healable, and tissue-adhesive hydrogel dressings involving nanoscale tannic acid/ferric ion complexes for combating bacterial infection and promoting wound healing. *ACS Applied Materials & Interfaces, 14*(38), 43010–43025. DOI: 10.1021/acsami.2c13283 PMID: 36108772

Young, A., & McNaught, C. E. (2011). The physiology of wound healing. *Surgery (Oxford), 29*(10), 475–479. DOI: 10.1016/j.mpsur.2011.06.011

Zayed, A., Abdelkareem, S., Talaat, N., Dayem, D. A., & Farag, M. A. (2025). Tannin in foods: Classification, Dietary Sources, and Processing Strategies to Minimize Anti-Nutrient Effects. *Food and Bioprocess Technology, 18*(11), 9221–9249. DOI: 10.1007/s11947-025-04020-3

Zekavat, O., Amanat, A., Karami, M., Paydar, S., Gramizadeh, B., & Zareian-Jahromi, M. (2016). Wound healing studies using Punica granatum peel: An animal experimental study. *Advances in Skin & Wound Care, 29*(5), 217–225. DOI: 10.1097/01.ASW.0000481116.16998.55 PMID: 27089150

Zhang, H., Chen, J., & Cen, Y. (2018). Burn wound healing potential of a polysaccharide from Sanguisorba officinalis L. in mice. *International Journal of Biological Macromolecules*, *112*, 862–867. DOI: 10.1016/j.ijbiomac.2018.01.214 PMID: 29425875

Zhang, H., Lin, X., Cao, X., Wang, Y., Wang, J., & Zhao, Y. (2024). Developing natural polymers for skin wound healing. *Bioactive Materials, 33*, 355-376. Zhang, J., Song, Q., Han, X., Zhang, Y., Zhang, Y., Zhang, X., Chu, X., Zhang, F., & Chu, L. (2017). Multi-targeted protection of acetaminophen-induced hepatotoxicity in mice by tannic acid. *International Immunopharmacology*, *47*, 95–105. DOI: 10.1016/j.intimp.2017.03.027 PMID: 28376392

Zuo, L., Wang, X., Cao, X., Chen, B., Shao, M., Yang, G., Fu, S., & Wang, L. (2023). Preparation and characterization of tannin-maltodextrin-polyvinyl alcohol hydrogel based on hydrogen bonding for wound healing. *Journal of the Mechanical Behavior of Biomedical Materials*, *145*, 105942. DOI: 10.1016/j.jmbbm.2023.105942 PMID: 37423009

Section 3
Traditional and Evidence-Based Herbal Therapies

Chapter 8
Healing Through the Ages:
Traditional Herbal Remedies

Priya Tiwari
http://orcid.org/0000-0003-0490-7884
Faculty of Pharmacy, Moradabad Educational Group of Institutions, Moradabad, India

Rajat Srivastava
http://orcid.org/0000-0002-7721-6710
BBS College of Pharmacy and Polytechnic, Prayagraj, India

ABSTRACT

The foundation for early medicine was established by ancient herbal practices, which still influence healthcare today. Ayurveda, Unani, and Traditional Chinese Medicine are examples of systems that have developed over centuries of spiritual, scientific, and cultural integration. Textual and archeobotanical records from Egypt, Greece, India, and Mesopotamia demonstrate the systematic usage of therapeutic plants. More than 80% of people still use herbal remedies today, particularly in rural regions. The preventive and holistic aspects of herbal therapy are reviving interest and bolstering traditional treatments. With an emphasis on clinical efficacy, illness management, prevention, and historical context, this study investigates the integration of traditional systems into contemporary care. Safety, regulation, standardisation, validation, and ethics are also examined. China, India, and Germany's integration models demonstrate the possibility of interdisciplinary, policy-driven cooperation.

DOI: 10.4018/979-8-3373-2908-6.ch008

1. ANCIENT HERBAL TRADITIONS

Introduction

Globally, medical knowledge has advanced astronomically over the past century. The overall mortality rate has declined, life expectancy has increased, numerous life-saving medications have been developed to combat infectious and chronic disease, and technological advancements expanded the capabilities of modern science. Despite these remarkable achievements, the benefits of modern medicine do not reach every corner of the globe. The World Health Organization (WHO), in the Declaration of Alma-Ata (1978), emphasized that poverty, illiteracy, and inadequate sanitation must be addressed to achieve the goal of "Health for All". Later, WHO introduced the "Health for All in the 21st Century" policy, focusing on health security, equity, increased life expectancy, and universal access to quality healthcare (Sen & Chakraborty, 2017). However, despite these advancements, the modern medical system continues to face challenges in accessibility and affordability, particularly in developing regions. A significant proportion of the global population still relies on traditional, complementary, or alternative medical systems, many of which predate allopathic medicine (Organization, 2004; Sen & Chakraborty, 2017). This raises critical questions: (1) Has the goal of "Health for All" been achieved? (2) Can it be achieved without integrating traditional herbal medicine into clinical practices?

Modern health challenges – including pollution, unhealthy lifestyles, and environmental degradation – have increased disease burden. Additionally, concerns about the side effects and misuse of allopathic medicines have renewed interest in traditional remedies. Recognizing this, WHO launched the "Traditional Medicine Strategy 2014-2023," emphasizing the integration of traditional medicine into healthcare systems to ensure safety, efficacy, and quality (Sen & Chakraborty, 2017).

EARLY USE OF MEDICINAL PLANTS IN PREHISTORIC SOCIETIES

Medicinal plants have been used since prehistoric times and formed the foundation of early healing practices. Early humans, closely connected to their natural environment, utilized plants for treating wounds, relieving pain, and performing ritualistic practices. This long-standing tradition represents one of humanity earliest engagements with pharmacology, bridging anthropology, ethnobotany, and medical history. Before the development of written language and formal healthcare systems, plant based remedies were essential for managing illnesses and injuries. This relationship between humans and nature laid the groundwork for modern herbal medicine. Early pharmacological knowledge shows a profound comprehension of nature's curative potential and is based on experimentation, observation, and communal experience.

Evidence of Early Medicinal Plant Use:

Archaeological evidence suggests that medicinal plants were used as early as 60,000 years ago. Fossilized pollen and plant residues in burial sites indicate intentional placement of plants, possibly for therapeutic or ritualistic purposes (Mariliyan Jen, 2022). Some of the earliest documented medicinal plants include:

- **Yarrow (Achillea millefolium)** – It treats wounds and stops bleeding by acting as an astringent.
- **Mallow (Malva neglecta)** – It is recognized for its colon-purifying properties and gastrointestinal advantages.
- **Rosemary (Rosmarinus officinalis)** – It is considered to possess antibacterial and cognitive-enhancing effects.
- **Birch Polypore (Piptoporus betulinus)** – In addition to being a laxative, it may also be used to treat infections (Mariliyan Jen, 2022).

The diversity of early herbal practices across civilizations is illustrated in Figure 1.

Figure 1. Herbal Medicine Practices Across Ancient Civilization including Ayurveda (India), Egyptian herbalism, Traditional Chinese Medicine and Greco-Roman Traditions

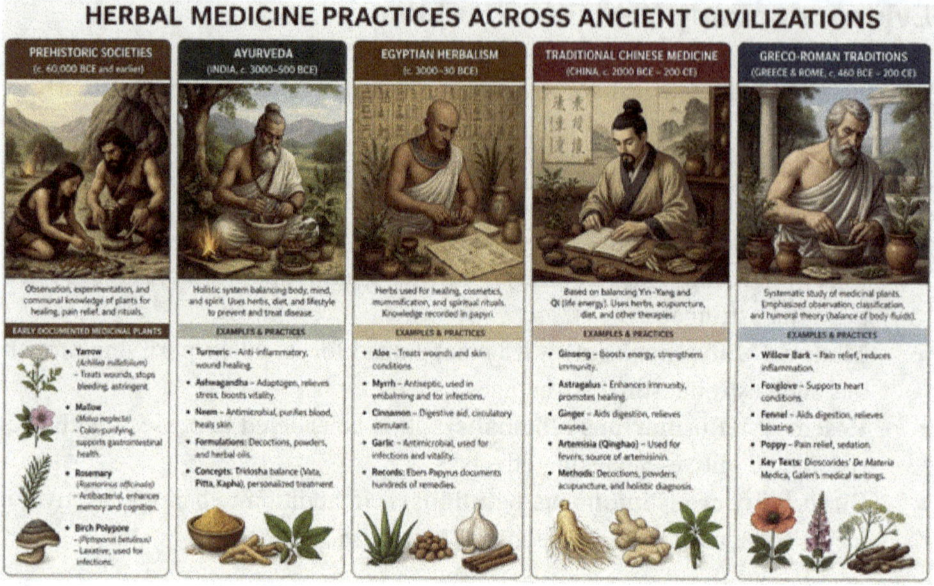

1.2 Herbal Practices in Ancient Civilizations

Ancient civilizations relied heavily on herbal medicine, reflecting a deep understanding of healing. These practices were closely intertwined with cultural, spiritual, and philosophical beliefs.

1.2.1 Indian society and traditional medicine

The prehistoric age marks the emergence of knowledge regarding the medicinal and toxicological effects of plants, minerals, and other substances. Evidence suggests that as early as 7000 BCE, medical procedures such as dentistry and trepanation were practiced on the Indian subcontinent. Archeobotanical findings indicate the use of medicinal plants in the Middle Gangetic region since the second millennium BCE, many of which are still used in Ayurvedic medicine (Sen & Chakraborty, 2017). India is home to diverse ethnic groups, each with distinct religions, beliefs, cultures, languages, and dialects. This diversity has contributed to the development of numerous medical systems. Indian society has historical relied on traditional medical systems; however, their prominence declined during British rule due to the introduction of allopathic medicines. Despite this, approximately 70% of India's rural population still depends on traditional medicine for primary health care (Mafuva & Marima-Matarira, 2014; Pandey et al., 2013).

- **Ayurveda:** Ayurveda is a comprehensive medical system native to India. Ayurveda, which combines the Sanskrit words Ayu (life) and Veda (knowledge or science), means 'knowledge of life'. Four Vedas, considered the oldest Indian literature (5000-1000 BC), offer knowledge about herbal medicines. Ayurveda was formed as a fully developed medical system (Chaudhury & Rafei, 2002). Both the Susruta Samhita, which focuses on surgery, and the Charaka Samhita, which focuses on internal medicine, were methodically composed and are regarded as fundamental Ayurvedic texts. Together, the essential information from Susruta Samhita and Charaka Samhita was gathered, and Astanga Sangraha and Astanga Hrdaya were also updated. Other ancient classics that contain minor Ayurvedic works include Madhava Nidana, which focuses on disease diagnosis, Bhava Prakasa, which focuses on additional plant and diet-related material, and Sarngadhara Samhita, which focuses on formulation and dosage form (Naswamy, 1981). Eight major clinical subdivisions comprised Ayurveda: internal medicine (Kayacikitsa), surgery (Salya Tantra), diseases of supraclavicular origin (Salakya), pediatrics, obstetrics, and gynecology (Kaumarabhrtya), psychiatry (Bhutavidya), toxicology (Agada Tantra), rejuvenation and geriatrics (Rasayana Tantra), and aphrodisiology and eugenics (Vajikarana) (Sen & Chakraborty, 2017).
- **Siddha:** The Siddha medical system is regarded as a magnificent accomplishment and a representation of Tamil culture, which has its roots in southern India. Siddha medicine was developed during the Indus Valley Civilization and was inspired by Dravidian culture. Siddha alchemy is thought to have been primarily inspired by Chinese alchemy, Taoism, and Taoist Patrology. It is thought that eighteen siddhar, a group of Tamil sages, created the system in antiquity. Despite having many similarities to Ayurveda, the Siddha medical system has its own philosophy, concept, holistic approach, and lifestyle-oriented methods (Karunamoorthi et al., 2012; Subbarayappa, 1997, 2001).
- **Unani:** The modern traditional medical systems of Egypt, Syria, Iran, Iraq, China, India, and several other eastern nations have been combined to form the Unani system. The latter originated in Greece and was developed in the Arab world. Unani medicine was brought to India by Arab and Persian colonists in the eleventh century, and throughout the Mughal era, the method gained acceptance and was enhanced (Ansari, 2006; Sen & Chakraborty, 2015).

- **Amchi:** Another long-standing, well-recorded traditional medical system is Amchi or Sowa-Rigpa, which was widely used in Tibet, Mongolia, Nepal, Bhutan, the Himalayan area of India, some regions of China, and the former Soviet Union. There are disagreements over the origin of Amchi medicine; some people think it came from India, while others claim it originated in the Tibetan region. Amchi and Ayurveda are quite similar, however this system also shows the impact of Tibetan folklore and Chinese traditional medicine (Gurmet, 2004).
- **Folk medicine:** For many members of rural, indigenous, and ethnic communities, uncodified folk medicine is just as important in maintaining health and curing illnesses as the codified traditional medical system. This kind of knowledge is passed down orally from ancestors and is not adequately documented. Around 25,000 potent plant-based medicines are employed by India's rural and ethnic populations, and nearly 8000 plant species are used in folk medicine (Sen & Chakraborty, 2015).

1.2.2 Mesopotamia: The Birth of Written Medicinal Knowledge

Mesopotamia, the early cradle of civilization, is where the oldest documented accounts of herbal therapy may be found. Some of the earliest pharmacological writings were produced by the Sumerians, Babylonians, and Assyrians; a large portion of these works addressed plant cures. These early pharmacopoeias described plants and their therapeutic properties, laying the groundwork for later, more sophisticated herbal remedy techniques.

One of the earliest examples of documented information about the therapeutic use of plants is the Sumerian Pharmacopoeia, a compilation of herbal cures and medicinal recipes. This book, which was written on clay tablets, included numerous plant treatments for ailments like infections, headaches, and digestive problems.

1.2.3 Ancient Egypt: The Integration of Herbs with Religion and Culture

Ancient Egypt is well known for having had a big impact on the development of herbal medicine. The Egyptians included plant-based medicine into their religious rites because they believed that gods and goddesses could cure. The Ebers Papyrus, one of the oldest surviving medical records, was created around 1500 BCE. More than 700 medical recipes are included in it, many of which call for plants like juniper, myrrh, and garlic. Many illnesses, such as wounds, gastrointestinal disorders, and infections, were treated using these plants. Herbal medicines were frequently utilized with ritual and magical practices in ancient Egypt. In addition to being utilized to treat medical conditions, plants were also believed to offer protection and spiritual healing. Ancient Egyptian medical traditions were firmly based on this integrated approach to medicine, which merged the body and the soul. This approach also influenced subsequent medical procedures.

1.2.4 Traditional Chinese Medicine (TCM)

With a history of more than 2,000 years, it is one of the most influential and enduring systems of herbal therapy in the world. With a focus on the movement of Qi (life force) throughout the body, TCM is based on the ideas of balance and harmony. based on the notion that illness arises from an imbalance in the body's energy and that disease can be cured and health improved by restoring harmony through acupuncture, herbal treatments, and food therapy.

One of the classics of Chinese materia medica is the Shennong Bencao Jing, also known as Divine Farmer's Materia Medica. The mythological Emperor Shennong is credited with writing this incredibly early book, which lists the medicinal benefits of hundreds of plants and arranges them based on their therapeutic roles. Several herbs, including ginkgo biloba, licorice, and ginseng, have been utilized in traditional Chinese medicine (TCM) for thousands of years and are still frequently found in herbal supplements today (Yadav et al., 2025).

1.2.5 Classical Herbalism in The Greco-Roman World

Some of the brightest minds in Western medicine organized and improved herbal medicine in ancient Greece and Rome. The use of plant cures and a natural approach to healing were highly valued by Hippocrates, who is considered the founder of modern medicine. The majority of modern herbal medicines had their roots in his medical philosophy, which encouraged the use of natural remedies and dietary changes.

Dioscorides' De Materia Medica, a book that enumerated over 600 plants and their therapeutic applications, is among the oldest and most significant ancient writings on medicinal herbs. For centuries, the writings of the Greek doctor and pharmacist Dioscorides were translated and disseminated throughout Europe and the Middle East. His thorough publications on herbs like mint, sage, and garlic made a significant contribution to the development of herbalism in the West.

Galen, a Roman physician, influenced medical practice for centuries by expanding on the works of Hippocrates and Dioscorides. Galen's use of herbal remedies, particularly for gastrointestinal and wound conditions, became a cornerstone of Roman medicine and influenced medieval medicine (Yadav et al., 2025).

1.3 Herbal medicine and its importance

In many ancient medical systems, plants are always the main source of a medication or treatment plan. Plant-based medications and products have been popular in recent years as a means of treating illnesses or as a cure, either by themselves or in conjunction with other substances. The World Health Organization states that many populations use herbs or herbal products for basic medical requirements. Herbs, herbal materials (such as plant parts or preparations), processed and completed herbal products, and active substances are all included in herbal medicine (Pan et al., 2014; Sen et al., 2010). The usage of herbal products has greatly increased in recent years as a result of germ resistance, the adverse effects of contemporary medications, and the failure of contemporary treatments for chronic illnesses. It is believed that traditional/folk medicine accounted for around 75% of the plant-based medicinal substances employed globally. About 70% of contemporary medications in India are derived from natural resources, and several synthetic counterparts have been created using plant-derived prototype chemicals (Pan et al., 2014; Sen et al., 2009, 2011) . According to reports, almost 60% of cancer medications now on the market or during study are derived from natural materials. Nowadays, plants constitute the source of almost 80% of antibacterial, immunosuppressive, cardiovascular, and anticancer medications. Of the 177 approved anticancer medications, about 70% are based on natural compounds or mimics. There are approximately 121 prescription medication entities in use worldwide, and about 25% of them are derived from plant sources. Over 100 medications based on natural products are undergoing clinical trials, and 13 medications of natural origin were approved in the US between 2005 and 2007. Additionally, it was calculated that 11% of the 252 medications listed in the WHO's essential medicine list are solely derived from plants (Pan et al., 2013; Wachtel-Galor & Benzie, 2012). Numerous plants are employed in Indian traditional medicine. An estimated 1200–1800 plants are used in Ayurveda, 500–900 in Siddha medicine, 400–700 in Unani medicine, and around 300 in Amchi medicine. Additionally, Indian traditional healers use over 7500 medicinal plants in various forms of medicine. About 526,573 and 902 plants were mentioned in the three main Ayurvedic texts, Charaka Samhita, Sushruta Samhita, and Astanga Hridaya (Debnath et al., 2015; Shankar & Majumdar, 1997).

2. MEDIEVAL AND RENAISSANCE HERBALISM

Herbalism, or the use of plants for therapeutic purposes, has been crucial to the development of medical study throughout human history. Herbal therapy was not only a common medical practice in the Middle Ages and Renaissance, but it also acted as a bridge between ancient traditions and the advancement of modern science. In medieval Europe (about the fifth to the fifteenth century), Christian monastic traditions, classical Greco-Roman literature, and Arab-Islamic study all had a significant influence on medical knowledge. "Herbals"—herbal manuals that featured illustrations and information of plants used for therapeutic purposes—were crucial in this field.

Herbalism is a folk medicine or traditional medical practice that uses plants and plant extracts. One of the primary forms of life on Earth are herbs and plants, which make up the majority of traditional materia medica worldwide. About 350,000 plant species are thought to exist today (including ferns, bryophytes, and seed plants), of which 287,655 species have been identified as of 2004. Herbs, herbal materials, herbal preparations, and completed herbal products that incorporate plant parts or other materials as active components are all considered herbal medicine (HM), also known as botanical medicine, phytomedicine, or phytotherapy. Plant parts such as seeds, berries, roots, leaves, fruits, bark, flowers, or even the entire plant are utilized in herbal therapy (Pan et al., 2014).

Both patients and scientists are interested in herbal medications for the validation of traditional medicine (TM) and for all areas of drug development from natural sources. Due to their accessibility and affordability, TM is utilized in many impoverished nations. Scientists worldwide view herbal species as a source of novel chemical entities and use them to extract substances like digoxin, morphine, taxol, atropine, and vinblastine (Mukherjee et al., 2022).

2.1 The Renaissance: A Rebirth of Herbal Medicine

Herbalism was given more attention throughout the renaissance (the 14th–17th century), particularly in the context of scientific research. A new understanding of plant medicine was facilitated by the discovery of ancient Greek and Roman writings and the development of botany as a scientific discipline. The Renaissance resurgence of herbalism was significantly influenced by notable individuals like Paracelsus and John Gerard. The utilization of chemically active plant derivatives was the emphasis of Swiss physician Paracelsus, who helped modern pharmacology eventually arise. One of the first systematic herbals written in English, The Herbal, was written by English herbalist John Gerard in 1597 and quickly became a standard text for herbal practitioners throughout Europe (Yadav et al., 2025).

3. INDIGENOUS AND FOLK MEDICINE

Across cultures, folk medicine has played a crucial role in accomplishing traditional health and well-being objectives as a fundamental part of Indigenous Knowledge Systems (IKS) (Saeed Anowar and Somnath Das, 2024). Folk medicine, which has its roots in centuries-old customs, includes holistic methods that combine cultural beliefs with medical applications, herbal cures, and spiritual healing (Pawar et al., 2023; Sanyal, 2023; Sarkar & Madhu, 2016). The efficacy of these methods is frequently confirmed by oral traditions, practical learning, and information transfer across generations. Theoretically, ethnomedicine and cultural epistemology—which highlight the importance of non-western health perspectives in addressing community health needs, are closely related to folk medicine. Indigenous societies build their knowledge through lived experiences and interactions with their surroundings, as demonstrated by theoretical frameworks like the constructivist method. These frameworks highlight the value of conventional therapeutic techniques in preserving the resilience and balance of society (Saeed Anowar and Somnath Das, 2024).

3.1 The role of folk medicine within Indigenous Knowledge Systems (IKS) in preserving and transmitting traditional healing practices and cultural values:

As an essential part of Indigenous Knowledge Systems (IKS), folk medicine is essential to maintaining and passing down cultural values and traditional healing methods. This goal is to investigate how traditional medicine preserves indigenous health practices and promotes cultural continuity. This review highlights the functions, significance, and impact of folk medicine in IKS by classifying important facets of its role using a content analytic approach, as shown in table 1.

Table 1. Key Themes in the Role of Folk Medicine in IKS.

Theme	Description	Reference
Preservation of Traditional Healing Practices	Folk medicine ensures the survival of ancient healing techniques, such as herbal remedies, spiritual healing, and holistic treatments, passed down through generations.	(Saeed Anowar and Somnath Das, 2024)
Transmission of Indigenous Knowledge	Knowledge transfer occurs through oral traditions, apprenticeship, and community-based practices, ensuring that younger generations inherit medicinal wisdom.	(Oyebode et al., 2016)
Cultural and Spiritual Significance	Many indigenous healing methods are deeply connected to spiritual beliefs, rituals, and ancestral heritage, reinforcing cultural identity	(Kimmerer, 2013; Saeed Anowar and Somnath Das, 2024)
Sustainability and Biodiversity Conservation	The use of local plants and natural resources in folk medicine promotes environmental sustainability and biodiversity conservation	(Berkes, 1999; Saeed Anowar and Somnath Das, 2024)
Integration with Modern Healthcare	Traditional medicine is increasingly being integrated into modern healthcare systems, offering complementary approaches to health and wellness.	(Saeed Anowar and Somnath Das, 2024)

3.2 Cultural and Spiritual Significance:

Many folk medicine techniques are more than just ways to heal physical illnesses; they have their roots in spiritual beliefs and cultural traditions. According to Warren et al. (2018), the goal of indigenous healing techniques is to restore spiritual harmony and balance, which are seen to be crucial for general wellbeing. This viewpoint is consistent with Indigenous Knowledge Systems' holistic orientation, which views sickness as a disturbance in the interdependent link between the environment, body, mind, and spirit. Kimmerer (2013) goes on to show how important ceremonies and rituals are to traditional healing because they not only activate therapeutic properties but also strengthen a community's feeling of cultural continuity and belonging. The purpose of these rituals, which could include chants, prayers, or symbolic offerings, is to enhance the healing process by connecting people with the spiritual world and calling forth ancestral wisdom. By strengthening social ties and preserving indigenous identity, such culturally entrenched techniques guarantee the survival of folk medicine traditions for many generations to come.

3.3 Sustainability and Biodiversity Conservation:

Through promoting sustainable harvesting and conservation initiatives among indigenous people, the use of natural resources in traditional medicine raises ecological consciousness. When gathering medicinal plants, traditional healers adhere to ethical standards, which include selective choosing, seasonal harvesting, and replanting to guarantee long-term availability (Gómez-Beloz, 2002). These methods aid in preserving biodiversity and avoiding overuse of important plants. Berkes (2018) goes on to explain that environmental stewardship is a fundamental component of indigenous medical traditions, highlighting the interdependence of nature and humans. The usage of medicinal plants is frequently governed by rituals and cultural taboos, maintaining the equilibrium and regeneration of ecosystems. Furthermore, as younger generations learn to appreciate and preserve their natural surroundings, information transfer across communities serves to further emphasize the significance of biodiversity protection. Indigenous organizations support the preservation of wider ecosystems and medicinal resources by incorporating ecological ideas into their healing methods.

3.4 The Contribution of Folk Medicine to Traditional Health and Well-being Goals:

With its roots in centuries of empirical knowledge and cultural legacy, folk medicine is an essential part of Indigenous Knowledge Systems (IKS). Folk medicine incorporates indigenous beliefs, rituals, and natural cures to promote holistic health and well-being, in contrast to modern biomedicine, which largely focuses on physical symptoms (Bodeker, 2019). This review examines the fundamental ideas, important practices, and social effects of folk medicine in order to assess its contribution to the attainment of traditional health objectives.

4. MODERN SCIENTIFIC VALIDATION OF HERBAL REMEDIES

The knowledge, abilities, and methods that indigenous cultures have accumulated over ages to preserve health and cure illnesses are collectively referred to as traditional medicine. This type of medicine has strong cultural roots and frequently uses natural cures made from plants, minerals, and animal products (Organization, 2018). Since traditional medicine is one of the oldest medical practices used by human communities, it has great historical relevance (Fabricant & Farnsworth, 2001). The rich legacy of ancestral wisdom has been preserved by these treatments and customs, which have been handed down through the years, frequently verbally or in written documents (Patwardhan et al., 2005).

4.1 Global Practices and Cultural Perspectives

Because traditional medicine comes from a variety of contexts and cultures, it differs greatly throughout the world. For example, Tai Chi, herbal medicine, and acupuncture are all part of Traditional Chinese Medicine (TCM), which aims to balance the body's energy (Qi) in order to support health (Wang & Zhang, 2017). In India, Ayurveda uses yoga, herbal remedies, and diet to promote the harmony of body, mind, and spirit. African traditional medicine frequently incorporates spiritual activities, community-based healing rituals, and the use of medicinal herbs (Abdullahi, 2011). Local plants and animals, as well as spiritual and communal aspects, are all included into the distinctive medicinal practices of indigenous civilizations in the Americas, Australia, and other places.

4.2 Integration with Modern Science

4.2.1 Rise of Interest in Traditional Remedies

Both the general public and the scientific community have become more interested in traditional medicines in recent decades. The limitations of traditional medicine, the rise in chronic illness, and the growing interest in natural and holistic remedies are some of the factors driving this renaissance (Ekor, 2014). In line with their desire for more individualized and natural healthcare solutions, people are increasingly looking for complementary and alternative medicine (CAM) choices. Furthermore, increased accessibility to traditional treatments from different cultures has been made possible by the globalization of knowledge, igniting interest in and admiration for these ancient customs (DR. Nitin Nama, 2022).

4.2.2 Benefits and Challenges of Combining Traditional and Modern Approaches

There are several advantages to combining modern science and traditional medicine. The possibility of finding novel medicinal drugs is one important benefit (Newman & Cragg, 2016). Many contemporary medications are made from substances that were first used in conventional treatments. Through scientific validation of these cures' efficacy and safety, we can add more effective medications to our pharmacopeia. Additionally, combining traditional and modern methods can improve patient treatment by offering a more comprehensive approach to health that addresses mental, emotional, and spiritual well-being in addition to physical problems.

4.3 The Science of Phytochemistry

The area of chemistry that examines compounds derived from plants is called phytochemistry. Known as phytochemicals, these plant-derived substances comprise a diverse range of bioactive compounds that are essential to the development, reproduction, and defense systems of plants. Alkaloids, flavonoids, terpenoids, glycosides, and numerous more chemical groups are all included in the broad category of phytochemicals. Investigating these chemicals' chemical structures, production, and methods of extraction and analysis is necessary to comprehend them.

4.3.1 Key Phytochemicals in Herbal Medicine

Phytochemicals are classified into several major groups based on their chemical structures and biological activities:

1. **Alkaloids:** These substances that contain nitrogen frequently have strong pharmacological effects. Quinine from cinchona bark (Cinchona spp.), caffeine from coffee (Coffea spp.), and morphine from opium poppies (Papaver somniferum) are a few examples (Barnes & Heinrich, 2004).
2. **Flavonoids:** The antioxidant qualities of these polyphenolic chemicals are well-known. Fruits, vegetables, and other plant-based diets contain them in large quantities. Well-known flavonoids include quercetin, which is abundant in apples and onions, and catechins, which are found in green tea (Middleton Jr et al., 2000).
3. **Terpenoids:** They are the largest class of phytochemicals and are also referred to as isoprenoids. These include substances with antimalarial qualities as artemisinin from sweet wormwood (Artemisia annua) and menthol from mint (Mentha spp.) (Newman & Cragg, 2016).
4. **Glycosides:** A sugar moiety is joined to a non-sugar moiety (aglycone) to form these molecules. Examples include the aspirin precursor salicin from willow bark (Salix spp.) and digoxin from foxglove (Digitalis spp.).
5. **Others:** Other bioactive substances that contribute to the therapeutic qualities of plants fall under this group, including tannins, saponins, and polyphenols.

The major classes of phytochemicals and their plant sources are summarized in Figure 2.

Figure 2. Major Phytochemicals and Their Source in Herbal Medicine

4.3.2 Mechanisms of Action

Phytochemicals exert their effects through various mechanisms of action, including:

1. **Enzyme Inhibition or Activation:** Numerous phytochemicals have the ability to alter biochemical pathways by blocking or activating particular enzymes. Alkaloids such as morphine, for instance, block pain signals by binding to opioid receptors.
2. **Antioxidant Activity:** By neutralizing free radicals, flavonoids and other polyphenols can lower oxidative stress and stop cellular damage. Preventing chronic illnesses like cancer and heart disease requires this activity.
3. **Anti-Inflammatory Effects:** Turmeric contains compounds called curcumin, which inhibits inflammatory cytokines and enzymes like cyclooxygenase-2 (COX-2) to produce anti-inflammatory effects (Gupta et al., 2013).
4. **Antimicrobial Action:** Numerous phytochemicals are helpful against a range of diseases because they have antibacterial qualities. For example, tea tree oil's terpenoids have broad-spectrum antibacterial properties (DR. Nitin Nama, 2022).

5. INTEGRATION OF HERBAL MEDICINE IN CONTEMPORARY HEALTHCARE

The World Health Organization (WHO) has identified and recognized one hundred different forms of traditional medical practices that are used worldwide. Acupuncture in China, magnetic treatment in France, Heilpraxis in Germany, herbalism in Sweden, shiatsu in Japan, and sowa rig-pa in Tibet and Bhutan are only a few examples of the global continuity of traditional medicine systems (Kala, 2017).

Phytotherapy, often known as herbal medicine, is the use of compounds obtained from plants for both preventative and therapeutic purposes. Herbal medicine, which has its roots in traditional healing systems like Traditional Chinese Medicine (TCM), Ayurveda, and other indigenous practices in Africa and Latin America, continues to be an essential source of healthcare for a sizable section of the world's population. Growing public interest in natural, holistic techniques and discontent with the shortcomings of traditional treatments, particularly in the management of chronic illnesses, have contributed to a renaissance in the use of herbal remedies in recent decades.

More than 80% of people worldwide receive some form of primary healthcare from traditional medicine, according to the World Health Organization (WHO, 2013). Herbal medicine's incorporation into modern healthcare signifies a revolutionary move toward more individualized, sustainable, and preventive healthcare-it is not only a synthesis of ancient and modern knowledge systems.

5.1 The Role of Herbal Remedies in Alternative and Complementary Medicine

5.1.1 Historical and Global Perspectives

One of the oldest forms of treatment used by humans, herbal medicine predates written history. Indian Ayurvedic literature like the "Charaka Samhita" and ancient Chinese books like the "Shennong Ben Cao Jing" emphasize the importance of plants in early medical systems (Bharat B Aggarwal, 2009; Organization, 2018). For millions of people throughout Africa and Latin America, traditional botanical knowledge remains their main source of medical care. The usage of herbal supplements has become more popular in Western nations; in the United States, up to 20% of adults regularly take at least one botanical product (Ekor, 2014).

5.1.2 Applications in Disease Management

Herbs have been extensively researched and used to treat chronic illnesses and diseases linked to lifestyle choices:

- Boswellia serrata and Curcuma longa, or turmeric, have anti-inflammatory properties for **arthritis**.
- **Diabetes:** Momordica charantia (bitter melon) and Gymnema sylvestre help regulate blood sugar levels.
- **Mental health:** mild-to-moderate depression can be effectively treated with Hypericum perforatum, also known as St. John's Wort (Linde et al., 2008).
- **Respiratory Conditions:** Glycyrrhiza glabra, or licorice, and eucalyptus are traditionally used to treat asthma and bronchitis.

Clinical trials and laboratory investigations are increasingly being used to validate these applications (Yuan et al., 2016).

5.1.3 Preventive and Integrative Health

Herbs are not only used for treatment but also for health maintenance and disease prevention. For example:

- Ginseng and Garlic are known to boost immunity.
- Echinacea is widely used to prevent colds.
- Ginger has proven efficacy in mitigating chemotherapy-induced nausea.

Integrative medicine models employ such botanicals alongside conventional treatments, enhancing patient outcomes and satisfaction.

5.1.4 Evidence Base and Regulation

Despite the fact that there is growing evidence to support various herbs, regional regulatory systems differ greatly. The Traditional Herbal Medicinal Products Directive (THMPD) in Europe mandates safety assurance and historical data. Herbal products are regulated as dietary supplements in the United States under the DSHEA (1994), which leads to gaps in dosage standardization and quality control (U.S. FDA., 1994). Although thorough research shows the medicinal potential of phytochemicals, such as that on curcumin (Bharat B Aggarwal, 2009), wider acceptance necessitates more robust data and supervision.

5.2 Challenges and Opportunities in Incorporating Traditional Herbal Treatments into Modern Medical Practices

5.2.1 Scientific Validation and Standardization

It can be difficult to incorporate traditional knowledge into evidence-based frameworks since it frequently lacks contemporary scientific validation:

- Complex interactions seen in herbal mixtures usually defy reductionist research methodologies.
- Consistency is impacted by species, cultivation, and processing variation (Bent, 2008). However, important discoveries such as the discovery of artemisinin from Artemisia annua (Tu, 2011) show how traditional treatments can be translated.

5.2.2 Models of Successful Integration

Different nations provide different models:

- **China:** In hospitals, herbal treatment is used in addition to conventional medication.
- **India:** AYUSH incorporates homeopathy, Siddha, Unani, and Ayurveda into public health systems.
- **Germany:** According to (Michael Heinrich, 2020), herbal medications are frequently prescribed, regulated, and covered by insurance.

These ideas show how policy frameworks might encourage traditional and modern activities to coexist and operate together.

5.2.3 Regulatory and Safety Frameworks

- For safety and effectiveness, standardization is essential. Among the difficulties are:
- Product adulteration (such as heavy metal contamination).
- Interactions between herbs and medications (e.g., St. John's Wort reducing SSRI serum levels).
- Differences in focus and quality (Ekor, 2014).

The goal of international initiatives like WHO's GACP and EMA's guidelines is to standardize quality assurance procedures.

5.2.4 Role of Healthcare Professionals

Medical practitioners must:

- Gather thorough patient histories, including information on herb use, for safe integration.
- Inform patients about the interactions between drugs and herbs.
- Work together with CAM professionals and herbalists.
- Encourage the integration of herbal pharmacology in courses for medical education.

To close the knowledge gap and guarantee safe, patient-centered care, training initiatives and interdisciplinary cooperation are essential.

5.3 Ethical and Sustainable Sourcing of Medicinal Plants

5.3.1 Ethical Harvesting and Indigenous Knowledge

Numerous therapeutic plants are obtained using traditional knowledge, frequently without recognition or payment.

- Cases of biopiracy, such as those involving turmeric and neem, have generated discussion around the world.
- In order to guarantee Access and Benefit Sharing (ABS) and safeguard community rights, the Nagoya Protocol (2010) was created (Convention on Biological Diversity, 2010).

Fair pay, informed permission, and the preservation of cultural intellectual property are all required by ethical sourcing.

5.3.2 Conservation and Biodiversity

Biodiversity is threatened by the global demand for botanicals;

- Overharvesting is threatening to wipe out species like Hoodia and goldenseal (Hamilton, 2004).
- Conservation initiatives encourage agroforestry and ex situ conservation operations and prioritize cultivation over wild collecting (Canter, 2005).

Fair Wild and other certifications encourage community stewardship and sustainable harvesting.

5.3.3 Supply Chain Transparency and Quality Control

Consumer safety and ethical integrity depend on a transparent supply chain;

- Traceability and GACP technologies, such as DNA barcoding, improve verification.
- A new instrument for supply chain transparency is blockchain technology.
- Analytical technologies (like HPLC) detect adulterants and guarantee pharmacological quality.

These procedures strengthen customer confidence and increase product dependability.

5.3.4 Future Directions and Global Collaboration

For herbal medicine to have a sustainable future,

- Local cultivation capacities must be strengthened.
- Promoting global collaboration on norms and regulations.
- Funding interdisciplinary studies that connect conservation, pharmacology, and ethnobotany (Schippmann et al., 2002).

Herbal medicine can continue to be a useful resource for upcoming generations if inclusive and environmentally responsible practices are implemented.

6. CONCLUSION

In conclusion, the progression of herbal medicine from prehistoric periods to the contemporary period underscores its lasting importance in human healthcare systems. Herbal medicine, grounded in ancient traditions and cultural practices, has functioned as a principal means of healing throughout civilisations such as India, China, Egypt, and the Greco-Roman world. Notwithstanding the significant progress in contemporary medical science, a considerable segment of the global populace persists in using traditional and herbal medicines owing to its accessibility, cost-effectiveness, and comprehensive approach to health. The amalgamation of ancient herbal wisdom with contemporary scientific validation has created novel opportunities for medication discovery, preventive healthcare, and chronic illness management. Phytochemicals sourced from plants have played a crucial role in the advancement of several life-saving pharmaceuticals, underscoring the scientific significance of traditional methods. Nonetheless, obstacles such the absence of standardisation, regulatory discrepancies, safety apprehensions, and ethical dilemmas pertaining to biodiversity and indigenous knowledge require resolution. Moreover, herbal medicine is essential for safeguarding cultural heritage, fostering sustainability, and aiding biodiversity conservation through the judicious use of natural resources. The rising global interest in natural and integrative healthcare necessitates enhanced interdisciplinary collaboration, rigorous research, and comprehensive policy frameworks to guarantee the safe, effective, and ethical application of herbal remedies. Ultimately, realising the objective of "Health for All" necessitates a balanced and inclusive healthcare strategy that integrates modern medicine with ancient herbal practices. By appreciating both scientific advancement and traditional knowledge, society may progress towards a more comprehensive, inclusive, and sustainable paradigm for world health.

REFERENCES

Abdullahi, A. A. (2011). Trends and challenges of traditional medicine in Africa. *African Journal of Traditional, Complementary, and Alternative Medicines : AJTCAM, 8*(5S). Advance online publication. DOI: 10.4314/ajtcam.v8i5S.5 PMID: 22754064

Ansari, A. A. (2006). Global status of Unani medicine. *The Proceeding of International Conclave on Traditional Medicine (16–17 November 2006), National Agriculture Science Complex, New Delhi. Available from:* Http://Www. Niscair. Res. in/Conclave/Downloadables/Plenary% *20Session, 203*.

Barnes, J., & Heinrich, M. (2004). *Fundamentals of pharmacognosy and phytotherapy*. Churchill Livingstone.

Bent, S. (2008). Herbal medicine in the United States: review of efficacy, safety, and regulation: grand rounds at University of California, San Francisco Medical Center. *Journal of General Internal Medicine, 23*, 854–859. DOI: 10.1007/s11606-008-0632-y PMID: 18415652

Berkes, F. (1999). Context of traditional ecological knowledge. *Sacred Ecology: Traditional Ecological Knowledge and Resource Management. Philadelphia.*

Bharat, B., & Aggarwal, K. B. H. (2009). Potential therapeutic effects of curcumin, the anti-inflammatory agent, against neurodegenerative, cardiovascular, pulmonary, metabolic, autoimmune and neoplastic diseases. *The International Journal of Biochemistry & Cell Biology, 1*(41), 40–59. DOI: 10.1016/j.biocel.2008.06.010 PMID: 18662800

Canter, P. H. (2005). Bringing medicinal plants into cultivation. *Focus on Alternative and Complementary Therapies, 10*(3), 167–168. DOI: 10.1211/fact.10.3.0002 PMID: 15780709

Chaudhury, R. R., & Rafei, U. M. (2002). *Traditional medicine in Asia*. World Health Organization Geneva.

Debnath, P. K., Banerjee, S., Debnath, P., Mitra, A., & Mukherjee, P. K. (2015). Ayurveda–opportunities for developing safe and effective treatment choices for the future. In *Evidence-based validation of herbal medicine* (pp. 427–454). Elsevier. DOI: 10.1016/B978-0-12-800874-4.00020-9

Ekor, M. (2014). The growing use of herbal medicines: Issues relating to adverse reactions and challenges in monitoring safety. *Frontiers in Pharmacology, 4*, 177. DOI: 10.3389/fphar.2013.00177 PMID: 24454289

Fabricant, D. S., & Farnsworth, N. R. (2001). The value of plants used in traditional medicine for drug discovery. *Environmental Health Perspectives, 109*(suppl 1), 69–75. PMID: 11250806

Gupta, S. C., Patchva, S., & Aggarwal, B. B. (2013). Therapeutic roles of curcumin: Lessons learned from clinical trials. *The AAPS Journal, 15*(1), 195–218. DOI: 10.1208/s12248-012-9432-8 PMID: 23143785

Gurmet, P. (2004). "Sowa-Rigpa": Himalayan art of healing. *Indian Journal of Traditional Knowledge, 3*, 212–218.

Hamilton, A. C. (2004). Medicinal plants, conservation and livelihoods. *Biodiversity and Conservation, 13*(8), 1477–1517. DOI: 10.1023/B:BIOC.0000021333.23413.42

Jen, M. (2022). Prehistoric Medicine: History, Procedures and Practices. *Anthropology; a Guide to First Degree Courses in UK Universities, 10*, 281.

Kala, C. P. (2017). Traditional health care systems and herbal medicines. *European Journal of Environment and Public Health, 1*(1), 3. DOI: 10.20897/ejeph.201703

Karunamoorthi, K., Jegajeevanram, K., Xavier, J., Vijayalakshmi, J., & Melita, L. (2012). Tamil traditional medicinal system-siddha: an indigenous health practice in the international perspectives. , *2*(2), 25–35.

Kimmerer, R. W. (2013). *Braiding sweetgrass: Indigenous wisdom, scientific knowledge and the teachings of plants*. Milkweed editions.

Linde, K., Berner, M. M., & Kriston, L. (2008). St John's wort for major depression. *Cochrane Database of Systematic Reviews, 4*. Advance online publication. DOI: 10.1002/14651858.CD000448.pub3 PMID: 18843608

Mafuva, C., & Marima-Matarira, H. T. (2014). *Towards professionalization of traditional medicine in Zimbabwe: A comparative analysis to the South African policy on traditional medicine and the Indian Ayurvedic system.*

Michael Heinrich, A. K. J. (2020). Ethnopharmacology and sustainable development: Reviewing recent progress in holistic research. *Journal of Ethnopharmacology, 257*, 112830.

Middleton, E.Jr, Kandaswami, C., & Theoharides, T. C. (2000). The effects of plant flavonoids on mammalian cells: Implications for inflammation, heart disease, and cancer. *Pharmacological Reviews, 52*(4), 673–751. DOI: 10.1016/S0031-6997(24)01472-8 PMID: 11121513

Mukherjee, P. K., Banerjee, S., Das Gupta, B., & Kar, A. (2022). Evidence-based validation of herbal medicine: Translational approach. In *Evidence-based validation of herbal medicine* (pp. 1–41). Elsevier. DOI: 10.1016/B978-0-323-85542-6.00025-1

Naswamy, V. (1981). Origin and development of Ayurveda. *Ancient Science of Life, 1*, 1–7. PMID: 22556454

Newman, D. J., & Cragg, G. M. (2016). Natural products as sources of new drugs from 1981 to 2014. *Journal of Natural Products, 79*(3), 629–661. DOI: 10.1021/acs.jnatprod.5b01055 PMID: 26852623

Nitin Nama, D. R. (2022). Modern Science Meets Traditional Medicine. In M. R. K. A. (Edit. DR. Harshita Jain, Dr. Nitin Nama, Dr. Girish Kumar Vyas (Ed.), *Herbal Remedies: Bridging Traditional Knowledge with Modern Sciences* (pp. 14–22). Career Point Ltd.

Organization, W. H. (2004). The world medicines situation. In *The world medicines situation.*

Organization, W. H. (2018). *WHO traditional medicine strategy 2014–2023. Geneva: WHO; 2013.*

Oyebode, O., Kandala, N. B., Chilton, P. J., & Lilford, R. J. (2016). Use of traditional medicine in middle-income countries: A WHO-SAGE study. *Health Policy and Planning, 31*(8), 984–991. DOI: 10.1093/heapol/czw022 PMID: 27033366

Pan, S.-Y., Litscher, G., Gao, S.-H., Zhou, S.-F., Yu, Z.-L., Chen, H.-Q., Zhang, S.-F., Tang, M.-K., Sun, J.-N., & Ko, K.-M. (2014). Historical perspective of traditional indigenous medical practices: The current renaissance and conservation of herbal resources. *Evidence-Based Complementary and Alternative Medicine : eCAM, 2014*(1), 525340. DOI: 10.1155/2014/525340 PMID: 24872833

Pan, S.-Y., Zhou, S.-F., Gao, S.-H., Yu, Z.-L., Zhang, S.-F., Tang, M.-K., Sun, J.-N., Ma, D.-L., Han, Y.-F., Fong, W.-F., & Ko, K.-M. (2013). New perspectives on how to discover drugs from herbal medicines: CAM′ S outstanding contribution to modern therapeutics. *Evidence-Based Complementary and Alternative Medicine : eCAM, 2013*(1), 627375. DOI: 10.1155/2013/627375 PMID: 23634172

Pandey, M. M., Rastogi, S., & Rawat, A. K. S. (2013). Indian traditional ayurvedic system of medicine and nutritional supplementation. *Evidence-Based Complementary and Alternative Medicine : eCAM, 2013*(1), 376327. DOI: 10.1155/2013/376327 PMID: 23864888

Patwardhan, B., Warude, D., Pushpangadan, P., & Bhatt, N. (2005). Ayurveda and traditional Chinese medicine: A comparative overview. *Evidence-Based Complementary and Alternative Medicine : eCAM*, *2*(4), 465–473. DOI: 10.1093/ecam/neh140 PMID: 16322803

Pawar, S., Pawade, K., Nipate, S., Balap, A., Pimple, B., Wagh, V., Kachave, R., & Gaikwad, A. (2023). Preclinical evaluation of the diabetic wound healing activity of phytoconstituents extracted from Ficus racemosa Linn. leaves. *Int. J. Exp. Res. Rev*, *32*, 365–377. DOI: 10.52756/ijerr.2023.v32.032

Saeed Anowar and Somnath Das. (2024). The role of Folk medicine in achieving the traditional goals through IKS: A Review. In D. M. D. and D. V. A. (eds. Dr. Somnath Das, Dr. Jayanta Kumar Das (Ed.), *Life as Basic Science: An Overview and Prospects for the Future Volume: 3* (pp. 164–179). International Academic Publishing House (IAPH), https://doi.org/https://doi.org/10.52756/lbsopf.2024.e03.007

Sanyal, R. (2023). Green Pharmacy : Unveiling the Healing Potential of Medicinal Plants. In D. A. C. S. & D. A. P. (eds. Mrs. Bhanumati Sarkar, Dr. (Professor) Surjyo Jyoti Biswas (Ed.), *The Basic Handbook of Indian Ethnobotany and Traditional Medicine [Volume: 2]* (pp. 104–116). International Academic Publishing House (IAPH). https://doi.org/https://doi.org/10.52756/bhietm.2023.e02.009

Sarkar, B., & Madhu, N. R. (2016). *Knowledge of traditional medicines and practices used for reproductive health problems. Review article. UGC-sponsored National Seminar on "Medicinal Plants, Health and Environment*. Nectar Publisher, Kolkata.

Schippmann, U., Leaman, D. J., & Cunningham, A. B. (2002). *Impact of cultivation and gathering of medicinal plants on biodiversity: global trends and issues*. Biodiversity and the Ecosystem Approach in Agriculture, Forestry and Fisheries.

Sen, S., & Chakraborty, R. (2015). Toward the integration and advancement of herbal medicine: A focus on traditional Indian medicine. *Botanics : Targets and Therapy*, •••, 33–44. DOI: 10.2147/BTAT.S66308

Sen, S., & Chakraborty, R. (2017). Revival, modernization and integration of Indian traditional herbal medicine in clinical practice: Importance, challenges and future. *Journal of Traditional and Complementary Medicine*, *7*(2), 234–244. DOI: 10.1016/j.jtcme.2016.05.006 PMID: 28417092

Sen, S., Chakraborty, R., & De, B. (2011). Challenges and opportunities in the advancement of herbal medicine: India's position and role in a global context. *Journal of Herbal Medicine*, *1*(3–4), 67–75. DOI: 10.1016/j.hermed.2011.11.001

Sen, S., Chakraborty, R., De, B., Ganesh, T., Raghavendra, H. G., & Debnath, S. (2010). Analgesic and anti-inflammatory herbs: A potential source of modern medicine. *International Journal of Pharmaceutical Sciences and Research, 1*(11), 32.

Sen, S., Chakraborty, R., De, B., & Mazumder, J. (2009). Plants and phytochemicals for peptic ulcer: An overview. *Pharmacognosy Reviews, 3*(6), 270.

Shankar, D., & Majumdar, B. (1997). Beyond the biodiversity convention: The challenges facing the biocultural heritage of India's medicinal plants. *Medicinal Plants for Forest Conservation and Health Care, 11*, 87–99.

Subbarayappa, B. V. (1997). Siddha medicine: An overview. *Lancet, 350*(9094), 1841–1844. DOI: 10.1016/S0140-6736(97)04223-2 PMID: 9428267

Subbarayappa, B. V. (2001). The roots of ancient medicine: An historical outline. *Journal of Biosciences, 26*(2), 135–143. DOI: 10.1007/BF02703637 PMID: 11426049

Tu, Y. (2011). The discovery of artemisinin (qinghaosu) and gifts from Chinese medicine. *Nature Medicine, 17*(10), 1217–1220. DOI: 10.1038/nm.2471 PMID: 21989013

U.S. FDA. (1994). *Dietary Supplement Health and Education Act (DSHEA).*

Wachtel-Galor, S., & Benzie, I. F. F. (2012). *Herbal medicine: an introduction to its history, usage, regulation, current trends, and research needs.*

Wang, W., & Zhang, T. (2017). Integration of traditional Chinese medicine and Western medicine in the era of precision medicine. *Journal of Integrative Medicine, 15*(1), 1–7. https://doi.org/https://doi.org/10.1016/S2095-4964(17)60314-5. DOI: 10.1016/S2095-4964(17)60314-5 PMID: 28088253

Yadav, A., Pandey, A. K., & Jaiswal, H. (2025). History of Ancient Herbal Medicine. *International Journal of Innovative Science and Research Technology, 10*(4), 2344–2347. DOI: 10.38124/ijisrt/25apr1442

Yuan, H., Ma, Q., Ye, L., & Piao, G. (2016). The traditional medicine and modern medicine from natural products. *Molecules (Basel, Switzerland), 21*(5), 559. DOI: 10.3390/molecules21050559 PMID: 27136524

Chapter 9
Phytoconstituents: A Promising Avenue for Diabetic Foot Ulcer Management

Prithiviraj Nagarajan
Aarupadai Veedu Medical College, Vinayaka Mission's Research Foundation, Puducherry, India

Gayathiri Ekambaram
http://orcid.org/0000-0002-7517-9151
Department of Plant Biology and Plant Biotechnology, Guru Nanak College, Velacherry, India

Arun Daniel Jayakumar
Aarupadai Veedu Medical College, Vinayaka Mission's Research Foundation, Puducherry, India

Prasad Thirumal
Aarupadai Veedu Medical College, Vinayaka Missions Research Foundation, Puducherry, India

Mani Rajarathinam
Aarupadai Veedu Medical College, Vinayaka Mission's Research Foundation, Puducherry, India

Deepa Shanmugam
Department of Obstetrics and Gynecology, Aarupadai Veedu Medical College, Vinayaka Mission's Research Foundation, Puducherry, India

Suganya Kannan
Central Research Laboratory, Vinayaka Mission's Research Foundation, Salem, India

Sankaranarayanan Nagarajan
Department of Chemistry, National Institute of Technology, Imphal, India

Satyavani Kaliamurthi
Aarupadai Veedu Medical College and Hospital, Vinayaka Mission's Research Foundation, Puducherry, India

Kirtik Ramadoss
Aarupadai Veedu Medical College, Vinayaka Mission's Research Foundation, Puducherry, India

DOI: 10.4018/979-8-3373-2908-6.ch009

ABSTRACT

Diabetic foot ulcers (DFUs) represent a major healthcare burden, driven by three interconnected pathophysiological barriers: chronic inflammation, impaired angiogenesis, and microbial infection. Conventional therapies address these challenges incompletely, driving interest in phytoconstituents, bioactive plant compounds with multi-target activity. This chapter examines evidence for three classes of phytoconstituents aligned with DFU pathophysiology includes anti-inflammatory, pro-angiogenic, and antimicrobial activities. Clinical evidence from Plantago major and Melilotus officinalis demonstrates accelerated epithelialization and wound closure compared to conventional dressings. However, translational barriers remain substantial. These limitations explain the gap between preclinical promise and clinical adoption. Addressing bioavailability through phytomedicines via chromatographic profiling and conducting rigorous pharmacokinetic studies are essential steps toward therapeutic translation.

1. INTRODUCTION

Diabetic foot ulcers (DFUs) are severe, costly complications of diabetes affecting 15% of the diabetic population at some point in their lifetime. Driven by prolonged hyperglycemia, DFUs result from three interconnected pathophysiological failures: peripheral neuropathy (loss of protective sensation), peripheral vascular disease (impaired blood supply), and immunosuppression (reduced infection resistance) (Dawi et al., 2025) . These deficits combine to create chronic, non-healing wounds in the lower extremities. Without intervention, DFUs progress from simple ulcers to severe infections, amputation, and death outcomes that carry substantial individual and societal costs (Jiang et al., 2024).

Conventional treatments wound debridement, pressure offloading, and antibiotics provide symptomatic relief but rarely achieve durable healing, particularly in chronic cases (Razavi et al., 2025). This therapeutic gap reflects a fundamental mismatch: DFU pathophysiology is multifactorial (inflammation, vascular insufficiency, infection), yet standard interventions address these issues independently and incompletely. The rising global diabetes burden and mounting healthcare expenditure on DFU management underscore the urgent need for novel, mechanism-directed therapies. (Tecilazich et al., 2013)

1.1. Overview of DFUs as a Complication of Diabetes

The pathophysiology of DFUs involves three primary obstacles to healing. First, chronic inflammation, sustained elevation of proinflammatory cytokines (TNF-α, IL-6) shifts wound healing from the proliferative phase into a pathological state that prevents tissue remodel oxidative stress from hyperglycemia overwhelms the body's antioxidant defenses, damaging fibroblasts and endothelial cells critical for wound repair (Thandavarayan et al., 2015; Xiong et al., 2025). Third, impaired angiogenesis the failure to form new blood vessels leaves wounds hypoxic and nutrient-starved, creating an environment permissive for opportunistic infection (Burgess et al., 2021).

Infection further complicates recovery. Multidrug-resistant organisms (MDROs) frequently colonize DFUs, establishing biofilms resistant to standard antibiotics, Additionally,the systemic comorbidities chronic kidney disease, cardiovascular disease impair immune function and reduce medication tolerance, narrowing therapeutic options (Matta-Gutiérrez et al., 2021). The cumulative result is wound chronicity: approximately 85% of lower-limb amputations are preceded by foot ulcers, with 19-34% lifetime risk for individuals with diabetes. Long-term management requires addressing not only the local wound environment but also the systemic metabolic and immunological defects underlying DFU pathogenesis. (Everett & Mathioudakis, 2018) Specifically, hyperglycemia-induced dysfunction in neutrophil chemotaxis and macrophage phagocytic activity compromises the innate immune response, allowing polymicrobial colonization to persist (Wang et al., 2025).

1.2. Importance of Exploring Alternative Therapies

Why conventional therapies fall short is instructive, standard antimicrobials target existing bacteria but do not address the underlying wound microenvironment the chronic inflammation, oxidative stress, and vascular insufficiency that perpetuate non-healing. (Whittam et al., 2014) Pressure offloading and debridement are essential but mechanistic: they reduce load and remove necrotic tissue yet do nothing to restore angiogenesis or suppress pathological inflammation (Halim et al., 2012).

Phytoconstituents bioactive plant compounds, offer a mechanistically different approach. Rather than single-target interventions, these molecules engage multiple healing pathways simultaneously (Yazarlu et al., 2021). Three classes are particularly relevant to DFU pathophysiology: Anti-inflammatory agents (curcumin, resveratrol, quercetin) suppress NF-κB signaling and reduce proinflammatory cytokines (Murdiana et al., 2024), facilitating the transition from inflammation to proliferation (Jiang et al., 2023).

Pro-angiogenic compounds (asiaticoside, madecassoside) upregulate VEGF and promote endothelial cell migration, restoring blood vessel formation (Li et al.,

2023) (Raguraman et al., 2021). Antimicrobial phytoconstituents (glycyrrhizic acid, berberine) disrupt bacterial cell membranes and prevent biofilm formation, reducing pathogenic burden (Ghosh & Gaba, 2013) (Zhou et al., 2021).

Compared to synthetic drugs, phytoconstituents carry potential advantages: lower toxicity, fewer systemic side effects, and the possibility of synergistic activity when combined with conventional therapies (Suganya et al., 2022). Clinical evidence from trials of Plantago major and Melilotus officinalis demonstrates accelerated epithelialization and reduced healing time compared to standard dressings (Ghanadian et al., 2022; Maver et al., 2019) .

However, translational barriers remain substantial. Poor oral bioavailability limits systemic effects; inconsistent extract standardization across suppliers reduces efficacy and reproducibility; and potential herb-drug interactions complicate integration into existing therapeutic regimens (Tie et al., 2025). These limitations explain the gap between preclinical promise and widespread clinical adoption. Addressing these barriers through advanced delivery systems, standardized extraction protocols, and rigorous pharmacokinetic studies is essential for clinical translation.

2. ROLE OF PHYTOCONSTITUENTS IN HEALING DIABETIC FOOT ULCERS

Wound healing is a tightly choreographed process with distinct phases: hemostasis, inflammation, proliferation, and remodeling. In healthy individuals, this sequence progresses over weeks to months. In diabetic patients, this process stalls inflammation becomes chronic, angiogenesis fails to initiate, and epithelialization plateaus, resulting in wounds that persist for months or years (Liu & Fang, 2025).

Phytoconstituents address this pathological stalling by modulating the specific biological pathways that malfunction in DFUs. This chapter examines three therapeutic strategies aligned with DFU pathophysiology: (1) resolving chronic inflammation through NF-κB pathway inhibition, (2) restoring angiogenesis through VEGF upregulation, and (3) controlling microbial burden while preventing biofilm-mediated treatment failure (Boulton et al., 2020; Lu et al., 2024). By targeting these mechanisms specifically, phytoconstituents can function as evidence-based adjunctive therapies not replacements for conventional care, but complements that address the underlying wound biology. DFUs healing, however, occurs with the defect of normal healing mechanisms, and phytoconstituents (bioactive compounds from a plant) have an important function in the healing process (Table 1). By modulating different biological pathways involved in wound healing, including inflammation, oxidative stress, angiogenesis, and infection control, these compounds can impede the healing process. In addition, because of their multifaceted roles, LR factor(s) could serve as adjunctive therapies in the management of DFU (Hegazi et al., 2024; Helal et al., 2024) (Figure 1).

Figure 1. Mechanism of Action of key- Phytoconstituents in DFU's healing

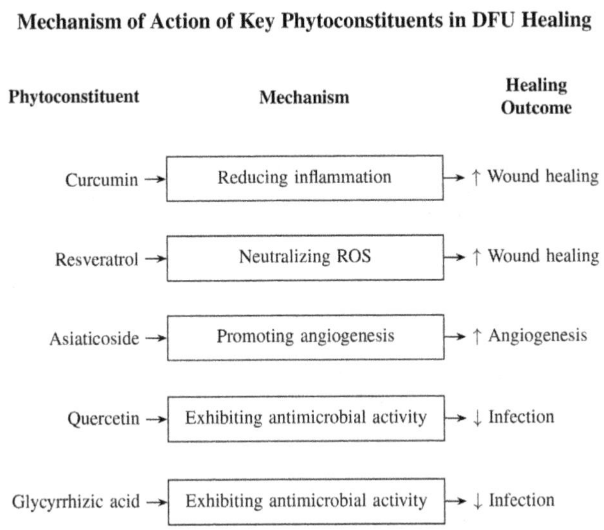

Table 1. The phytoconstituents that aid in the healing process of diabetic foot ulcers (DFUs).

Phytoconstituent	Mechanism of Action	Targeted Healing Process	Formulation Used	Reference
Curcumin	Anti-inflammatory, antioxidant	Reduces inflammation, promotes fibroblast proliferation	Topical formulations, oral supplements	**Zhang et al., 2023**
Resveratrol	Antioxidant, anti-inflammatory	Reduces oxidative stress, accelerates collagen synthesis	Oral supplements, topical gel	Fu et al., 2025
Quercetin	Antioxidant, anti-inflammatory	Scavenges ROS, reduces inflammatory cytokines	Oral capsules, topical formulations	Chittasupho et al., 2021
Asiaticoside	Angiogenesis promoter	Stimulates VEGF production, promotes endothelial cell migration	Topical hydrogel	Liu et al., 2021

continued on following page

Table 1. Continued

Phytoconstituent	Mechanism of Action	Targeted Healing Process	Formulation Used	Reference
Glycyrrhizic Acid	Antimicrobial, anti-inflammatory	Reduces microbial load, modulates immune response	Hydrogel dressing, topical ointment	Wang et al., 2024
Berberine	Antimicrobial, anti-inflammatory	Inhibits bacterial growth, reduces cytokine production	Oral, topical formulations	Gawali, 2021
Aloe Vera	Anti-inflammatory, antioxidant	Reduces inflammation, enhances collagen synthesis	Gel, topical creams	Singh et al., 2020
Centella Asiatica	Angiogenesis promotion, collagen synthesis	Stimulates fibroblast activity, promotes angiogenesis	Topical ointment, oral supplements	Lee et al., 2022
Ginger	Anti-inflammatory, antioxidant	Reduces oxidative stress, modulates inflammation	Topical gel, oral powder	Tan et al., 2021
Allium Sativum (Garlic)	Antimicrobial, anti-inflammatory	Inhibits bacterial growth, reduces inflammatory cytokines	Oral supplements, topical paste	Al-Knani et al., 2023

2.1. Anti-inflammatory Effects

Chronic inflammation is the primary barrier to effective wound healing in DFUs, In diabetic wounds, prolonged inflammatory responses prevent progression through normal healing phases, locking wounds into a chronic, non-healing state (Yadav et al., 2025). Curcumin and resveratrol suppress this pathological inflammation by reducing proinflammatory cytokines (TNF-α, IL-6) and blocking NF-κB signaling (Gonzales & Orlando, 2008; Jagiełło et al., 2023). In vitro and animal models demonstrate that curcumin downregulates inflammatory mediators and accelerates wound closure (Kumari et al., 2022). Beyond direct cytokine suppression, phytoconstituents can shift macrophage polarization toward the M2 phenotype, which actively promotes tissue repair and resolves inflammation. However, achieving these effects systemically requires adequate bioavailability a significant limitation, as curcumin and resveratrol are poorly absorbed orally and require advanced delivery systems to reach therapeutic concentrations at the wound site (Alyoussef et al., 2021; Omidian et al., 2023; Sideek et al., 2022).

2.2. Antioxidant Properties

Oxidative stress impairs DFU healing when reactive oxygen species (ROS) overwhelm endogenous antioxidant defenses, causing cellular damage and fibroblast dysfunction. Resveratrol and quercetin—potent antioxidant phytoconstituents neutralize free radicals and induce expression of endogenous antioxidant enzymes (superoxide dismutase, catalase), restoring cellular homeostasis at the wound site (Cui & Wang, 2025; McKay et al., 2023). Studies show that resveratrol and quercetin reduce ROS levels and enhance wound healing in diabetic animals by increasing fibroblast proliferation and collagen synthesis (Kaleci & Koyutürk, 2020; Nusantoro et al., 2024). Mechanistically, resveratrol activates the Nrf2 pathway, a key transcription factor that coordinates antioxidant gene expression and cellular repair responses, further reducing oxidative stress(Cui & Wang, 2025). Despite these mechanisms, clinical translation remains limited: poor solubility and low oral bioavailability restrict systemic delivery, and extract standardization inconsistencies make reproducibility difficult in clinical trials.

2.3. Angiogenesis

Impaired angiogenesis the failure to form new blood vessels is a critical bottleneck in DFU healing, Without new capillariesthe wounds remain hypoxic and nutrient-deprived, preventing epithelialization and tissue remodeling.(Rai et al., 2022) Asiaticoside from Centella asiatica upregulates VEGF, the primary angiogenic growth factor, promoting endothelial cell migration and capillary formation(Gohil et al., 2010; Li et al., 2022). Berberine and ferulic acid enhance these effects by directly stimulating endothelial cell survival and tube formation, These compounds address a fundamental DFU healing deficit restoring blood perfusion to ischemic tissues. (Lin et al., 2009; Qiao et al., 2024), However, delivering sufficient concentrations of these compounds to promote sustained angiogenesis remains challenging; most preclinical studies use direct wound injection or topical application, limiting their applicability to oral or systemic administration in clinical practice.

2.4. Antimicrobial Activity

Infection transforms DFU management from a healing challenge into a survival concern. When bacteria form biofilms, conventional antibiotics fail, and multidrug resistance escalates rapidly(Boulton et al., 2020; Ramirez-Acuña et al., 2019). Glycyrrhizic acid from licorice possesses potent antimicrobial properties, reducing microbial load at the wound site and lowering biofilm production by Staphylococcus aureus, a common DFU pathogen (Oyama et al., 2016) it also provides complementary antimicrobial activity against biofilm-forming bacteria, offering an alternative or adjunctive approach to antibiotics in cases of resistance, These compounds target a real clinical need. Yet their effectiveness depends on achieving adequate local concentrations a challenge for compounds with poor bioavailability and limited wound penetration. (Bayramov & Neff, 2016; Sevgi et al., 2014). Furthermore, standardization of antimicrobial potency across different plant extracts remains inconsistent, complicating dose optimization (Tirant et al., 2024).

2.5. Wound Healing and Tissue Regeneration

Phytoconstituents enhance multiple aspects of tissue repair essential for wound closure, Polyphenols particularly those in green tea and grapes improve fibroblast proliferation and collagen deposition, strengthening tissue structure and integrity (Guimarães et al., 2021). Combinations of embelin, berberine, and ferulic acid accelerate wound closure and epithelialization in diabetic rats by promoting collagen synthesis and granulation tissue formation, enhancing both functional and aesthetic outcomes.

Despite these promising mechanisms, three translational barriers limit clinical adoption. First, bioavailability: most phytoconstituents exhibit poor solubility and low oral absorption, resulting in insufficient therapeutic concentrations at the wound site without advanced delivery systems (Kumar et al., 2023; Sivamani et al., 2012). Second, standardization: variability in plant extract composition across suppliers driven by differences in growing conditions, harvest timing, and extraction methods makes reproducible dosing and clinical trial design difficult (Sivamani et al., 2012). Third, mechanism definition: while animal models demonstrate efficacy, the specific doses, duration of treatment, and optimal compound combinations for human DFU therapy remain undefined (Ning et al., 2022). Further research on pharmacokinetics, molecular mechanisms, and combination strategies is essential for clinical translation (Figure 2).

Figure 2. Phases of wound healing and phytoconstituent influence

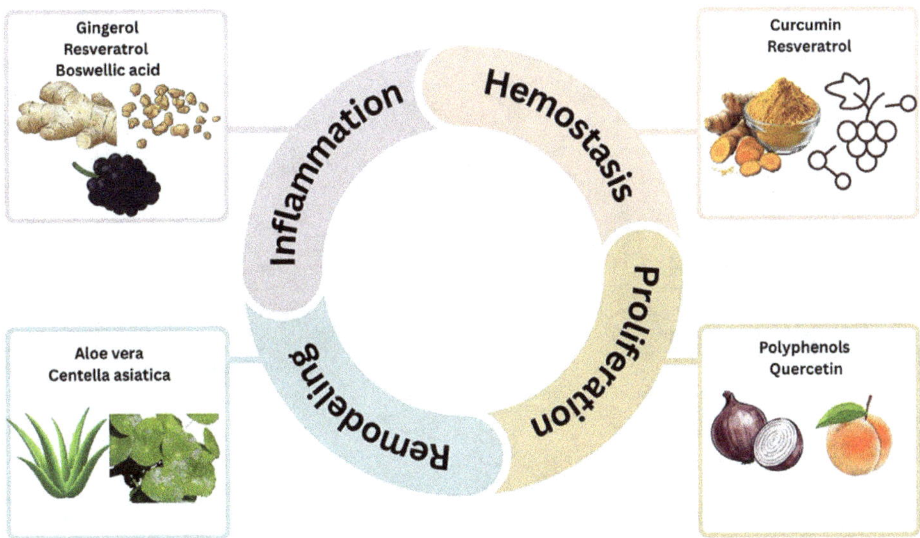

3. PATHOPHYSIOLOGY OF DIABETIC FOOT ULCERS: CHRONIC INFLAMMATION

Chronic inflammation is the central pathophysiological feature of DFUs, delaying or preventing tissue recovery. Unlike acute wounds that progress through distinct healing phases, diabetic foot ulcers become trapped in a chronic inflammatory state due to dysregulated local and systemic factors (Mohsin et al., 2024). This prolonged inflammatory environment disrupts immune response, impairs cellular function, and blocks tissue repair processes. Understanding these mechanisms is essential for targeting therapeutic interventions, including phytoconstituent-based approaches (Patil et al., 2019). (Figure 3).

Figure 3. Antioxidant and Anti-inflammatory pathway in DFS

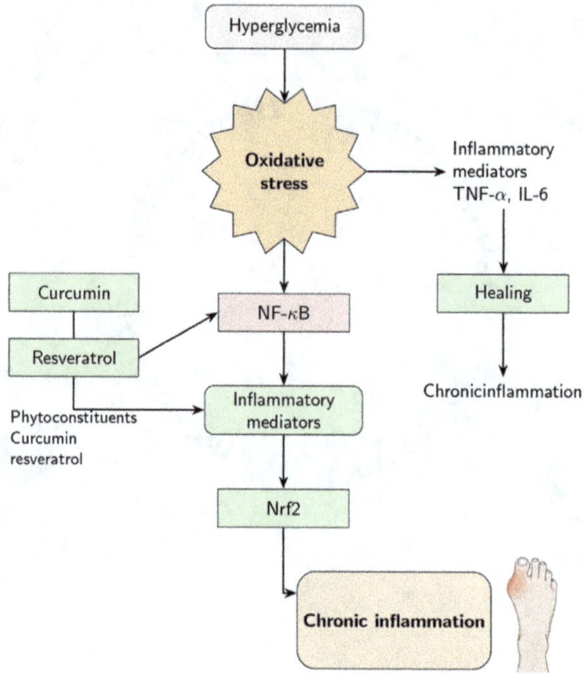

3.1. Prolonged Inflammatory Response

In DFUs, chronic inflammatory processes are driven by sustained elevation of proinflammatory cytokines TNF-α, IL-1, IL-6, and IFN-γ which prevent the transition to the wound healing resolution phase, These cytokines play dual roles: while essential for initiating acute immune responses, their chronic elevation becomes pathological (Nirenjen et al., 2023; Raziyeva et al., 2021) Elevated IL-6 and TNF-α impair fibroblast function and suppress collagen production, directly compromising the structural rebuild phase of healing (Nirenjen et al., 2023).

A key pathological feature is aberrant macrophage polarization. In healthy wounds, macrophages shift from pro-inflammatory M1 phenotype to anti-inflammatory M2 phenotype, promoting tissue repair through secretion of growth factors (VEGF, TGF-β) (Li et al., 2021). However, DFUs fail to complete this transition—macrophages remain locked in the M1 state, releasing ROS and matrix-degrading enzymes (MMPs) that perpetuate tissue destruction, immune cells become dysfunctional: neutrophils fail to effectively neutralize their own ROS production, and dendritic cells and T lymphocytes become dysregulated, creating a self-perpetuating inflammatory environment (Herrero-Cervera et al., 2022; Mittal et al., 2013). Hyperglycemia exacerbates this dysfunction by directly impairing immune cell function, reducing their capacity to resolve the ongoing inflammatory process (Rendra et al., 2018).

3.2. Effects of Chronic Inflammation on Wound Healing

Blocked Progression Through Healing Phases. In healthy wounds, inflammation naturally resolves, allowing progression to the proliferative phase where fibroblasts synthesize collagen and endothelial cells form new blood vessels (Almadani et al., 2021). In DFUs, this transition is blocked. Chronic elevation of TNF-α and IL-1β suppresses fibroblast activity and collagen deposition, trapping wounds in the inflammatory phase, this stalled progression is a defining feature of chronicity (Baltzis et al., 2014; Burgess et al., 2021).

Dysregulated Protease Activity. DFUs exhibit elevated matrix metalloproteinase activity, particularly MMP-9, which degrades extracellular matrix components faster than they can be rebuilt (Burgess et al., 2021). The balance between MMPs and their tissue inhibitors (TIMPs) is severely disrupted, exacerbating tissue destruction and delaying healing. This imbalance, coupled with chronic inflammation, creates a vicious cycle: degradation exceeds synthesis, preventing tissue accumulation (Schultz et al., 2011). This persistent proteolytic activity, coupled with diminished growth factor levels, directly impedes the migration and proliferation of keratinocytes and fibroblasts, crucial for re-epithelialization and granulation tissue formation (Zhu et al., 2023).

Angiogenesis Suppression and Biofilm Formation. Chronic inflammatory cytokines (IL-6, TNF-α) suppress VEGF expression, preventing endothelial cell migration and new vessel formation. Poor blood supply limits oxygen delivery, perpetuating hypoxia and impeding healing (Kumar et al., 2015). Simultaneously, the inflammatory microenvironment promotes bacterial biofilm formation biofilms provide pathogens with sanctuary from antibiotics and the immune system, establishing chronic infection, Repetitive infections reinforce inflammation, creating a self-sustaining pathological loop (Chen & Wen, 2011; Pai et al., 2023).

Oxidative Stress Accumulation. The prolonged inflammatory state generates excessive ROS, which overwhelms antioxidant defenses. ROS damages DNA, proteins, and lipids, disrupting cellular function and further impairing healing (Mittal et al., 2013; Polaka et al., 2022). This oxidative burden, combined with insufficient antioxidant capacity in diabetic tissues, perpetuates both inflammation and cellular damage.

3.3. Impaired Angiogenesis and Vascular Dysfunction

Angiogenesis formation of new blood vessels from existing vasculature is critical for wound healing. Without new capillaries, wounds remain hypoxic and nutrient-starved, preventing tissue regeneration. DFUs exhibit profound angiogenic failure due to multiple converging mechanisms (Burgess et al., 2021; Okonkwo & DiPietro, 2017).

VEGF Pathway Suppression. High glucose levels damage endothelial cells, impairing their capacity to proliferate and migrate processes essential for capillary formation. More critically, inflammatory cytokines and oxidative stress suppress VEGF expression, reducing the primary pro-angiogenic signal (Caldwell et al., 2005). When VEGF is insufficient, endothelial cells cannot be recruited to form new vessels, and angiogenesis fails, DFUs also overexpress anti-angiogenic factors (thrombospondin 1, endostatin) while simultaneously downregulating pro-angiogenic factors, creating a hostile environment for vascular growth (Liu et al., 2021) .

Nitric Oxide Deficiency. Nitric oxide (NO) is essential for vascular tone regulation and endothelial function. In diabetes, endothelial NO synthase (eNOS) activity is reduced, lowering NO levels(Spinetti et al., 2008). This leads to vasoconstriction, decreased blood flow, and impaired angiogenesis . Low NO also exacerbates inflammation and oxidative stress, creating a secondary impediment to vascular regeneration(Calver et al., 1992; Yang et al., 2024).

Peripheral Arterial Disease (PAD). Diabetic patients frequently develop atherosclerotic narrowing or obstruction of blood vessels, reducing macrovascular perfusion to the affected limb(Rajendran et al., 2013). Combined with endothelial dysfunction at the microvascular level, this significantly worsens DFU outcomes by limiting oxygen and nutrient delivery to already compromised tissues. Poor baseline blood supply makes it nearly impossible to support the metabolic demands of active wound healing(Moura et al., 2019).

3.4. Defective Tissue Regeneration

Even when inflammation temporarily subsides, fibroblasts in the diabetic microenvironment become dysfunctional. Prolonged exposure to inflammatory cytokines, hyperglycemia, and altered extracellular matrix components forces fibroblasts into a less regenerative phenotype a phenomenon termed fibroblast plasticity(Khamaisi et al., 2016) . This altered phenotype reduces collagen synthesis and impairs the structural restoration necessary for wound closure(Dasari et al., 2021).

Hypoxia further disrupts healing through impaired HIF (hypoxia-inducible factor) signaling. Under normoxic conditions, HIF mediates the cellular response to hypoxia and upregulates angiogenic pathways(Hong et al., 2014). In DFUs, combined hyperglycemia and chronic hypoxia depress HIF signaling, preventing the wound from activating its own pro-healing programs(Botusan et al., 2008). The result is a wound incapable of self-rescue remaining hypoxic and metabolically depleted, unable to generate the proliferation and differentiation signals needed for tissue regeneration(Botusan et al., 2008; Ruthenborg et al., 2014).

Additionally, insufficient nutrient and growth factor delivery via inadequate angiogenesis starves fibroblasts of the substrates needed for collagen synthesis . This creates a coupled vicious cycle:

poor angiogenesis → inadequate nutrients → impaired fibroblast function → failed collagen deposition → persistent wound.(Brown et al., 2018; Johnson & Wilgus, 2014)

3.5. Clinical Evidence and Therapeutic Targeting

Current clinical approaches attempt to address these mechanisms through emerging therapies. Exosome-based therapies from adipose-derived stem cells (ADSCs) with enhanced angiogenic signaling have accelerated wound healing in diabetic rats(Ren et al., 2022; Zhang et al., 2022) . Specific exosomes promote VEGF-dependent angiogenesis and upregulate tissue repair genes. Stromal vascular fraction (SVF) cell therapy has demonstrated clinical efficacy(Shukla et al., 2020): SVF injections increase blood flow and accelerate wound closure in patients with chronic DFUs, improving vascularization and tissue healing (Bi et al., 2019).

Combination gas therapy addresses hypoxia and NO deficiency directly. Recent work demonstrates that simultaneous delivery of nitric oxide and oxygen via hydrogel scaffolds promotes endothelial cell proliferation, migration, and tube formation, offering a mechanistically targeted approach to restore angiogenesis in diabetic mice (Jen et al., 2011; Phelps & García, 2008)). These strategies represent advances in targeting specific pathophysiological deficits rather than generic wound care.

4. OXIDATIVE STRESS IN DIABETIC FOOT ULCERS

Oxidative stress the imbalance between reactive oxygen species (ROS) production and antioxidant capacity is a central pathophysiological driver of DFU chronicity(Deng et al., 2021). Unlike Section 3's focus on inflammation, this section isolates the oxidative stress pathway and its tissue-level consequences.

4.1. Imbalance between ROS and Antioxidants

Chronic hyperglycemia drives persistent ROS production, primarily from mitochondria but also from inflammatory cells (neutrophils, macrophages) in the diabetic microenvironment. Under normal conditions, endogenous antioxidant enzymes superoxide dismutase (SOD), catalase (CAT), and glutathione peroxidase (GPx) neutralize ROS and maintain cellular homeostasis(Teodoro et al., 2019; Volpe et al., 2018). However, in diabetes, these antioxidant systems become overwhelmed by excessive ROS generation, resulting in unchecked oxidative damage to proteins, lipids, and DNA(Johansen et al., 2005) .This oxidative imbalance amplifies pathological inflammation.

Hyperglycemia-induced ROS activates the NF-κB pathway, driving sustained elevation of proinflammatory cytokines that prevent wound healing progression (Giacco & Brownlee, 2010; Volpe et al., 2018). ROS also triggers lipid peroxidation, generating malondialdehyde (MDA) a marker of oxidative damage which disrupts cell membranes and exacerbates tissue injury (Elajaili et al., 2025) (Hussain et al., 2023). Additionally, ROS suppresses fibroblast collagen synthesis, directly impairing the structural repair phase of wound healing (Kumar et al., 2023). By simultaneously driving inflammation, damaging cell membranes, and blocking collagen deposition, oxidative stress creates a multi-pronged impediment to healing (Dunnill et al., 2015; Wang et al., 2022).

Translational Challenge: Systemic antioxidant therapy faces a fundamental barrier delivering sufficient concentrations of exogenous antioxidants to the wound site while maintaining bioactivity (Comino-Sanz et al., 2021; Zhang et al., 2023). Most synthetic antioxidants exhibit poor tissue penetration and rapid clearance, limiting their therapeutic window.(Whittam et al., 2014)

4.2. Tissue Damage Induced by Oxidative Stress

Mitochondrial Dysfunction and Cellular Energy Depletion. Mitochondria are both the primary source and primary target of ROS in diabetic environments. Excessive ROS damages mitochondrial membranes and impairs oxidative phosphorylation, reducing ATP production in metabolically active cells particularly fibroblasts and endothelial cells required for wound repair (Kaludercic & Lisa, 2020; Zhang et al., 2023) . ATP depletion decreases cell viability, increases apoptosis, and leaves wound-healing cells unable to perform their synthetic and migratory functions (Hanna et al., 1992).

Extracellular Matrix Degradation. ROS and its byproducts directly damage structural ECM proteins, ROS degrades collagen and fibronectin essential components for tissue remodeling while simultaneously overwhelming the capacity for new

ECM synthesis (Egea et al., 2020; Martins et al., 2021) . This imbalance between degradation and synthesis prolongs the inflammatory phase of wound healing and prevents organized tissue remodeling. The consequence is poor ulcer closure and disorganized, weak scar tissue formation (Diller & Tabor, 2022).

Endothelial Dysfunction and Angiogenesis Suppression. As detailed in Section 3, ROS-induced endothelial dysfunction reduces VEGF signaling and impairs angiogenesis, creating a coupled mechanism:

Oxidative stress → endothelial damage → reduced VEGF → impaired vessel formation → persistent hypoxia (Xu et al., 2022). This vicious cycle hypoxia increases ROS production, which further impairs angiogenesis—locks wounds in an ischemic, non-healing state (Guo et al., 2025; Hunt et al., 2024; Hutchings et al., 2021).

Clinical Implication: Addressing oxidative stress alone is insufficient; wounds require simultaneous restoration of antioxidant capacity, resolution of inflammation, and restoration of angiogenesis (Guan et al., 2021; Wang et al., 2022). Single-target antioxidant interventions have shown limited clinical efficacy, suggesting the need for multi-mechanistic approaches precisely the rationale for phytoconstituent combinations discussed in Section 6.structure of the ECM, an important constituent of the healing process (Zhang et al., 2021). ROS degradation of collagen and fibronectin components of the ECM impairs tissue remodeling and prolongs the inflammatory phase of wound healing (Ukaegbu et al., 2025) . Therefore, ulcer closure is poor because tissue formation is not well organized.

5. INFECTION IN DIABETIC FOOT ULCERS: CLINICAL EVIDENCE AND THERAPEUTIC CHALLENGES

While preceding sections focused on chronic inflammation and oxidative stress, both are exacerbated catastrophically by infection. Infection is the leading complication in DFUs, increasing treatment duration, amputation risk, and mortality (Boulton et al., 2020; Rodríguez-Rodríguez et al., 2021). Understanding infection pathophysiology is essential because therapeutic approaches must address not only baseline wound healing deficits, but also infection-driven complications that often render standard antibiotics ineffective (Burgess et al., 2021). This section integrates clinical evidence on infection mechanisms and barriers to treatment.

5.1. Infection Pathophysiology in DFUs

5.1.1. Compromised Immune Response

Hyperglycemia compromises the immune system's ability to prevent and resolve infection. Neutrophils—the first-line immune cells to migrate to injury sites exhibit reduced activity in diabetes, losing the capacity to migrate effectively and phagocytose pathogens(Chávez-Reyes et al., 2021) . Additionally, macrophages and T-cells become dysregulated, further reducing immune competence and making diabetic patients vulnerable to infections. This immunocompromised state is compounded by poor vascular perfusion peripheral vascular disease is common in DFUs and reduces immune cell delivery to the wound site (Burgess et al., 2021) noted in (Section 5.2).

5.1.2. Biofilm Formation and Antibiotic Resistance

Pathogenic bacteria particularly Pseudomonas aeruginosa and Staphylococcus aureus readily form biofilms in diabetic wounds (Afonso et al., 2021). Biofilms are complex bacterial communities encased in protective polysaccharide matrices that shield pathogens from both immune cells and antibiotics (Afonso et al., 2021; Pouget et al., 2020). Traditional antibiotics fail to penetrate biofilms, rendering biofilm-associated bacteria notoriously resistant to treatment and requiring aggressive or alternative therapeutics (Scherr et al., 2014).

5.1.3. Polymicrobial Infections and Treatment Complexity

DFU infections are typically polymicrobial multiple bacterial species coexist within single wounds. Diverse microbial populations exhibit varying antibiotic sensitivities, complicating infection control strategies (Ramirez-Acuña et al., 2019), Microenvironment not only delays healing but perpetuates inflammation, establishing a vicious cycle where persistent infection maintains chronic inflammatory state and prevents wound progression (Raziyeva et al., 2021; Uberoi et al., 2024) (detailed in Section 5.2).

5.2. How Infection Derails Wound Healing

The infected state promotes a "wound chronicity phenotype"(Boulton et al., 2020; Rodríguez-Rodríguez et al., 2021), a self-perpetuating state of chronic inflammation (Raziyeva et al., 2021; Uberoi et al., 2024) that prevents wounds from progressing toward healing while sustaining biofilms of polymicrobial communities (Afonso et al., 2021; Ramirez-Acuña et al., 2019)—these perpetuate infection, resist antibiotics (Pouget et al., 2020; Scherr et al., 2014), and inhibit tissue regeneration. Tissue necrosis from infection exacerbates this cycle by increasing bacterial burden through protease release (Demidova-Rice et al., 2012; McCarty & Percival, 2013) and further compromising regenerative capacity. Furthermore, the persistent inflammatory response induced by chronic infection, often characterized by polymicrobial biofilms, promotes local plasma leakage and cellular senescence, thereby hindering optimal wound repair (Pereira et al., 2017).

5.2.1. Multidrug-Resistant Organisms (MDROs): An Emerging Crisis

DFU infections increasingly involve multidrug-resistant organisms (MDROs), which substantially complicate treatment outcomes (López-Díez et al., 2021) . MDROs are ineffective against standard antibiotics and require prolonged, specialized treatment protocols. This resistance problem fundamentally alters the clinical landscape: conventional antimicrobial strategies lose efficacy, necessitating novel approaches including alternative therapeutics such as antimicrobial phytoconstituents (Ashraf et al., 2023; Khare et al., 2021) discussed in Section 6.

6. PHYTOCONSTITUENTS IN DFU TREATMENT

Due to their biological activities, they have attracted a multiplicity of phytoconstituents as potential therapeutic agents for DFUs (Khare et al., 2021). The five bioactive compounds referred to in these plants participate in wound healing by increasing fibroblast activity, collagen synthesis, and other wound healing mechanisms, such as anti-inflammatory actions, antibacterial activity, and increased blood circulation to the wound site. These phytochemicals play a major role in DFU treatment (Ahmadian et al., 2021; Ambika & Nair, 2018).

6.1. Polyphenols

In vitro studies of fibroblasts have shown that polyphenols, especially flavonoids and phenolic acids, improve fibroblast proliferation and migration, which are of key importance in wound healing (Herman et al., 2023; Niu et al., 2025) . Fibroblast activity is also promoted by polyphenols, resulting in collagen deposition and extracellular matrix synthesis. For instance, the well-recognized polyphenol resveratrol has been shown to enhance fibroblast function and accelerate tissue repair in diabetic wounds (Addis et al., 2020; Kunkemoeller & Kyriakides, 2017). Furthermore, it was also proved that polyphenols like quercetin and curcumin can promote fibroblast migration and proliferation, which result in better wound healing in diabetic ulcers

(Ding et al., 2022; Pignet et al., 2021) . DFUs tissue regeneration is promoted by collagen synthesis and re-epithelization by Carica papaya polyphenol-rich extract. it showed that Carica papaya polyphenol-rich extracts effectively aid in wound healing by enhancing collagen synthesis and speeding up re-epithelialization (Hakim et al., 2019; Li et al., 2023). These compounds modulate inflammatory cytokines (suppressing TNF-α and IL-6), collagen degradation, and tissue remodeling (Samavati et al., 2025).

6.2. Flavonoids

Flavonoids are strong immune response modulators in the wound microenvironment owing to their strong anti-inflammatory effects (Zulkefli et al., 2023). Quercetin, kaempferol, and luteolin also suppress the release of proinflammatory cytokines (TNF-α, IL-6, and IL-1β), which delays the wound healing process (Papadopoulou et al., 2004). Flavonoids also limit inflammation caused by nuclear factor kappa B (NF-κB), a significant mediator of inflammatory pathway regulators that manage inflammation in DFUs . Flavonoids are potent antioxidants and anti-inflammatory substances. Due to ROS scavenging, oxidative stress, a major predisposing factor for delayed wound healing in diabetes, is reduced. Adding quercetin and rutin, a class of flavonoids, to a wound gel increases the levels of antioxidants in the body, such as superoxide dismutase and catalase, to protect the wound site from oxidative damage (Jin & Arroo, 2023). It neutralizes the antioxidants to prevent tissue damage and promote healing.

Flavonoids modulate the immune response in the wound microenvironment through potent anti-inflammatory effects (Zulkefli et al., 2023). Quercetin, kaempferol, and luteolin suppress proinflammatory cytokines and inhibit nuclear factor kappa B, a key mediator of inflammation in DFUs (Papadopoulou et al., 2004). As potent antioxidants, they scavenge ROS to reduce oxidative stress—a major factor delaying diabetic wound healing—and elevate endogenous antioxidants like superoxide dismutase and catalase. Incorporating quercetin and rutin into wound gels protects tissues from oxidative damage and promotes healing (Devi et al., 2021; Khan et al., 2019). Furthermore, flavonoids have demonstrated antimicrobial properties against various pathogens commonly found in DFUs, contributing to infection control (Chanu et al., 2023).

6.3. Alkaloids

Berberine and morphine are examples of alkaloids that have proven to be highly antimicrobial and are therefore useful in preventing and treating DFU infections (Herman et al., 2023). Diabetic wounds pose a significant risk of infection; thus, the ability of alkaloids to prevent microbial growth is vital for treating such ulcers (Thawabteh et al., 2024). Since they have been proven to be active against a plethora of bacteria, such as Staphylococcus aureus and Pseudomonas aeruginosa, common pathogens in DFUs (Moses et al., 2023; Oliveira et al., 2022), Berberis vulgaris alkaloids can serve as an example, and it also aid in minimizing the microbial load in infected wounds due to their antimicrobial properties (El-Zahar et al., 2022). Clinical studies using alkaloids from plants, such as Sophora interrupta, have shown a decrease in the amount of bacterial biofilms and an increase in the healing environment (Li et al., 2016; Mathi et al., 2015), Alkaloid synergism with antibiotics may abbreviate and minimize the period of infection and its severity, which are imperative in the wound healing process that must be expedited in DFUs (Manosalva et al., 2016).

6.4. Terpenoids

Some terpenoids and specific compounds, such as asiaticoside and ursolic acid, promote angiogenesis, the formation of new blood vessels, which plays an important role in healing DFU (Min et al., 2024). Terpenoids stimulate the formation of new blood vessels at the wound site, increasing blood flow and providing oxygen and nutrients to aid in wound regeneration (Min et al., 2024). (Setiawati et al., 2024) conducted a study where asiaticoside of Centella asiatica has been found to enhance expression of vascular endothelial growth factor (VEGF), a potent mediator of angiogenesis. Terpenoids are also known to improve blood circulation, which is important in DFUs that are frequently characterized by poor vascularization and delayed healing. They further mentioned that Ginkgo biloba and Commiphora wightii terpenoids can enhance the growth of new capillaries and improve endothelial function, leading to an improved supply of nutrients and oxygen to the healing tissues (Mohan et al., 2017; Wang et al., 2022). In addition to improving the speed of wound healing, it reduces the occurrence of these problems.

7. INFLAMMATION AND ANTI-INFLAMMATORY PHYTOCONSTITUENTS

Chronic inflammation in diabetic foot ulcers (DFUs) impedes healing and damages tissue. Three phytoconstituents curcumin, resveratrol, and quercetin—modulate the NF-κB pathway and reduce proinflammatory cytokines, accelerating the transition from inflammatory to proliferative healing phases(Chaturvedi et al., 2022; Shukla et al., 2012).

Curcumin and NF-κB Inhibition

Curcumin inhibits NF-κB activation by preventing NF-κB p65 translocation to the nucleus, suppressing expression of TNF-α and IL-6 (Kenawy et al., 2022) . In diabetic wound models, curcumin increases fibroblast activity and collagen synthesis while mitigating oxidative stress (Kumari et al., 2022) . Resveratrol works synergistically with curcumin, further reducing NF-κB signaling and proinflammatory cytokine production. Both compounds promote re-epithelialization and collagen deposition, accelerating wound closure (Chittasupho et al., 2021; Zhang et al., 2022).

Quercetin's Anti-inflammatory Mechanism

Quercetin, a flavonoid with potent antioxidant properties, inhibits NF-κB activation and reduces TNF-α and IL-6 level. It promotes fibroblast migration and collagen deposition, facilitating tissue repair (Chittasupho et al., 2021; Kant et al., 2020) . The combined use of quercetin and curcumin enhances oxidative stress reduction and cell proliferation (Chittasupho et al., 2021) .

Resveratrol and Oxidative Stress Management

Beyond NF-κB inhibition, resveratrol directly scavenges ROS and upregulates the body's endogenous antioxidant defense system, It activates the Nrf2/ARE signaling pathway, enhancing expression of superoxide dismutase (SOD), catalase, and glutathione peroxidase (GPx) enzymes that detoxify ROS (Farkhondeh et al., 2020) . This dual mechanism strengthens cellular defenses and improves mitochondrial function, supporting energy production needed for wound healing (Li et al., 2019) . Resveratrol also increases glutathione levels and enhances mitochondrial biogenesis, reducing oxidative damage and improving cellular repair mechanisms (Pignet et al., 2021) (Huang et al., 2019)(Zhang, 2023).

Glycyrrhizic Acid: Antimicrobial and Anti-inflammatory Properties

Glycyrrhizic acid, a triterpenoid saponin from Glycyrrhiza glabra (licorice), provides dual therapeutic benefits for infected DFUs (Richard, 2021) It exhibits antimicrobial activity against Staphylococcus aureus and Pseudomonas aeruginosa through multiple mechanisms: disrupting bacterial cell wall integrity, causing intracellular leakage, and activating phagocytic immune responses. Importantly, glycyrrhizic acid prevents biofilm formation, addressing a major barrier to conventional antibiotic treatment (Hazlett et al., 2019; Long et al., 2012)t .

When incorporated into hydrogel delivery systems, glycyrrhizic acid demonstrates synergistic antimicrobial and anti-inflammatory effects. It reduces bacterial load while suppressing chronic inflammation by shifting macrophages toward anti-inflammatory phenotypes, facilitating transition from inflammatory to proliferative healing phases (Nascimento & Araújo, 2022; Oyama et al., 2016). Additionally, it enhances collagen synthesis and extracellular matrix components, promoting re-epithelialization and accelerating wound closure. Glycyrrhizic acid also induces angiogenesis through VEGF expression, improving vascular supply to healing tissues (Aly et al., 2023; Hung et al., 2017) .

8. ANGIOGENESIS AND PHYTOCONSTITUENTS THAT PROMOTE VASCULARIZATION

Angiogenesis, the formation of new blood vessels from pre-existing vessels, is an important component of wound healing. Importantly, it provides sufficient oxygen and nutrients to regenerating tissues during the healing process. Poor tissue regeneration and delayed healing are consequences of poor angiogenesis, which is observed in chronic wounds, including DFUs (Okonkwo et al., 2020). Moreover, the effects of the phytoconstituent 'asiaticoside' have been reported for angiogenesis, stimulating angiogenesis, and upregulating crucial pro-angiogenic factors such as VEGF that promote angiogenesis, increase blood flow, and tissue repair. These natural compounds have great potential as therapeutic agents for promoting wound healing in ischemic tissues and chronic wounds. (Raguraman et al., 2021)

8.1. Importance of Angiogenesis in Wound Healing

Since angiogenesis is an essential step in restoring oxygen and nutrient supply to regenerating tissues through the formation of new blood vessels, it is not surprising that angiogenesis of posttraumatic tissue lesions is usually observed between 24 and 72 h after the lesion. During the proliferative phase of wound healing, endothelial cells migrate, proliferate, and form new capillary networks to reestablish blood circulation in damaged tissues (Mamun et al., 2024; Shams et al., 2022; Shi et al., 2023). The inability to form new blood vessels impedes ischemia and chronic wound healing processes. Patients with diabetes suffer from angiogenesis, which is often diminished when blood sugar and circulation are low, wounds do not close in a timely manner, and have the potential to become infected. Studies on wound healing have shown that targeting angiogenesis has a highly positive impact on healing (Okonkwo et al., 2020; Rai et al., 2022). For example, Wang et al. (2023) found that therapeutic strategies to activate angiogenesis promote the regenerative potential of ischemic wounds, indicating that vasculature is a key factor in proper wound healing. The main stimulator of endothelial cell proliferation and the formation of new blood vessels is increased expression of VEGF, a key angiogenic factor. Angiogenesis promotion is especially crucial in DFUs because angiogenesis is impaired by poor vascular supply and oxidative stress in the ulcer (Raguraman et al., 2021; Rai et al., 2022).

8.2. Role of Asiaticoside in VEGF Upregulation

Asiaticoside, a phytoconstituent of Centella asiatica, is a triterpenoid glycoside with angiogenesis-promoting properties. Asiaticoside upregulates VEGF, which is responsible for blood vessel formation and angiogenesis (Kumar et al., 2024). It activates endothelial cell proliferation and migration to form new capillaries at the wound site. In wound tissues where asiaticoside significantly elevated VEGF levels, it also improved wound healing in diabetic rats by fostering blood vessel formation in the wound tissues (Hou et al., 2016; Wang et al., 2017) . To explore the potential of asiaticoside as a therapeutic agent for improving angiogenesis in chronic wounds with reduced angiogenesis, the effect of asiaticoside, either alone or in combination with an allergen, on VEGF expression and angiogenesis was investigated using a chicken CAM model. In addition, asiaticoside affects angiogenesis, particularly without mediating upregulated VEGF (Hou et al., 2016; Shukla et al., 1999). Similarly, it has been demonstrated to influence other signalling pathways involved in blood vessel formation (for instance, the PI3K/Akt pathway). Hou et. al. (2016) shows that TUBE is a key process in endothelial cell survival, migration, and tube formation in order to form new blood vessels. In addition to its influence on VEGF expression, activation of these pathways enables asiaticoside to assist the endothelium in performing its function better in aiding wound healing and tissue regeneration (Hou et al., 2016; Lu et al., 2004; Nowwarote et al., 2012; Yuliati et al., 2015).

8.3 Improved Blood Flow and Tissue Regeneration

In addition to phytoconstituents such as asiaticoside, asiatic acid, and madecassic acid, it (supercritical extract) promotes angiogenesis to foster blood flow to the wound site for tissue regeneration (Azmin et al., 2016; Hou et al., 2016). The difficulty with impaired blood supply is that, as in any resting tissue, and in chronic wounds in particular, hypoxia prevents the proper functioning of key cells required for wound healing (e.g., fibroblasts and keratinocytes)(Shi et al., 2023) (Martin et al., 2010). Asiaticoside aids in oxygen and nutrient delivery and supports cellular processes (such as collagen synthesis and re-epithelialization) by stimulating new blood vessel formation. In addition to increasing blood flow, asiaticoside strengthens other aspects of tissue regeneration (Hou et al., 2016; Kumar et al., 2024). For instance, in the study done by Liu et al. (2021), their results showed that asiaticoside treatment increased collagen deposition and improved tissue structure in diabetic wound models. This is highly important because wound closure and extracellular matrix restoration, which provide structural support for tissue regeneration, are necessary tasks (Anand et al., 2022; Nie et al., 2019). Patients with diabetes and chronic wounds are characterized by impaired circulation, and asiaticoside has an effective therapeutic role, as it can stimulate angiogenesis, counteract collagen synthesis, and increase blood flow. In addition, asiaticoside diminishes oxidative stress, which is the main cause of defective angiogenesis in diabetic wounds (Anand et al., 2022; Nie et al., 2019). By enhancing the antioxidant defense system, asiaticoside creates a favorable environment for endothelial cell function and blood vessel formation, thereby promoting tissue regeneration (Anand et al., 2022; Mu et al., 2025; Nie et al., 2019; Razif et al., 2025) (Wang et al., 2023).

Clinical Evidence Supporting Phytoconstituents in DFU Treatment

Diabetic foot ulcers (DFUs) are a major complication in patients with diabetes, and their management remains challenging, requiring novel therapeutic approaches. Conventional treatments involving debridement, antibiotics, and wound dressings are frequently used, sometimes to no avail if the process is accelerated and infection is prevented (Baig et al., 2022; Burgess et al., 2021). Wound healing can be accelerated, wound size can be reduced, and healing time can be decreased in the resolution of chronic wounds, and in the first trials, promising clinical efficacy was shown with the medicinal plants Melilotus officinalis and Plantago major, which makes them a valuable adjunct to traditional wound treatment methods (Ghanadian et al., 2022; Kartini et al., 2021) (Table 2).

Table 2. Phytoconstituents with their corresponding mode of action with clinical evidences.

Plant Name	Phytoconstituent	Mechanism of Action	Clinical Evidence	Reference
Melilotus officinalis	Coumarins	Antimicrobial, anti-inflammatory	Reduces infection risk and inflammation, promotes wound healing	Tabassum & Raj, 2020
Plantago major	Flavonoids, Glycosides	Reduces inflammation, accelerates epithelialization	Significant reduction in wound size, increased healing rate in diabetic ulcers	Ghanadian et al., 2022
Centella Asiatica	Asiaticoside, Madecassoside	Stimulates collagen synthesis, promotes angiogenesis	Enhances wound closure and collagen deposition in diabetic wounds	Liu et al., 2021
Aloe Vera	Polysaccharides, Anthraquinones	Anti-inflammatory, antimicrobial, promotes tissue regeneration	Significant improvement in wound healing and reduction in infection	Singh et al., 2020
Glycyrrhiza glabra	Glycyrrhizic acid	Antimicrobial, anti-inflammatory	Reduces microbial load, improves tissue regeneration in diabetic wounds	Wang et al., 2024
Curcuma longa	Curcumin	Anti-inflammatory, antioxidant	Reduces inflammation and accelerates collagen synthesis in chronic wounds	Zhang et al., 2023
Allium sativum	Allicin	Antimicrobial, anti-inflammatory	Inhibits bacterial growth and enhances wound healing in diabetic ulcers	Al-Knani et al., 2023
Ginkgo biloba	Flavonoids, Terpenoids	Antioxidant, improves circulation	Improves blood flow and reduces oxidative stress in diabetic wounds	(Tan, 2021)
Berberis vulgaris	Berberine	Antimicrobial, anti-inflammatory	Reduces bacterial load, modulates inflammatory response in DFUs	Gawali, 2021
Silybum marianum	Silymarin	Antioxidant, anti-inflammatory	Protects tissues from oxidative damage and promotes wound healing	Soni et al., 2020

One well-studied plant is Plantago major (broadleaf plantain), which exhibits strong wound-healing activity, especially in diabetic ulcers (Kartini et al., 2021). A recent clinical trial showed that the application of 10% hydroalcoholic extract of Plantago Major had a significant effect on the healing of diabetic foot and pressure ulcers (Najafian et al., 2018). The results showed that wounds treated with Plantago major gel experienced a statistically significant decrease in wound area compared to those treated with conventional dressings. After the first week, the proportion of reduction was 64.9% compared to 33.1% in the control group, and 86.8% the following week compared to 52.9% in the control group. Moreover, 64% of the patients with completely healed wounds in the experimental group with Plantago major were more pronounced than those in the control group (20.45%) (Ghanadian et al. 2022).

The high flavonoid, tannin, and glycoside contents of Plantago major confer anti-inflammatory, antimicrobial, and antioxidant properties; hence, the therapeutic effects of this plant are thought to be due to these(Adom et al., 2017). This agent decreases properties that lower infection risk, promotes collagen synthesis, and accelerates the process of wound bed re-epithelialization, making this medicine promising for the treatment of DFU. Among others, it contains Melilotus officinalis, or sweet clover, which is used for wound healing. Tabassum and Raj (2020) studied the antimicrobial activity of Melilotus officinalis extracts against pathogens involved in DFUs. In the second study, Melilotus extracts proved very successful in killing multidrug-resistant Staphylococcus aureus and Pseudomonas aeruginosa, the two major pathogens of diabetic wounds(Kamarudin et al., 2016). The antimicrobial effect contributes to reducing the microbial load in primary wounds and is related to the theme of healing in a balanced environment. In addition, the anti-inflammatory properties of Melilotus officinalis support its use as a wound-healing agent. Several studies have shown that Melilotus has been used to decrease levels of proinflammatory cytokines of the TNF-α and IL-6, and also has positive effects on fast healing of DFU, since it is the most common complication to inhibit the healing of DFU(Famurewa et al., 2019) . Considering their antimicrobial and anti-inflammatory effects, along with their ability to reduce wound size, Melilotus officinalis may be a suitable candidate for treating DFUs.

Epithelialization is an important part of re-epithelization, which involves covering the wound surface with new epithelial cells. The solidification process was accelerated by Plantago major and Melilotus officinalis(El-Sherbeni & Negm, 2023; Kartini et al., 2021). The application of Plantago major gel, according to Ghanadian et al. (2022) clinical trial, enhanced re-epithelialization, reduced erythema (redness), and reduced time to heal over the wound skin barrier function compared to the control group(Kartini et al., 2021; Najafian et al., 2018). Similarly, Melilotus officinalis has been reported to increase epithelial cell migration and proliferation, leading to faster wound closure in diabetic DFUs. A reduction in wound size is an essential indication of successful treatment, but this was demonstrated to be effective for both Plantago major and Melilotus officinalis Ghanadian et al., et al reported that the topical application of Plantago major extract significantly decreased the size of the wound after just two weeks of treatment(Ghanadian et al., 2022; Kartini et al., 2021; Liu et al., 2018) . In the case of Melilotus officinalis, the antibacterial and anti-inflammatory properties of the plant reduce the bacterial load and inflammation, leading to wound contraction and a decrease in size (Ghanadian et al., 2022; Liu et al., 2018). Finally, the medicinal plants reduced the overall DFU healing time. In addition, Plantago major extract facilitated the healing process, with a complete wound closure rate of 64% in patients treated with the extract, whereas only 20.45% of the control group had complete wound closure(Mustafa et al., 2019; Najafian et al., 2018) (Ghanadian et al., 2022). In addition, Melilotus officinalis has been shown to enhance healing time by encouraging faster tissue regeneration and decreasing the duration of the inflammatory healing phase. Other botanicals, such as *Calendula officinalis* and *Moringa oleifera*, have also demonstrated efficacy in accelerating DFU healing by reducing inflammation and promoting tissue granulation (Raghav et al., 2022; Walach et al., 2017).

9. CHALLENGES AND LIMITATIONS IN PHYTOCONSTITUENT APPLICATIONS

Phytoconstituents of medicinal plants have led to promising solutions for the treatment of different ailments, including chronic wounds such as diabetic foot ulcers (DFUs). However, the clinical use of these traps is associated with numerous challenges and limitations(Herman et al., 2023; Sivamani et al., 2012). The main challenges in the utilization of these herbs are the problems associated with bioavailability, standardization of extracts, and the possibility of herb–drug interactions. Resolving these issues would provide an impetus for improving the therapeutic efficacy of phytoconstituents and their safe and acceptable use in clinical settings(Fasinu et al., 2012; Gurley et al., 2012).

9.1. Bioavailability Issues

However, there is a challenge in obtaining the clinical phytoconstituents themselves; their bioavailability is very low. Poor absorption by the body is a function of molecular size, hydrophilicity, and low lipid solubility, and many active compounds in medicinal plants are poorly absorbed by the body. Therefore, they may not be able to achieve maximal therapeutic doses in the blood, and hence, the medications may be ineffective(Setia et al., 2024). Therefore, phytosomes could be considered an advanced drug delivery system for enhanced absorption and bioavailability of phytoconstituents to overcome this limitation of herbal formulations. Phytoconstituents with phospholipids as added lipophilicity and permeability across biological membranes have recently been developed in the form of phytosomes(Gaikwad et al., 2021; Kumar et al., 2017). Currently, this concept is being used to enhance the bioavailability of poorly absorbed plant compounds. For instance, phytosome forms of Ginkgo biloba, green tea, and grape seed extracts have been proven to be absorbed more effectively and act better than natural extracts(Kumar et al., 2017; Şenol et al., 2025)). This advancement in drug delivery is essential to broaden the clinical application of phytoconstituents in the treatment of DFUs, considering the need for prolonged administration and to ensure good bioavailability of phytoconstituents to be used in combating such chronic conditions(Ahmadian et al., 2021; Azmin et al., 2016). In addition, the best formulation needs to be developed for herbal compounds possessing phenolic compounds (polyphenols) and terpenoids, which are insoluble in aqueous systems and anti-absorbable. Phytochemicals with higher solubility and absorption have been used in strategies such as nanoemulsions, lipid-based formulations, and cyclodextrin inclusion complexes(Sreenathkumar, 2021; Yang et al., 2020) ((Tan, 2021)). Nevertheless, additional studies are required to confirm that this bioavailability is consistently realized in other population groups.

9.2. Extract Standardization Challenges

Another challenge in the clinical use of plant phytoconstituents is the standardization of plant extracts. Due to the variability in the preparation of herbal medicines, the manifestation of their therapeutic effects is also variable. Because different plant species are used under different cultivation conditions, at different times of harvesting, and using different methods for the extraction of active compounds, there is no uniformity in the amount of active compounds, and hence, the outcomes are unpredictable in the clinic(Hossain et al., 2022). Standardization ensures the consistency of the efficacy and safety of a given herbal extract and guarantees that it contains a certain amount of active ingredient. For example, the extraction method and plant origin determine the concentration of active compounds (flavonoids and glycosides) in the wound-healing plant Plantago major(Adom et al., 2017; Kartini et al., 2021). Chromatographic techniques and molecular profiling are recent studies that standardize the extraction of herbs, considering bioactive compounds that are important for wound healing(Kartini et al., 2021; Negi et al., 2012) . Standardization of the precise level of all medicinal plants in a complex matrix, such as individual plants, and the environmental effects on the chemistry of the plant seem challenging(Ghosh & Gaba, 2013; Maver et al., 2019). Moreover, the extraction process is often standardized and based on well-developed analytical techniques (i.e., HPLC and MS), which are generally expensive and resource-intensive(Azmin et al., 2016; Sun et al., 2025). Therefore, there is a need for cheaper and more easily accessible means of standardizing herbal extracts for clinical and commercial uses.

9.3. Herb-Drug Interactions

The most important concern regarding the use of phytoconstituents as adjuncts to conventional pharmaceuticals is herb–drug interaction. Many of these medicinal plants contain bioactive compounds that can interact with drug-metabolizing enzymes, particularly CYPs, which alter drug metabolism and subsequently reduce the efficacy and/or increase toxicity(Cho & Yoon, 2015; Fasinu et al., 2012; Gowder,

2013). For example, CYP3A4 is responsible for the metabolism of the majority of pharmaceutical drugs, and Ginkgo biloba and St. John's Wort interact with it, while the CYP2D6 enzyme, Ginkgo biloba interacts with this as well. Interactions between food and drugs can increase or decrease drug concentrations in the bloodstream, altering the expected therapeutic outcome or causing adverse effects(Abad et al., 2010; Ohnishi et al., 2003; Saxena et al., 2008) . Reviewing the propolis herb-drug interactions, it was reported that propolis can inhibit certain CYP450 enzymes, leading to the alteration of the pharmacokinetics of the drugs administered concurrently(R. Hossain et al., 2022; S. Hossain et al., 2022) . For example, some reactions can be beneficial, such as enhancing the therapeutic potency of certain chemotherapeutic agents, whereas others can be detrimental, leading to toxicity or therapeutic failure. There is a potential for herb-drug interactions; therefore, patients taking conventional drugs alongside herbal products require careful monitoring(Knecht et al., 2020; Okem et al., 2022). The potential for such interactions suggests that patients should inform their healthcare providers about their use of herbal supplements, and healthcare providers should consider these adverse interactions when prescribing medications, as per clinical guidelines(Hermann & Richter, 2012; Tirona & Bailey, 2006). However, further research is needed to elucidate the mechanisms underlying herb drug interactions, especially for phytoconstituents with complex pharmacokinetics.

CONCLUSION

Medicinal plants are potential sources of phytoconstituents with significant therapeutic potential for the treatment of diabetic foot ulcers (DFUs). Among the various groups of molecules with healing properties are polyphenols, flavonoids, alkaloids, and terpenoids, which offer multiple actions to overcome the hurdles of healing, such as chronic inflammation, oxidative effects, infection, and suppressed angiogenesis. Curcumin, resveratrol, and asiaticoside phytoconstituents can modulate the inflammatory pathway, neutralize ROS, induce angiogenesis, and exert antimicrobial effects, all of which are important for DFU management. Although these effects seem promising, problems with bioavailability, maintaining extract standardization, and herb-drug interactions exist. Consequently, there is a need for continued research to fully understand the mechanism(s) of action of these compounds, optimize their clinical use, and guarantee the safety and efficacy of their administration for DFUs treatment. However, further clinical validation is necessary based on well-designed trials to establish whether phytoconstituents can be used as treatments for chronic wounds and, if so, whether they are reliable and effective treatments, leading to improved patient outcomes and reduced healthcare burden.

Funding: This chapter received no external funding.

Institutional Review Board Statement: Not applicable
Informed Consent Statement: Not applicable
Data Availability Statement: All the relevant data included in the manuscript.

ACKNOWLEDGMENTS

The authors express their sincere gratitude Aarupadai Veedu Medical College & Hospital, Vinayaka Mission's Research Foundation (Deemed to be University), Puducherry, Karaikal, and Guru Nanak College (Autonomous), Chennai 600042,

Conflicts of Interest: No conflicts of interest.

REFERENCES

Al-Knani, Z. M. I., Ashoor, A. S., & Hussein, A. N. (2023). Evaluation of antimicrobial activity of Allium sativum and its synergistic effects with antibiotics in the treatment of diabetic foot ulcers. *Journal of Ethnopharmacology*, *280*, 114309. DOI: 10.1016/j.jep.2023.114309

Abad, M. J., Bedoya, L. M., & Bermejo, P. (2010). An Update on Drug Interactions with the Herbal Medicine Ginkgo biloba. *Current Drug Metabolism*, *11*(2), 171–181. DOI: 10.2174/138920010791110818 PMID: 20359286

Addis, R., Cruciani, S., Santaniello, S., Bellu, E., Sarais, G., Ventura, C., Maioli, M., & Pintore, G. (2020). Fibroblast Proliferation and Migration in Wound Healing by Phytochemicals: Evidence for a Novel Synergic Outcome. *International Journal of Medical Sciences*, *17*(8), 1030–1042. DOI: 10.7150/ijms.43986 PMID: 32410832

. Adom, M. B., Taher, M., Mutalabisin, M. F., Amri, M. S., Kudos, M. B. A., Sulaiman, M. W. A. W., Sengupta, P., & Darnis, D. S. (2017). Chemical constituents and medical benefits of Plantago major [Review of *Chemical constituents and medical benefits of Plantago major*]. *Biomedicine & Pharmacotherapy, 96*, 348. Elsevier BV. DOI: 10.1016/j.biopha.2017.09.152

Afonso, A., Oliveira, D., Saavedra, M. J., Borges, A., & Simões, M. (2021). Biofilms in Diabetic Foot Ulcers: Impact, Risk Factors and Control Strategies. *International Journal of Molecular Sciences*, *22*(15), 8278. DOI: 10.3390/ijms22158278 PMID: 34361044

Ahmadian, R., Bahramsoltani, R., Marques, A. M., Rahimi, R., & Farzaei, M. H. (2021). Medicinal Plants as Efficacious Agents for Diabetic Foot Ulcers: A Systematic Review of Clinical Studies. *Wounds : a Compendium of Clinical Research and Practice*, *33*(6), 207–218. DOI: 10.25270/wnds/2021.207218 PMID: 34357879

Almadani, Y., Vorstenbosch, J., Davison, P., & Murphy, A. (2021). Wound Healing: A Comprehensive Review. *Seminars in Plastic Surgery*, *35*(3), 141–144. DOI: 10.1055/s-0041-1731791 PMID: 34526860

Aly, S. H., Elissawy, A. M., Mahmoud, A. M. A., El-Tokhy, F. S., Mageed, S. S. A., Almahli, H., Al-Rashood, S. T., Binjubair, F. A., Hassab, M. A. E., Eldehna, W. M., & Singab, A. N. B. (2023). Synergistic Effect of Sophora japonica and Glycyrrhiza glabra Flavonoid-Rich Fractions on Wound Healing: In Vivo and Molecular Docking Studies. *Molecules (Basel, Switzerland)*, *28*(7), 2994. DOI: 10.3390/molecules28072994 PMID: 37049756

Alyoussef, A., El-Gogary, R. I., Ahmed, R. F., Ahmed-Farid, O. A., Bakeer, R. M., & Nasr, M. (2021). The beneficial activity of curcumin and resveratrol loaded in nanoemulgel for healing of burn-induced wounds. *Journal of Drug Delivery Science and Technology*, *62*, 102360. DOI: 10.1016/j.jddst.2021.102360

. Ambika, A. P., & Nair, S. N. (2018). Wound Healing Activity of Plants from the Convolvulaceae Family [Review of *Wound Healing Activity of Plants from the Convolvulaceae Family*]. *Advances in Wound Care, 8*(1), 28. Mary Ann Liebert, Inc. DOI: 10.1089/wound.2017.0781

Anand, S., Rajinikanth, P. S., Arya, D. K., Pandey, P., Gupta, R., Sankhwar, R., & Chidambaram, K. (2022). Multifunctional Biomimetic Nanofibrous Scaffold Loaded with Asiaticoside for Rapid Diabetic Wound Healing. *Pharmaceutics, 14*(2), 273. DOI: 10.3390/pharmaceutics14020273 PMID: 35214006

Ashraf, M. V., Pant, S., Khan, M. A. H., Shah, A. A., Siddiqui, S., Jeridi, M., Alhamdi, H. W., & Ahmad, S. (2023). Phytochemicals as Antimicrobials: Prospecting Himalayan Medicinal Plants as Source of Alternate Medicine to Combat Antimicrobial Resistance. *Pharmaceuticals (Basel, Switzerland)*, *16*(6), 881. DOI: 10.3390/ph16060881 PMID: 37375828

Azmin, S. N. H. M., Manan, Z. A., Alwi, S. R. W., Chua, L. S., Mustaffa, A. A., & Yunus, N. A. (2016). Herbal Processing and Extraction Technologies. *Separation and Purification Reviews*, *45*(4), 305–320. DOI: 10.1080/15422119.2016.1145395

. Baig, M. S., Banu, A., Zehravi, M., Rana, R., Burle, S. S., Khan, S. L., Islam, F., Siddiqui, F. A., Massoud, E. E. S., Rahman, Md. H., & Cavalu, S. (2022). An Overview of Diabetic Foot Ulcers and Associated Problems with Special Emphasis on Treatments with Antimicrobials [Review of *An Overview of Diabetic Foot Ulcers and Associated Problems with Special Emphasis on Treatments with Antimicrobials*]. *Life, 12*(7), 1054. Multidisciplinary Digital Publishing Institute. DOI: 10.3390/life12071054

Baltzis, D., Eleftheriadou, I., & Veves, A. (2014). Pathogenesis and Treatment of Impaired Wound Healing in Diabetes Mellitus: New Insights. *Advances in Therapy*, *31*(8), 817–836. DOI: 10.1007/s12325-014-0140-x PMID: 25069580

. Bayramov, D. F., & Neff, J. A. (2016). Beyond conventional antibiotics — New directions for combination products to combat biofilm [Review of *Beyond conventional antibiotics—New directions for combination products to combat biofilm*]. *Advanced Drug Delivery Reviews, 112*, 48. Elsevier BV. DOI: 10.1016/j.addr.2016.07.010

Bi, H., Li, H., Zhang, C., Mao, Y., Nie, F., Xing, Y., Sha, W., Wang, X., Irwin, D. M., & Tan, H. (2019). Stromal vascular fraction promotes migration of fibroblasts and angiogenesis through regulation of extracellular matrix in the skin wound healing process. *Stem Cell Research & Therapy, 10*(1), 302. DOI: 10.1186/s13287-019-1415-6 PMID: 31623669

Botusan, I. R., Sunkari, V. G., Savu, O., Catrina, A. I., Grünler, J., Lindberg, S., Pereira, T., Ylä-Herttuala, S., Poellinger, L., Brismar, K., & Catrina, S. (2008). Stabilization of HIF-1α is critical to improve wound healing in diabetic mice. *Proceedings of the National Academy of Sciences of the United States of America, 105*(49), 19426–19431. DOI: 10.1073/pnas.0805230105 PMID: 19057015

Boulton, A. J. M., Armstrong, D. G., Hardman, M. J., Malone, M., Embil, J. M., Attinger, C. E., Lipsky, B. A., Aragón-Sánchez, J., Kwong, H., Schultz, G. S., & Kirsner, R. S. (2020). Diagnosis and Management of Diabetic Foot Infections. *ADA Clinical Compendia, 2020*(1), 1–24. DOI: 10.2337/db2020-01 PMID: 32105420

. Brown, M. S., Ashley, B. K., & Koh, A. (2018). Wearable Technology for Chronic Wound Monitoring: Current Dressings, Advancements, and Future Prospects [Review of *Wearable Technology for Chronic Wound Monitoring: Current Dressings, Advancements, and Future Prospects*]. *Frontiers in Bioengineering and Biotechnology, 6*. Frontiers Media. DOI: 10.3389/fbioe.2018.00047

Burgess, J. L., Wyant, W. A., Abujamra, B. A., Kirsner, R. S., & Jozic, I. (2021). Diabetic Wound-Healing Science. *Medicina, 57*(10), 1072. DOI: 10.3390/medicina57101072 PMID: 34684109

Caldwell, R. B., Bartoli, M., Behzadian, M. A., El-Remessy, A. B., Al-Shabrawey, M., Platt, D., Liou, G. I., & Caldwell, R. W. (2005). Vascular Endothelial Growth Factor and Diabetic Retinopathy: Role of Oxidative Stress. *Current Drug Targets, 6*(4), 511–524. DOI: 10.2174/1389450054021981 PMID: 16026270

Calver, A., Collier, J., & Vallance, P. (1992). Inhibition and stimulation of nitric oxide synthesis in the human forearm arterial bed of patients with insulin-dependent diabetes. *The Journal of Clinical Investigation, 90*(6), 2548–2554. DOI: 10.1172/JCI116149 PMID: 1469103

Chanu, N. R., Gogoi, P., Barbhuiya, P. A., Dutta, P. P., Pathak, M. P., & Sen, S. (2023). Natural Flavonoids as Potential Therapeutics in the Management ofDiabetic Wound: A Review. *Current Topics in Medicinal Chemistry*, *23*(8), 690–710. DOI: 10.2174/1568026623666230419102140 PMID: 37114791

Chaturvedi, S., Agrawal, S., Garg, A., & Rastogi, V. (2022). Potential of Nano-encapsulated Quercetin Topical Formulations in the Management of Diabetic Foot Ulcer. *Revista Brasileira de Farmacognosia*, *33*(3), 484–501. DOI: 10.1007/s43450-022-00345-8

. Chávez-Reyes, J., Escárcega-González, C. E., Chavira-Suárez, E., León-Buitimea, A., Vázquez-León, P., Morones-Ramírez, J. R., Villalón, C. M., Quintanar-Stephano, A., & Marichal-Cancino, B. A. (2021). Susceptibility for Some Infectious Diseases in Patients With Diabetes: The Key Role of Glycemia [Review of *Susceptibility for Some Infectious Diseases in Patients With Diabetes: The Key Role of Glycemia*]. *Frontiers in Public Health, 9*. Frontiers Media. DOI: 10.3389/fpubh.2021.559595

. Chen, L., & Wen, Y. (2011). The role of bacterial biofilm in persistent infections and control strategies [Review of *The role of bacterial biofilm in persistent infections and control strategies*]. *International Journal of Oral Science, 3*(2), 66. Springer Nature. DOI: 10.4248/IJOS11022

Chittasupho, C., Manthaisong, A., Okonogi, S., Tadtong, S., & Samee, W. (2021). Effects of Quercetin and Curcumin Combination on Antibacterial, Antioxidant, In Vitro Wound Healing and Migration of Human Dermal Fibroblast Cells. *International Journal of Molecular Sciences*, *23*(1), 142. DOI: 10.3390/ijms23010142 PMID: 35008566

Cho, H., & Yoon, I. (2015). Pharmacokinetic Interactions of Herbs with Cytochrome P450 and P-Glycoprotein. *Evidence-Based Complementary and Alternative Medicine : eCAM*, *1*, 1–10. Advance online publication. DOI: 10.1155/2015/736431 PMID: 25632290

Comino-Sanz, I. M., López-Franco, M. D., Castro, B., & Pancorbo-Hidalgo, P. L. (2021). The Role of Antioxidants on Wound Healing: A Review of the Current Evidence. *Journal of Clinical Medicine*, *10*(16), 3558. DOI: 10.3390/jcm10163558 PMID: 34441854

. Cui, Q., & Wang, H. (2025). Resveratrol in Dermatological Therapy: A Critical Review of Mechanisms, Delivery Innovations, and Clinical Frontiers [Review of *Resveratrol in Dermatological Therapy: A Critical Review of Mechanisms, Delivery Innovations, and Clinical Frontiers*]. *Clinical Cosmetic and Investigational Dermatology*, 2229. Dove Medical Press. DOI: 10.2147/CCID.S543849

Dasari, N., Jiang, A., Skochdopole, A. J., Chung, J., Reece, E. M., Vorstenbosch, J., & Winocour, S. (2021). Updates in Diabetic Wound Healing, Inflammation, and Scarring. *Seminars in Plastic Surgery, 35*(3), 153–158. DOI: 10.1055/s-0041-1731460 PMID: 34526862

. Dawi, J., Tumanyan, K., Tomas, K., Misakyan, Y., Gargaloyan, A., Gonzalez, E. R., Hammi, M., Tомас, C., & Venketaraman, V. (2025). Diabetic Foot Ulcers: Pathophysiology, Immune Dysregulation, and Emerging Therapeutic Strategies [Review of *Diabetic Foot Ulcers: Pathophysiology, Immune Dysregulation, and Emerging Therapeutic Strategies*]. *Biomedicines, 13*(5), 1076. Multidisciplinary Digital Publishing Institute. DOI: 10.3390/biomedicines13051076

Demidova-Rice, T. N., Hamblin, M. R., & Herman, I. M. (2012). Acute and Impaired Wound Healing. *Advances in Skin & Wound Care, 25*(7), 304–314. DOI: 10.1097/01.ASW.0000416006.55218.d0 PMID: 22713781

Deng, L., Du, C., Song, P., Chen, T., Rui, S., Armstrong, D. G., & Deng, W. (2021). The Role of Oxidative Stress and Antioxidants in Diabetic Wound Healing. *Oxidative Medicine and Cellular Longevity, 2021*(1), 8852759. DOI: 10.1155/2021/8852759 PMID: 33628388

. Devi, S., Kumar, V., Singh, S. K., Dubey, A. K., & Kim, J.-J. (2021). Flavonoids: Potential Candidates for the Treatment of Neurodegenerative Disorders [Review of *Flavonoids: Potential Candidates for the Treatment of Neurodegenerative Disorders*]. *Biomedicines, 9*(2), 99. Multidisciplinary Digital Publishing Institute. DOI: 10.3390/biomedicines9020099

Diller, R. B., & Tabor, A. J. (2022). The Role of the Extracellular Matrix (ECM) in Wound Healing: A Review. *Biomimetics, 7*(3), 87. DOI: 10.3390/biomimetics7030087 PMID: 35892357

Ding, Y., Yang, P., Li, S., Zhang, H., Ding, X., & Tan, Q. (2022). Resveratrol accelerates wound healing by inducing M2 macrophage polarisation in diabetic mice. *Pharmaceutical Biology, 60*(1), 2328–2337. DOI: 10.1080/13880209.2022.2149821 PMID: 36469602

Dunnill, C., Patton, T., Brennan, J., Barrett, J. N., Dryden, M., Cooke, J., Leaper, D., & Georgopoulos, N. T. (2015). Reactive oxygen species (ROS) and wound healing: The functional role of ROS and emerging ROS-modulating technologies for augmentation of the healing process. *International Wound Journal, 14*(1), 89–96. DOI: 10.1111/iwj.12557 PMID: 26688157

Egea, G., Jiménez-Altayó, F., & Campuzano, V. (2020). Reactive Oxygen Species and Oxidative Stress in the Pathogenesis and Progression of Genetic Diseases of the Connective Tissue. *Antioxidants*, *9*(10), 1013. DOI: 10.3390/antiox9101013 PMID: 33086603

Elajaili, H., Lyttle, B. D., Lewis, C., Bardill, J. R., Dee, N., Seal, S., Nozik-Grayck, E., Liechty, K. W., & Zgheib, C. (2025). Increased ROS and Persistent Pro-Inflammatory Responses in a Diabetic Wound Healing Model (db/db): Implications for Delayed Wound Healing. *International Journal of Molecular Sciences*, *26*(10), 4884. DOI: 10.3390/ijms26104884 PMID: 40430024

. El-Sherbeni, S. A., & Negm, W. A. (2023). The wound healing effect of botanicals and pure natural substances used in in vivo models [Review of *The wound healing effect of botanicals and pure natural substances used in in vivo models*]. *Inflammopharmacology, 31*(2), 755. Springer Nature. DOI: 10.1007/s10787-023-01157-5

El-Zahar, K. M., Al-Jamaan, M. E., Al-Mutairi, F. R., Al-Hudiab, A. M., Al-Einzi, M. S., & Mo, A. A.-Z. M. (2022). Antioxidant, Antibacterial, and Antifungal Activities of the Ethanolic Extract Obtained from Berberis vulgaris Roots and Leaves. *Molecules (Basel, Switzerland)*, *27*(18), 6114. DOI: 10.3390/molecules27186114 PMID: 36144846

Everett, E., & Mathioudakis, N. (2018). Update on management of diabetic foot ulcers. *Annals of the New York Academy of Sciences*, *1411*(1), 153–165. DOI: 10.1111/nyas.13569 PMID: 29377202

Fu, X., Chen, H., Zhang, X., & Zhang, Z. (2025). Resveratrol: A potent antioxidant and anti-inflammatory compound in the treatment of chronic wounds. *Antioxidants*, *14*(3), 210. DOI: 10.3390/antiox14030210

Famurewa, A. C., Maduagwuna, E. K., Folawiyo, A. M., Besong, E. E., Eteudo, A. N., Famurewa, O. A., & Ejezie, F. E. (2019). Antioxidant, anti-inflammatory, and antiapoptotic effects of virgin coconut oil against antibiotic drug gentamicin-induced nephrotoxicity via the suppression of oxidative stress and modulation of iNOS/NF-κB/caspase-3 signaling pathway in Wistar rats. *Journal of Food Biochemistry*, *44*(1). Advance online publication. DOI: 10.1111/jfbc.13100 PMID: 31721240

Farkhondeh, T., Llorens, S., Pourbagher-Shahri, A. M., Ashrafizadeh, M., & Samarghandian, S. (2020). The therapeutic effect of resveratrol: Focusing on the Nrf2 signaling pathway. *Biomedicine and Pharmacotherapy*, *127*, 110234. DOI: 10.1016/j.biopha.2020.110234 PMID: 32559855

Fasinu, P. S., Bouic, P., & Rosenkranz, B. (2012). An Overview of the Evidence and Mechanisms of Herb–Drug Interactions. *Frontiers in Pharmacology, 3*. Advance online publication. DOI: 10.3389/fphar.2012.00069 PMID: 22557968

Gawali, V. (2021). Antihyperglycemic and antimicrobial activity of Berberis vulgaris in the treatment of diabetic foot ulcers. *Phytotherapy Research, 35*(5), 1863–1872. DOI: 10.1002/ptr.6907

. Gaikwad, A. R., Ahire, K. D., Gosavi, A. A., Salunkhe, K. S., & Khalkar, A. (2021). Gaikwad Abhijeet R. Phytosome as a Novel Drug Delivery System for Bioavailability Enhancement of Phytoconstituents and its Applications: A Review [Review of *Gaikwad Abhijeet R. Phytosome as a Novel Drug Delivery System for Bioavailability Enhancement of Phytoconstituents and its Applications: A Review*]. *Journal of Drug Delivery and Therapeutics, 11*(3), 138. Society of Pharmaceutical Tecnocrats. DOI: 10.22270/jddt.v11i3.4847

Ghanadian, M., Soltani, R., Homayouni, A., Khorvash, F., Jouabadi, S. M., & Abdollahzadeh, M. (2022). The Effect of Plantago major Hydroalcoholic Extract on the Healing of Diabetic Foot and Pressure Ulcers: A Randomized Open-Label Controlled Clinical Trial. *The International Journal of Lower Extremity Wounds, 23*(3), 475–481. DOI: 10.1177/15347346211070723 PMID: 35044254

. Ghosh, P., & Gaba, A. (2013). Phyto-Extracts in Wound Healing [Review of *Phyto-Extracts in Wound Healing*]. *Journal of Pharmacy & Pharmaceutical Sciences, 16*(5), 760. Canadian Society for Pharmaceutical Sciences. DOI: 10.18433/J3831V

Giacco, F., & Brownlee, M. (2010). Oxidative Stress and Diabetic Complications. *Circulation Research, 107*(9), 1058–1070. DOI: 10.1161/CIRCRESAHA.110.223545 PMID: 21030723

Gohil, K., Patel, J., & Gajjar, A. (2010). Pharmacological review on Centella asiatica: A potential herbal cure-all. *Indian Journal of Pharmaceutical Sciences, 72*(5), 546. DOI: 10.4103/0250-474X.78519 PMID: 21694984

Gonzales, A. M., & Orlando, R. A. (2008). Curcumin and resveratrol inhibit nuclear factor-kappaB-mediated cytokine expression in adipocytes. *Nutrition & Metabolism, 5*(1), 17. DOI: 10.1186/1743-7075-5-17 PMID: 18549505

Gowder, S. J. (2013). New Insights into Toxicity and Drug Testing. In *InTech eBooks*. DOI: 10.5772/55886

Guan, Y., Niu, H., Liu, Z., Dang, Y., Shen, J., Zayed, M. A., Ma, L., & Guan, J. (2021). Sustained oxygenation accelerates diabetic wound healing by promoting epithelialization and angiogenesis and decreasing inflammation. *Science Advances*, *7*(35), eabj0153. Advance online publication. DOI: 10.1126/sciadv.abj0153 PMID: 34452918

Guimarães, I., Baptista-Silva, S., Pintado, M., & Oliveira, A. L. (2021). Polyphenols: A Promising Avenue in Therapeutic Solutions for Wound Care. *Applied Sciences (Basel, Switzerland)*, *11*(3), 1230. DOI: 10.3390/app11031230

Guo, Y., Niu, W., Jiao, H., Li, Y., Xu, C., Zhou, X., & Wang, J. (2025). Crosstalk between oxidative stress and inflammatory pathways: Natural therapeutic approaches for diabetic wound healing. *World Journal of Diabetes*, *16*(11), 111400. DOI: 10.4239/wjd.v16.i11.111400 PMID: 41278434

Gurley, B. J., Fifer, E. K., & Gardner, Z. (2012). Pharmacokinetic Herb-Drug Interactions (Part 2): Drug Interactions Involving Popular Botanical Dietary Supplements and Their Clinical Relevance [Review of *Pharmacokinetic Herb-Drug Interactions (Part 2): Drug Interactions Involving Popular Botanical Dietary Supplements and Their Clinical Relevance*]. *Planta Medica*, *78*(13), 1490. Thieme Medical Publishers (Germany). DOI: 10.1055/s-0031-1298331

Hakim, R. F., Fakhrurrazi, F., & Dinni, . (2019). Effect of Carica papaya Extract toward Incised Wound Healing Process in Mice (Mus musculus) Clinically and Histologically. *Evidence-Based Complementary and Alternative Medicine : eCAM*, *1*, 1–5. Advance online publication. DOI: 10.1155/2019/8306519 PMID: 31827564

Halim, A. S., Khoo, T. L., & Saad, A. Z. M. (2012). Wound bed preparation from a clinical perspective. *Indian Journal of Plastic Surgery : Official Publication of the Association of Plastic Surgeons of India*, *45*(2), 193–202. DOI: 10.4103/0970-0358.101277 PMID: 23162216

Hanna, P., Kochi, S. K., & Collier, R. J. (1992). Biochemical and physiological changes induced by anthrax lethal toxin in J774 macrophage-like cells. *Molecular Biology of the Cell*, *3*(11), 1269–1277. DOI: 10.1091/mbc.3.11.1269 PMID: 1457831

Hazlett, L. D., Ekanayaka, S. A., McClellan, S. A., & Francis, R. (2019). Glycyrrhizin Use for Multi-Drug Resistant Pseudomonas aeruginosa: In Vitro and In Vivo Studies. *Investigative Ophthalmology & Visual Science*, *60*(8), 2978. DOI: 10.1167/iovs.19-27200 PMID: 31311033

Hegazi, S., Aly, R., Mesilhy, R., & Aljohary, H. (2024). Diabetic Foot Ulcer Wound Healing and Tissue Regeneration: Signaling Pathways and Mechanisms. In *IntechOpen eBooks*. IntechOpen. DOI: 10.5772/intechopen.1004267

Helal, M. A., Elwakeel, H., Sorogy, M. E., Abdelmaksoud, M. A., & Abdelgawad, M. S. (2024). Effectiveness of autologous leucocyte and platelet-rich fibrin (L-PRF) in management of chronic diabetic foot ulcers. *The Egyptian Journal of Surgery*, *43*(4), 1540–1547. DOI: 10.21608/EJSUR.2024.297983.1104

Herman, A., Herman, A. P., Herman, A. P., & Herman, A. P. (2023). Herbal Products and Their Active Constituents for Diabetic Wound Healing—Preclinical and Clinical Studies. *Systematic Reviews in Pharmacy*, *15*(1), 281. DOI: 10.3390/pharmaceutics15010281 PMID: 36678910

Hermann, R., & von Richter, O. (2012). Clinical Evidence of Herbal Drugs As Perpetrators of Pharmacokinetic Drug Interactions [Review of *Clinical Evidence of Herbal Drugs As Perpetrators of Pharmacokinetic Drug Interactions*]. *Planta Medica, 78*(13), 1458. Thieme Medical Publishers (Germany). DOI: 10.1055/s-0032-1315117

Herrero-Cervera, A., Soehnlein, O., & Kenne, E. (2022). Neutrophils in chronic inflammatory diseases. *Cellular & Molecular Immunology*, *19*(2), 177–191. DOI: 10.1038/s41423-021-00832-3 PMID: 35039631

. Hong, W. X., Hu, M. S., Esquivel, M., Liang, G., Rennert, R. C., McArdle, A., Paik, K. J., Duscher, D., Gurtner, G. C., Lorenz, H. P., & Longaker, M. T. (2014). The Role of Hypoxia-Inducible Factor in Wound Healing [Review of *The Role of Hypoxia-Inducible Factor in Wound Healing*]. *Advances in Wound Care, 3*(5), 390. Mary Ann Liebert, Inc. DOI: 10.1089/wound.2013.0520

Hossain, C. M., Gera, M., & Ali, K. A. (2022). CURRENT STATUS AND CHALLENGES OF HERBAL DRUG DEVELOPMENT AND REGULATORY ASPECT: A GLOBAL PERSPECTIVE. *Asian Journal of Pharmaceutical and Clinical Research*, *31*, 31–41. Advance online publication. DOI: 10.22159/ajpcr.2022.v15i12.46134

Hossain, R., Quispe, C., Khan, R. A., Saikat, A. S. M., Ray, P., Ongalbek, D., Yeskaliyeva, B., Jain, D., Smeriglio, A., Trombetta, D., Kiani, R., Kobarfard, F., Mojgani, N., Saffarian, P., Ayatollahi, S. A., Sarkar, C. K., Islam, M. T., Keriman, D., Uçar, A., & Cho, W. C. (2022). Propolis: An update on its chemistry and pharmacological applications. *Chinese Medicine*, *17*(1), 100. Advance online publication. DOI: 10.1186/s13020-022-00651-2 PMID: 36028892

Hossain, S., Yousaf, M., Liu, Y., Chang, D., & Zhou, X. (2022). An Overview of the Evidence and Mechanism of Drug–Herb Interactions Between Propolis and Pharmaceutical Drugs. *Frontiers in Pharmacology*, *13*, 876183. DOI: 10.3389/fphar.2022.876183 PMID: 35444531

Hou, Q., Li, M., Lu, Y., Liu, D., & Li, C.-C. (2016). Burn wound healing properties of asiaticoside and madecassoside. *Experimental and Therapeutic Medicine*, *12*(3), 1269–1274. DOI: 10.3892/etm.2016.3459 PMID: 27588048

Huang, X., Sun, J., Chen, G., Niu, C., Wang, Y., Congcong, Z., Sun, J., Huang, H., Huang, S., Liang, Y., Shen, Y., Cong, W., Jin, L., & Zhu, Z. (2019). Resveratrol Promotes Diabetic Wound Healing via SIRT1-FOXO1-c-Myc Signaling Pathway-Mediated Angiogenesis. *Frontiers in Pharmacology*, *10*, 421. DOI: 10.3389/fphar.2019.00421 PMID: 31068817

Hung, C., Hsiao, C., Hsieh, W.-H., Li, H., Tsai, Y., Lin, C., Chang, H., & Wu, N. (2017). 18ß-glycyrrhetinic acid derivative promotes proliferation, migration and aquaporin-3 expression in human dermal fibroblasts. *PLoS One*, *12*(8), e0182981. Advance online publication. DOI: 10.1371/journal.pone.0182981 PMID: 28813533

Hunt, M., Torres, M. S., Bachar-Wikström, E., & Wikström, J. D. (2024). Cellular and molecular roles of reactive oxygen species in wound healing. *Communications Biology*, *7*(1), 1534. DOI: 10.1038/s42003-024-07219-w PMID: 39562800

Hutchings, G., Kruszyna, Ł., Nawrocki, M. J., Strauss, E., Bryl, R., Spaczyńska, J., Perek, B., Jemielity, M., Mozdziak, P., Kempisty, B., Nowicki, M., & Krasiński, Z. (2021). Molecular Mechanisms Associated with ROS-Dependent Angiogenesis in Lower Extremity Artery Disease. *Antioxidants*, *10*(5), 735. DOI: 10.3390/antiox10050735 PMID: 34066926

Jagiełło, K., Uchańska, O., Matyja, K., Jackowski, M., Wiatrak, B., Kubasiewicz–Ross, P., & Karuga-Kuźniewska, E. (2023). Supporting the Wound Healing Process—Curcumin, Resveratrol and Baicalin in In Vitro Wound Healing Studies. *Pharmaceuticals (Basel, Switzerland)*, *16*(1), 82. DOI: 10.3390/ph16010082 PMID: 36678579

Jen, M. C., Serrano, M. C., van Lith, R., & Ameer, G. A. (2011). Polymer-Based Nitric Oxide Therapies: Recent Insights for Biomedical Applications. *Advanced Functional Materials*, *22*(2), 239–260. DOI: 10.1002/adfm.201101707 PMID: 25067935

. Jiang, P., Li, Q., Yanhong, L., Luo, F., Che, Q., Lu, Z., Yang, S., Yang, Y., Chen, X., & Cai, Y. (2023). Current status and progress in research on dressing management for diabetic foot ulcer [Review of *Current status and progress in research on dressing management for diabetic foot ulcer*]. *Frontiers in Endocrinology, 14*. Frontiers Media. DOI: 10.3389/fendo.2023.1221705

Jiang, X., Zeng, Y., Li, C., Wang, K., & Yu, D. (2024). Enhancing diabetic wound healing: Advances in electrospun scaffolds from pathogenesis to therapeutic applications. *Frontiers in Bioengineering and Biotechnology*, *12*, 1354286. Advance online publication. DOI: 10.3389/fbioe.2024.1354286 PMID: 38375451

Johansen, J. S., Harris, A. K., Rychly, D. J., & Ergul, A. (2005). Oxidative stress and the use of antioxidants in diabetes: Linking basic science to clinical practice [Review of *Oxidative stress and the use of antioxidants in diabetes: Linking basic science to clinical practice*]. *Cardiovascular Diabetology, 4*(1). BioMed Central. DOI: 10.1186/1475-2840-4-5

Johnson, K. E., & Wilgus, T. A. (2014). Vascular Endothelial Growth Factor and Angiogenesis in the Regulation of Cutaneous Wound Repair [Review of *Vascular Endothelial Growth Factor and Angiogenesis in the Regulation of Cutaneous Wound Repair*]. *Advances in Wound Care, 3*(10), 647. Mary Ann Liebert, Inc. DOI: 10.1089/wound.2013.0517

Kaleci, B., & Koyutürk, M. (2020). Efficacy of resveratrol in the wound healing process by reducing oxidative stress and promoting fibroblast cell proliferation and migration. *Dermatologic Therapy*, *33*(6). Advance online publication. DOI: 10.1111/dth.14357 PMID: 32996685

Kaludercic, N., & Lisa, F. D. (2020). Mitochondrial ROS Formation in the Pathogenesis of Diabetic Cardiomyopathy. *Frontiers in Cardiovascular Medicine*, *7*, 12. DOI: 10.3389/fcvm.2020.00012 PMID: 32133373

Kamarudin, N. A., Markom, M., & Latip, J. (2016). Effects of Solvents and Extraction Methods on Herbal Plants *Phyllanthus niruri, Orthosiphon stamineus* and *Labisia pumila. Indian Journal of Science and Technology*, *9*(21). Advance online publication. DOI: 10.17485/ijst/2016/v9i21/95235

Kant, V., Jangir, B. L., Kumar, V., Nigam, A., & Sharma, V. (2020). Quercetin accelerated cutaneous wound healing in rats by modulation of different cytokines and growth factors. *Growth Factors (Chur, Switzerland)*, *38*(2), 105–119. DOI: 10.1080/08977194.2020.1822830 PMID: 32957814

Kartini, K., Wati, N. I., Gustav, R., Wahyuni, R., Anggada, Y. F., Hidayani, R., Raharjo, A., Islamie, R., & Putra, S. E. D. (2021). Wound healing effects of Plantago major extract and its chemical compounds in hyperglycemic rats. *Food Bioscience*, *41*, 100937. DOI: 10.1016/j.fbio.2021.100937

Kenawy, E., Kamoun, E. A., Ghaly, Z. S., Shokr, A. M., El-Meligy, M. A., & Mahmoud, Y. A.-G. (2022). Novel Physically Cross-Linked Curcumin-Loaded PVA/Aloe vera Hydrogel Membranes for Acceleration of Topical Wound Healing: In Vitro and In Vivo Experiments. *Arabian Journal for Science and Engineering, 48*(1), 497–514. DOI: 10.1007/s13369-022-07283-6

Khamaisi, M., Katagiri, S., Keenan, H. A., Park, K., Maeda, Y., Li, Q., Qi, W., Thomou, T., Eschuk, D., Tellechea, A., Veves, A., Huang, C., Orgill, D. P., Wagers, A. J., & King, G. L. (2016). PKCδ inhibition normalizes the wound-healing capacity of diabetic human fibroblasts. *The Journal of Clinical Investigation, 126*(3), 837–853. DOI: 10.1172/JCI82788 PMID: 26808499

Khan, H., Ullah, H., Aschner, M., Cheang, W. S., & Akkol, E. K. (2019). Neuroprotective Effects of Quercetin in Alzheimer's Disease [Review of *Neuroprotective Effects of Quercetin in Alzheimer's Disease*]. *Biomolecules, 10*(1), 59. Multidisciplinary Digital Publishing Institute. DOI: 10.3390/biom10010059

Khare, T., Anand, U., Dey, A., Assaraf, Y. G., Chen, Z., Liu, Z., & Kumar, V. (2021). Exploring Phytochemicals for Combating Antibiotic Resistance in Microbial Pathogens [Review of *Exploring Phytochemicals for Combating Antibiotic Resistance in Microbial Pathogens*]. *Frontiers in Pharmacology, 12*. Frontiers Media. DOI: 10.3389/fphar.2021.720726

Knecht, K. T., Kinder, D. H., & Stockert, A. (2020). Biologically-Based Complementary and Alternative Medicine (CAM) Use in Cancer Patients: The Good, the Bad, the Misunderstood [Review of *Biologically-Based Complementary and Alternative Medicine (CAM) Use in Cancer Patients: The Good, the Bad, the Misunderstood*]. *Frontiers in Nutrition, 6*. Frontiers Media. DOI: 10.3389/fnut.2019.00196

Kumar, A., Kumar, B., Singh, S. K., Kaur, B., & Singh, S. (2017). A REVIEW ON PHYTOSOMES: NOVEL APPROACH FOR HERBAL PHYTOCHEMICALS [Review of *A REVIEW ON PHYTOSOMES: NOVEL APPROACH FOR HERBAL PHYTOCHEMICALS*]. *Asian Journal of Pharmaceutical and Clinical Research, 10*(10), 41. Innovare Academic Sciences. DOI: 10.22159/ajpcr.2017.v10i10.20424

Kumar, M., Keshwania, P., Chopra, S., Mahmood, S., & Bhatia, A. (2023). Therapeutic Potential of Nanocarrier-Mediated Delivery of Phytoconstituents for Wound Healing: Their Current Status and Future Perspective. *AAPS PharmSciTech, 24*(6), 155. DOI: 10.1208/s12249-023-02616-6 PMID: 37468691

Kumar, M., Kumar, D., Mahmood, S., Singh, V., Chopra, S., Hilles, A. R., & Bhatia, A. (2024). Nanotechnology-driven wound healing potential of asiaticoside: a comprehensive review [Review of *Nanotechnology-driven wound healing potential of asiaticoside: a comprehensive review*]. *RSC Pharmaceutics, 1*(1), 9. Royal Society of Chemistry. DOI: 10.1039/D3PM00024A

Kumar, P., Kumar, S., Udupa, E. P., Kumar, U., Rao, P., & Honnegowda, T. (2015). Role of angiogenesis and angiogenic factors in acute and chronic wound healing. *Plastic and Aesthetic Research, 2*(5), 243. DOI: 10.4103/2347-9264.165438

Kumar, S., Balachandran, V., Njoo, L., Anastasopoulos, A., & Tsvetkov, Y. (2023). Language Generation Models Can Cause Harm: So What Can We Do About It? *An Actionable Survey, 3299*, 3299–3321. Advance online publication. DOI: 10.18653/v1/2023.eacl-main.241

Kumari, A., Raina, N., Wahi, A., Goh, K. W., Sharma, P., Nagpal, R., Jain, A., Ming, L. C., & Gupta, M. (2022). Wound-Healing Effects of Curcumin and Its Nanoformulations: A Comprehensive Review. *Pharmaceutics, 14*(11), 2288. DOI: 10.3390/pharmaceutics14112288 PMID: 36365107

Kunkemoeller, B., & Kyriakides, T. R. (2017). Redox Signaling in Diabetic Wound Healing Regulates Extracellular Matrix Deposition [Review of *Redox Signaling in Diabetic Wound Healing Regulates Extracellular Matrix Deposition*]. *Antioxidants and Redox Signaling, 27*(12), 823. Mary Ann Liebert, Inc. DOI: 10.1089/ars.2017.7263

Li, H., Beg, O. U., Rafie, A. R., Kanwal, S., Ovalle-Cisneros, A., Faison, M. O., & Siddiqui, R. A. (2023). Characterization of Green and Yellow Papaya (Carica papaya) for Anti-Diabetic Activity in Liver and Myoblast Cells and Wound-Healing Activity in Fibroblast Cells. *Nutrients, 15*(8), 1929. DOI: 10.3390/nu15081929 PMID: 37111148

Liu, Y., Xie, J., & Yang, Z. (2021). Asiaticoside promotes angiogenesis and wound healing by upregulating VEGF expression in diabetic rats. *Journal of Ethnopharmacology, 267*, 113642. DOI: 10.1016/j.jep.2020.113642

Li, M., Hou, Q., Zhong, L., Zhao, Y., & Fu, X. (2021). Macrophage Related Chronic Inflammation in Non-Healing Wounds [Review of *Macrophage Related Chronic Inflammation in Non-Healing Wounds*]. *Frontiers in Immunology, 12*. Frontiers Media. DOI: 10.3389/fimmu.2021.681710

Li, M., Yu, H., Pan, H., Zhou, X., Ruan, Q., Kong, D., Chu, Z., Li, H., Huang, J., Huang, X., Chau, A., Xie, W., Ding, Y., & Yao, P. (2019). Nrf2 Suppression Delays Diabetic Wound Healing Through Sustained Oxidative Stress and Inflammation. *Frontiers in Pharmacology*, *10*, 1099. Advance online publication. DOI: 10.3389/fphar.2019.01099 PMID: 31616304

Li, X., Guan, C., He, Y., Wang, Y., Liu, X., & Zhou, X. (2016). Effects of Total Alkaloids ofSophora alopecuroideson Biofilm Formation inStaphylococcus epidermidis. *BioMed Research International*, *1*, 1–10. Advance online publication. DOI: 10.1155/2016/4020715 PMID: 27413745

Lee, S., Lee, J., & Lee, H. (2022). The role of Centella Asiatica in the treatment of chronic wounds and its effect on collagen synthesis. *Journal of Wound Care*, *31*(6), 350–355. DOI: 10.12968/jowc.2022.31.6.350

Lin, C.-M., Chiu, J., Wu, I.-H., Baowei, W., Pan, C.-M., & Chen, Y. (2009). Ferulic acid augments angiogenesis via VEGF, PDGF and HIF-1α. *The Journal of Nutritional Biochemistry*, *21*(7), 627–633. DOI: 10.1016/j.jnutbio.2009.04.001 PMID: 19443196

Liu, Y., Gong, P., Xiao, F., Shao, S., Zhao, D., Yan, M., & Yang, X. (2018). Chemical Constituents and Antioxidant, Anti-Inflammatory and Anti-Tumor Activities of Melilotus officinalis (Linn.) Pall. *Molecules (Basel, Switzerland)*, *23*(2), 271. DOI: 10.3390/molecules23020271 PMID: 29382154

Liu, Y., Liu, Y., Deng, J., Li, W., & Nie, X. (2021). Fibroblast Growth Factor in Diabetic Foot Ulcer: Progress and Therapeutic Prospects [Review of *Fibroblast Growth Factor in Diabetic Foot Ulcer: Progress and Therapeutic Prospects*]. *Frontiers in Endocrinology, 12*. Frontiers Media. DOI: 10.3389/fendo.2021.744868

Liu, Z., & Fang, Y. (2025). Wound healing and signaling pathways. *Open Life Sciences*, *20*(1), 20251166. DOI: 10.1515/biol-2025-1166 PMID: 40917783

Long, D. R., Mead, J., Hendricks, J. M., Hardy, M. E., & Voyich, J. M. (2013). 18β-Glycyrrhetinic acid inhibits methicillin-resistant Staphylococcus aureus survival and attenuates virulence gene expression. Antimicrobial agents and chemotherapy, 57(1), 241-247.

López-Díez, R., Egaña-Gorroño, L., Senatus, L. M., Shekhtman, A., Ramasamy, R., & Schmidt, A. M. (2021). Diabetes and Cardiovascular Complications: The Epidemics Continue [Review of *Diabetes and Cardiovascular Complications: The Epidemics Continue*]. *Current Cardiology Reports, 23*(7), 74. Springer Science+Business Media. DOI: 10.1007/s11886-021-01504-4

Lu, L., Ying, K., Wei, S.-M., Fang, Y., Liu, Y., Lin, H., Ma, L., & Mao, Y. (2004). Asiaticoside induction for cell-cycle progression, proliferation and collagen synthesis in human dermal fibroblasts. *International Journal of Dermatology, 43*(11), 801–807. DOI: 10.1111/j.1365-4632.2004.02047.x PMID: 15533060

Lu, Y., Zhao, D., Cao, G., Yin, S., Liu, C., Song, R., Ma, J., Sun, R., Wu, Z., Liu, J., Wu, P., & Wang, Y. (2024). Research progress on and molecular mechanism of vacuum sealing drainage in the treatment of diabetic foot ulcers. *Frontiers in Surgery, 11*, 1265360. Advance online publication. DOI: 10.3389/fsurg.2024.1265360 PMID: 38464666

Mamun, A. A., Shao, C., Geng, P., Wang, S., & Xiao, J. (2024). Recent advances in molecular mechanisms of skin wound healing and its treatments. *Frontiers in Immunology, 15*, 1395479. Advance online publication. DOI: 10.3389/fimmu.2024.1395479 PMID: 38835782

Manosalva, L., Mutis, A., Urzúa, A., Fajardo, V., & Quiróz, A.Forst and Study of Synergism with Ampicillin and Cephalothin. (2016). Antibacterial Activity of Alkaloid Fractions from Berberis microphylla G. Forst and Study of Synergism with Ampicillin and Cephalothin. *Molecules (Basel, Switzerland), 21*(1), 76. DOI: 10.3390/molecules21010076 PMID: 26760994

Martins, S. G., Zilhão, R., Þorsteinsdóttir, S., & Carlos, A. R. (2021). Linking Oxidative Stress and DNA Damage to Changes in the Expression of Extracellular Matrix Components [Review of *Linking Oxidative Stress and DNA Damage to Changes in the Expression of Extracellular Matrix Components*]. *Frontiers in Genetics, 12*. Frontiers Media. DOI: 10.3389/fgene.2021.673002

Mathi, P., Bokka, V., & Botlagunta, M. (2015). MEDICINAL USES AND BIOLOGICAL ACTIVITIES OF SOPHORA INTERRUPTA BEDD-A REVIEW. *International Journal of Pharmaceutics*, *5*(15), 265. https://www.pharmascholars.com/articles/medicinal-uses-and-biological-activities-of-sophora-interrupta-bedda-review.pdf

Matta-Gutiérrez, G., García-Morales, E., García-Álvarez, Y., Álvaro-Afonso, F. J., Molines-Barroso, R. J., & Lázaro-Martínez, J. L. (2021). The Influence of Multidrug-Resistant Bacteria on Clinical Outcomes of Diabetic Foot Ulcers: A Systematic Review. *Journal of Clinical Medicine, 10*(9), 1948. DOI: 10.3390/jcm10091948 PMID: 34062775

Maver, T., Kurečič, M., Smrke, D., Kleinschek, K. S., & Maver, U. (2019). Plant-Derived Medicines with Potential Use in Wound Treatment. In *IntechOpen eBooks*. IntechOpen., DOI: 10.5772/intechopen.72813

. McCarty, S., & Percival, S. L. (2013). Proteases and Delayed Wound Healing [Review of *Proteases and Delayed Wound Healing*]. *Advances in Wound Care, 2*(8), 438. Mary Ann Liebert, Inc. DOI: 10.1089/wound.2012.0370

McKay, T. B., Emmitte, K. A., German, C., & Karamichos, D. (2023). Quercetin and Related Analogs as Therapeutics to Promote Tissue Repair. *Bioengineering (Basel, Switzerland)*, *10*(10), 1127. DOI: 10.3390/bioengineering10101127 PMID: 37892857

Min, D.-H., Yu, Y.-B., Kim, T., Kim, H., & Lee, S. (2024). Pharmacological effects of pentacyclic triterpenoids isolated from Centella asiatica. *Horticulture, Environment and Biotechnology*, *65*(2), 189–197. DOI: 10.1007/s13580-023-00561-8

Mittal, M., Siddiqui, M. R., Tran, K. A., Reddy, S. P., & Malik, A. B. (2013). Reactive Oxygen Species in Inflammation and Tissue Injury. *Antioxidants & Redox Signaling*, *20*(7), 1126–1167. DOI: 10.1089/ars.2012.5149 PMID: 23991888

Mohan, Ch., Naresh, B., Kumar, B. K., Reddy, V. D., Manjula, P., Keerthi, B., Sreekanth, D., Manzelat, S. F., & Cherku, P. D. (2017). Micropropagation studies and phytochemical analysis of the endangered tree Commiphora wightii. *Journal of Applied Research on Medicinal and Aromatic Plants*, *6*, 70–79. DOI: 10.1016/j.jarmap.2017.02.004

Mohsin, F., Javaid, S., Tariq, M. H., & Mustafa, M. (2024). Molecular immunological mechanisms of impaired wound healing in diabetic foot ulcers (DFU), current therapeutic strategies and future directions. *International Immunopharmacology*, *139*, 112713. DOI: 10.1016/j.intimp.2024.112713 PMID: 39047451

Moses, R. L., Prescott, T. A. K., Mas-Claret, E., Steadman, R., Moseley, R., & Sloan, A. J. (2023). Evidence for Natural Products as Alternative Wound-Healing Therapies. *Biomolecules*, *13*(3), 444. DOI: 10.3390/biom13030444 PMID: 36979379

. Moura, J., Madureira, P., Leal, E. C., Fonseca, A. C., & Carvalho, E. (2019). Immune aging in diabetes and its implications in wound healing [Review of *Immune aging in diabetes and its implications in wound healing*]. *Clinical Immunology, 200*, 43. Elsevier BV. DOI: 10.1016/j.clim.2019.02.002

Mu, X., Chen, J., Zhu, H., Deng, J., Wu, X., He, W., Ye, P., Gu, R., Wu, Y., Han, F. Y., & Nie, X. (2025). Asiaticoside–nitric oxide synergistically accelerate diabetic wound healing by regulating key metabolites and SRC/STAT3 signaling. *Burns and Trauma*, *13*, tkaf009. Advance online publication. DOI: 10.1093/burnst/tkaf009 PMID: 40066291

Murdiana, H. E., Murwanti, R., Fakhrudin, N., & Ikawati, Z. (2024). Multi-target mechanism of polyherbal extract to treat diabetic foot ulcer based on network pharmacology and molecular docking. *Journal of Herbmed Pharmacology*, *13*(2), 289–299. DOI: 10.34172/jhp.2024.49362

Mustafa, I., Chin, N. L., Fakurazi, S., & Arulselvan, P. (2019). Comparison of Phytochemicals, Antioxidant and Anti-Inflammatory Properties of Sun-, Oven- and Freeze-Dried Ginger Extracts. *Foods*, *8*(10), 456. DOI: 10.3390/foods8100456 PMID: 31590464

Najafian, Y., Khorasani, Z. M., Najafi, M. N., Hamedi, S. S., Mahjour, M., & Feyzabadi, Z. (2018). Efficacy of Aloe vera/ Plantago Major Gel in Diabetic Foot Ulcer: A Randomized Double-Blind Clinical Trial. *Current Drug Discovery Technologies*, *16*(2), 223–231. DOI: 10.2174/1570163815666180115093007 PMID: 29336265

do Nascimento, M. H. M., & de Araújo, D. R. (2022). Exploring the Pharmacological Potential of Glycyrrhizic Acid: From Therapeutic Applications to Trends in Nanomedicine. *Future Pharmacology*, *2*(1), 1–15. DOI: 10.3390/futurepharmacol2010001

Negi, A., Sharma, N., & Singh, M. F. (2012). Formulation and Evaluation of an Herbal Anti-Inflammatory Gel Containing Eupatorium Leaves Extract. *Journal of Pharmacognosy and Phytochemistry*, *1*(4), 112. https://www.phytojournal.com/archives/2012/vol1issue4/PartA/16.1.pdf

Nie, X., Zhang, H., Shi, X., Zhao, J., Chen, Y., Wu, F., Yang, J., & Li, X. (2019). Asiaticoside nitric oxide gel accelerates diabetic cutaneous ulcers healing by activating Wnt/β-catenin signaling pathway. *International Immunopharmacology*, *79*, 106109. DOI: 10.1016/j.intimp.2019.106109 PMID: 31865242

Ning, S., Zang, J., Zhang, B., Feng, X., & Qiu, F. (2022). Botanical Drugs in Traditional Chinese Medicine With Wound Healing Properties [Review of *Botanical Drugs in Traditional Chinese Medicine With Wound Healing Properties*]. *Frontiers in Pharmacology, 13*. Frontiers Media. DOI: 10.3389/fphar.2022.885484

Nirenjen, S., Narayanan, J., Tamilanban, T., Subramaniyan, V., Chitra, V., Fuloria, N. K., Wong, L. S., Ramachawolran, G., Sekar, M., Gupta, G., Fuloria, S., Chinni, S. V., & Selvaraj, S. (2023). Exploring the contribution of pro-inflammatory cytokines to impaired wound healing in diabetes. *Frontiers in Immunology*, *14*, 1216321. DOI: 10.3389/fimmu.2023.1216321 PMID: 37575261

Nowwarote, N., Osathanon, T., Jitjaturunt, P., Manopattanasoontorn, S., & Pavasant, P. (2012). Asiaticoside Induces Type I Collagen Synthesis and Osteogenic Differentiation in Human Periodontal Ligament Cells. *Phytotherapy Research*, *27*(3), 457–462. DOI: 10.1002/ptr.4742 PMID: 22628285

Nusantoro, A. P., Kuntaman, K., & Perdanakusuma, D. S. (2024). Synergistic Interaction between Topical Application of Allicin and Quercetin Enhances Diabetic Wound Healing Via Inflammatory in Wistar Rats. *Sains Malaysiana, 53*(2), 397–408. DOI: 10.17576/jsm-2024-5302-13

Ohnishi, N., Kusuhara, M., Yoshioka, M., Kuroda, K., Soga, A., Nishikawa, F., Koishi, T., Nakagawa, M., Hori, S., Matsumoto, T., Yamashita, M., Ohta, S., Takara, K., & Yokoyama, T. (2003). Studies on Interactions between Functional Foods or Dietary Supplements and Medicines. I. Effects of Ginkgo biloba Leaf Extract on the Pharmacokinetics of Diltiazem in Rats. *Biological & Pharmaceutical Bulletin, 26*(9), 1315–1320. DOI: 10.1248/bpb.26.1315 PMID: 12951478

Okem, A., Henstra, C., Lambert, M., & Hayeshi, R. (2022). A review of the pharmacodynamic effect of chemo-herbal drug combinations therapy for cancer treatment [Review of *A review of the pharmacodynamic effect of chemo-herbal drug combinations therapy for cancer treatment*]. *Medicine in Drug Discovery, 17*, 100147. Elsevier BV. DOI: 10.1016/j.medidd.2022.100147

Okonkwo, U. A., Chen, L., Ma, D., Haywood, V. A., Barakat, M., Urao, N., & DiPietro, L. A. (2020). Compromised angiogenesis and vascular Integrity in impaired diabetic wound healing. *PLoS One, 15*(4), e0231962. Advance online publication. DOI: 10.1371/journal.pone.0231962 PMID: 32324828

Okonkwo, U. A., & DiPietro, L. A. (2017). Diabetes and Wound Angiogenesis [Review of *Diabetes and Wound Angiogenesis*]. *International Journal of Molecular Sciences, 18*(7), 1419. Multidisciplinary Digital Publishing Institute. DOI: 10.3390/ijms18071419

Oliveira, D., Borges, A., Saavedra, M. J., Borges, F., & Simões, M. (2022). Screening of Natural Molecules as Adjuvants to Topical Antibiotics to Treat Staphylococcus aureus from Diabetic Foot Ulcer Infections. *Antibiotics (Basel, Switzerland), 11*(5), 620. DOI: 10.3390/antibiotics11050620 PMID: 35625264

Omidian, H., Wilson, R. L., & Chowdhury, S. D. (2023). Enhancing Therapeutic Efficacy of Curcumin: Advances in Delivery Systems and Clinical Applications. *Gels (Basel, Switzerland), 9*(8), 596. DOI: 10.3390/gels9080596 PMID: 37623051

Oyama, K., Kawada-Matsuo, M., Oogai, Y., Hayashi, T., Nakamura, N., & Komatsuzawa, H. (2016). Antibacterial Effects of Glycyrrhetinic Acid and Its Derivatives on Staphylococcus aureus. *PLoS One, 11*(11), e0165831. Advance online publication. DOI: 10.1371/journal.pone.0165831 PMID: 27820854

Pai, L., Patil, S., Liu, S., & Wen, F. (2023). A growing battlefield in the war against biofilm-induced antimicrobial resistance: Insights from reviews on antibiotic resistance. *Frontiers in Cellular and Infection Microbiology, 13*, 1327069. Advance online publication. DOI: 10.3389/fcimb.2023.1327069 PMID: 38188636

Papadopoulou, A., Green, R., & Frazier, R. A. (2004). Interaction of Flavonoids with Bovine Serum Albumin: A Fluorescence Quenching Study. *Journal of Agricultural and Food Chemistry, 53*(1), 158–163. DOI: 10.1021/jf048693g PMID: 15631523

Patil, K. R., Mahajan, U. B., Unger, B. S., Goyal, S. N., Belemkar, S., Surana, S. J., Ojha, S., & Patil, C. R. (2019). Animal Models of Inflammation for Screening of Anti-inflammatory Drugs: Implications for the Discovery and Development of Phytopharmaceuticals [Review of *Animal Models of Inflammation for Screening of Anti-inflammatory Drugs: Implications for the Discovery and Development of Phytopharmaceuticals*]. *International Journal of Molecular Sciences, 20*(18), 4367. Multidisciplinary Digital Publishing Institute. DOI: 10.3390/ijms20184367

Pereira, S. G., Moura, J., Carvalho, E., & Empadinhas, N. (2017). Microbiota of Chronic Diabetic Wounds: Ecology, Impact, and Potential for Innovative Treatment Strategies [Review of *Microbiota of Chronic Diabetic Wounds: Ecology, Impact, and Potential for Innovative Treatment Strategies*]. *Frontiers in Microbiology, 8*. Frontiers Media. DOI: 10.3389/fmicb.2017.01791

Phelps, E. A., & García, A. J. (2008). Update on Therapeutic Vascularization Strategies [Review of *Update on Therapeutic Vascularization Strategies*]. *Regenerative Medicine, 4*(1), 65. Future Medicine. DOI: 10.2217/17460751.4.1.65

Pignet, A.-L., Schellnegger, M., Hecker, A., Kohlhauser, M., Kotzbeck, P., & Kamolz, L. (2021). Resveratrol-Induced Signal Transduction in Wound Healing. *International Journal of Molecular Sciences, 22*(23), 12614. DOI: 10.3390/ijms222312614 PMID: 34884419

Polaka, S., Katare, P., Pawar, B., Vasdev, N., Gupta, T., Rajpoot, K., Sengupta, P., & Tekade, R. K. (2022). Emerging ROS-Modulating Technologies for Augmentation of the Wound Healing Process. *ACS Omega, 7*(35), 30657–30672. DOI: 10.1021/acsomega.2c02675 PMID: 36092613

Pouget, C., Dunyach-Remy, C., Pantel, A., Schuldiner, S., Sotto, A., & Lavigne, J. (2020). Biofilms in Diabetic Foot Ulcers: Significance and Clinical Relevance. *Microorganisms, 8*(10), 1580. DOI: 10.3390/microorganisms8101580 PMID: 33066595

Qiao, W., Niu, L., Jiang, W., Lu, L., & Liu, J. (2024). Berberine ameliorates endothelial progenitor cell function and wound healing in vitro and in vivo via the miR-21-3p/RRAGB axis for venous leg ulcers. *Regenerative Therapy, 26*, 458–468. DOI: 10.1016/j.reth.2024.06.011 PMID: 39100534

Raghav, A., Dugbartey, G. J., Alornyo, K. K., Long, I., Zhu, L., Xiao, Y., Xiao, Y., Jiang, Y., Prathap, M., & Gittes, G. K. (2022). Diabetic Foot - Recent Advances. In *IntechOpen eBooks*. IntechOpen., DOI: 10.5772/intechopen.102211

Raguraman, R., Srivastava, A., Munshi, A., & Ramesh, R. (2021). Therapeutic approaches targeting molecular signaling pathways common to diabetes, lung diseases and cancer [Review of *Therapeutic approaches targeting molecular signaling pathways common to diabetes, lung diseases and cancer*]. *Advanced Drug Delivery Reviews, 178*, 113918. Elsevier BV. DOI: 10.1016/j.addr.2021.113918

Rai, V., Moellmer, R., & Agrawal, D. K. (2022). Stem Cells and Angiogenesis: Implications and Limitations in Enhancing Chronic Diabetic Foot Ulcer Healing. *Cells, 11*(15), 2287. DOI: 10.3390/cells11152287 PMID: 35892584

Rajendran, P., Rengarajan, T., Thangavel, J., Nishigaki, Y., Sakthisekaran, D., Sethi, G., & Nishigaki, I. (2013). The Vascular Endothelium and Human Diseases [Review of *The Vascular Endothelium and Human Diseases*]. *International Journal of Biological Sciences, 9*(10), 1057. Ivyspring International Publisher. DOI: 10.7150/ijbs.7502

Ramirez-Acuña, J. M., Cardenas-Cadena, S. A., Marquez-Salas, P. A., Garza-Veloz, I., Perez-Favila, A., Cid-Báez, M. A., Flores-Morales, V., & Martínez-Fierro, M. L. (2019). Diabetic Foot Ulcers: Current Advances in Antimicrobial Therapies and Emerging Treatments. *Antibiotics (Basel, Switzerland), 8*(4), 193. DOI: 10.3390/antibiotics8040193 PMID: 31652990

Razavi, Z. S., Sharafshadehi, S. A., Yousefi, M. H., Javaheri, F., Barghani, M. R. R., Afkhami, H., & Heidari, F. (2025). Application of novel strategies in chronic wound management with focusing on pressure ulcers: New perspective. *Archives of Dermatological Research, 317*(1), 320. DOI: 10.1007/s00403-024-03790-8 PMID: 39888392

Razif, R., Fadilah, N. I. M., Ahmad, H., Hao, D. L. Q., Maarof, M., & Fauzi, M. B. (2025). Asiaticoside-Loaded Multifunctional Bioscaffolds for Enhanced Hyperglycemic Wound Healing. *Biomedicines, 13*(2), 277. DOI: 10.3390/biomedicines13020277 PMID: 40002691

Raziyeva, K., Kim, Y., Zharkinbekov, Z., Kassymbek, K., Jimi, S., & Saparov, A. (2021). Immunology of Acute and Chronic Wound Healing. *Biomolecules, 11*(5), 700. DOI: 10.3390/biom11050700 PMID: 34066746

Ren, S., Chen, J., Guo, J., Liu, Y., Xiong, H., Jing, B., Yang, X., Li, G., Kang, Y., Wang, C., Xu, X., Liu, Z., Zhang, M., Xiang, K., Li, C., Li, Q., Machens, H., & Chen, Z. (2022). Exosomes from Adipose Stem Cells Promote Diabetic Wound Healing through the eHSP90/LRP1/AKT Axis. *Cells*, *11*(20), 3229. DOI: 10.3390/cells11203229 PMID: 36291096

Rendra, E., Riabov, V., Mossel, D. M., Sevastyanova, T., Harmsen, M. C., & Kzhyshkowska, J. (2018). Reactive oxygen species (ROS) in macrophage activation and function in diabetes [Review of *Reactive oxygen species (ROS) in macrophage activation and function in diabetes*]. *Immunobiology, 224*(2), 242. Elsevier BV. DOI: 10.1016/j.imbio.2018.11.010

Richard, S. A. (2021). Exploring the Pivotal Immunomodulatory and Anti-Inflammatory Potentials of Glycyrrhizic and Glycyrrhetinic Acids. *Mediators of Inflammation*, *2021*(1), 6699560. DOI: 10.1155/2021/6699560 PMID: 33505216

Rodríguez-Rodríguez, N., Martínez-Jiménez, I., García-Ojalvo, A., Mendoza-Marí, Y., Guillén, G., Armstrong, D., & Berlanga-Acosta, J. (2021). Wound Chronicity, Impaired Immunity and Infection in Diabetic Patients [Review of Wound Chronicity, Impaired Immunity and Infection in Diabetic Patients]. *MEDICC Review*, *24*(1), 44. DOI: 10.37757/MR2021.V23.N3.8 PMID: 34653116

Ruthenborg, R. J., Ban, J.-J., Wazir, A., Takeda, N., & Kim, J. (2014). Regulation of Wound Healing and Fibrosis by Hypoxia and Hypoxia-Inducible Factor-1. *Molecules and Cells*, *37*(9), 637–643. DOI: 10.14348/molcells.2014.0150 PMID: 24957212

Singh, B., Yadav, A., & Sharma, N. (2020). The efficacy of aloe vera in wound healing and its application in the management of burns. *Burns and Trauma*, *8*, 47. DOI: 10.1186/s41038-020-00222-x

Samavati, S. S., Kashanian, S., Derakhshankhah, H., Rabiei, M., Sajadimajd, S., Fakhri, S., & Rashidi, Z. (2025). Accelerated wound healing through tannin-rich Jaft extract, concentration-dependent efficacy and mechanistic insights from Quercus brantii ointment formulations. *Scientific Reports*, *15*(1), 29004. Advance online publication. DOI: 10.1038/s41598-025-13832-4 PMID: 40781116

Saxena, A., Tripathi, K. P., Roy, S., Khan, F., & Sharma, A. (2008). Pharmacovigilance: Effects of herbal components on human drugs interactions involving Cytochrome P450. *Bioinformation*, *3*(5), 198–204. DOI: 10.6026/97320630003198 PMID: 19255634

Scherr, T. D., Heim, C. E., Morrison, J. M., & Kielian, T. (2014). Hiding in Plain Sight: Interplay between Staphylococcal Biofilms and Host Immunity [Review of *Hiding in Plain Sight: Interplay between Staphylococcal Biofilms and Host Immunity*]. *Frontiers in Immunology, 5*. Frontiers Media. DOI: 10.3389/fimmu.2014.00037

Schultz, G. S., Davidson, J. M., Kirsner, R. S., Börnstein, P., & Herman, I. M. (2011). Dynamic reciprocity in the wound microenvironment [Review of *Dynamic reciprocity in the wound microenvironment*]. *Wound Repair and Regeneration, 19*(2), 134. Wiley. DOI: 10.1111/j.1524-475X.2011.00673.x

Şenol, F. S., Tuğcu-Demiröz, F., Ulutaş, O. K., Erdoğan, İ., & Orhan, I. E. (2025). Phytosomes-Unraveling the Unique Properties of Plant-Derived Nanotechnological Drug Delivery Systems: A Review [Review of *Phytosomes-Unraveling the Unique Properties of Plant-Derived Nanotechnological Drug Delivery Systems: A Review*]. *Current Medicinal Chemistry, 32*(15), 3088. Bentham Science Publishers. DOI: 10 .2174/0109298673319759250116104648

Setia, A., Vallamkonda, B., Challa, R. R., Mehata, A. K., Badgujar, P., & Muthu, M. S. (2024). Herbal Theranostics: Controlled, Targeted Delivery and Imaging of Herbal Molecules [Review of *Herbal Theranostics: Controlled, Targeted Delivery and Imaging of Herbal Molecules*]. *Nanotheranostics, 8*(3), 344. Ivyspring International Publisher. DOI: 10.7150/ntno.94987

Setiawati, A., Maharani, B. A., Sari, P. A. P., Widyantara, K. A., Saputra, B. W., Febriansah, R., & Dwiastuti, R. (2024). Deciphering the molecular pathway of an asiaticosiderich fraction of Centella asiatica as an anti-melanogenesis agent. *Journal of Herbmed Pharmacology, 13*(2), 269–279. DOI: 10.34172/jhp.2024.49332

Sevgi, M., Toklu, A., Vecchio, D., & Hamblin, M. R. (2014). Topical Antimicrobials for Burn Infections – An Update [Review of *Topical Antimicrobials for Burn Infections – An Update*]. *Recent Patents on Anti-Infective Drug Discovery, 8*(3), 161. Bentham Science Publishers. DOI: 10.2174/1574891X08666131112143447

Shams, F., Moravvej, H., Hosseinzadeh, S., Mostafavi, E., Bayat, H., Kazemi, B., Bandehpour, M., Rostami, E., Rahimpour, A., & Moosavian, H. (2022). Overexpression of VEGF in dermal fibroblast cells accelerates the angiogenesis and wound healing function: In vitro and in vivo studies. *Scientific Reports, 12*(1), 18529. Advance online publication. DOI: 10.1038/s41598-022-23304-8 PMID: 36323953

Shi, Z., Yao, C., Shui, Y., Li, S., & Yan, H. (2023). Research progress on the mechanism of angiogenesis in wound repair and regeneration [Review of *Research progress on the mechanism of angiogenesis in wound repair and regeneration*]. *Frontiers in Physiology, 14*, 1284981. Frontiers Media. DOI: 10.3389/fphys.2023.1284981

Shukla, A., Rasik, A. M., & Dhawan, B. N. (1999). Asiaticoside-induced elevation of antioxidant levels in healing wounds. *Phytotherapy Research, 13*(1), 50–54. DOI: 10.1002/(SICI)1099-1573(199902)13:1<50::AID-PTR368>3.0.CO;2-V PMID: 10189951

Shukla, L., Yuan, Y., Shayan, R., Greening, D. W., & Karnezis, T. (2020). Fat Therapeutics: The Clinical Capacity of Adipose-Derived Stem Cells and Exosomes for Human Disease and Tissue Regeneration [Review of *Fat Therapeutics: The Clinical Capacity of Adipose-Derived Stem Cells and Exosomes for Human Disease and Tissue Regeneration*]. *Frontiers in Pharmacology, 11*. Frontiers Media. DOI: 10.3389/fphar.2020.00158

Shukla, S. D., Bhatnagar, M., & Khurana, S. (2012). Critical Evaluation of Ayurvedic Plants for Stimulating Intrinsic Antioxidant Response. *Frontiers in Neuroscience, 6*. Advance online publication. DOI: 10.3389/fnins.2012.00112 PMID: 22855669

Sideek, S. A., El-Nassan, H. B., Fares, A., ElMeshad, A. N., & Elkasabgy, N. A. (2022). Different Curcumin-Loaded Delivery Systems for Wound Healing Applications: A Comprehensive Review. *Pharmaceutics, 15*(1), 38. DOI: 10.3390/pharmaceutics15010038 PMID: 36678665

Sivamani, R. K., Brian, R., Wehrli, L. N., & Maverakis, E. (2012). Phytochemicals and Naturally Derived Substances for Wound Healing [Review of *Phytochemicals and Naturally Derived Substances for Wound Healing*]. *Advances in Wound Care, 1*(5), 213. Mary Ann Liebert, Inc. DOI: 10.1089/wound.2011.0330

Spinetti, G., Kraenkel, N., Emanueli, C., & Madeddu, P. (2008). Diabetes and vessel wall remodelling: from mechanistic insights to regenerative therapies [Review of *Diabetes and vessel wall remodelling: from mechanistic insights to regenerative therapies*]. *Cardiovascular Research, 78*(2), 265. Oxford University Press. DOI: 10.1093/cvr/cvn039

Sreenathkumar, S. (2021). Current Updates on Global Phytoceuticals and Novel Phyto Drug Delivery System in Herbal Medicine. In *IntechOpen eBooks*. IntechOpen., DOI: 10.5772/intechopen.97671

Suganya, T., Packiavathy, I. A. S. V., Aseervatham, G. S. B., Carmona, A., Rashmi, V., Mariappan, S., Devi, N. R., & Ananth, D. A. (2022). Tackling Multiple-Drug-Resistant Bacteria With Conventional and Complex Phytochemicals [Review of *Tackling Multiple-Drug-Resistant Bacteria With Conventional and Complex Phytochemicals*]. *Frontiers in Cellular and Infection Microbiology, 12*. Frontiers Media. DOI: 10.3389/fcimb.2022.883839

Sun, S., Yu, Y., Jo, Y., Han, J. H., Xue, Y., Cho, M., Bae, S.-J., Ryu, D., Park, M.-J., Ha, K., & Zhuang, S. (2025). Impact of extraction techniques on phytochemical composition and bioactivity of natural product mixtures. *Frontiers in Pharmacology, 16*, 1615338. DOI: 10.3389/fphar.2025.1615338 PMID: 40808686

Soni, H., Khare, S., & Singh, G. (2020). Silymarin for diabetic foot ulcer healing: Mechanisms and therapeutic application. *Journal of Wound Care, 29*(5), 270–278. DOI: 10.12968/jowc.2020.29.5.270 PMID: 32421483

Tabassum, A., & Raj, M. (2020). Plant extracts as potential antibacterial agents against pathogens causing diabetic foot ulcers. *Journal of Ethnopharmacology, 262*, 113145. DOI: 10.1016/j.jep.2020.113145

Tecilazich, F., Dinh, T. L., & Veves, A. (2013). Emerging drugs for the treatment of diabetic ulcers [Review of *Emerging drugs for the treatment of diabetic ulcers*]. *Expert Opinion on Emerging Drugs, 18*(2), 207. Taylor & Francis. DOI: 10.1517/14728214.2013.802305

Teodoro, J. S., Nunes, S., Rolo, A. P., Reis, F., & Palmeira, C. M. (2019). Therapeutic Options Targeting Oxidative Stress, Mitochondrial Dysfunction and Inflammation to Hinder the Progression of Vascular Complications of Diabetes. *Frontiers in Physiology, 9*, 1857. DOI: 10.3389/fphys.2018.01857 PMID: 30705633

Thandavarayan, R. A., Garikipati, V. N. S., Joladarashi, D., Babu, S. S., Jeyabal, P., Verma, S. K., Mackie, A. R., Khan, M., Arumugam, S., Watanabe, K., Kishore, R., & Krishnamurthy, P. (2015). Sirtuin-6 deficiency exacerbates diabetes-induced impairment of wound healing. *Experimental Dermatology, 24*(10), 773–778. DOI: 10.1111/exd.12762 PMID: 26010430

Thawabteh, A. M., Ghanem, A. W., AbuMadi, S., Thaher, D., Jaghama, W., Karaman, R., Scrano, L., & Bufo, S. A. (2024). Antibacterial Activity and Antifungal Activity of Monomeric Alkaloids. *Toxins, 16*(11), 489. DOI: 10.3390/toxins16110489 PMID: 39591244

Tie, Y., Liu, H., Zhang, T., Meng, T., & Liang, Q. (2025). Natural products alleviate viral pneumonia by modulating inflammatory and oxidative-stress pathways [Review of *Natural products alleviate viral pneumonia by modulating inflammatory and oxidative-stress pathways*]. *Frontiers in Pharmacology, 16*. Frontiers Media. DOI: 10.3389/fphar.2025.1657829

Tirant, M., Tirant, H., & Wollina, U. (2024). Herbal Bioactive Compounds for Skin Infections and Inflammatory Conditions. *Open Access Macedonian Journal of Medical Sciences, 12*, 1–44. DOI: 10.3889/oamjms.2024.11888

Tirona, R. G., & Bailey, D. G. (2006). Herbal product–drug interactions mediated by induction. *British Journal of Clinical Pharmacology, 61*(6), 677–681. DOI: 10.1111/j.1365-2125.2006.02684.x PMID: 16722828

. Uberoi, A., McCready-Vangi, A. R., & Grice, E. A. (2024). The wound microbiota: microbial mechanisms of impaired wound healing and infection [Review of *The wound microbiota: microbial mechanisms of impaired wound healing and infection*]. *Nature Reviews Microbiology, 22*(8), 507. Nature Portfolio. DOI: 10.1038/s41579-024-01035-z

Ukaegbu, K., Allen, E. P., & Svoboda, K. K. H. (2025). Reactive Oxygen Species and Antioxidants in Wound Healing: Mechanisms and Therapeutic Potential [Review of *Reactive Oxygen Species and Antioxidants in Wound Healing: Mechanisms and Therapeutic Potential*]. *International Wound Journal, 22*(5). Wiley. DOI: 10.1111/iwj.70330

Volpe, C. M. O., Villar-Delfino, P. H., dos Anjos, P. M. F., & Nogueira-Machado, J. A. (2018). Cellular death, reactive oxygen species (ROS) and diabetic complications. *Cell Death & Disease, 9*(2), 119. Advance online publication. DOI: 10.1038/s41419-017-0135-z PMID: 29371661

Walach, H., Huber, B., Bassler, D., Naumann, J., Bismarck, D., Schneider, M., Müller, E., Walach, H., Mangelsdorf, I., & Mutter, J. (2017). Journal Club. *Complementary Medicine Research, 24*(3), 135–138. DOI: 10.1159/000477783 PMID: 28641284

Wang, Y., Zhang, Z., & Li, Y. (2024). Glycyrrhizic acid-based hydrogel for improving wound healing in diabetic ulcers. *Journal of Materials Chemistry. B, 12*(7), 2125–2132. DOI: 10.1039/D4TB01572J PMID: 38230414

Wang, C., Lou, Y., Tong, M., Zhang, L., Zhang, Z., Feng, Y., Shi, L., Xu, H., & Mao, C. (2017). Asperosaponin VI promotes angiogenesis and accelerates wound healing in rats via up-regulating HIF-1α/VEGF signaling. *Acta Pharmacologica Sinica, 39*(3), 393–404. DOI: 10.1038/aps.2017.161 PMID: 29219948

Wang, G., Yang, F., Zhou, W., Xiao, N., Luo, M., & Tang, Z. (2022). The initiation of oxidative stress and therapeutic strategies in wound healing. *Biomedicine and Pharmacotherapy, 157*, 114004. DOI: 10.1016/j.biopha.2022.114004 PMID: 36375308

Wang, L., Xia, K., Han, L., Zhang, M., Fan, J., Liu, S., Liao, A., Wang, W., & Guo, J. (2022). Local Administration of Ginkgolide B Using a Hyaluronan-Based Hydrogel Improves Wound Healing in Diabetic Mice. *Frontiers in Bioengineering and Biotechnology, 10*, 898231. Advance online publication. DOI: 10.3389/fbioe.2022.898231 PMID: 35694224

Wang, Q., Liu, C., An, J., Liu, C., Wang, Y., & Cai, Y. (2025). Mechanisms of microbial infection and wound healing in diabetic foot ulcer: Pathogenicity in the inflammatory-proliferative phase, chronicity, and treatment strategies. *Frontiers in Endocrinology, 16*, 1657928. DOI: 10.3389/fendo.2025.1657928 PMID: 41103650

Whittam, A. J., Maan, Z. N., Duscher, D., Wong, V. W., Barrera, J. A., Januszyk, M., & Gurtner, G. C. (2014). Challenges and Opportunities in Drug Delivery for Wound Healing [Review of *Challenges and Opportunities in Drug Delivery for Wound Healing*]. *Advances in Wound Care, 5*(2), 79. Mary Ann Liebert, Inc. DOI: 10.1089/wound.2014.0600

Xiong, Y., Knoedler, S., Alfertshofer, M., Kim, B., Jiang, D., Liu, G., Rinkevich, Y., & Mi, B. (2025). Mechanisms and therapeutic opportunities in metabolic aberrations of diabetic wounds: A narrative review. *Cell Death & Disease, 16*(1), 341. DOI: 10.1038/s41419-025-07583-3 PMID: 40280905

Yadav, P., Singh, M. P., Vinayagam, R., & Shukla, P. (2025). Therapies and delivery systems for diabetic wound care: Current insights and future directions. *Frontiers in Pharmacology, 16*, 1628252. DOI: 10.3389/fphar.2025.1628252 PMID: 40766758

Yang, B., Dong, Y., Wang, F., & Zhang, Y. (2020). Nanoformulations to Enhance the Bioavailability and Physiological Functions of Polyphenols. *Molecules (Basel, Switzerland), 25*(20), 4613. DOI: 10.3390/molecules25204613 PMID: 33050462

Yang, D., Wang, M., Zhang, C., & Wang, Y. (2024). Endothelial dysfunction in vascular complications of diabetes: A comprehensive review of mechanisms and implications. *Frontiers in Endocrinology, 15*, 1359255. DOI: 10.3389/fendo.2024.1359255 PMID: 38645427

Yazarlu, O., Iranshahi, M., Kashani, H. R. K., Reshadat, S., Habtemariam, S., Iranshahy, M., & Hasanpour, M. (2021). Perspective on the application of medicinal plants and natural products in wound healing: A mechanistic review [Review of *Perspective on the application of medicinal plants and natural products in wound healing: A mechanistic review*]. *Pharmacological Research, 174*, 105841. Elsevier BV. DOI: 10.1016/j.phrs.2021.105841

Yuliati, L., Mardliyati, E., Bramono, K., & Freisleben, H. J. (2015). Asiaticoside induces cell proliferation and collagen synthesis in human dermal fibroblasts. *Universa Medicina, 34*(2), 96. DOI: 10.18051/UnivMed.2015.v34.96-103

Zhang, L., Wang, X., & Si, H. (2022). Synergistic anti-inflammatory effects and mechanisms of the combination of resveratrol and curcumin in human vascular endothelial cells and rodent aorta. *The Journal of Nutritional Biochemistry, 108*, 109083. DOI: 10.1016/j.jnutbio.2022.109083 PMID: 35691595

Zhang, X., Liu, S., & Wang, J. (2023). Curcumin and its role in wound healing and inflammation modulation. *Journal of Ethnopharmacology*, *289*, 115131. DOI: 10.1016/j.jep.2023.115131

Zhang, W., Chen, L., Xiong, Y., Panayi, A. C., Abududilibaier, A., Hu, Y., Yu, C., Zhou, W., Sun, Y., Liu, M., Xue, H., Hu, L., Yan, C., Xie, X., Lin, Z., Cao, F., Mi, B., & Liu, G. (2021). Antioxidant Therapy and Antioxidant-Related Bionanomaterials in Diabetic Wound Healing. *Frontiers in Bioengineering and Biotechnology*, *9*, 707479. DOI: 10.3389/fbioe.2021.707479 PMID: 34249895

Zhang, W., Liu, L., Cheng, H., Zhu, J., Li, X., Ye, S., & Li, X. (2023). Hydrogel-based dressings designed to facilitate wound healing. *Materials Advances*, *5*(4), 1364–1394. DOI: 10.1039/D3MA00682D

Zhang, Y., Bai, X., Shen, K., Luo, L., Zhao, M., Xu, C., Jia, Y., Xiao, D., Li, Y., Gao, X., Tian, C., Wang, Y., & Hu, D. (2022). Exosomes Derived from Adipose Mesenchymal Stem Cells Promote Diabetic Chronic Wound Healing through SIRT3/SOD2. *Cells*, *11*(16), 2568. DOI: 10.3390/cells11162568 PMID: 36010644

Zhang, Z., Huang, Q., Zhao, D., Lian, F., Li, X., & Qi, W. (2023). The impact of oxidative stress-induced mitochondrial dysfunction on diabetic microvascular complications. *Frontiers in Endocrinology*, *14*, 1112363. Advance online publication. DOI: 10.3389/fendo.2023.1112363 PMID: 36824356

Zhou, P., Chrepa, V., Karoussis, I. K., Pikos, M. A., & Kotsakis, G. A. (2021). Cytocompatibility Properties of an Herbal Compound Solution Support In vitro Wound Healing. *Frontiers in Physiology*, *12*, 653661. Advance online publication. DOI: 10.3389/fphys.2021.653661 PMID: 33868021

Zhu, Y., Lu, J., Wang, S., Xu, D., Wu, M., Xian, S., Zhang, W., Tong, X., Liu, Y., Huang, J., Jiang, L., Guo, X., Xie, S., Gu, M., Jin, S., Ma, Y., Huang, R., Xiao, S., & Ji, S. (2023). Mapping intellectual structure and research hotspots in the field of fibroblast-associated DFUs: A bibliometric analysis. *Frontiers in Endocrinology*, *14*, 1109456. Advance online publication. DOI: 10.3389/fendo.2023.1109456 PMID: 37124747

Zulkefli, N., Zahari, C. N. M. C., Sayuti, N. H., Kamarudin, A. A., Saad, N., Hamezah, H. S., Bunawan, H., Baharum, S. N., Mediani, A., Ahmed, Q. U., Ismail, A. F. H., & Sarian, M. N. (2023). Flavonoids as Potential Wound-Healing Molecules: Emphasis on Pathways Perspective. *International Journal of Molecular Sciences*, *24*(5), 4607. DOI: 10.3390/ijms24054607 PMID: 36902038

Section 4

Modern Approaches and Future Perspectives

Chapter 10
Wound Healing by Phytochemicals:
Insights From Molecular Docking and Computational Analysis

Raghunath Satpathy
http://orcid.org/0000-0001-5296-8492
Gangadhar Meher University, Sambalpur, India

ABSTRACT

Wound healing is a complex molecular process that aims to repair the tissue damaged due to the cause of an injury. Chronic and delayed wound healing processes pose significant clinical challenges for the patient. The synthetic drug possesses several side effects when taken for a long time. However, the phytochemicals derived from medicinal plants have gained attention and are eventually used to treat wounds. Due to the large molecular diversity in the phytochemicals, it is difficult to screen the key compound by using traditional experimental methods, as they are expensive and time-consuming in nature. Therefore, computational approaches, particularly molecular docking, provide a better understanding of the mechanisms through which phytochemicals interact with important biological targets involved in wound healing. The mechanistic action of various proteins in the wound-healing process is highlighted in this chapter along with the procedure and implementation of molecular docking studies used to screen potential phytochemicals are discussed.

INTRODUCTION

The skin, which makes up a large portion of the body and serves as an immunological barrier as well as a protective outer layer. In addition to maintaining a normal

DOI: 10.4018/979-8-3373-2908-6.ch010

body temperature, it protects septicity, microbial contagion, and mechanical pressure (Kwiecien et al., 2019; Kumar and Kothari, 2021; Kumar et al., 2022). Wounds occur when the skin is injured physically, chemically, thermally, infectiously, or immunologically, disrupting the continuity of cells. Wound healing is the process of maintaining functional tissue integrity. Injuries have occurred as a result of either a sequential loss of function in any layer of the skin, leading to tissue disruption, or the loss of the histological composition of the skin tissue. Wounds allow various microbial agents, such as bacteria, viruses, or other foreign substances, to enter the body. Local microbial infections are the cause of skin wound inflammation. Additionally, septicemia—a potentially fatal condition—a widespread systemic infection was discovered. Therefore, more attention is required to identify simple and effective ways to take care of skin wounds so that they heal properly (Strodtbeck, 2001; Kumar et al., 2004; Blume-Peytavi et al., 2016). An effective wound-healing process assists in preventing bleeding, eliminating microbial infections in wounds, and promoting proper wound healing without complications or abnormalities. Also, in this process, several cellular and extracellular pathways are involved to repair, grow, and regenerate damaged skin tissue (Percival, 2002; Grice and Segre, 2011; Richmond and Harris, 2014; Jones, 2015). This wound-healing process consists of stages like hemostasis, inflammation, proliferation, and remodeling (Guo et al., 2010; Gonzalez et al., 2016). Chronic wounds and delayed wound healing, which are frequently made worse by oxidative stress, infections, and inflammatory reactions, continue to be major clinical challenges despite tremendous advancements in modern medicine (Dinh et al., 2011).

The objective of this chapter is to present a thorough analysis of the wound healing process, covering its main molecular targets as well as the function of phytochemicals in facilitating tissue repair. The chapter will also discuss the use and benefits of computational methods, specifically molecular docking, in phytochemical screening and assessment. The evaluation of the binding affinities of these compounds and the mechanistic interactions with matrix metalloproteinases (MMPs), which are essential for extracellular matrix remodeling during wound healing.

MOLECULAR BASIS OF THE WOUND HEALING PROCESS

The molecular basis of wound healing is achieved by adequate activation and infiltration of inflammatory cells such as neutrophils and macrophages, which generate pro-inflammatory cytokines (tumor necrosis factor α (TNF-α) and interleukin-I (IL-I)). Subsequently, these inflammatory cytokines lead to the activation of growth factors like transforming growth factor (TGF)-β and several fibroblast growth factors, causing proliferation in the wound site. However, aging, obesity, and endocrine disorders like diabetes mellitus can all impair these natural healing processes. Although the right amount of each cytokine is necessary for healing, excessive fibroblast proliferation brought on by inflammatory cytokine levels can lead to hypertrophic scarring, which can seriously deform the skin (Guo and Dipietro, 2010; Portou et al., 2015).

All these normal wound processes consist of four interconnected stages, as illustrated in Figure 1. The initial stage is hemostasis, which is followed by wound remodeling, inflammation, and proliferation. Several proteins are involved in the first step, including A stable blood clot forms at the site of injury as a result of thrombin's conversion of fibrinogen into fibrin. In a similar vein, activated platelets release Platelet-Derived Growth Factor (PDGF), which draws immune cells like neutrophils and macrophages to the wound site and starts the inflammatory phase. When blood vessels are injured, they release tissue factor (TF), which starts the coagulation cascade. This process eventually makes thrombin for clotting blood.

When blood vessels are injured, they release tissue factor (TF), which starts the coagulation cascade. This process eventually makes thrombin for the clotting of blood.

Figure 1. Sequential Phases in the Wound Healing Process

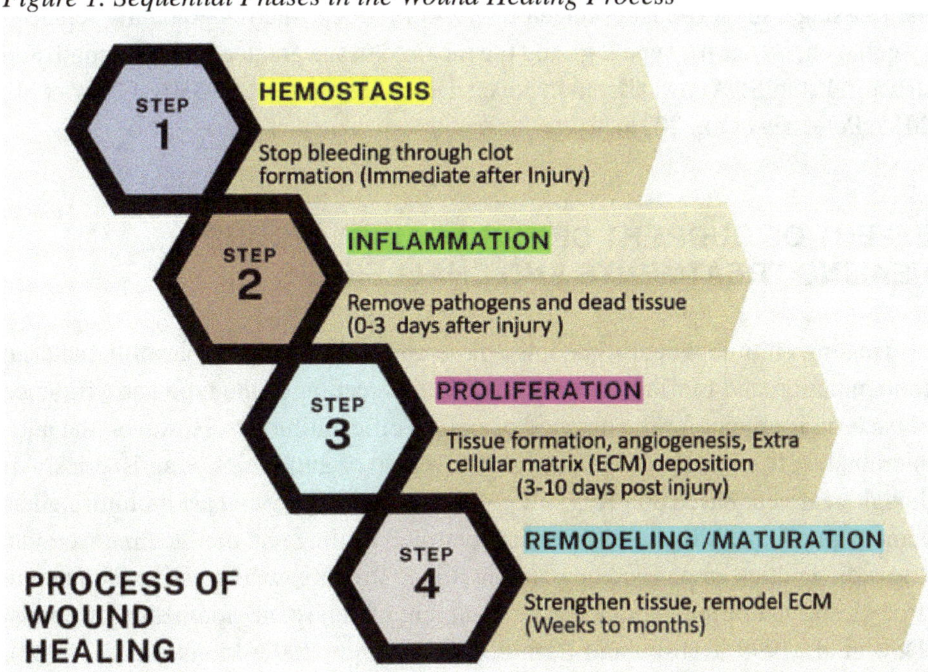

Tumor necrosis factor-alpha (TNF-α), a pro-inflammatory cytokine, is crucial for initiating and sustaining the inflammatory response in the initial phases of wound healing. Similarly, the interleukins, such as IL-1, IL-6, and IL-8, are crucial signaling molecules that promote the recruitment and activation of leukocytes to enhance immune surveillance and response at the site of injury. The breakdown of damaged extracellular matrix (ECM) components is done by enzymes known as metalloproteinases (MMPs) that facilitate tissue remodeling and create space for the growth of new tissue. During the process as a biomarker for infection and tissue damage, C-reactive protein (CRP) levels increase in response to inflammation. Vascular Endothelial Growth Factor (VEGF) provides enough oxygen and nutrients. The production of extracellular matrix and tissue repair depend on fibroblasts, which are stimulated to proliferate and migrate by fibroblast growth factor-2 (FGF-2).

TGF-β, or transforming growth factor-beta, is essential for encouraging the production of collagen and other matrix proteins, which support the strength and structural integrity of tissues. To restore the skin barrier, keratinocytes must migrate and proliferate, which is supported by keratinocyte growth factor (KGF).

Important structural proteins called collagen types I and III help create the new extracellular matrix (ECM), which gives the regenerating tissue mechanical support.

The adhesive glycoproteins fibronectin and laminin work together to create a scaffold that directs cell migration and arranges the healing tissue's structure. In the last stage, the Type III collagen that was first deposited is gradually replaced by collagen, primarily Type I, giving the repaired tissue greater tensile strength and structural stability (Powanda and Moyer, 1981; Schreml et al., 2010; Caley et al., 2015; Potekaev et al., 2021).

EFFECT OF CURRENT DRUGS AGAINST WOUND HEALING TREATMENTS AND CHALLENGES

Treating chronic wounds becomes more complicated due to harmful bacterial contamination and biofilm formation. Hence, detection of the type and virulence of bacteria is crucial for the targeted use of specific antibiotics. However, the antimicrobial resistance may arise due to the overuse of antibiotics or antiseptics. Although treatment based on pH is still poorly understood, wound pH monitoring has demonstrated the ability to differentiate between healing and non-healing wounds. These difficulties emphasize how urgent it is to investigate alternative therapeutic strategies as viable substitutes for the treatment of persistent, non-healing wounds. (Dow et al., 1999; Robson and Barbul, 2006; Gethin, 2007; Monika et al., 2022). Current medical and surgical treatments face limitations such as low efficacy, high costs, and associated side effects (Guo & DiPietro, 2010; Siddiqui & Bernstein, 2010; Tottoli et al., 2020), and some examples of drugs are given in Table 1.

Table 1. Common Drugs Used in Wound Healing with Mechanisms of Action and Side Effects

S. N	Drug/ Compound	Mechanism of Action	Common Side Effects
1	Neomycin	Inhibits bacterial protein synthesis	Allergic contact dermatitis, rash
2	Mupirocin	Inhibits bacterial isoleucyl tRNA synthetase	Burning, itching, irritation
3	Silver sulfadiazine	Disrupts bacterial DNA and membranes	Leukopenia, skin discoloration, hypersensitivity
4	Hydrocortisone	Reduces inflammatory cytokines	Skin thinning, delayed wound healing
5	Ibuprofen	Inhibits COX enzyme	Gastric irritation, renal effects, delayed healing (chronic use)
6	Acetaminophen	Central analgesic, weak COX inhibitor	Hepatotoxicity (high doses)
7	Tramadol	μ-opioid receptor agonist	Nausea, drowsiness, dependency

PHYTOCHEMICALS AS HEALING AGENTS

The chronic and unhealed wound is a serious health concern that brings severe economic burdens, physical pain, and psychological pressure to patients. Recently, phytochemicals derived from medicinal plants have gained attention due to their antioxidant, anti-inflammatory, antimicrobial, and regenerative properties (Thangapazham et al., 2016; Vitale et al., 2022). Due to the large diversity in phytochemicals, identification of specific compounds in wound healing requires in-silico-based screening approaches.

Figure 2. Plant-Based Mechanisms Contributing to Wound Healing

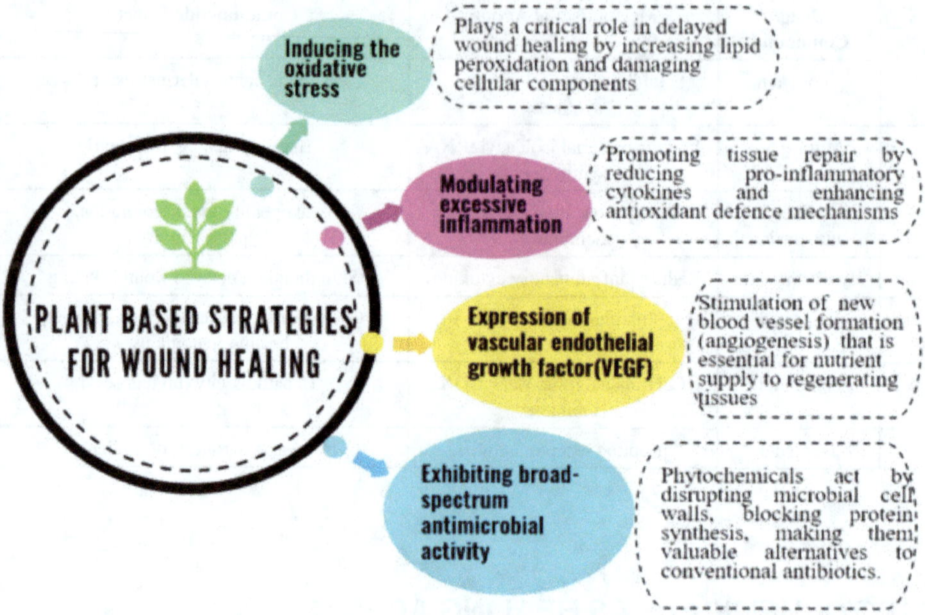

Figure 2 shows the main biological mechanisms by which compounds derived from plants aid in the healing of wounds. By minimizing lipid peroxidation and cellular damage, these strategies (1) reduce oxidative stress, which can delay healing; (2) modulate excessive inflammation by suppressing pro-inflammatory cytokines and enhancing antioxidant defense; (3) stimulate the expression of vascular endothelial growth factor (VEGF), which increases angiogenesis and nutrient delivery to regenerating tissues; and (4) exhibit broad-spectrum antimicrobial activity by breaking down microbial cell walls and preventing protein synthesis. When combined, these processes promote tissue repair and present viable plant-based substitutes for traditional wound care treatments. In recent years, there have been numerous herbal wound-healing compounds available on the market all over the world. Also, a growing number of studies have been carried out by researchers to clarify the potential activity of these compounds in promoting wound-healing activity (Ning et al., 2022).

Computational methods, specifically molecular docking and MD simulation, provide valuable insights into the mechanisms through which phytochemicals interact with key biological targets involved in wound healing (Haq et al., 2021; Shady et al., 2022; Thomas et al., 2024).

Several major targets have been predicted for wound-healing activity of the body with which the active phytochemical can interact (Schreml et al., 2010; Palmieri et al., 2017). Some of them are matrix metalloproteinases (MMPs), and pro-inflammatory cytokines such as TNF-α and IL-6 are key cytokines involved in prolonged inflammation, angiogenic and growth factor modulation such as VEGF, and other pathogenic proteins and enzymes. However, targeting these wound-healing molecules by using the phytochemicals may face several issues like *low bioavailability, lack of standardized dosing, and so* on. Some of the observed pharmacological properties have been described by researchers (Azab et al., 2016; Aziz et al., 2018; Satpathy, 2020a; Satpathy, 2021; Gonfa et al., 2023; Satpathy, 2024a; Satpathy, 2024b). According to some research studies, different phytoextracts have shown their positive activity when acting synergistically in the wound-healing mechanism. Typical phytochemicals such as flavonols, flavanones, isoflavones, flavonolignans, proanthocyanidins, β-glucans, bromelain, and curcumin are widely used for treatment for wound-repairing purposes (Shah & Amini-Nik, 2017;Attallah et al., 2021; Alherz et al., 2022; El-Sherbeni & Negm, 2023). Table 2 lists some of the key phytochemicals and the plants that provide them for the wound-healing process.

Table 2. Phytochemical Constituents of Medicinal Plants and Their Role in Wound Healing Mechanisms

S.N	Plant Name	Plant Part Used	Major Phytochemicals	Activity in the Wound healing process	References
1	*Aloe vera*	Gel, Leaves	Aloin, Emodin, Polysaccharides	Enhances fibroblast proliferation, collagen synthesis	(Arcaya et al., 2025)
2	*Curcuma longa*	Rhizome	Curcumin, Flavonoids, Alkaloids	Anti-inflammatory, granulation tissue formation	(Irham & Hardiyant, 2021)
3	*Azadirachta indica*	Leaves, Bark, Oil	Nimbidin, Azadirachtin, Tannins	Antiseptic, antimicrobial, enhances wound contraction	(Boukeloua et al., 2023)
4	*Centella asiatica*	Leaves	Asiaticoside, Madecassoside, Triterpenoids	Stimulates collagen production	(Diniz et al., 2023)
5	*Matricaria chamomilla*	Flowers	Apigenin, Bisabolol, Chamazulene	Reduces inflammation, antimicrobial	(Shivananda Nayak et al., 2007)
6	*Aegle marmelos*	Leaves, Bark	Tannins, Alkaloids, Flavonoids	Astringent, promotes epithelialization	(Arunachalam et al., 2012)

continued on following page

Table 2. Continued

S.N	Plant Name	Plant Part Used	Major Phytochemicals	Activity in the Wound healing process	References
7	*Punica granatum*	Peel, Seeds	Ellagic Acid, Punicalagins, Flavonoids	Antioxidant, promotes re-epithelialization	(Nayak et al., 2013)
8	*Eucalyptus globulus*	Leaves, Oil	Cineole, Tannins, Terpenes	Antiseptic, reduces swelling	(Mumtaz et al., 2022)
9	*Psidium guajava*	Leaves	Tannins, Quercetin, Lycopene	Antibacterial, helps in clot formation	(Bilal et al., 2024)
10	*Terminalia arjuna*	Bark	Tannins, Saponins, Flavonoids	Enhances fibroblast migration	(Chaudhari & Mengi, 2006)
11	*Melia azedarach*	Leaves, Bark	Limonoids, Quercetin, Terpenoids	Antibacterial, promotes cell regeneration	(Vijaya et al., 2012)
12	*Mimosa pudica*	Leaves, Roots	Tannins, Alkaloids, Flavonoids	Clot-promoting, antimicrobial	(Kokane et al., 2009)
13	*Cassia alata*	Leaves, Flowers	Anthraquinones, Flavonoids, Tannins	Antimicrobial, promotes collagen production	(Keng et al., 2024)
14	*Bacopa monnieri*	Leaves	Bacosides, Saponins, Alkaloids	Stimulates fibroblast activity, reduces oxidative stress	(Ghosh et al., 2007)
15	*Hemidesmus indicus*	Root, Leaves	Steroidal Saponins, Flavonoids	Collagen synthesis, antimicrobial	(Ganesan et al., 2012)
16	*Acorus calamus*	Rhizome	Asarones, Flavonoids, Tannins	Antiseptic, anti-inflammatory	(Jain et al., 2010)
17	*Piper nigrum*	Seeds	Piperine, Alkaloids, Flavonoids	Antimicrobial, enhances collagen formation	(Wong & Ling, 2014)
18	*Cymbopogon citratus*	Leaves, Essential Oil	Citral, Flavonoids, Terpenoids	Antimicrobial, collagen formation	(Hasibuan et al., 2023)
19	*Butea monosperma*	Flowers, Bark	Isoflavonoids, Tannins, Alkaloids	Promotes wound contraction, antimicrobial activity	(Muralidhar et al., 2013)

The extraction process of phytochemicals in a typical experimental approach consists of the preparation of the extract by Soxhlet extraction followed by GC-MS profiling with solvents like methanol and petroleum ether. These extracts are evaluated using both in vitro (cytotoxicity and scratch assays) and in vivo models. Animal wound models, which allow for direct assessment of wound healing progression, are commonly created using anesthetized rodents. Hematoxylin and eosin staining is used in histological evaluations to provide microscopic information about histological information (tissue regeneration and epithelialization). The healing process can be measured using macroscopic wound closure measurements (İğci & Aytac, 2020; Premarathna et al., 2021; Teymoorian et al., 2024). However, traditional studies involving animal models and histological studies only reveal phenotypic outcomes, not the specific molecular interactions. In addition, these methods are resource-intensive and often time-consuming and require skilled personnel.

TARGETING MATRIX METALLOPROTEINASES IN WOUND HEALING

The enzyme matrix metalloproteinases (MMPs), involved in the wound healing process, are particularly helpful in the breakdown of extracellular matrix (ECM) components. However, excessive tissue damage, delayed healing, and impaired re-epithelialization in chronic wounds can result from their overactivity, particularly MMP-2 and MMP-9. Due to this imbalance, some important growth factors are also degraded. Therefore, controlling MMP activity is crucial for proper healing (Stacey et al., 2002; Xue et al., 2006; Kandhwal et al., 2022). Functionally, the MMPs are generated as pro-MMPs, which are inactive precursors that are blocked from the catalytic site by a regulatory segment. When this inhibitory region is eliminated, either by proteolytic enzymes or by outside cues like oxidative stress, activation takes place. Once activated, MMPs hydrolyze peptide bonds in ECM proteins using a water molecule bound to zinc, allowing tissue remodeling. A hemopexin-like domain is one of the extra domains that some MMPs have that aid in interacting with other proteins in the wound environment or identifying particular substrates. They are released by a variety of cells, including immune cells, fibroblasts, and keratinocytes. Tissue inhibitors of metalloproteinases (TIMPs), which attach to MMPs and stop unchecked ECM degradation, strictly regulate their activity. Therefore, gaining knowledge of MMPs' structural dynamics is crucial for understanding how these enzymes work and how to potentially target them therapeutically in wound care (Parks et al., 2004; Nagase et al., 2006; Page-McCaw et al., 2007).

The following figure 3 shows the solved three-dimensional structures of various human matrix metalloproteinases (MMPs) and their inhibitor-bound complexes, based on Protein Data Bank (PDB) entries. These enzymes are crucial for extracellular matrix remodeling and are involved in wound healing, inflammation, angiogenesis, and cancer metastasis.

Figure 3. Selected crystal structure of human metalloproteinases involved in the wound healing process with their PDB ID (Source: www. https://www.rcsb.org/)

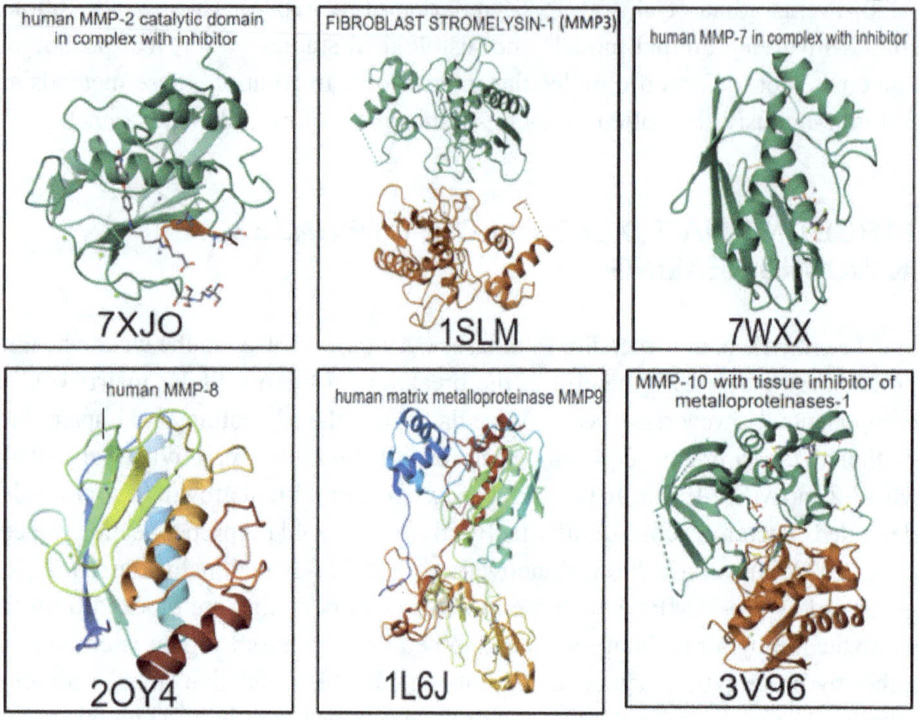

MOLECULAR DOCKING STUDY OF PHYTOCHEMICALS WITH THE TARGET MATRIX METALLOPROTEINASES (MMPS)

In wound healing research, molecular docking is an efficient computational technique that has several advantages over conventional experimental techniques due to its cost- and time-effectiveness. Docking simulations enable quick virtual screening of a large number of phytochemicals against particular protein targets linked to wound healing. In a docking program, a module usually consists of two types of algorithms. The first one is the search algorithm that investigates various ligand positions and assesses how well the ligand fits. The search algorithm simulates various ligand movements and shapes in an attempt to identify the optimal binding conformation. During this process, the protein may be regarded as flexible or rigid, depending on the technique. While some docking techniques allow the ligand to move freely while the protein remains fixed, others give the ligand and protein flexibility. Table 3 shows some of the popular docking programs based on the nature of these algorithms. Further, the strength and effectiveness of the ligand binding to the receptor are estimated using a scoring algorithm. These assist in determining the most likely binding pose and evaluating a large number of compounds to choose the most promising ones for additional research. Docking is therefore helpful in virtual screening, which allows for the rapid screening of thousands of compounds to find promising ones (Morris & Lim-Wilby, 2008; Fan et al., 2019; Paggi et al., 2024).

Figure 4. Workflow of Molecular Docking Simulation

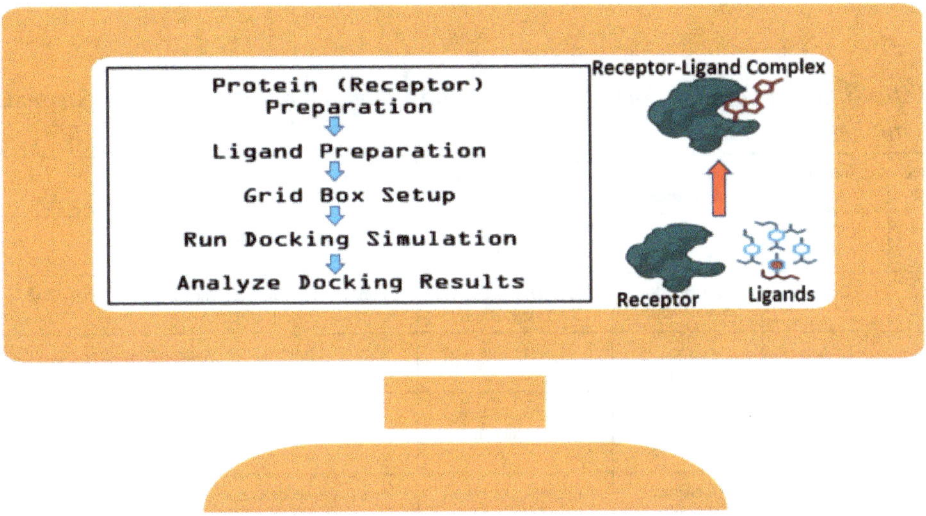

Performing the molecular docking involves several key steps (Pagadala et al., 2017; Satpathy, 2020b) and is shown in Figure 4.

1. **Receptor and Ligand 3D Structure Preparation**
2. Both the protein and the ligand three-dimensional (3D) structures are typically retrieved from databases such as the Protein Data Bank (PDB) and PubChem. Sometimes computationally derived 3D structures (models) can also be used as receptors. Then the partial charges are assigned and hydrogen atoms are added.
3. **Establishing the**
4. A grid box is created to indicate the area of the protein where docking will occur. Predicted pockets or known active sites are used to determine the grid box's center and size.
5. **Docking Simulation**
6. Search algorithms are used by docking software to investigate various ligand-protein binding scenarios. It creates a variety of binding positions and assesses them using energy scoring functions.

7. **Scoring and Ranking**
8. The binding affinity score of each pose reflects its likelihood, with the most probable binding conformations corresponding to poses with the lowest energy values.
9. **Visualization and Analysis**
10. Tools like PyMOL or Discovery Studio are used to visualize the results and analyze interactions like binding orientation, hydrophobic contacts, and hydrogen bonds.

Table 3. Some popular docking programs based on the nature of these algorithms with their availability

S. N	Software	Scoring Function	Search Algorithm	Availability
1	AutoDock Vina	Empirical Free Energy Function	Lamarckian Genetic Algorithm	https://autodock.scripps.edu/
2	Dock 6	Grid-based + Force Field	Anchor-and-Grow, Systematic	https://dock.compbio.ucsf.edu/DOCK_6/index.htm
3	LeDock	Empirical + Force Field	Stochastic	https://www.lephar.com/software
4	FlexX	Incremental Construction + Empirical	Fragment-based Incremental	
5	GOLD	Genetic Algorithm + Empirical Scoring	Genetic Algorithm	https://www.ccdc.cam.ac.uk/solutions/software/gold/
6	Glide	Empirical + Force Field (SP/XP)	Exhaustive + Hierarchical	https://www.schrodinger.com/platform/products/glide/
7	MOE-Dock	Empirical, Force Field	Multiple (Systematic/Stochastic)	https://www.chemcomp.com/en/Products.htm
8	SwissDock	CHARMM Force Field	Exhaustive	https://www.swissdock.ch/
9	rDock	Empirical, Knowledge-Based	Genetic + Stochastic	https://rdock.github.io/
10	RosettaLigand	Knowledge-based + Energy Functions	Monte Carlo + Minimization	https://meilerlab.org/rosetta-ligand/
11	MZ dock	Genetic algorithm	Smina	https://sourceforge.net/projects/mzdock/

continued on following page

Table 3. Continued

S. N	Software	Scoring Function	Search Algorithm	Availability
12	PatchDock	Geometric Shape Complementarity	Surface Patch Matching	https://bioinfo3d.cs.tau.ac.il/PatchDock/
13	HADDOCK (High Ambiguity Driven protein-protein Docking)	Biophysical and Knowledge-based	Data-driven Docking	https://rascar.science.uu.nl/haddock2.4/

By using the docking program, the phytochemicals (ligands) can be screened for their binding affinity with the respective MMP receptors involved in the wound healing process. Table 4 presents a curated list of phytochemicals derived from medicinal plants that have demonstrated inhibitory effects on specific matrix metalloproteinases (MMPs).

Table 4. List of phytochemicals that show their efficacy against MMPs evaluated by molecular docking with experimental study

S. N	Phytochemical	Plant Source	Target MMP(s)	Reference
1	quercetin, isoquercetin, isorhamnetin, and isorhamnetin 3-O glycoside	*Calendula officinalis*	MMP-8) and MMP-9	(Belal et al., 2022)
2	licoflavone A, pinostrobin	Natural sources	MMP-1 and MMP-2	(Arda et al., 2024)
3	alliospiroside A	*Cissus rotundifolia*	MMP -2 and MMP -9	(Mohan et al., 2022)
4	4H1Q revealed that compounds 1, 3, 5 and 6 exhibited maximum interactions with the target protein,	*Symplocos racemosa* Roxb	MMP -2	(Rafey et al., 2024)
5	3,3-Dimethyl-1,2,4-cyclopentanetrione	*Chromolaena Odorata*	MMP-9	(Hazian et al., 2022)
6	luteolin 7-*O*-*β*-D-glucopyranoside (4)	*Glechoma hederacea* L.	MMP-8	(El-Aasr et al., 2023)
7	Linoleic acid, Kaempferol-7-O-Glucoside, Betulin	*Hygrophila auriculata*	MMPs (Elastase, gelatinase, and collagenase.)	(Deepika & Santhy, 2023)
8	trans-3-indole acrylic acid	*Anredera cordifolia* (Ten.) Steenis	MMP1	(Dewi et al., 2025)
9	1-Tetracosanol	*Eupatorium glandulosum*	MMP-9	(Shalini et al., 2023)

continued on following page

Table 4. Continued

S. N	Phytochemical	Plant Source	Target MMP(s)	Reference
10	anthraquinones	*Rumex crispus* L.	MMPs	(Uzun et al., 2020)
11	(-)-epicatechin and proanthocyanidin B2	Natural source	MMP-1	(Lee et al., 2020)
12	Tagitinin F	Natural sources	MMP-1 and MMP-2	(Silva et al., 2021)
13	nordentatin, dentatin, clausenidin, and xanthoxyletin	*Clausena excavata*	MMP-9.	(Afriza et al., 2020)
14	wogonin	*Radix Scutellariae*	MMP-9	(Hong et al., 2018)
15	lignan-related molecules	*Olea europaea* L. Cv. Arbequina	MMP-1	(Al-Warhi et al., 2022)
16	Isochiapin B	*Calendula officinalis* L.	MMP-9	(Salama et al., 2024)
17	Naringin	natural sources	MMP-2 and MMP-9	(Aroui et al., 2016)
18	friedelin and lupeol,	*Holoptelea integrifolia* (Roxb.) Planch.	MMP-9	(Somwong et al., 2022)
19	myricetin and quinic acid	*Helianthemum canum* (L.) Baumg	MMPs	(Küpeli Akkol et al., 2023)
20	kaempferol.	*Moringa oleifera* Lam.	MPs (matrix metalloproteinases) subtype I and II,	(Shady et al., 2022)
21	Sophoraflavonoloside, sophoraflavanone G, genistein 7,4′-di-O- -D-glucopyranoside and kaempferol	*Glycyrrhiza glabra* and *Sophora japonica*	MMP-8	(Aly et al., 2023)
22	quercetin-3,4′-*O*-diglucoside, alliospiroside C, and alliospiroside D	*Allium cepa Solanum tuberosum Allium sativum*	MMP-13	(Salem et al., 2023)

PHYTOCHEMICALS IN TARGETING BIOFILM AND ANTIBIOTIC RESISTANCE IN WOUND MANAGEMENT

The major cause of non-healing of chronic wounds is due to persistent microbial infections and the formation of biofilms. These biofilms consist of organized microbial communities on the extracellular polymeric matrix that protects the pathogens from host immune defenses. Recently, phytochemicals have gained attention as useful alternatives to conventional antibiotics due to their capability in disrupting multiple bacterial targets (Ashraf et al., 2023; Banerjee et al., 2024). The phytochemicals act in different ways, such as the disruption of biofilm formation by interfering with quorum sensing (the chemical communication system of bacteria) used to regulate the biofilm formation process (Almuhanna et al., 2024; Banerjee et al., 2024). In addition to this, some plant-derived phenolic compounds weaken the biofilm matrix by breaking down the polysaccharide and protein components of the extracellular polymeric substance (Suganya et al., 2022). It has been established that the phytochemicals, such as tannins, can block the early stages of biofilm formation by changing bacterial surface properties and preventing strong attachment to wound tissues. Sometimes the phytochemicals also enhance the healing process by modulating the multidrug resistance mechanism of the microbes, particularly through their action on bacterial efflux systems (El-Sapagh et al., 2023). Natural compounds like piperine and berberine can inhibit these efflux pumps, allowing antibiotics to remain inside bacterial cells for longer and improving their effectiveness (Jhanji et al., 2020). In addition to efflux pump inhibition, some terpenoids and essential oil components damage the bacterial cell membrane by increasing its fluidity and permeability (Andrade-Ochoa et al., 2021).Molecular docking and dynamics studies indicate that *Andrographis echioides* extracts show strong antibacterial effects and support wound healing similar to standard treatments. Stable binding with *Staphylococcus aureus* proteins and important wound-healing targets, including GSK-3β and the TGF-β receptor kinase, supports these findings (Dhamodiran et al., 2024). The phytochemical curcumin has also shown strong antibiofilm activity against *Staphylococcus aureus*. Subsequent docking studies confirmed its interaction with key biofilm-regulating proteins, suggesting its potential at the molecular level (Gao et al., 2023). Another effective approach involves blocking quorum sensing signalling. The compounds, such as squalene and sorbitol from medicinal plant extracts, have been shown to disrupt this mechanism. This phenomenon of reduction in biofilms has been observed in the case of pathogens causing diabetic foot ulcers (Mashamba et al., 2022). Antibiofilm activity can be further enhanced by targeting the extracellular matrix. Recent docking and molecular dynamics studies identified glucosamine as a stable inhibitor of matrix-forming enzymes in mixed *Pseudomonas aeruginosa–Candida albicans* biofilms (Putri et al., 2025). In addition, docking studies also showed that inhibition of bacterial DNA gyrase is a key mechanism in blocking microbial biofilm activity. As evidence, the neem-derived phytochemicals exhibited strong antibiofilm activity (Mahmoud et al., 2024). In summary, these results indicate that phytochemicals control wound biofilms through several complementary mechanisms. However, the molecular docking and dynamics simulations help explain these effects adequately and support their application in wound-care therapy.

FUTURE DIRECTIONS

The complicated roles of growth factors and other related proteins make it difficult to develop medications that effectively heal wounds because imbalances in their function can have negative side effects. Hence, proper dosing and controlled delivery are crucial to ensure therapeutic action against wound healing. The strategies, such as gene therapy and RNA interference, may offer hope for safer and more effective wound-healing solutions in the future (Karukonda et al., 2000; Zielins et al., 2015).

However, some future directions can be proposed based on current insights:

- Developments in virtual screening pipelines, like integrated machine learning models, could enable the quick testing of thousands of phytochemicals against a variety of wound-related targets, including microbial enzymes, growth factor receptors, cytokines, and MMPs.
- Mapping of how a single phytochemical can affect interconnected targets of the wound-healing process like TNF-α, VEGFR2, and MMP-9 by using network pharmacology.
- Molecular docking assumes rigid receptor-ligand interactions, limiting its accuracy in dynamic biological systems like wound healing. Integrating molecular dynamics and MM-GBSA calculations can better capture protein flexibility and improve predictions of ligand stability under stress conditions.
- Two main hurdles to the healing of chronic wounds are polymicrobial biofilms and an unbalanced wound microbiome. Future docking methods should focus on this aspect.
- Phytochemicals can be optimized through structure-based design or combined with nanoparticles to enhance their bioavailability and wound-healing efficacy.

CONCLUSION

The biological process of wound healing is complex and includes angiogenesis, tissue remodeling, inflammation, and cell proliferation. Because little is known about the underlying molecular mechanisms, treating both acute and chronic wounds is still difficult in modern medicine. The medicinal plants contain bioactive substances like flavonoids, alkaloids, triterpenoids, carotenoids, and polyphenols, that significantly used in promoting wound healing. However, evaluating the potential plant compound on an experimental basis is a tedious job. The molecular docking method is used to evaluate the binding profile of these phytochemicals with particular wound-related proteins. Researchers can find promising compounds more quickly and effectively by combining this computational technique with experiments. Future research should focus on combined methods such as docking studies with molecular dynamics, customized animal models, and nanotechnology to improve the design of plant-based treatments for a better outcome.

REFERENCES

Afriza, D., Orienty, F. N., & Ayu, W. P. (2020). Molecular docking analysis of the interactions between MMP-9 protein and four coumarin compounds (nordentatin, dentatin, calusenidin and xanthoxyletin). *Journal of International Dental and Medical Research, 13*(4), 1286–1292.

Al-Warhi, T., Elmaidomy, A. H., Maher, S. A., Abu-Baih, D. H., Selim, S., Albqmi, M., Al-Sanea, M. M., Alnusaire, T. S., Ghoneim, M. M., Mostafa, E. M., Hussein, S., El-Damasy, A. K., Saber, E. A., Elrehany, M. A., Sayed, A. M., Othman, E. M., El-Sherbiny, M., & Abdelmohsen, U. R. (2022). The wound-healing potential of *Olea europaea* L. cv. Arbequina leaves extract: An integrated in vitro, in silico, and in vivo investigation. *Metabolites, 12*(9), 791. DOI: 10.3390/metabo12090791 PMID: 36144197

Alherz, F. A., Negm, W. A., Elekhnawy, E., El-Masry, T. A., Haggag, E. M., Alqahtani, M. J., & Hussein, I. A. (2022). Silver nanoparticles prepared using *Encephalartos laurentianus* de Wild leaf extract have inhibitory activity against *Candida albicans* clinical isolates. *Journal of Fungi (Basel, Switzerland), 8*(10), 1005. DOI: 10.3390/jof8101005 PMID: 36294570

Almuhanna, Y., Alshalani, A., AlSudais, H., Alanazi, F., Alissa, M., & Joseph, B. (2024). Antibacterial, antibiofilm, and wound healing activities of rutin and quercetin and their interaction with gentamicin on excision wounds in diabetic mice. *Biology (Basel), 13*(9), 676. DOI: 10.3390/biology13090676 PMID: 39336103

Aly, S. H., Elissawy, A. M., Mahmoud, A. M. A., El-Tokhy, F. S. E., Mageed, S. S. A., Almahli, H., Al-Rashood, S. T., Binjubair, F. A., Hassab, M. A. E., Eldehna, W. M., & Singab, A. E. B. (2023). Synergistic effect of *Sophora japonica* and *Glycyrrhiza glabra* flavonoid-rich fractions on wound healing: In vivo and molecular docking studies. *Molecules (Basel, Switzerland), 28*(7), 2994. DOI: 10.3390/molecules28072994 PMID: 37049756

Andrade-Ochoa, S., Chacón-Vargas, K. F., Sánchez-Torres, L. E., Rivera-Chavira, B. E., Nogueda-Torres, B., & Nevárez-Moorillón, G. V. (2021). Differential antimicrobial effect of essential oils and their main components: Insights based on the cell membrane and external structure. *Membranes, 11*(6), 405. DOI: 10.3390/membranes11060405 PMID: 34071618

Arda, A. G., Syaifie, P. H., Ramadhan, D., Jauhar, M. M., Nugroho, D. W., Kaswati, N. M. N., Noviyanto, A., Safihtri, M., Rochman, N. T., Andrianto, D., & Mardliyati, E. (2024). Activity of propolis compounds as potential MMP1 and MMP2 inhibitors by in silico studies in wound healing application. *Journal of Pharmacy & Pharmacognosy Research*, *12*(2), 264–285. DOI: 10.56499/jppres23.1719_12.2.264

Aroui, S., Aouey, B., Chtourou, Y., Meunier, A. C., Fetoui, H., & Kenani, A. (2016). Naringin suppresses cell metastasis and the expression of matrix metalloproteinases (MMP-2 and MMP-9) via the inhibition of ERK-P38-JNK signaling pathway in human glioblastoma. *Chemico-Biological Interactions*, *244*, 195–203. DOI: 10.1016/j.cbi.2015.12.011 PMID: 26721195

Arunachalam, K. D., Subhashini, S., & Annamalai, S. K. (2012). Wound healing and antigenotoxic activities of *Aegle marmelos* with relation to its antioxidant properties. *Journal of Pharmacy Research*, *5*(3), 1492–1502.

Ashraf, M. V., Pant, S., Khan, M. A. H., Shah, A. A., Siddiqui, S., Jeridi, M., & Ahmad, S. (2023). Phytochemicals as antimicrobials: Prospecting Himalayan medicinal plants as source of alternate medicine to combat antimicrobial resistance. *Pharmaceuticals (Basel, Switzerland)*, *16*(6), 881. DOI: 10.3390/ph16060881 PMID: 37375828

Attallah, N. G. M., Negm, W. A., Elekhnawy, E., Elmongy, E. I., Altwaijry, N., El-Haroun, H., El-Masry, T. A., & El-Sherbeni, S. A. (2021). Elucidation of phytochemical content of *Cupressus macrocarpa* leaves in vitro and in vivo anti-bacterial effect against methicillin resistant *Staphylococcus aureus* clinical isolates. *Antibiotics (Basel, Switzerland)*, *10*(8), 890. DOI: 10.3390/antibiotics10080890 PMID: 34438940

Azab, A., Nassar, A., & Azab, A. N. (2016). Anti-inflammatory activity of natural products. *Molecules (Basel, Switzerland)*, *21*(10), 1321. DOI: 10.3390/molecules21101321 PMID: 27706084

Aziz, N., Kim, M. Y., & Cho, J. Y. (2018). Anti-inflammatory effects of luteolin: A review of in vitro, in vivo, and in silico studies. *Journal of Ethnopharmacology*, *225*, 342–358. DOI: 10.1016/j.jep.2018.05.019 PMID: 29801717

Banerjee, D., Biswas, P., Mazumder, K., Palai, S., Hossain, C. M., Karmakar, S., & Biswas, K. (2024). Exploration of phytochemicals as anti-biofilm agents against pathogenic bacteria: Their potential and challenges. *Infectious Disorders Drug Targets*, *25*(6), e18715265324950. Advance online publication. DOI: 10.2174/0118715265324950241204182204 PMID: 39757612

Belal, A., Elanany, M. A., Raafat, M., Hamza, H. T., & Mehany, A. B. M. (2022). *Calendula officinalis* phytochemicals for the treatment of wounds through matrix metalloproteinases-8 and 9 (MMP-8 and MMP-9): In silico approach. *Natural Product Communications*, *17*(5), 1–7. DOI: 10.1177/1934578X221098848

Bilal, K., Mehboob, F., Akhtar, N., Mirza, I. A., Okla, M. K., Dar, M. J., Saleh, I. A., Zomot, N., & Fatima, H. (2024). Wound healing, antioxidant and antibacterial activities of polyphenols of *Psidium guajava* L. leaves. *South African Journal of Botany*, *165*, 538–551. DOI: 10.1016/j.sajb.2023.12.026

Blume-Peytavi, U., Tan, J., Tennstedt, D., Boralevi, F., Fabbrocini, G., Torrelo, A., Soares-Oliveira, R., Haftek, M., Rossi, A. B., Thouvenin, M. D., Mangold, J., Galliano, M. F., Hernandez-Pigeon, H., Aries, M. F., Rouvrais, C., Bessou-Touya, S., Duplan, H., Castex-Rizzi, N., Mengeaud, V., & Dauskardt, R. (2016). Fragility of epidermis in newborns, children and adolescents. *Journal of the European Academy of Dermatology and Venereology*, *30*(9), 1634. DOI: 10.1111/jdv.13636 PMID: 27062556

Boukeloua, A., Kouadria, M., Bendif, H., Plavan, G., Alsalamah, S. A., Alghonaim, M. I., Boufahja, F., & Abd-Elkader, O. H. (2023). Physicochemical analysis and wound healing activity of *Azadirachta indica* (A. Juss.) fruits. *Processes (Basel, Switzerland)*, *11*(6), 1692. DOI: 10.3390/pr11061692

Caley, M. P., Martins, V. L. C., & O'Toole, E. A. (2015). Metalloproteinases and wound healing. *Advances in Wound Care*, *4*(4), 225–234. DOI: 10.1089/wound.2014.0581 PMID: 25945285

Chaudhari, M., & Mengi, S. (2006). Evaluation of phytoconstituents of *Terminalia arjuna* for wound healing activity in rats. *Phytotherapy Research*, *20*(9), 799–805. DOI: 10.1002/ptr.1857 PMID: 16835874

Deepika, E., & Santhy, K. S. (2023). Unravelling the mode of action of wound healing efficiency with the compounds screened from *Hygrophila auriculata* by molecular docking studies. *Journal of Plant Science Research*, *2*, 1–9.

Dewi, K. A., Kusumaningrum, S., Prasetya, N. B. A., Kusumastuti, S. A., Rismana, E., Ngatinem, N., Nuralih, N., & Wulansari, M. T. (2025). Potential of binahong (*Anredera cordifolia* (Ten.) Steenis) leaf extract as a wound healing agent that inhibits matrix metalloproteinases (MMPs) using in silico and stimulates NIH-3T3 cell proliferation by in vitro assay. *Farmatsiia (Sofia)*, *72*, 1–15. DOI: 10.3897/pharmacia.72.e135811

Dhamodiran, M., Chinnaperumal, K., Venkatesan, G., Alshiekheid, M. A., & Suseem, S. R. (2024). Isolation, structural elucidation of bioactive compounds and their wound-healing ability, antibacterial and In silico molecular docking applications. *Environmental Research*, *252*, 119023. DOI: 10.1016/j.envres.2024.119023 PMID: 38685295

Dinh, T., Elder, S., & Veves, A. (2011). Delayed wound healing in diabetes: Considering future treatments. *Diabetes Management (London, England)*, *1*(5), 509–519. DOI: 10.2217/dmt.11.44

Diniz, L. R. L., Calado, L. L., Duarte, A. B. S., & de Sousa, D. P. (2023). *Centella asiatica* and its metabolite asiatic acid: Wound healing effects and therapeutic potential. *Metabolites*, *13*(2), 276. DOI: 10.3390/metabo13020276 PMID: 36837896

Dow, G., Browne, A., & Sibbald, R. G. (1999). Infection in chronic wounds: Controversies in diagnosis and treatment. *Ostomy/Wound Management*, *45*(1), 23. PMID: 10655866

El-Aasr, M., Nohara, T., Ikeda, T., Abu-Risha, S. E., Elekhnawy, E., Tawfik, H. O., Shoeib, N., & Attia, G. (2023). LC-MS/MS metabolomics profiling of *Glechoma hederacea* L. methanolic extract; in vitro antimicrobial and in vivo with in silico wound healing studies on *Staphylococcus aureus* infected rat skin wound. *Natural Product Research*, *37*(10), 1730–1734. DOI: 10.1080/14786419.2022.2108421 PMID: 35942893

El-Sapagh, S., El-Shenody, R., Pereira, L., & Elshobary, M. (2023). Unveiling the potential of algal extracts as promising antibacterial and antibiofilm agents against multidrug-resistant Pseudomonas aeruginosa: In vitro and in silico studies including molecular docking. *Plants*, *12*(18), 3324. DOI: 10.3390/plants12183324 PMID: 37765485

El-Sherbeni, S. A., & Negm, W. A. (2023). The wound healing effect of botanicals and pure natural substances used in in vivo models. *Inflammopharmacology*, *31*(2), 755–772. DOI: 10.1007/s10787-023-01157-5 PMID: 36811778

Fan, J., Fu, A., & Zhang, L. (2019). Progress in molecular docking. *Quantitative Biology*, *7*(2), 83–89. DOI: 10.1007/s40484-019-0172-y

Ganesan, S., Parasuraman, S., Maheswaran, S. U., & Gnanasekar, N. (2012). Wound healing activity of *Hemidesmus indicus* formulation. *Journal of Pharmacology & Pharmacotherapeutics*, *3*(1), 66–67. DOI: 10.4103/0976-500X.92516 PMID: 22368424

Gao, K., Zhang, B., & Zhao, F. (2023). Antibiofilm effect of curcumin against Staphylococcus aureus surface wound biofilm–associated infection: In vitro and in silico. *Applied Biochemistry and Biotechnology*, *195*(9), 5329–5337. DOI: 10.1007/s12010-022-03844-5 PMID: 35166997

Gethin, G. (2007). The significance of surface pH in chronic wounds. *Wounds UK*, *3*(1), 52–56.

Ghosh, T., Maity, T. K., Dash, D. K., & Boss, A. (2007). A study on wound healing activity of *Bacopa monnieri* Linn. aerial parts. *Oriental Pharmacy and Experimental Medicine*, *7*(2), 150–156. DOI: 10.3742/OPEM.2007.7.2.150

Gonfa, Y. H., Tessema, F. B., Bachheti, A., Rai, N., Tadesse, M. G., Nasser Singab, A. N., Chaubey, K. K., & Bachheti, R. K. (2023). Anti-inflammatory activity of phytochemicals from medicinal plants and their nanoparticles: A review. *Current Research in Biotechnology*, *6*, 100152. DOI: 10.1016/j.crbiot.2023.100152

Gonzalez, A. C. O., Costa, T. F., Andrade, Z. A., & Medrado, A. R. A. P. (2016). Wound healing – A literature review. *Anais Brasileiros de Dermatologia*, *91*(5), 614–620. DOI: 10.1590/abd1806-4841.20164741 PMID: 27828635

Grice, E. A., & Segre, J. A. (2011). The skin microbiome. *Nature Reviews. Microbiology*, *9*(4), 244–253. DOI: 10.1038/nrmicro2537 PMID: 21407241

Guo, S., & DiPietro, L. A. (2010). Factors affecting wound healing. *Journal of Dental Research*, *89*(3), 219–229. DOI: 10.1177/0022034509359125 PMID: 20139336

Haq, M. M., Chowdhury, M. A. R., Tayara, H., Abdelbaky, I., Islam, M. S., Chong, K. T., & Jeong, S. (2021). A report on multi-target anti-inflammatory properties of phytoconstituents from *Monochoria hastata* (Family: Pontederiaceae). *Molecules (Basel, Switzerland)*, *26*(23), 7397. DOI: 10.3390/molecules26237397 PMID: 34885978

Hasibuan, P. F., Tambunan, A. Z., & Mutia, M. S. (2023). Anti-inflammatory potential of lemongrass leaves in wound healing. *Jurnal Prima Medika Sains*, *5*(2), 126–129. DOI: 10.34012/jpms.v5i2.4402

Hazian, N. Q., Mokhtar, N. A., Khairudin, N. A., & Yousif, R. H.Nur Qamarina HazianNur Ainun MokhtarNurulbahiyah Ahmad Khairudin. (2022). In silico molecular docking simulation of *Chromolaena odorata* phytoconstituents against matrix metalloproteinase-9 (MMP-9). *Journal of Research in Nanoscience and Nanotechnology*, *7*(1), 1–5. DOI: 10.37934/jrnn.7.1.16

Hong, M., Cheng, H., Song, L., Wang, W., Wang, Q., Xu, D., & Xing, W. (2018). Wogonin suppresses the activity of matrix metalloproteinase-9 and inhibits migration and invasion in human hepatocellular carcinoma. *Molecules (Basel, Switzerland), 23*(2), 384. DOI: 10.3390/molecules23020384 PMID: 29439451

İğci, B. K., & Aytaç, Z. (2020). An investigation on the in vitro wound healing activity and phytochemical composition of *Hypericum pseudolaeve* N. Robson growing in Turkey. *Turkish Journal of Pharmaceutical Sciences, 17*(6), 610–619. DOI: 10.4274/tjps.galenos.2019.80037 PMID: 33389950

Irham, W. H., & Hardiyanti, R. (2021). Wound healing bioactivity of *Curcuma longa* Linn. *Rasayan Journal of Chemistry, 14*(4), 2386–2391. DOI: 10.31788/RJC.2021.1446345

Jain, N., Jain, R., Jain, A., Jain, D. K., & Chandel, H. S. (2010). Evaluation of wound-healing activity of *Acorus calamus* Linn. *Natural Product Research, 24*(6), 534–541. DOI: 10.1080/14786410802531782 PMID: 20182947

Jhanji, R., Bhati, V., Singh, A., & Kumar, A. (2020). Phytomolecules against bacterial biofilm and efflux pump: An in silico and in vitro study. *Journal of Biomolecular Structure & Dynamics, 38*(18), 5500–5512. DOI: 10.1080/07391102.2019.1704884 PMID: 31847713

Jones, M. L. (2015). A short history of the development of wound care dressings. *British Journal of Healthcare Assistants, 9*(10), 482–485. DOI: 10.12968/bjha.2015.9.10.482

Kandhwal, M., Behl, T., Singh, S., Sharma, N., Arora, S., Bhatia, S., Al-Harrasi, A., Sachdeva, M., & Bungau, S. (2022). Role of matrix metalloproteinase in wound healing. *American Journal of Translational Research, 14*(7), 4391–4405. PMID: 35958464

Karukonda, S. R., Flynn, T. C., Boh, E. E., McBurney, E. I., Russo, G. G., & Millikan, L. E. (2000). The effects of drugs on wound healing: Part 1. *International Journal of Dermatology, 39*(4), 250–257. DOI: 10.1046/j.1365-4362.2000.00948.x PMID: 10809970

Keng, J. W., Lee, S. K., Sang, S. H., Liew, K. B., Teo, S. S., Mossadeq, W. M. S. M., & Chew, Y. L. (2024). *Cassia alata* and its phytochemicals: A promising natural strategy in wound recovery. *Sci Mossadeq, W. M. S. M., & Chew, Y. L., 6*(2), 34.

Kokane, D. D., More, R. Y., Kale, M. B., Nehete, M. N., Mehendale, P. C., & Gadgoli, C. H. (2009). Evaluation of wound healing activity of root of *Mimosa pudica*. *Journal of Ethnopharmacology*, *124*(2), 311–315. DOI: 10.1016/j.jep.2009.04.038 PMID: 19397984

Kumar, P., & Kothari, V. (2021). *Wound healing research: Current trends and future directions*. Springer. DOI: 10.1007/978-981-16-2677-7

Kumar, S., Kumar, A., Kumar, N., Singh, P., Singh, T. U., Singh, B. R., Gupta, P. K., & Thakur, V. K. (2022). In vivo therapeutic efficacy of *Curcuma longa* extract loaded ethosomes on wound healing. *Veterinary Research Communications*, *46*(4), 1033–1049. DOI: 10.1007/s11259-022-09952-1 PMID: 35796857

Kumar, S., Wong, P. F., & Leaper, D. J. (2004). What is new in wound healing? *Turkish Journal of Medical Sciences*, *34*(3), 147–160.

Küpeli Akkol, E., Kosar, M., Baldemir, A., Şeker Karatoprak, G., Demirpolat, E., Yerer Aycan, M. B., Süntar, I., & Ilgün, S. (2023). The wound-healing potential of the endemic plant *Helianthemum canum* (L.) Baumg.: Preclinical studies supported with phytochemical profiling. *Chemistry & Biodiversity*, *20*(12), e202301529. DOI: 10.1002/cbdv.202301529 PMID: 37955210

Kwiecień, K., Zegar, A., Jung, J., Brzoza, P., Kwitniewski, M., Godlewska, U., Grygier, B., Kwiecińska, P., Morytko, A., & Cichy, J. (2019). Architecture of antimicrobial skin defense. *Cytokine & Growth Factor Reviews*, *49*, 70–84. DOI: 10.1016/j.cytogfr.2019.08.001 PMID: 31473081

Lee, K. E., Bharadwaj, S., Yadava, U., & Kang, S. G. (2020). Computational and in vitro investigation of (-)-epicatechin and proanthocyanidin B2 as inhibitors of human matrix metalloproteinase 1. *Biomolecules*, *10*(10), 1379. DOI: 10.3390/biom10101379 PMID: 32998374

Mahmoud, G. A. E., Rashed, N. M., El-Ganainy, S. M., & Salem, S. H. (2024). Unveiling the neem (Azadirachta indica) effects on biofilm formation of food-borne bacteria and the potential mechanism using a molecular docking approach. *Plants*, *13*(18), 2669. DOI: 10.3390/plants13182669 PMID: 39339644

Mashamba, T. G., Adeosun, I. J., Baloyi, I. T., Tshikalange, E. T., & Cosa, S. (2022). Quorum sensing modulation and inhibition in biofilm forming foot ulcer pathogens by selected medicinal plants. *Heliyon*, *8*(4), e09303. DOI: 10.1016/j.heliyon.2022.e09303 PMID: 35520625

Mohan, S., Prabhakaran, V. S., & Narayanaswamy, R. (2022). In silico analysis of *Cissus rotundifolia* constituents as human neutrophil elastase (HNE), matrix metalloproteinases (MMP 2 and MMP 9), and tyrosinase inhibitors. *Applied Biochemistry and Biotechnology*, *194*(1), 232–245. DOI: 10.1007/s12010-021-03758-8 PMID: 34800282

Monika, P., Chandraprabha, M. N., Rangarajan, A., Waiker, P. V., & Chidambara Murthy, K. N. (2022). Challenges in healing wound: Role of complementary and alternative medicine. *Frontiers in Nutrition*, *8*, 791899. DOI: 10.3389/fnut.2021.791899 PMID: 35127787

Morris, G. M., & Lim-Wilby, M. (2008). Molecular docking. In Kukol, A. (Ed.), *Molecular Modeling of Proteins* (pp. 365–382). Humana Press., DOI: 10.1007/978-1-59745-177-2_19

Mumtaz, R., Zubair, M., Khan, M. A., Muzammil, S., & Siddique, M. H. (2022). Extracts of *Eucalyptus alba* promote diabetic wound healing by inhibiting α-glucosidase and stimulating cell proliferation. *Evidence-Based Complementary and Alternative Medicine : eCAM*, *4953105*, 1–12. Advance online publication. DOI: 10.1155/2022/4953105 PMID: 35463094

Muralidhar, A., Babu, K. S., Sankar, T. R., Reddanna, P., & Latha, J. (2013). Wound healing activity of flavonoid fraction isolated from the stem bark of *Butea monosperma* (Lam.) in albino Wistar rats. *European Journal of Experimental Biology*, *3*(6), 1–6.

Nagase, H., Visse, R., & Murphy, G. (2006). Structure and function of matrix metalloproteinases and TIMPs. *Cardiovascular Research*, *69*(3), 562–573. DOI: 10.1016/j.cardiores.2005.12.002 PMID: 16405877

Nayak, B. S., Raju, S. S., & Rao, A. V. C. (2007). Wound healing activity of *Matricaria recutita* L. extract. *Journal of Wound Care*, *16*(7), 298–302. DOI: 10.12968/jowc.2007.16.7.27061 PMID: 17708380

Nayak, S. B., Rodrigues, V., Maharaj, S., & Bhogadi, V. S. (2013). Wound healing activity of the fruit skin of *Punica granatum*. *Journal of Medicinal Food*, *16*(9), 857–861. DOI: 10.1089/jmf.2012.0229 PMID: 24044494

Ning, S., Zang, J., Zhang, B., Feng, X., & Qiu, F. (2022). Botanical drugs in traditional Chinese medicine with wound healing properties. *Frontiers in Pharmacology*, *13*, 885484. DOI: 10.3389/fphar.2022.885484 PMID: 35645789

Pagadala, N. S., Syed, K., & Tuszynski, J. (2017). Software for molecular docking: A review. *Biophysical Reviews*, *9*(2), 91–102. DOI: 10.1007/s12551-016-0247-1 PMID: 28510083

Page-McCaw, A., Ewald, A. J., & Werb, Z. (2007). Matrix metalloproteinases and the regulation of tissue remodelling. *Nature Reviews. Molecular Cell Biology*, *8*(3), 221–233. DOI: 10.1038/nrm2125 PMID: 17318226

Paggi, J. M., Pandit, A., & Dror, R. O. (2024). The art and science of molecular docking. *Annual Review of Biochemistry*, *93*(1), 389–410. DOI: 10.1146/annurev-biochem-030222-120000 PMID: 38594926

Palmieri, B., Vadalà, M., & Laurino, C. (2017). Review of the molecular mechanisms in wound healing: New therapeutic targets? *Journal of Wound Care*, *26*(12), 765–775. DOI: 10.12968/jowc.2017.26.12.765 PMID: 29244975

Panares Arcaya, P. J., Claro Capilitan, L. M. D., Decano, Z. R., Guyo Dugaduga, C. D., Lamasan Hernane, A. G., Calo, J. R. D., & Paspe Saldo, I. J. (2025). Phytochemical analysis and wound healing potential of *Aloe vera* (*Aloe massawana*) peel extract. *Asian Journal of Biological and Life Sciences*, *14*(1), 147–152. DOI: 10.5530/ajbls.20251440

Parks, W. C., Wilson, C. L., & López-Boado, Y. S. (2004). Matrix metalloproteinases as modulators of inflammation and innate immunity. *Nature Reviews. Immunology*, *4*(8), 617–629. DOI: 10.1038/nri1418 PMID: 15286728

Percival, N. J. (2002). Classification of wounds and their management. *Surgical Infections*, *20*(5), 114–117.

Portou, M. J., Baker, D., Abraham, D., & Tsui, J. (2015). The innate immune system, toll-like receptors and dermal wound healing: A review. *Vascular Pharmacology*, *71*, 31–36. DOI: 10.1016/j.vph.2015.02.007 PMID: 25869514

Potekaev, N. N., Borzykh, O. B., Medvedev, G. V., Pushkin, D. V., Petrova, M. M., Petrov, A. V., Dmitrenko, D. V., Karpova, E. I., Demina, O. M., & Shnayder, N. A. (2021). The role of extracellular matrix in skin wound healing. *Journal of Clinical Medicine*, *10*(24), 5947. DOI: 10.3390/jcm10245947 PMID: 34945243

Powanda, M. C., & Moyer, E. D. (1981). Plasma proteins and wound healing. *Surgery, Gynecology & Obstetrics*, *153*(5), 749–755. PMID: 7027503

Putri, R. F., Abdjan, M. I., Santosaningsih, D., Ni, Y. L., Putri, N., Khairunnisa, F., & Baktir, A. (2025). molecular dynamic and docking simulation in exploring a potential extracellular matrix-based antibiofilm: Fighting the resistant Pseudomonas aeruginosa-Candida albicans. *Engineering and Science*, *34*, 1386. DOI: 10.30919/es1386

Rafey, H. A., Amin, A., Ross, S. A., El-Shazly, M., Zahid, M. A., Niaz, S. I., Ul Mahmood, F., & Ullah, H. (2024). Multiple integrated computational approach to analyse wound healing potential of *Symplocos racemosa* bark as matrix metalloproteinase inhibitors. *Natural Product Research*, •••, 1–10. DOI: 10.1080/14786419.2024.2345678 PMID: 38497294

Richmond, J. M., & Harris, J. E. (2014). Immunology and skin in health and disease. *Cold Spring Harbor Perspectives in Medicine*, *4*(12), a015339. DOI: 10.1101/cshperspect.a015339 PMID: 25452424

Robson, M. C., & Barbul, A. (2006). Guidelines for the best care of chronic wounds. *Wound Repair and Regeneration*, *14*(6), 647–648. DOI: 10.1111/j.1524-475X.2006.00173.x PMID: 17199830

Salama, A., El-Kassaby, M. I., Refaat, A., & Mohasib, R. M. (2024). GC–MS and molecular docking analyses of phytochemicals from *Calendula officinalis* L. hexane extract and evaluation of its antioxidant and wound healing properties in rats. *Egyptian Journal of Chemistry*, *67*(13), 1037–1058. DOI: 10.21608/ejchem.2024.328645.10642

Salem, M. A., Mohamed, O. G., Mosalam, E. M., Elberri, A. I., Abdel-Bar, H. M., Hassan, M., Al-Karmalawy, A. A., Tripathi, A., Ezzat, S. M., & Abo Mansour, H. E. (2023). Investigation of the phytochemical composition, antioxidant, antibacterial, anti-osteoarthritis, and wound healing activities of selected vegetable waste. *Scientific Reports*, *13*(1), 13034. DOI: 10.1038/s41598-023-38591-y PMID: 37563154

Satpathy, R. (2020a). Therapeutic importance and application of boswellic acid from the plant *Boswellia serrata*. In Singh, A., Singh, P., & Bithel, N. (Eds.), *Advanced Pharmacological Uses of Medicinal Plants and Natural Products* (pp. 302–315). IGI Global., DOI: 10.4018/978-1-7998-2094-9.ch016

Satpathy, R. (2020b). Application of molecular docking methods on endocrine disrupting chemicals: A review. *Journal of Applied Biotechnology Reports*, *7*(2), 74–80.

Satpathy, R. (2021). Application of some medicinal plants and their constituents in the treatment of diabetes mellitus. In Hussain, A., & Behl, S. (Eds.), *Treating Endocrine and Metabolic Disorders with Herbal Medicines* (pp. 32–47). IGI Global., DOI: 10.4018/978-1-7998-4808-0.ch002

Satpathy, R. (2024a). Application of molecular modeling techniques to investigate phytochemicals as prospective anti-malarial agents. In Tripathi, R. K. P., & Tiwari, S. (Eds.), *Converging Pharmacy Science and Engineering in Computational Drug Discovery* (pp. 116–139). IGI Global., DOI: 10.4018/979-8-3693-2897-2.ch006

Satpathy, R. (2024b). Computational approaches to discover novel phytochemical inhibitors against novel coronavirus (SARS-CoV-2). In *Computational Biology in Drug Discovery and Repurposing* (pp. 323–336). Apple Academic Press. DOI: 10.1201/9781003455424-15

Schreml, S., Szeimies, R. M., Prantl, L., Landthaler, M., & Babilas, P. (2010). Wound healing in the 21st century. *Journal of the American Academy of Dermatology, 63*(5), 866–881. DOI: 10.1016/j.jaad.2009.10.048 PMID: 20576319

Shady, N. H., Mostafa, N. M., Fayez, S., Abdel-Rahman, I. M., Maher, S. A., Zayed, A., Saber, E. A., Khowdiary, M. M., Elrehany, M. A., Alzubaidi, M. A., Altemani, F. H., Shawky, A. M., & Abdelmohsen, U. R. (2022). Mechanistic wound healing and antioxidant potential of *Moringa oleifera* seeds extract supported by metabolic profiling, in silico network design, molecular docking, and in vivo studies. *Antioxidants, 11*(9), 1743. DOI: 10.3390/antiox11091743 PMID: 36139817

Shah, A., & Amini-Nik, S. (2017). The role of phytochemicals in the inflammatory phase of wound healing. *International Journal of Molecular Sciences, 18*(5), 1068. DOI: 10.3390/ijms18051068 PMID: 28509885

Shalini, R., Chandrasekar, M. J. N., Nanjan, M. J., Madhunapantula, S. V., Karnik, M., Selvaraj, J., & Ganesh, G. N. K. (2023). 1-Tetracosanol isolated from the leaves of *Eupatorium glandulosum* accelerates wound healing by expressing inflammatory cytokines and matrix metalloproteinase. *Journal of Ethnopharmacology, 315*, 116654. DOI: 10.1016/j.jep.2023.116654 PMID: 37225028

Silva, L. P., Santos, E. C., Borges, B. A., Veloso, M. P., Chagas-Paula, D. A., Gonçalves, R. V., & Novaes, R. D. (2021). Tagitinin F has anti-inflammatory, anti-nociceptive and anti-matrix metalloproteinase properties: An in silico, in vitro and in vivo study. *Pharmacological Research, 164*, 105303. DOI: 10.1016/j.phrs.2020.105303 PMID: 33212290

Somwong, K., Lertpatipanpong, P., Nimlamool, W., Panya, A., Tragoolpua, Y., Yongsawas, R., Gritsanapan, W., Pandith, H., & Baek, S. J. (2022). Effect of *Holoptelea integrifolia* (Roxb.) Planch. N-hexane extract and its bioactive compounds on wound healing and anti-inflammatory activity. *Molecules (Basel, Switzerland), 27*(23), 8540. DOI: 10.3390/molecules27238540 PMID: 36500633

Stacey, M. C., Wysocki, S. J., Wallace, H. J., Yeoh, S., & Davies, M. J. (2002). Matrix metalloproteinases in wound healing. *ANZ Journal of Surgery, 72*(S1), A118.

Strodtbeck, F. (2001). Physiology of wound healing. *Newborn and Infant Nursing Reviews; NAINR, 1*(1), 43–52. DOI: 10.1053/nbin.2001.23176

Suganya, T., Packiavathy, I. A. S. V., Aseervatham, G. S. B., Carmona, A., Rashmi, V., Mariappan, S., & Ananth, D. A. (2022). Tackling multiple-drug-resistant bacteria with conventional and complex phytochemicals. *Frontiers in Cellular and Infection Microbiology*, *12*, 883839. DOI: 10.3389/fcimb.2022.883839 PMID: 35846771

Teymoorian, S. K., Nouri, H., & Moghimi, H. (2024). In-vivo and in-vitro wound healing and tissue repair effect of *Trametes versicolor* polysaccharide extract. *Scientific Reports*, *14*(1), 3796. DOI: 10.1038/s41598-024-54565-0 PMID: 38360911

Thangapazham, R. L., Sharad, S., & Maheshwari, R. K. (2016). Phytochemicals in wound healing. *Advances in Wound Care*, *5*(5), 230–241. DOI: 10.1089/wound.2013.0505 PMID: 27134766

Thomas, A., Shinde, S., Wavhale, R., Jadhav, P., Tambe, S., Lokhande, K., & Chitlange, S. (2024). In-silico screening of phytomolecules against multiple targets for wound management. *In Silico Pharmacology*, *12*(1), 19. DOI: 10.1007/s40203-024-00194-4 PMID: 38550524

Uzun, M., Guvenalp, Z., Kazaz, C., & Demirezer, L. O. (2020). Matrix metalloproteinase inhibitor and sunscreen effective compounds from *Rumex crispus* L.: Isolation, identification, bioactivity and molecular docking study. *Phytochemical Analysis*, *31*(6), 818–834. DOI: 10.1002/pca.2948 PMID: 32488908

Vijaya, V. T., Srinivasan, D., & Sengottuvelu, S. (2012). Wound healing potential of *Melia azedarach* L. leaves in alloxan induced diabetic rats. *Global Journal of Research on Medicinal Plants and Indigenous Medicine*, *1*(7), 265.

Vitale, S., Colanero, S., Placidi, M., Di Emidio, G., Tatone, C., Amicarelli, F., & D'Alessandro, A. M. (2022). Phytochemistry and biological activity of medicinal plants in wound healing: An overview of current research. *Molecules (Basel, Switzerland)*, *27*(11), 3566. DOI: 10.3390/molecules27113566 PMID: 35684503

Wong, C. M., & Ling, J. J. (2014). In vitro study of wound healing potential in black pepper (*Piper nigrum* L.). *Pharmaceutical and Biosciences Journal*, 5–9.

Xue, M., Le, N. T. V., & Jackson, C. J. (2006). Targeting matrix metalloproteases to improve cutaneous wound healing. *Expert Opinion on Therapeutic Targets*, *10*(1), 143–155. DOI: 10.1517/14728222.10.1.143 PMID: 16441234

Zielins, E. R., Brett, E. A., Luan, A., Hu, M. S., Walmsley, G. G., Paik, K., Senarath-Yapa, K., Atashroo, D. A., Wearda, T., Lorenz, H. P., Wan, D. C., & Longaker, M. T. (2015). Emerging drugs for the treatment of wound healing. *Expert Opinion on Emerging Drugs*, *20*(2), 235–246. DOI: 10.1517/14728214.2015.1018176 PMID: 25704608

Compilation of References

Abad, M. J., Bedoya, L. M., & Bermejo, P. (2010). An Update on Drug Interactions with the Herbal Medicine Ginkgo biloba. *Current Drug Metabolism, 11*(2), 171–181. DOI: 10.2174/138920010791110818 PMID: 20359286

Abbas, F., Ke, Y., Yu, R., Yue, Y., Amanullah, S., Jahangir, M. M., & Fan, Y. (2017). Volatile terpenoids: Functions, biosynthesis, modulation and genetic manipulation. *Planta, 246*, 803–816. DOI: 10.1007/s00425-017-2749-x PMID: 28803364

Abbas, M., Hussain, T., Arshad, M., Ansari, A. R., Irshad, A., Nisar, J., Hussain, F., Masood, N., Nazir, A., & Iqbal, M. (2019). Wound healing potential of curcumin cross-linked chitosan/polyvinyl alcohol. *International Journal of Biological Macromolecules, 140*, 871–876. DOI: 10.1016/j.ijbiomac.2019.08.153 PMID: 31437503

Abd Jalil, M. A., Kasmuri, A. R., & Hadi, H.AbdJalil. (2017). Stingless bee honey, the natural wound healer: A review. *Skin Pharmacology and Physiology, 30*(2), 66–75. DOI: 10.1159/000458416 PMID: 28291965

Abdullahi, A. A. (2011). Trends and challenges of traditional medicine in Africa. *African Journal of Traditional, Complementary, and Alternative Medicines : AJTCAM, 8*(5S). Advance online publication. DOI: 10.4314/ajtcam.v8i5S.5 PMID: 22754064

Abedanzadeh, M., Abolmaali, S. S., Heidari, R., Aalaei, E., Kaviani, M., Dara, M., Mohammadi, S., Azarpira, N., & Amaddon, A. M. (2024). Photo-crosslinked hyaluronic acid hydrogels designed for simultaneous delivery of mesenchymal stem cells and tannic acid: Advancing towards scarless wound healing. *International Journal of Biological Macromolecules, 281*, 136394. DOI: 10.1016/j.ijbiomac.2024.136394 PMID: 39406324

Abolarin, P. O., & Owoyele, B. V. (2024). Tannic acid inhibits pain mediators, inflammation and oxidative stress in mice exposed to glyphosate-based herbicide. *Environmental Analysis, Health and Toxicology, 39*(2), e2024019. DOI: 10.5620/eaht.2024019 PMID: 39054833

Abu-Izneid, T., Rauf, A., Khalil, A. A., Olatunde, A., Khalid, A., Alhumaydhi, F. A., Aljohani, A. S. M., Sahab Uddin, M., Heydari, M., Khayrullin, M., Shariati, M. A., Aremu, A. O., Alafnan, A., & Rengasamy, K. R. R. (2020). Nutritional and Health Beneficial Properties of Saffron (Crocus sativus L): A Comprehensive Review. *Critical Reviews in Food Science and Nutrition*, *62*(10), 2683–2706. DOI: 10.1080/10408398.2020.1857682 PMID: 33327732

Aćimović, M., Tešević, V., Smiljanić, K. T., Cvetković, M., Stanković, J., Kiprovski, B., & Sikora, V. (2020). Hydrolates: Chemical composition and applications. *Advanced Technologies*, *9*(2), 54–70.

Addis, R., Cruciani, S., Santaniello, S., Bellu, E., Sarais, G., Ventura, C., Maioli, M., & Pintore, G. (2020). Fibroblast Proliferation and Migration in Wound Healing by Phytochemicals: Evidence for a Novel Synergic Outcome. *International Journal of Medical Sciences*, *17*(8), 1030–1042. DOI: 10.7150/ijms.43986 PMID: 32410832

Adukwu, E. C., Bowles, M., Edwards-Jones, V., & Bone, H. (2016). Antimicrobial and cytotoxic evaluation of lemongrass essential oil and citral. *Applied Microbiology and Biotechnology*, *100*, 9619–9627. DOI: 10.1007/s00253-016-7807-y PMID: 27562470

Afonso, A., Oliveira, D., Saavedra, M. J., Borges, A., & Simões, M. (2021). Biofilms in Diabetic Foot Ulcers: Impact, Risk Factors and Control Strategies. *International Journal of Molecular Sciences*, *22*(15), 8278. DOI: 10.3390/ijms22158278 PMID: 34361044

Afriza, D., Orienty, F. N., & Ayu, W. P. (2020). Molecular docking analysis of the interactions between MMP-9 protein and four coumarin compounds (nordentatin, dentatin, calusenidin and xanthoxyletin). *Journal of International Dental and Medical Research*, *13*(4), 1286–1292.

Agyare, C., Akindele, A. J., & Steenkamp, V. (2019). Natural products and/or isolated compounds on wound healing. *Evidence-Based Complementary and Alternative Medicine : eCAM*, *4594965*, 1–3. Advance online publication. DOI: 10.1155/2019/4594965 PMID: 31186659

Ahmadian, R., Bahramsoltani, R., Marques, A. M., Rahimi, R., & Farzaei, M. H. (2021). Medicinal Plants as Efficacious Agents for Diabetic Foot Ulcers: A Systematic Review of Clinical Studies. *Wounds : a Compendium of Clinical Research and Practice*, *33*(6), 207–218. DOI: 10.25270/wnds/2021.207218 PMID: 34357879

Ahmed, O. M., Mohamed, T., Moustafa, H., Hamdy, H., Ahmed, R. R., & Aboud, E. (2018). Quercetin and low level laser therapy promote wound healing process in diabetic rats via structural reorganization and modulatory effects on inflammation and oxidative stress. *Biomedicine and Pharmacotherapy*, *101*, 58–73. DOI: 10.1016/j.biopha.2018.02.040 PMID: 29477473

Ajmal, G., Bonde, G. V., Thokala, S., Mittal, P., Khan, G., Singh, J., Pandey, V. K., & Mishra, B. (2019). Ciprofloxacin HCl and quercetin functionalized electrospun nanofiber membrane: Fabrication and its evaluation in full thickness wound healing. *Artificial Cells, Nanomedicine, and Biotechnology*, *47*(1), 228–240. DOI: 10.1080/21691401.2018.1548475 PMID: 30688107

Akin, B., Akgul, B., Tasdurmazli, S., Abamor, E. S., Ozbek, T., Ozcelik, B., & Ozmen, M. M. (2024). Tannic Acid Incorporated Antibacterial Polyethylene Glycol Based Hydrogel Sponges for Management of Wound Infections. *Macromolecular Bioscience*, *24*(8), 2400101. DOI: 10.1002/mabi.202400101 PMID: 38748813

Akintoye, O. O., Ajibare, A. J., Fabunmi, O. A., Asuku, A. O., & Tajudeen, R. A. (2021). Effect of Honey on Oxidative Stress in Brain Tissue of Sleep Deprived Rats. *Nigerian Journal of Basic and Applied Science.*, *29*(1), 83–89. DOI: 10.4314/njbas.v29i1.10

Alam, M. M., Greco, A., Rajabimashhadi, Z., & Corcione, C. E. (2024). Green extraction of lignin and applications. Cleaner Materials, 100253.

Alam, W., Hussain, Y., Ahmad, S., Ali, A., & Khan, H. (2023). Neuroprotective effects of essential oils. In *Phytonutrients and neurological disorders* (pp. 305–333). Academic Press. DOI: 10.1016/B978-0-12-824467-8.00006-1

Albahri, G., Badran, A., Hijazi, A., Daou, A., Baydoun, E., Nasser, M., & Merah, O. (2023). Therapeutic wound healing bioactivities of medicinal plants. *Life (Chicago, Ill.)*, *13*(2), 317. DOI: 10.3390/life13020317 PMID: 36836674

Alherz, F. A., Negm, W. A., Elekhnawy, E., El-Masry, T. A., Haggag, E. M., Alqahtani, M. J., & Hussein, I. A. (2022). Silver nanoparticles prepared using encephalartos laurentianus de wild leaf extract have inhibitory activity against candida albicans clinical isolates. *Journal of Fungi (Basel, Switzerland)*, *8*(10), 1005. DOI: 10.3390/jof8101005 PMID: 36294570

Ali, S., Ekbbal, R., Salar, S., Ali, S. A., Jaiswal, A. K., Singh, M., Yadav, D. K., & Kumar, S. (2023). Quality standards and pharmacological interventions of natural oils. *ACS Omega*, *8*(43), 39945–39963. DOI: 10.1021/acsomega.3c05241 PMID: 37953833

Aljaafari, M. N., AlAli, A. O., Baqais, L., Alqubaisy, M., AlAli, M., Molouki, A., Ong-Abdullah, J., Abushelaibi, A., Lai, K. S., & Lim, S. H. (2021). Therapeutic applications of essential oils. *Molecules (Basel, Switzerland), 26*(3), 628. DOI: 10.3390/molecules26030628 PMID: 33530290

Al-Khamis, A., McCallum, I., King, P. M., & Bruce, J. (2010). Healing by primary versus secondary intention after surgical treatment for pilonidal sinus. *Cochrane Database of Systematic Reviews, 2010*(1), CD006213. Advance online publication. DOI: 10.1002/14651858.CD006213.pub3 PMID: 20091589

Al-Knani, Z. M. I., Ashoor, A. S., & Hussein, A. N. (2023). Evaluation of antimicrobial activity of Allium sativum and its synergistic effects with antibiotics in the treatment of diabetic foot ulcers. *Journal of Ethnopharmacology, 280*, 114309. DOI: 10.1016/j.jep.2023.114309

Allen, T. M., & Cullis, P. R. (2013). Liposomal drug delivery systems: From concept to clinical applications. *Advanced Drug Delivery Reviews, 65*(1), 36–48. DOI: 10.1016/j.addr.2012.09.037 PMID: 23036225

Almadani, Y. H., Vorstenbosch, J., Davison, P. G., & Murphy, A. M. (2021). Wound healing: a comprehensive review. In *Seminars in plastic surgery* (Vol. 35, No. 03, pp. 141-144). Thieme Medical Publishers, Inc. DOI: 10.1055/s-0041-1731791

Almeida, J., Benvegnú, D., Boufleur, N., Reckziegel, P., Barcelos, R., Coradini, K., de Carvalho, L. M., Bürger, M. E., & Beck, R. C. R. (2012). Hydrogels containing rutin intended for cutaneous administration: Efficacy in wound healing in rats. *Drug Development and Industrial Pharmacy, 38*(7), 792–799. DOI: 10.3109/03639045.2011.628676 PMID: 22066462

Almuhanna, Y., Alshalani, A., AlSudais, H., Alanazi, F., Alissa, M., & Joseph, B. (2024). Antibacterial, antibiofilm, and wound healing activities of rutin and quercetin and their interaction with gentamicin on excision wounds in diabetic mice. *Biology (Basel), 13*(9), 676. DOI: 10.3390/biology13090676 PMID: 39336103

Al-Nahari, A. A., Almasaudi, S. B., Abd El-Ghany, E. S. M., Barbour, E., Al Jaouni, S. K., & Harakeh, S. (2015). Antimicrobial activities of Saudi honey against Pseudomonas aeruginosa. *Saudi Journal of Biological Sciences, 22*(5), 521–525. DOI: 10.1016/j.sjbs.2015.04.006 PMID: 26288553

Al-Obaidi, J. R., Halabi, M. F., AlKhalifah, N. S., Asanar, S., Al-Soqeer, A. A., & Attia, M. F. (2017). A review on plant importance, biotechno logical aspects, and cultivation challenges of jojoba plant. *Biological Research, 50*(1), 25. DOI: 10.1186/s40659-017-0131-x PMID: 28838321

Alonso, G. L., Zalacain, A., & Carmona, M. (2012). Saffron. In *Handbook of Herbs and Spices* (pp. 469–498). Elsevier. DOI: 10.1533/9780857095671.469

Alotibi, I. A., Harakeh, S. M., Al-Mamary, M., Mariod, A. A., Al-Jaouni, S. K., Al-Masaud, S., Alharbi, M. G., & Al-Hindi, R. R. (2018). Floral markers and biological activity of Saudi honey. *Saudi Journal of Biological Sciences*, *25*(7), 1369–1374. DOI: 10.1016/j.sjbs.2018.05.021 PMID: 30505183

Al-Rawaf, H. A., Gabr, S. A., & Alghadir, A. H. (2019). Molecular Changes in Diabetic Wound Healing following Administration of Vitamin D and Ginger Supplements: Biochemical and Molecular Experimental Study. Evid.-. *Evidence-Based Complementary and Alternative Medicine : eCAM*, *4352470*, 1–13. Advance online publication. DOI: 10.1155/2019/4352470 PMID: 31428171

Al-Roujayee, A. S. (2017). Naringenin improves the healing process of thermally-induced skin damage in rats. *The Journal of International Medical Research*, *45*(2), 570–582. DOI: 10.1177/0300060517692483 PMID: 28415935

Al-Samydai, A., Al Qaraleh, M., Alshaer, W., Al-Halaseh, L. K., Issa, R., & Alshaikh, F. (2022). Preparation, Characterization, Wound Healing, and Cytotoxicity Assay of PEGylated Nanophytosomes Loaded with 6-Gingerol. *Nutrients*, *14*(23), 5170. DOI: 10.3390/nu14235170 PMID: 36501201

Al-Warhi, T., Elmaidomy, A. H., Maher, S. A., Abu-Baih, D. H., Selim, S., Albqmi, M., Al-Sanea, M. M., Alnusaire, T. S., Ghoneim, M. M., Mostafa, E. M., Hussein, S., El-Damasy, A. K., Saber, E. A., Elrehany, M. A., Sayed, A. M., Othman, E. M., El-Sherbiny, M., & Abdelmohsen, U. R. (2022). The wound-healing potential of *Olea europaea* L. cv. Arbequina leaves extract: An integrated in vitro, in silico, and in vivo investigation. *Metabolites*, *12*(9), 791. DOI: 10.3390/metabo12090791 PMID: 36144197

Alyoussef, A., El-Gogary, R. I., Ahmed, R. F., Ahmed-Farid, O. A., Bakeer, R. M., & Nasr, M. (2021). The beneficial activity of curcumin and resveratrol loaded in nanoemulgel for healing of burn-induced wounds. *Journal of Drug Delivery Science and Technology*, *62*, 102360. DOI: 10.1016/j.jddst.2021.102360

Aly, S. H., Elissawy, A. M., Mahmoud, A. M. A., El-Tokhy, F. S., Mageed, S. S. A., Almahli, H., Al-Rashood, S. T., Binjubair, F. A., Hassab, M. A. E., Eldehna, W. M., & Singab, A. N. B. (2023). Synergistic Effect of Sophora japonica and Glycyrrhiza glabra Flavonoid-Rich Fractions on Wound Healing: In Vivo and Molecular Docking Studies. *Molecules (Basel, Switzerland)*, *28*(7), 2994. DOI: 10.3390/molecules28072994 PMID: 37049756

Amanat, S., Taymouri, S., Varshosaz, J., Minaiyan, M., & Talebi, A. (2020). Carboxymethyl cellulose-based wafer enriched with resveratrol-loaded nanoparticles for enhanced wound healing. *Drug Delivery and Translational Research*, *10*(5), 1241–1254. DOI: 10.1007/s13346-020-00711-w PMID: 31981141

Ambreen, M., & Mirza, S. A. (2020). Evaluation of anti-inflammatory and wound healing potential of tannins isolated from leaf callus cultures of Achyranthes aspera and Ocimum basilicum. *Pakistan Journal of Pharmaceutical Sciences*, *33*(1), 361–36. PMID: 32122869

Amini, S. M., & Akbari, A. (2019). Metal nanoparticles synthesis through natural phenolic acids. *IET Nanobiotechnology / IET*, *13*(8), 771–777. DOI: 10.1049/iet-nbt.2018.5386 PMID: 31625516

Anand, S., Rajinikanth, P. S., Arya, D. K., Pandey, P., Gupta, R., Sankhwar, R., & Chidambaram, K. (2022). Multifunctional Biomimetic Nanofibrous Scaffold Loaded with Asiaticoside for Rapid Diabetic Wound Healing. *Pharmaceutics*, *14*(2), 273. DOI: 10.3390/pharmaceutics14020273 PMID: 35214006

Andrade-Ochoa, S., Chacón-Vargas, K. F., Sánchez-Torres, L. E., Rivera-Chavira, B. E., Nogueda-Torres, B., & Nevárez-Moorillón, G. V. (2021). Differential antimicrobial effect of essential oils and their main components: Insights based on the cell membrane and external structure. *Membranes*, *11*(6), 405. DOI: 10.3390/membranes11060405 PMID: 34071618

Ansari, A. A. (2006). Global status of Unani medicine. *The Proceeding of International Conclave on Traditional Medicine (16–17 November 2006), National Agriculture Science Complex, New Delhi. Available from:* Http://Www. Niscair. Res. in/Conclave/Downloadables/Plenary% *20Session, 203*.

Arbenz, A., & Avérous, L. (2015). Chemical modification of tannins to elaborate aromatic biobased macromolecular architectures. *Green Chemistry*, *17*(5), 2626–2646. DOI: 10.1039/C5GC00282F

Arda, A. G., Syaifie, P. H., Ramadhan, D., Jauhar, M. M., Nugroho, D. W., Kaswati, N. M. N., Noviyanto, A., Safihtri, M., Rochman, N. T., Andrianto, D., & Mardliyati, E. (2024). Activity of propolis compounds as potential MMP1 and MMP2 inhibitors by in silico studies in wound healing application. *Journal of Pharmacy & Pharmacognosy Research*, *12*(2), 264–285. DOI: 10.56499/jppres23.1719_12.2.264

Arif, N., Baumann, L., & Felcht, M. (2024). Combination of secondary intention healing and primary closure to reconstruct large facial defects. *Journal der Deutschen Dermatologischen Gesellschaft*, *22*(10), 1344–1349. DOI: 10.1111/ddg.15481 PMID: 39167560

Armstrong, D. G., Boulton, A. J. M., & Bus, S. A. (2017). Diabetic foot ulcers and recurrence. *The New England Journal of Medicine, 376*(24), 2367–2375. DOI: 10.1056/NEJMra1615439 PMID: 28614678

Aroui, S., Aouey, B., Chtourou, Y., Meunier, A. C., Fetoui, H., & Kenani, A. (2016). Naringin suppresses cell metastasis and the expression of matrix metalloproteinases (MMP-2 and MMP-9) via the inhibition of ERK-P38-JNK signaling pathway in human glioblastoma. *Chemico-Biological Interactions, 244*, 195–203. DOI: 10.1016/j.cbi.2015.12.011 PMID: 26721195

Arring, N. M., Millstine, D., Marks, L. A., & Nail, L. M. (2018). Ginseng as a treatment for fatigue: A systematic review. *Journal of Alternative and Complementary Medicine (New York, N.Y.), 24*(7), 624–633. DOI: 10.1089/acm.2017.0361 PMID: 29624410

Arunachalam, K. D., Subhashini, S., & Annamalai, S. K. (2012). Wound healing and antigenotoxic activities of *Aegle marmelos* with relation to its antioxidant properties. *Journal of Pharmacy Research, 5*(3), 1492–1502.

Asadi, S. Y., Parsaei, P., Karimi, M., Ezzati, S., Zamiri, A., Mohammadizadeh, F., & Rafieian-Kopaei, M. (2013). Effect of green tea (Camellia sinensis) extract on healing process of surgical wounds in rat. *International Journal of Surgery, 11*(4), 332–337. DOI: 10.1016/j.ijsu.2013.02.014 PMID: 23459184

Asfour, M. H., Elmotasem, H., Mostafa, D. M., & Salama, A. A. (2017). Chitosan based Pickering emulsion as a promising approach for topical application of rutin in a solubilized form intended for wound healing: In vitro and in vivo study. *International Journal of Pharmaceutics, 534*(1-2), 325–338. DOI: 10.1016/j.ijpharm.2017.10.044 PMID: 29074391

Ashour, M., Wink, M., & Gershenzon, J. (2010). Biochemistry of terpenoids: Monoterpenes, sesquiterpenes and diterpenes. In M. Wink (Ed.), Biochemistry of Plant Secondary Metabolism (Annual Plant Reviews, Vol. 40, pp. 258–303). Wiley.

Ashraf, M. V., Pant, S., Khan, M. A. H., Shah, A. A., Siddiqui, S., Jeridi, M., Alhamdi, H. W., & Ahmad, S. (2023). Phytochemicals as Antimicrobials: Prospecting Himalayan Medicinal Plants as Source of Alternate Medicine to Combat Antimicrobial Resistance. *Pharmaceuticals (Basel, Switzerland), 16*(6), 881. DOI: 10.3390/ph16060881 PMID: 37375828

Asuku, A. O., Ayinla, M. T., Ajibare, A. J., & Olajide, T. S. (2024). Mercury chloride causes cognitive impairment, oxidative stress and neuroinflammation in male Wistar rats: The potential protective effect of 6-gingerol-rich fraction of *Zingiber officinale* via regulation of antioxidant defense system and reversal of pro-inflammatory markers increase. Brain Research, 148741: 1826, ISSN 0006-8993, https://doi.org/ DOI: 10.1016/j.brainres.2023.148741

Asuku, A. O., Ajibare, A. J., Ajagun, E. J., Babatunde, S. S., Olajide, T. S., Oyerinde, T. O., Ogungbangbe, G. O., & Odugbesan, O. A. (2024). Herbal Extracts in Antiaging Products. In Yingngam, B. (Ed.), *Eco-friendly Skin Solutions for Natural Cosmeceuticals*. IGI Publisher., DOI: 10.4018/979-8-3693-5473-5.ch012

Atkin, L. (2019). Chronic wounds: The challenges of appropriate management. *British Journal of Community Nursing*, *24*(Sup9, Suppl. 9), S26–S32. DOI: 10.12968/bjcn.2019.24.Sup9.S26 PMID: 31479336

Attallah, N. G. M., Negm, W. A., Elekhnawy, E., Elmongy, E. I., Altwaijry, N., El-Haroun, H., El-Masry, T. A., & El-Sherbeni, S. A. (2021). Elucidation of phytochemical content of *Cupressus macrocarpa* leaves in vitro and in vivo anti-bacterial effect against methicillin resistant *Staphylococcus aureus* clinical isolates. *Antibiotics (Basel, Switzerland)*, *10*(8), 890. DOI: 10.3390/antibiotics10080890 PMID: 34438940

Augoye, O., Sobowale, A., & Muyiwa-Ajayi, T. P. (2024). Sustainable process design in essential oil extraction.

Awad, H. M., Boersma, M. G., Boeren, S., van der Woude, H., van Zanden, J., van Bladeren, P. J., Vervoort, J., & Rietjens, I. M. C. M. (2002). Identification of o-quinone/quinone methide metabolites of quercetin in a cellular in vitro system. *FEBS Letters*, *520*(1-3), 30–34. DOI: 10.1016/S0014-5793(02)02754-0 PMID: 12044865

Ayinla, M. T., & Asuku, A. O. (2025). The neurotoxic effects of lead acetate and the abrogating actions of 6-gingerol-rich extract of ginger via modulation of antioxidant defence system, pro-inflammatory markers, and apoptotic cascade. *Naunyn-Schmiedeberg's Archives of Pharmacology*, *398*(7), 9341–9356. Advance online publication. DOI: 10.1007/s00210-025-03873-x PMID: 39937255

Azab, A., Nassar, A., & Azab, A. N. (2016). Anti-inflammatory activity of natural products. *Molecules (Basel, Switzerland)*, *21*(10), 1321. DOI: 10.3390/molecules21101321 PMID: 27706084

Azeredo, J., Azevedo, N. F., Briandet, R., Cerca, N., Coenye, T., Costa, A. R., Desvaux, M., Di Bonaventura, G., Hébraud, M., Jaglic, Z., Kačániová, M., Knøchel, S., Lourenço, A., Mergulhão, F., Meyer, R. L., Nychas, G., Simões, M., Tresse, O., & Sternberg, C. (2017). Critical review on biofilm methods. *Critical Reviews in Microbiology, 43*(3), 313–351. DOI: 10.1080/1040841X.2016.1208146 PMID: 27868469

Aziz, N., Kim, M. Y., & Cho, J. Y. (2018). Anti-inflammatory effects of luteolin: A review of in vitro, in vivo, and in silico studies. *Journal of Ethnopharmacology, 225*, 342–358. DOI: 10.1016/j.jep.2018.05.019 PMID: 29801717

Aziz, Z. A., Ahmad, A., Setapar, S. H., Karakucuk, A., Azim, M. M., & Lokhat, D.. (2018). Essential oils: Extraction techniques and therapeutic potential. *Current Drug Metabolism, 19*(13), 1100–1110. DOI: 10.2174/1389200219666180723144850 PMID: 30039757

Azmin, S. N. H. M., Manan, Z. A., Alwi, S. R. W., Chua, L. S., Mustaffa, A. A., & Yunus, N. A. (2016). Herbal Processing and Extraction Technologies. *Separation and Purification Reviews, 45*(4), 305–320. DOI: 10.1080/15422119.2016.1145395

Babbar, N., Oberoi, H. S., & Sandhu, S. K. (2015). Nutraceutical potential of bioactives from fruit residues. *Critical Reviews in Food Science and Nutrition, 55*(3), 319–337. DOI: 10.1080/10408398.2011.653734 PMID: 24915390

Badylak, S. F. (2002). The extracellular matrix as a scaffold for tissue reconstruction. *Seminars in Cell & Developmental Biology, 13*(5), 377–383. DOI: 10.1016/S1084952102000940 PMID: 12324220

Bagher, Z., Ehterami, A., Safdel, M. H., Khastar, H., Semiari, H., Asefnejad, A., Davachi, S. M., Mirzaii, M., & Salehi, M. (2020). Wound healing with alginate/chitosan hydrogel containing hesperidin in rat model. *Journal of Drug Delivery Science and Technology, 55*, 101379. DOI: 10.1016/j.jddst.2019.101379

Bahadur, S., & Fatima, S. (2024). Essential oils of medicinal plants and wound healing activities. *Current Pharmaceutical Biotechnology, 25*(14), 1818–1834. DOI: 10.2174/0113892010282605231218064053 PMID: 38310452

Bai, F., Wang, Y., Zhang, S., Wang, Y., Zhang, J., Cao, J., & Sun, L. (2020). Caffeoyl substitution changes the inhibition mode of tartaric acid against α-amylase: Analysis of the enzyme inhibition by four caffeic and tartaric acid derivates. *Lebensmittel-Wissenschaft + Technologie, 133*, 109942. DOI: 10.1016/j.lwt.2020.109942

Baliga, M. S., Latheef, L., Haniadka, R., Fazal, F., Chacko, J., & Arora, R. (2013). Bioactive Food as Dietary Interventions for Arthritis and Related Inflammatory Diseases. Elsevier Inc.; Amsterdam, The Netherlands: 2013. Ginger (Zingiber officinale Roscoe) in the Treatment and Prevention of Arthritis; pp. 529–544.

Baltzis, D., Eleftheriadou, I., & Veves, A. (2014). Pathogenesis and Treatment of Impaired Wound Healing in Diabetes Mellitus: New Insights. *Advances in Therapy*, *31*(8), 817–836. DOI: 10.1007/s12325-014-0140-x PMID: 25069580

Banerjee, D., Biswas, P., Mazumder, K., Palai, S., Hossain, C. M., Karmakar, S., & Biswas, K. (2024). Exploration of phytochemicals as anti-biofilm agents against pathogenic bacteria: Their potential and challenges. *Infectious Disorders Drug Targets*, *25*(6), e18715265324950. Advance online publication. DOI: 10.2174/0118715265324950241204182204 PMID: 39757612

Baptista-Silva, S., Borges, S., Ramos, O. L., Pintado, M., & Sarmento, B. (2020). Essential oils as therapeutic agents. *The Journal of Essential Oil Research*, *32*(4), 279–295. DOI: 10.1080/10412905.2020.1746698

Baradaran Rahimi, V., & Askari, V. R. (2022). Immunomodulatory effects of β-caryophyllene. *BioFactors (Oxford, England)*, *48*(4), 857–882. DOI: 10.1002/biof.1869 PMID: 35648433

Barman, P. K., & Koh, T. J. (2020). Macrophage dysregulation and impaired skin wound healing in diabetes. *Frontiers in Cell and Developmental Biology*, *8*, 528. Advance online publication. DOI: 10.3389/fcell.2020.00528 PMID: 32671072

Barnes, J., & Heinrich, M. (2004). *Fundamentals of pharmacognosy and phytotherapy*. Churchill Livingstone.

Baron, J. M., Glatz, M., & Proksch, E. (2020). Optimal Support of Wound Healing: New Insights. *Dermatology (Basel, Switzerland)*, *236*(6), 593–600. DOI: 10.1159/000505291 PMID: 31955162

Basak, S., Ghosh, N., & Saha, S. (2021). Anti-inflammatory mechanisms of polyphenols and their application in medicinal chemistry: A review. *Medicinal Chemistry (Shariqah, United Arab Emirates)*, *17*(1), 20–35. DOI: 10.2174/1573406415666210715110612

Belal, A., Elanany, M. A., Raafat, M., Hamza, H. T., & Mehany, A. B. M. (2022). *Calendula officinalis* phytochemicals for the treatment of wounds through matrix metalloproteinases-8 and 9 (MMP-8 and MMP-9): In silico approach. *Natural Product Communications*, *17*(5), 1–7. DOI: 10.1177/1934578X221098848

Belay, A., & Gholap, A. V. (2009). Characterization and determination of chlorogenic acids (CGA) in coffee beans by UV-VIS spectroscopy. *African Journal of Pure and Applied Chemistry*, *3*(11), 234–240.

Belwal, T., Chemat, F., Venskutonis, P. R., Cravotto, G., Jaiswal, D. K., Bhatt, I. D., Devkota, H. P., & Luo, Z. (2020). Scaling-up non-conventional extraction techniques. *Trends in Analytical Chemistry*, *127*, 115895. DOI: 10.1016/j.trac.2020.115895

Benedek, B., Kopp, B., & Melzig, M. F. (2007). Achillea millefolium L. s.l.–Is the anti-inflammatory activity mediated by protease inhibition? *Journal of Ethnopharmacology*, *113*(2), 312–317. DOI: 10.1016/j.jep.2007.06.014 PMID: 17689902

Bent, S. (2008). Herbal medicine in the United States: review of efficacy, safety, and regulation: grand rounds at University of California, San Francisco Medical Center. *Journal of General Internal Medicine*, *23*, 854–859. DOI: 10.1007/s11606-008-0632-y PMID: 18415652

Berkes, F. (1999). Context of traditional ecological knowledge. *Sacred Ecology: Traditional Ecological Knowledge and Resource Management. Philadelphia.*

Bharat, B., & Aggarwal, K. B. H. (2009). Potential therapeutic effects of curcumin, the anti-inflammatory agent, against neurodegenerative, cardiovascular, pulmonary, metabolic, autoimmune and neoplastic diseases. *The International Journal of Biochemistry & Cell Biology*, *1*(41), 40–59. DOI: 10.1016/j.biocel.2008.06.010 PMID: 18662800

Bhattacharya, A., & Hossain, S. (2021). Nutritional support in wound healing: The role of polyphenolic-rich diets. *Nutrition Research Reviews*, *34*(2), 146–158. DOI: 10.1017/S0954422421000014

Bi, H., Li, H., Zhang, C., Mao, Y., Nie, F., Xing, Y., Sha, W., Wang, X., Irwin, D. M., & Tan, H. (2019). Stromal vascular fraction promotes migration of fibroblasts and angiogenesis through regulation of extracellular matrix in the skin wound healing process. *Stem Cell Research & Therapy*, *10*(1), 302. DOI: 10.1186/s13287-019-1415-6 PMID: 31623669

Bilal, K., Mehboob, F., Akhtar, N., Mirza, I. A., Okla, M. K., Dar, M. J., Saleh, I. A., Zomot, N., & Fatima, H. (2024). Wound healing, antioxidant and antibacterial activities of polyphenols of *Psidium guajava* L. leaves. *South African Journal of Botany*, *165*, 538–551. DOI: 10.1016/j.sajb.2023.12.026

Binsuwaidan, R., Elekhnawy, E., Elseady, W. S., Keshk, W. A., Shoeib, N. A., Attallah, N. G., Mokhtar, F. A., Abd El Hadi, S. R., Ahmed, E., Magdeldin, S., & Negm, W. A. (2022). Anti-bacterial activity and wound heal ing potential of Cycas thouarsii R. Br n-butanol fraction in diabetic rats supported with phytochemical profiling. *Biomedicine and Pharmacotherapy*, *155*, 113763. DOI: 10.1016/j.biopha.2022.113763 PMID: 36182739

Blume-Peytavi, U., Tan, J., Tennstedt, D., Boralevi, F., Fabbrocini, G., Torrelo, A., Soares-Oliveira, R., Haftek, M., Rossi, A. B., Thouvenin, M. D., Mangold, J., Galliano, M. F., Hernandez-Pigeon, H., Aries, M. F., Rouvrais, C., Bessou-Touya, S., Duplan, H., Castex-Rizzi, N., Mengeaud, V., & Dauskardt, R. (2016). Fragility of epidermis in newborns, children and adolescents. *Journal of the European Academy of Dermatology and Venereology*, *30*(9), 1634. DOI: 10.1111/jdv.13636 PMID: 27062556

Boakye, Y. D., Agyare, C., & Hensel, A. (2016). Anti-infective properties and time-kill kinetics of Phyllanthus muellerianus and its major constituent, geraniin. *Medicinal Chemistry (Sunnyvale, Calif.)*, *6*(2), 2. DOI: 10.4172/2161-0444.1000332

Bodas, K., & Shinde, V. (2021). Healing of wounds: A detailed review on models, biomarkers, biochemical and other wound assessment parameters. *Inflammation*, *9*(3), 2069–2085.

Botusan, I. R., Sunkari, V. G., Savu, O., Catrina, A. I., Grünler, J., Lindberg, S., Pereira, T., Ylä-Herttuala, S., Poellinger, L., Brismar, K., & Catrina, S. (2008). Stabilization of HIF-1α is critical to improve wound healing in diabetic mice. *Proceedings of the National Academy of Sciences of the United States of America*, *105*(49), 19426–19431. DOI: 10.1073/pnas.0805230105 PMID: 19057015

Boukeloua, A., Kouadria, M., Bendif, H., Plavan, G., Alsalamah, S. A., Alghonaim, M. I., Boufahja, F., & Abd-Elkader, O. H. (2023). Physicochemical analysis and wound healing activity of *Azadirachta indica* (A. Juss.) fruits. *Processes (Basel, Switzerland)*, *11*(6), 1692. DOI: 10.3390/pr11061692

Boulton, A. J. M., Armstrong, D. G., Hardman, M. J., Malone, M., Embil, J. M., Attinger, C. E., Lipsky, B. A., Aragón-Sánchez, J., Kwong, H., Schultz, G. S., & Kirsner, R. S. (2020). Diagnosis and Management of Diabetic Foot Infections. *ADA Clinical Compendia*, *2020*(1), 1–24. DOI: 10.2337/db2020-01 PMID: 32105420

Brem, H., Stojadinovic, O., Diegelmann, R. F., Entero, H., Lee, B., Pastar, I., Golinko, M., Rosenberg, H., & Tomic-Canic, M. (2007). Molecular markers in patients with chronic wounds to guide surgical debridement. *Molecular Medicine (Cambridge, Mass.)*, *13*(1–2), 30–39. DOI: 10.2119/2006-00054.Brem PMID: 17515955

Brewer, D. B. (2006). Max Schultze (1865), G. Bizzozero (1882) and the discovery of the platelet. *British Journal of Haematology, 133*(3), 251–258. DOI: 10.1111/j.1365-2141.2006.06036.x PMID: 16643426

Brudzynski, K., & Maldonado-Alvarez, L. (2015). Polyphenol-protein complexes and their consequences for the redox activity, structure and function of honey. A current view and new hypothesis—A review. *Polish Journal of Food and Nutrition Sciences, 65*(2), 71–80. DOI: 10.1515/pjfns-2015-0030

Burgess, J. L., Wyant, W. A., Abujamra, B. A., Kirsner, R. S., & Jozic, I. (2021). Diabetic Wound-Healing Science. *Medicina, 57*(10), 1072. DOI: 10.3390/medicina57101072 PMID: 34684109

Burud, A., Galatage, S., Manjappa, A., Salawi, A., Nadaf, S., Holam, M., Harale, S., Kumbhar, R., Peram, M. R., & Suryadevara, V. (2024). Sericin stabilized emulgel for improving therapeutic efficacy of quercetin in treatment of diabetic wound healing. *Journal of Dispersion Science and Technology*, 1–21. Advance online publication. DOI: 10.1080/01932691.2024.2444980

Caldwell, R. B., Bartoli, M., Behzadian, M. A., El-Remessy, A. B., Al-Shabrawey, M., Platt, D., Liou, G. I., & Caldwell, R. W. (2005). Vascular Endothelial Growth Factor and Diabetic Retinopathy: Role of Oxidative Stress. *Current Drug Targets, 6*(4), 511–524. DOI: 10.2174/1389450054021981 PMID: 16026270

Caley, M. P., Martins, V. L. C., & O'Toole, E. A. (2015). Metalloproteinases and wound healing. *Advances in Wound Care, 4*(4), 225–234. DOI: 10.1089/wound.2014.0581 PMID: 25945285

Calver, A., Collier, J., & Vallance, P. (1992). Inhibition and stimulation of nitric oxide synthesis in the human forearm arterial bed of patients with insulin-dependent diabetes. *The Journal of Clinical Investigation, 90*(6), 2548–2554. DOI: 10.1172/JCI116149 PMID: 1469103

Campos, I., Geiger, J. A., Santos, A. C., Carlos, V., & Jacinto, A. (2010). Genetic screen in Drosophila melanogaster uncovers a novel set of genes required for embryonic epithelial repair. *Genetics, 184*(1), 129–140. DOI: 10.1534/genetics.109.110288 PMID: 19884309

Canter, P. H. (2005). Bringing medicinal plants into cultivation. *Focus on Alternative and Complementary Therapies, 10*(3), 167–168. DOI: 10.1211/fact.10.3.0002 PMID: 15780709

Caponio, G. R., Lippolis, T., Tutino, V., Gigante, I., De Nunzio, V., Milella, R. A., Gasparro, M., & Notarnicola, M. (2022). Nutraceuticals: Focus on anti-inflammatory, anti-cancer, antioxidant properties in gastrointestinal tract. *Antioxidants*, *11*(7), 1274. DOI: 10.3390/antiox11071274 PMID: 35883765

Caputo, L., Smeriglio, A., Trombetta, D., Cornara, L., Trevena, G., & Valussi, M.. (2020). Chemical composition and biological activities of Leptospermum petersonii and Eucalyptus gunnii essential oils. *Frontiers in Microbiology*, *11*, 409. DOI: 10.3389/fmicb.2020.00409 PMID: 32351456

Carrasco, S., González, L., Tapia, M., Urbano, B. F., Aguayo, C., & Fernández, K. (2024). Enhancing alginate hydrogels as possible wound-healing patches: The synergistic impact of reduced graphene oxide and tannins on mechanical and adhesive properties. *Polymers*, *16*(8), 1081. DOI: 10.3390/polym16081081 PMID: 38675000

Carvalho, M. T. B., Araújo-Filho, H. G., Barreto, A. S., Quintans-Júnior, L. J., Quintans, J. S., & Barreto, R. S. S. (2021). Wound healing properties of flavonoids: A systematic review highlighting the mechanisms of action. *Phytomedicine*, *90*, 153636. DOI: 10.1016/j.phymed.2021.153636 PMID: 34333340

Castangia, I., Nácher, A., Caddeo, C., Valenti, D., Fadda, A. M., Díez-Sales, O., Ruiz-Saurí, A., & Manconi, M. (2014). Fabrication of quercetin and curcumin bionanovesicles for the prevention and rapid regeneration of full-thickness skin defects on mice. *Acta Biomaterialia*, *10*(3), 1292–1300. DOI: 10.1016/j.actbio.2013.11.005 PMID: 24239901

Cavanagh, H. M. A., & Miller, J. J. (2019). Essential oils: A new perspective on antimicrobial activity. *International Journal of Aromatherapy*, *19*(1-2), 1–7.

Cedillo-Cortezano, M., Martínez-Cuevas, L. R., López, J. A. M., Barrera López, I. L., Escutia-Pérez, S., & Petricevich, V. L. (2024). Use of medicinal plants in the process of wound healing: A literature review. *Pharmaceuticals (Basel, Switzerland)*, *17*(3), 303. DOI: 10.3390/ph17030303 PMID: 38543089

Cerdá, B., Cerón, J. J., Tomás-Barberán, F. A., & Espín, J. C. (2003). Repeated oral administration of high doses of the pomegranate ellagitannin punicalagin to rats for 37 days is not toxic. *Journal of Agricultural and Food Chemistry*, *51*(11), 3493–3501. DOI: 10.1021/jf020842c PMID: 12744688

Chanu, N. R., Gogoi, P., Barbhuiya, P. A., Dutta, P. P., Pathak, M. P., & Sen, S. (2023). Natural Flavonoids as Potential Therapeutics in the Management ofDiabetic Wound: A Review. *Current Topics in Medicinal Chemistry*, *23*(8), 690–710. DOI: 10.2174/1568026623666230419102140 PMID: 37114791

Charles, C. A., Tomic-Canic, M., Vincek, V., Nassiri, M., Stojadinovic, O., Eaglstein, W. H., & Kirsner, R. S. (2008). A gene signature of nonhealing venous ulcers: Potential diagnostic markers. *Journal of the American Academy of Dermatology, 59*(5), 758–771. DOI: 10.1016/j.jaad.2008.07.018 PMID: 18718692

Chaturvedi, S., Agrawal, S., Garg, A., & Rastogi, V. (2022). Potential of Nano-encapsulated Quercetin Topical Formulations in the Management of Diabetic Foot Ulcer. *Revista Brasileira de Farmacognosia, 33*(3), 484–501. DOI: 10.1007/s43450-022-00345-8

Chaudhari, M., & Mengi, S. (2006). Evaluation of phytoconstituents of *Terminalia arjuna* for wound healing activity in rats. *Phytotherapy Research, 20*(9), 799–805. DOI: 10.1002/ptr.1857 PMID: 16835874

Chaudhury, R. R., & Rafei, U. M. (2002). *Traditional medicine in Asia*. World Health Organization Geneva.

Cheesman, M. J., Alcorn, S., Verma, V., & Cock, I. E. (2021). An assessment of the growth inhibition profiles of Hamamelis virginiana L. extracts against Streptococcus and Staphylococcus spp. *Journal of Traditional and Complementary Medicine, 11*(5), 457–465. DOI: 10.1016/j.jtcme.2021.03.002 PMID: 34522640

Chelu, M. (2024). Hydrogels with essential oils: Recent advances. *Gels (Basel, Switzerland), 10*(10), 636. DOI: 10.3390/gels10100636 PMID: 39451288

Chen, G., He, L., Zhang, P., Zhang, J., Mei, X., Wang, D., Zhang, Y., Ren, X., & Chen, Z. (2020). Encapsulation of green tea polyphenol nanospheres in PVA/alginate hydrogel for promoting wound healing of diabetic rats by regulating PI3K/AKT pathway. *Materials Science and Engineering C, 110*, 110686. DOI: 10.1016/j.msec.2020.110686 PMID: 32204114

Cheng, C., Peng, X., Xi, L., Wan, C., Shi, S., Wang, Y., & Yu, X. (2022). An agar–polyvinyl alcohol hydrogel loaded with tannic acid with efficient hemostatic and antibacterial capacity for wound dressing. *Food & Function, 13*(18), 9622–9634. DOI: 10.1039/D2FO02251F PMID: 36004684

Cheng, H. A., Drinnan, C. T., Pleshko, N., & Fisher, O. Z. (2015). Pseudotannins self-assembled into antioxidant complexes. *Soft Matter, 11*(39), 7783–7791. DOI: 10.1039/C5SM01224D PMID: 26313262

Cheng, M., Zhang, L., Zhang, H., Li, X., Wang, Y., Xia, F., Wang, B., Cai, R., Guo, Z., Zhang, Y., Ji, Y., Sun, C., Feng, X., Lei, L., Yang, Y., Han, W., & Gu, J. (2018). An ointment consisting of the phage lysin LysGH15 and apigenin for decolonization of methicillin-resistant Staphylococcus aureus from skin wounds. *Viruses*, *10*(5), 244. DOI: 10.3390/v10050244 PMID: 29734776

Chen, X., Wang, M., Xu, X., Liu, J., Mei, B., Fu, P., Zhao, D., & Sun, L. (2017). Panax ginseng total protein promotes proliferation and secretion of collagen in NIH/3T3 cells by activating extracellular signal-related kinase path way. *Journal of Ginseng Research*, *41*(3), 411–418. DOI: 10.1016/j.jgr.2017.02.001 PMID: 28701885

Chen, X., Zhang, Q., Wang, Y., Meng, J., Wu, M., Xu, H., Du, L., & Yang, X. (2023). Fabrication and characterization of electrospun poly (caprolactone)/tannic acid scaffold as an antibacterial wound dressing. *Polymers*, *15*(3), 593. DOI: 10.3390/polym15030593 PMID: 36771894

Chen, Y., Tian, L., Yang, F., Tong, W., Jia, R., Zou, Y., Yin, L., Li, L., He, C., Liang, X., Ye, G., Lv, C., Song, X., & Yin, Z. (2019). Tannic Acid Accelerates Cutaneous Wound Healing in Rats Via Activation of the ERK 1/2 Signaling Pathways. *Advances in Wound Care*, *8*(7), 341–354. DOI: 10.1089/wound.2018.0853 PMID: 31737421

Chhabra, S., Chhabra, N., Kaur, A., & Gupta, N. (2017). Wound healing concepts in clinical practice of OMFS. *Journal of Maxillofacial and Oral Surgery*, *16*(4), 403–423. DOI: 10.1007/s12663-016-0880-z PMID: 29038623

Chiangnoon, R., Samee, W., Uttayarat, P., Jittachai, W., Ruksiriwanich, W., Sommano, S. R., Athikomkulchai, S., & Chittasupho, C. (2022). Phytochemical analysis, antioxidant, and wound healing activity of *Pluchea indica* branch extract nanoparticles. *Molecules (Basel, Switzerland)*, *27*(3), 635. DOI: 10.3390/molecules27030635 PMID: 35163900

Chircov, C., Miclea, I. I., Grumezescu, V., & Grumezescu, A. M. (2021). Essential oils for bone repair and regeneration—Mechanisms and applications. *Materials (Basel)*, *14*(8), 1867. DOI: 10.3390/ma14081867 PMID: 33918697

Chittasupho, C., Manthaisong, A., Okonogi, S., Tadtong, S., & Samee, W. (2021). Effects of Quercetin and Curcumin Combination on Antibacterial, Antioxidant, In Vitro Wound Healing and Migration of Human Dermal Fibroblast Cells. *International Journal of Molecular Sciences*, *23*(1), 142. DOI: 10.3390/ijms23010142 PMID: 35008566

Cho, H., & Yoon, I. (2015). Pharmacokinetic Interactions of Herbs with Cytochrome P450 and P-Glycoprotein. *Evidence-Based Complementary and Alternative Medicine : eCAM, 1*, 1–10. Advance online publication. DOI: 10.1155/2015/736431 PMID: 25632290

Choi, S. S., & Diehl, A. M. (2009). Epithelial-to-mesenchymal transitions in liver. *Hepatology (Baltimore, Md.), 50*(6), 2007–2013. DOI: 10.1002/hep.23196 PMID: 19824076

Cho, J.-G., Lee, M.-K., Lee, J.-W., Park, H.-J., Lee, D.-Y., Lee, Y.-H., Yang, D.-C., & Baek, N.-I. (2010). Physicochemical Characterization and NMR Assignments of Ginsenosides Rb1, Rb2, Rc, and Rd Isolated from Panax ginseng. *Journal of Ginseng Research, 34*(2), 113–121. DOI: 10.5142/jgr.2010.34.2.113

Choudhary, A., Kant, V., Jangir, B. L., & Joshi, V. (2020). Quercetin loaded chitosan tripolyphosphate nanoparticles accelerated cutaneous wound healing in Wistar rats. *European Journal of Pharmacology, 880*, 173172. DOI: 10.1016/j.ejphar.2020.173172 PMID: 32407724

Chu, C., Deng, J., Man, Y., & Qu, Y. (2017). Green Tea Extracts Epigallocatechin-3-Gallate for Different Treatments. *BioMed Research International, 2017*(2017), 5615647. DOI: 10.1155/2017/5615647 PMID: 28884125

Coleridge-Smith, P. D. (2009). Leg ulcer treatment. *Journal of Vascular Surgery, 49*(3), 804–808. DOI: 10.1016/j.jvs.2009.01.003 PMID: 19268785

Coleridge-Smith, P., Lok, C., & Ramelet, A.-A. (2005). Venous leg ulcer: A meta-analysis of adjunctive therapy with micronized purified flavonoid fraction. *European Journal of Vascular and Endovascular Surgery, 30*(2), 198–208. DOI: 10.1016/j.ejvs.2005.04.017 PMID: 15936227

Comino-Sanz, I. M., López-Franco, M. D., Castro, B., & Pancorbo-Hidalgo, P. L. (2021). The Role of Antioxidants on Wound Healing: A Review of the Current Evidence. *Journal of Clinical Medicine, 10*(16), 3558. DOI: 10.3390/jcm10163558 PMID: 34441854

Cosme, F., Aires, A., Pinto, T., Oliveira, I., Vilela, A., & Gonçalves, B. (2025). A comprehensive review of bioactive tannins in foods and beverages: Functional properties, health benefits, and sensory qualities. *Molecules (Basel, Switzerland), 30*(4), 800. DOI: 10.3390/molecules30040800 PMID: 40005115

Costa, P., Grosso, C., Gonçalves, S., Andrade, P. B., Valentão, P., Bernardo-Gil, M. G., & Romano, A. (2012). Recovery of bioactives from Lavandula viridis by supercritical fluid extraction. *Food Chemistry*, *135*(1), 112–121. DOI: 10.1016/j.foodchem.2012.04.108 PMID: 26434270

Cotterell, A., Griffin, M., Downer, M. A., Parker, J. B., Wan, D., & Longaker, M. T. (2024). Understanding wound healing in obesity. *World Journal of Experimental Medicine*, *14*(1), 86898. DOI: 10.5493/wjem.v14.i1.86898 PMID: 38590299

Cox-Georgian, D., Ramadoss, N., Dona, C., & Basu, C. (2019). Therapeutic and medicinal uses of terpenes. In *Medicinal plants: From farm to pharmacy* (pp. 333–359). Springer. DOI: 10.1007/978-3-030-31269-5_15

Criollo-Mendoza, M. S., Contreras-Angulo, L. A., Leyva-López, N., Gutiérrez-Grijalva, E. P., Jiménez-Ortega, L. A., & Heredia, J. B. (2023). Wound healing properties of natural products: Mechanisms of action. *Molecules (Basel, Switzerland)*, *28*(2), 598. DOI: 10.3390/molecules28020598 PMID: 36677659

da Silva Carneiro, R. E., de Almeida da Costa, M. H., Lima Verde Santos, J. Z., Teixeira Uchoa, V., da Silva Lopes, L. U., & Mendes, A. N. (2025). Therapeutic actions of terpenes in neurodegenerative disorders and ECM remodeling. *Biocell*, *49*(1). Advance online publication. DOI: 10.32604/biocell.2024.058405

Da Silva, R. F., Carneiro, C. N., de Sousa, C. B., Gomez, F. J., Espino, M., Boiteux, J., Fernández, M. D., Silva, M. F., & Dias, F. D. (2022). Green analytical extraction of medicinal plant bioactives. *Microchemical Journal*, *175*, 107184. DOI: 10.1016/j.microc.2022.107184

Damrongrungruang, T., Paphangkorakit, J., Limsitthichaikoon, S., Khampaenjiraroch, B., Davies, M. J., Sungthong, B., & Priprem, A. (2021). Anthocyanin complex niosome gel accelerates oral wound healing: In vitro and clinical studies. *Nanomedicine; Nanotechnology, Biology, and Medicine*, *37*, 102423. DOI: 10.1016/j.nano.2021.102423 PMID: 34214683

Daniels, C., Isaacs, Z., Finelli, R., & Leisegang, K. (2022). The Efficacy of Zingiber officinale on Dyslipidaemia, Blood Pressure, and In-flammation as Cardiovascular Risk Factors: A Systematic Review. *Clinical Nutrition ESPEN*, *51*, 72–82. DOI: 10.1016/j.clnesp.2022.08.031 PMID: 36184251

Dargère, S., Dube, M., & Gervais, M. (2020). Probiotic treatment improves wound healing in diabetic mice. *Microorganisms*, *8*(6), 835. Advance online publication. DOI: 10.3390/microorganisms8060835 PMID: 32498344

Das, A. K., Islam, M. N., Faruk, M. O., Ashaduzzaman, M., & Dungani, R. (2020). Review on tannins: Extraction processes, applications and possibilities. *South African Journal of Botany*, *135*, 58–70. DOI: 10.1016/j.sajb.2020.08.008

Dasari, N., Jiang, A., Skochdopole, A. J., Chung, J., Reece, E. M., Vorstenbosch, J., & Winocour, S. (2021). Updates in Diabetic Wound Healing, Inflammation, and Scarring. *Seminars in Plastic Surgery*, *35*(3), 153–158. DOI: 10.1055/s-0041-1731460 PMID: 34526862

Dash, S., Behera, P. M., Mandal, U., Nayak, R., Parida, S., & Mahalik, G. (2021). Potential medicinal plants of Centurion University of Technology and Management and their medicinal uses. *Journal of Medicinal Plants Studies*, *9*(3), 251–258. DOI: 10.22271/plants.2021.v9.i3c.1303

Dashtian, K., Kamalabadi, M., Ghoorchian, A., Ganjali, M. R., & Rahimi-Nasrabadi, M. (2024). Integrated supercritical fluid extraction of essential oils. *Journal of Chromatography. A*, *1723*, 465240. DOI: 10.1016/j.chroma.2024.465240 PMID: 39154494

Davis, M. E., Chen, Z. G., & Golomb, H. M. (2008). Policy forum: Nanotechnology. *Science*, *319*(5869), 1673–1675.

de Macedo, L. M., dos Santos, É. M., Militão, L., Tundisi, L. L., Ataide, J. A., Souto, E. B., & Mazzola, P. G. (2020). Rosemary (Rosmarinus officinalis L., Syn Salvia rosmarinus Spenn.) and Its Topical Applications: A Review. *Plants*, *2020*(9), 651. DOI: 10.3390/plants9050651 PMID: 32455585

Debnath, P. K., Banerjee, S., Debnath, P., Mitra, A., & Mukherjee, P. K. (2015). Ayurveda–opportunities for developing safe and effective treatment choices for the future. In *Evidence-based validation of herbal medicine* (pp. 427–454). Elsevier. DOI: 10.1016/B978-0-12-800874-4.00020-9

Deepika, E., & Santhy, K. S. (2023). Unravelling the mode of action of wound healing efficiency with the compounds screened from *Hygrophila auriculata* by molecular docking studies. *Journal of Plant Science Research*, *2*, 1–9.

Demidova-Rice, T. N., Hamblin, M. R., & Herman, I. M. (2012). Acute and impaired wound healing: Pathophysiology and current methods for drug delivery, part 1: Normal and chronic wounds: Biology, causes, and approaches to care. *Advances in Skin & Wound Care*, *25*(7), 304–314. https://pubmed.ncbi.nlm.nih.gov/22713781/ . DOI: 10.1097/01.ASW.0000416006.55218.d0 PMID: 22713781

Deng, L., Du, C., Song, P., Chen, T., Rui, S., Armstrong, D. G., & Deng, W. (2021). The Role of Oxidative Stress and Antioxidants in Diabetic Wound Healing. *Oxidative Medicine and Cellular Longevity*, *2021*(1), 8852759. DOI: 10.1155/2021/8852759 PMID: 33628388

DeRouen, M. C., & Oro, A. E. (2009). The primary cilium: A small yet mighty organelle. *The Journal of Investigative Dermatology*, *129*(2), 264–265. DOI: 10.1038/jid.2008.404 PMID: 19148215

Dev, S. S., Farghadani, R., Abidin, S. A. Z., Othman, I., & Naidu, R. (2023). Flavonoids as receptor tyrosine kinase inhibitors in lung cancer. *Journal of Functional Foods*, *110*, 105845. Advance online publication. DOI: 10.1016/j.jff.2023.105845

Dewi, K. A., Kusumaningrum, S., Prasetya, N. B. A., Kusumastuti, S. A., Rismana, E., Ngatinem, N., Nuralih, N., & Wulansari, M. T. (2025). Potential of binahong (*Anredera cordifolia* (Ten.) Steenis) leaf extract as a wound healing agent that inhibits matrix metalloproteinases (MMPs) using in silico and stimulates NIH-3T3 cell proliferation by in vitro assay. *Farmatsiia (Sofia)*, *72*, 1–15. DOI: 10.3897/pharmacia.72.e135811

Dhamodiran, M., Chinnaperumal, K., Venkatesan, G., Alshiekheid, M. A., & Suseem, S. R. (2024). Isolation, structural elucidation of bioactive compounds and their wound-healing ability, antibacterial and In silico molecular docking applications. *Environmental Research*, *252*, 119023. DOI: 10.1016/j.envres.2024.119023 PMID: 38685295

Diller, R. B., & Tabor, A. J. (2022). The Role of the Extracellular Matrix (ECM) in Wound Healing: A Review. *Biomimetics*, *7*(3), 87. DOI: 10.3390/biomimetics7030087 PMID: 35892357

Ding, Y., Yang, P., Li, S., Zhang, H., Ding, X., & Tan, Q. (2022). Resveratrol accelerates wound healing by inducing M2 macrophage polarisation in diabetic mice. *Pharmaceutical Biology*, *60*(1), 2328–2337. DOI: 10.1080/13880209.2022.2149821 PMID: 36469602

Dinh, T., Elder, S., & Veves, A. (2011). Delayed wound healing in diabetes: Considering future treatments. *Diabetes Management (London, England)*, *1*(5), 509–519. DOI: 10.2217/dmt.11.44

Diniz, L. R. L., Calado, L. L., Duarte, A. B. S., & de Sousa, D. P. (2023). *Centella asiatica* and its metabolite asiatic acid: Wound healing effects and therapeutic potential. *Metabolites*, *13*(2), 276. DOI: 10.3390/metabo13020276 PMID: 36837896 do Nascimento, M. H. M., & de Araújo, D. R. (2022). Exploring the Pharmacological Potential of Glycyrrhizic Acid: From Therapeutic Applications to Trends in Nanomedicine. *Future Pharmacology, 2*(1), 1–15. DOI: 10.3390/futurepharmacol2010001

Doerge, D. R., & Chang, H. C. (2002). Inactivation of thyroid peroxidase by soy isoflavones, in vitro and in vivo. *Journal of Chromatography. B, Analytical Technologies in the Biomedical and Life Sciences, 777*(1-2), 269–279. DOI: 10.1016/S1570-0232(02)00214-3 PMID: 12270219

Dong, G., Liu, H., Yu, X., Zhang, X., Lu, H., Zhou, T., & Cao, J. (2018). Antimicrobial and anti-biofilm activity of tannic acid against Staphylococcus aureus. *Natural Product Research, 32*(18), 2225–2228. DOI: 10.1080/14786419.2017.1366485 PMID: 28826250

Doughty, D., Ramundo, J., Bonham, P., Beitz, J., Erwin-Toth, P., Anderson, R., & Rolstad, B. S. (2006). Issues and challenges in staging of pressure ulcers. *Journal of Wound, Ostomy, and Continence Nursing, 33*(2), 125–130. DOI: 10.1097/00152192-200603000-00004 PMID: 16572010

Dou, Q. P. (2019). Tea in Health and Disease. *Nutrients, 2019*(11), 929. DOI: 10.3390/nu11040929 PMID: 31027187

Dow, G., Browne, A., & Sibbald, R. G. (1999). Infection in chronic wounds: Controversies in diagnosis and treatment. *Ostomy/Wound Management, 45*(8), 23–42. PMID: 10655866

Duda-Chodak, A., & Tarko, T. (2023). Possible side effects of polyphenols and their interactions with medicines. *Molecules (Basel, Switzerland), 28*(6), 2536. DOI: 10.3390/molecules28062536 PMID: 36985507

Du, L., Feng, X., Xiang, X., & Jin, Y. (2016). Wound healing effect of an in situ forming hydrogel loading curcumin-phospholipid complex. *Current Drug Delivery, 13*(1), 76–82. DOI: 10.2174/1567201813666151202195437 PMID: 26634789

Dunnick, J. K., & Hailey, J. R. (1992). Toxicity and carcinogenicity studies of quercetin, a natural component of foods. *Toxicological Sciences, 19*(3), 423–431. DOI: 10.1093/toxsci/19.3.423 PMID: 1459373

Dunnill, C., Patton, T., Brennan, J., Barrett, J., Dryden, M., Cooke, J., Leaper, D., & Georgopoulos, N. T. (2017). Reactive oxygen species in wound healing. *International Wound Journal, 14*(1), 89–96. DOI: 10.1111/iwj.12557 PMID: 26688157

Duran, N.. (2019). Regulatory challenges for nanomedicine: The demand for clarity in the regulatory landscape. *Nanomedicine (London), 14*(4), 400–402.

Durazzo, A., Lucarini, M., Souto, E. B., Cicala, C., Caiazzo, E., Izzo, A. A., & Santini, A. (2019). Polyphenols: A concise overview on the chemistry, occurrence, and human health. *Phytotherapy Research*, *33*(9), 2221–2243. DOI: 10.1002/ptr.6419 PMID: 31359516

Dwivedi, J., Sachan, P., Wal, P., Wal, A., & Rai, A. K. (2024). Current state and future perspective of diabetic wound healing treatment: Present evidence from clinical trials. *Current Diabetes Reviews*, *20*(5), 45–65. DOI: 10.2174/1573399820666230828091708 PMID: 37641999

Dwivedi, J., Sachan, P., Wal, P., Wal, A., Vippamakula, S., Shankara, R. K., & Sharma, M. C. (2025). Clinical evidence of traditional medicines in modulating the immune response and diabetic wound healing. *Current Topics in Medicinal Chemistry*, *25*. Advance online publication. DOI: 10.2174/0115680266359349250707102203 PMID: 40676793

Earl, J. E., Aubert, J. F., Michael, E. O., & Edwards, D. J. (2025). Effects of Soil, Light, and Temperature on Freshwater Tannin Concentrations. *Hydrobiology*, *4*(1), 2. DOI: 10.3390/hydrobiology4010002

Egea, G., Jiménez-Altayó, F., & Campuzano, V. (2020). Reactive Oxygen Species and Oxidative Stress in the Pathogenesis and Progression of Genetic Diseases of the Connective Tissue. *Antioxidants*, *9*(10), 1013. DOI: 10.3390/antiox9101013 PMID: 33086603

Ekor, M. (2014). The growing use of herbal medicines: Issues relating to adverse reactions and challenges in monitoring safety. *Frontiers in Pharmacology*, *4*, 177. DOI: 10.3389/fphar.2013.00177 PMID: 24454289

El-Aasr, M., Nohara, T., Ikeda, T., Abu-Risha, S. E., Elekhnawy, E., Tawfik, H. O., Shoeib, N., & Attia, G. (2023). LC-MS/MS metabolomics profiling of *Glechoma hederacea* L. methanolic extract; in vitro antimicrobial and in vivo with in silico wound healing studies on *Staphylococcus aureus* infected rat skin wound. *Natural Product Research*, *37*(10), 1730–1734. DOI: 10.1080/14786419.2022.2108421 PMID: 35942893

Elajaili, H., Lyttle, B. D., Lewis, C., Bardill, J. R., Dee, N., Seal, S., Nozik-Grayck, E., Liechty, K. W., & Zgheib, C. (2025). Increased ROS and Persistent Pro-Inflammatory Responses in a Diabetic Wound Healing Model (db/db): Implications for Delayed Wound Healing. *International Journal of Molecular Sciences*, *26*(10), 4884. DOI: 10.3390/ijms26104884 PMID: 40430024

Elbahnasawy, A. S., Valeeva, E. R., El-Sayed, E. M., & Rakhimov, I. I. (2019). Impact of thyme and rosemary on osteoporosis prevention in rats. *Journal of Nutrition and Metabolism, 2019*, 1431384. DOI: 10.1155/2019/1431384 PMID: 31049223

El-Mabood, A., El-Sayed, A., Zaher, N. A., & Ali, H. E. (2014). Topical tannic acid application decreases posthemorrhoidectomy pain: A novel idea. *The Egyptian Journal of Surgery*, *33*(4), 213. DOI: 10.4103/1110-1121.147586

El-Refaie, W. M., Elnaggar, Y. S., El-Massik, M. A., & Abdallah, O. Y. (2015). Novel curcumin-loaded gel-core hyaluosomes with promising burn-wound healing potential: Development, in vitro appraisal and in vivo studies. *International Journal of Pharmaceutics*, *486*(1-2), 88–98. DOI: 10.1016/j.ijpharm.2015.03.052 PMID: 25818063

El-Saber Batiha, G., Alkazmi, L. M., Wasef, L. G., Beshbishy, A. M., Nadwa, E. H., & Rashwan, E. K. (2020). Syzygium aromaticum (Myrtaceae): Traditional uses, bioactive constituents, pharmacological and toxicological activities. *Biomolecules*, *10*(2), 202. DOI: 10.3390/biom10020202 PMID: 32019140

El-Sapagh, S., El-Shenody, R., Pereira, L., & Elshobary, M. (2023). Unveiling the potential of algal extracts as promising antibacterial and antibiofilm agents against multidrug-resistant Pseudomonas aeruginosa: In vitro and in silico studies including molecular docking. *Plants*, *12*(18), 3324. DOI: 10.3390/plants12183324 PMID: 37765485

El-Zahar, K. M., Al-Jamaan, M. E., Al-Mutairi, F. R., Al-Hudiab, A. M., Al-Einzi, M. S., & Mo, A. A.-Z. M. (2022). Antioxidant, Antibacterial, and Antifungal Activities of the Ethanolic Extract Obtained from Berberis vulgaris Roots and Leaves. *Molecules (Basel, Switzerland)*, *27*(18), 6114. DOI: 10.3390/molecules27186114 PMID: 36144846

Eo, H., Lee, H.-J., & Lim, Y. (2016). Ameliorative effect of dietary genistein on diabetes induced hyper-inflammation and oxidative stress during early stage of wound healing in alloxan induced diabetic mice. *Biochemical and Biophysical Research Communications*, *478*(3), 1021–1027. DOI: 10.1016/j.bbrc.2016.07.039 PMID: 27431618

Eseberri, I., Trepiana, J., Léniz, A., Gómez-García, I., Carr-Ugarte, H., González, M., & Portillo, M. P. (2022). Variability in the beneficial effects of phenolic compounds: A review. *Nutrients*, *14*(9), 1925. DOI: 10.3390/nu14091925 PMID: 35565892

Eteraf-Oskouei, T., & Najafi, M. (2013). Traditional and modern uses of natural honey in human diseases: A review. *Iranian Journal of Basic Medical Sciences.*, *16*(6), 731–742. PMID: 23997898

European Medicines Agency. (2011). Guideline on specifications for herbal substances and preparations.

Everett, E., & Mathioudakis, N. (2018). Management of diabetic foot ulcers. *Annals of the New York Academy of Sciences*, *1411*(1), 153–165. DOI: 10.1111/nyas.13569 PMID: 29377202

Fabricant, D. S., & Farnsworth, N. R. (2001). The value of plants used in traditional medicine for drug discovery. *Environmental Health Perspectives*, *109*(suppl 1), 69–75. PMID: 11250806

Fagot, D., Pham, D. M., Laboureau, J., Planel, E., Guerin, L., Nègre, C., Donovan, M., & Bernard, B. A. (2018). Crocin, a Natural Molecule with Potentially Beneficial Effects against Skin Ageing. *International Journal of Cosmetic Science*, *40*(4), 388–400. DOI: 10.1111/ics.12472 PMID: 29893408

Falanga, V., Isseroff, R. R., Soulika, A. M., Romanelli, M., Margolis, D., Kapp, S., Granick, M., & Harding, K. (2022). Chronic wounds. *Nature Reviews. Disease Primers*, *8*(1), 50. DOI: 10.1038/s41572-022-00377-3 PMID: 35864102

Famurewa, A. C., Maduagwuna, E. K., Folawiyo, A. M., Besong, E. E., Eteudo, A. N., Famurewa, O. A., & Ejezie, F. E. (2019). Antioxidant, anti-inflammatory, and antiapoptotic effects of virgin coconut oil against antibiotic drug gentamicin-induced nephrotoxicity via the suppression of oxidative stress and modulation of iNOS/NF-κB/caspase-3 signaling pathway in Wistar rats. *Journal of Food Biochemistry*, *44*(1). Advance online publication. DOI: 10.1111/jfbc.13100 PMID: 31721240

Fan, J., Fu, A., & Zhang, L. (2019). Progress in molecular docking. *Quantitative Biology*, *7*(2), 83–89. DOI: 10.1007/s40484-019-0172-y

Farahpour, M. R., Pirkhezr, E., Ashrafian, A., & Sonboli, A. (2020). Accelerated Healing by Topical Administration of Salvia officinalis Essential Oil on Pseudomonas Aeruginosa and Staphylococcus Aureus Infected Wound Model. *Biomedicine and Pharmacotherapy*, *128*, 110120. DOI: 10.1016/j.biopha.2020.110120 PMID: 32460189

Farha, A. K., Yang, Q.-Q., Kim, G., Li, H.-B., Zhu, F., Liu, H.-Y., Gan, R.-Y., & Corke, H. (2020). Tannins as an alternative to antibiotics. *Food Bioscience*, *38*, 100751. DOI: 10.1016/j.fbio.2020.100751

Farkhondeh, T., Llorens, S., Pourbagher-Shahri, A. M., Ashrafizadeh, M., & Samarghandian, S. (2020). The therapeutic effect of resveratrol: Focusing on the Nrf2 signaling pathway. *Biomedicine and Pharmacotherapy*, *127*, 110234. DOI: 10.1016/j.biopha.2020.110234 PMID: 32559855

Fasinu, P. S., Bouic, P., & Rosenkranz, B. (2012). An Overview of the Evidence and Mechanisms of Herb–Drug Interactions. *Frontiers in Pharmacology, 3*. Advance online publication. DOI: 10.3389/fphar.2012.00069 PMID: 22557968

Fatemi, M. J., Nikoomaram, B., Rahimi, A. A. K., Talayi, D., Taghavi, S., & Ghavami, Y. (2014). Effect of green tea on the second degree burn wounds in rats. *Indian Journal of Plastic Surgery : Official Publication of the Association of Plastic Surgeons of India, 47*(03), 370–374. DOI: 10.4103/0970-0358.146593 PMID: 25593422

Fathi, M.. (2019). Nano-encapsulation of bioactive compounds improves bioavailability and efficacy. *Food Chemistry, 276*, 817–828.

Fentahun, N., Anteneh, Y., & Menber, Y. (2021). Malnutrition in the Outcome of Wound Healing at Public Hospitals in Bahir Dar City, Northwest Ethiopia: A Prospective Cohort Study. *Journal of Nutrition and Metabolism, 8824951*, 1–7. Advance online publication. DOI: 10.1155/2021/8824951 PMID: 33680509

Fernández, K., Llanquileo, A., Bustos, M., Aedo, V., Ruiz, I., Carrasco, S., Tapia, M., Pereira, M., Meléndrez, M. F., Aguayo, C., & Atanase, L. I. (2023). Self-Assembled CNF/rGO/Tannin Composite: Study of the Physicochemical and Wound Healing Properties. *Polymers, 15*(12), 2752. DOI: 10.3390/polym15122752 PMID: 37376399

Ferreira, A., Lisboa, P., Oliveira, K., Lima, L., Barros, I., & Carvalho, D. (2002). Inhibition of thyroid type 1 deiodinase activity by flavonoids. *Food and Chemical Toxicology, 40*(7), 913–917. DOI: 10.1016/S0278-6915(02)00064-9 PMID: 12065212

Fife, C. E., & Carter, M. J. (2012). Wound care outcomes and associated cost among patients treated in US outpatient wound centers: Data from the US wound registry. *Wounds : a Compendium of Clinical Research and Practice, 24*(1), 10–17. PMID: 25875947

Figueiredo, A. C., Barroso, J. G., Pedro, L. G., & Scheffer, J. J. (2008). Factors affecting secondary metabolite production in plants. *Flavour and Fragrance Journal, 23*(4), 213–226. DOI: 10.1002/ffj.1875

Fikru, A., Makonnen, E., Eguale, T., Debella, A., & Mekonnen, G. A. (2012). Evaluation of in vivo wound healing activity of methanol extract of Achyranthes aspera L. *Journal of Ethnopharmacology, 143*(2), 469–474. DOI: 10.1016/j.jep.2012.06.049 PMID: 22771316

Finnson, K. W., McLean, S., Di Guglielmo, G. M., & Philip, A. (2013). Dynamics of transforming growth factor beta signaling in wound healing and scarring. *Advances in Wound Care, 2*(5), 195–214. DOI: 10.1089/wound.2013.0429 PMID: 24527343

Fraga-Corral, M., Otero, P., Cassani, L., Echave, J., Garcia-Oliveira, P., Carpena, M., Chamorro, F., Lourenço-Lopes, C., Prieto, M. A., & Simal-Gandara, J. (2021). Traditional Applications of Tannin Rich Extracts Supported by Scientific Data: Chemical Composition, Bioavailability and Bioaccessibility. *Foods*, *10*(2), 251. DOI: 10.3390/foods10020251 PMID: 33530516

Freedman, B. R., Hwang, C., Talbot, S., Hibler, B., Matoori, S., & Mooney, D. J. (2023). Breakthrough treatments for accelerated wound healing. *Science Advances*, *9*(20), eade7007. Advance online publication. DOI: 10.1126/sciadv.ade7007 PMID: 37196080

Friedman, S. L. (2008). Mechanisms of hepatic fibrogenesis. *Gastroenterology*, *134*(6), 1655–1669. DOI: 10.1053/j.gastro.2008.03.003 PMID: 18471545

Friedrich, R. B., Kann, B., Coradini, K., Offerhaus, H. L., Beck, R. C., & Windbergs, M. (2015). Skin penetration behavior of lipid-core nanocapsules for simultaneous delivery of resveratrol and curcumin. *European Journal of Pharmaceutical Sciences*, *78*, 204–213. DOI: 10.1016/j.ejps.2015.07.018 PMID: 26215463

Friedstat, J., Brown, D. A., & Levi, B. (2017). Chemical, Electrical, and Radiation Injuries. *Clinics in Plastic Surgery*, *44*(3), 657–669. DOI: 10.1016/j.cps.2017.02.021 PMID: 28576255

Frykberg, R. G., & Banks, J. (2015). Challenges in the treatment of chronic wounds. *Advances in Wound Care*, *4*(9), 560–582. DOI: 10.1089/wound.2015.0635 PMID: 26339534

Fuchs, T. A., Abed, U., Goosmann, C., Hurwitz, R., Schulze, I., Wahn, V., Weinrauch, Y., Brinkmann, V., & Zychlinsky, A. (2007). Novel cell death program leads to neutrophil extracellular traps. *The Journal of Cell Biology*, *176*(2), 231–241. DOI: 10.1083/jcb.200606027 PMID: 17210947

Fui, L. W., Lok, M. P., Govindasamy, V., Yong, T. K., Lek, T. K., & Das, A. K. (2019). Understanding the multifaceted mechanisms of diabetic wound healing and therapeutic application of stem cells conditioned medium in the healing process. *Journal of Tissue Engineering and Regenerative Medicine*, *13*(12), 2218–2233. DOI: 10.1002/term.2966 PMID: 31648415

Fu, J., Huang, J., Lin, M., Xie, T., & You, T. (2020). Quercetin promotes diabetic wound healing via switching macrophages from M1 to M2 polarization. *The Journal of Surgical Research*, *246*, 213–223. DOI: 10.1016/j.jss.2019.09.011 PMID: 31606511

Fujita, K., Kuge, K., Ozawa, N., Sahara, S., Zaiki, K., Nakaoji, K., Hamada, K., Takenaka, Y., Tanahashi, T., Tamai, K., Kaneda, Y., & Maeda, A. (2015). Cinnamtannin B-1 promotes migration of mesenchymal stem cells and accelerates wound healing in mice. *PLoS One*, *10*(12), e0144166. DOI: 10.1371/journal.pone.0144166 PMID: 26657737

Furie, B., & Furie, B. C. (2008). Mechanisms of thrombus formation. *The New England Journal of Medicine*, *359*(9), 938–949. DOI: 10.1056/NEJMra0801082 PMID: 18753650

Fu, X., Chen, H., Zhang, X., & Zhang, Z. (2025). Resveratrol: A potent antioxidant and anti-inflammatory compound in the treatment of chronic wounds. *Antioxidants*, *14*(3), 210. DOI: 10.3390/antiox14030210

Gadekar, R., Saurabh, M., Thakur, G., & Saurabh, A. (2012). Study of formulation, characterisation and wound healing potential of transdermal patches of curcumin. *Asian Journal of Pharmaceutical and Clinical Research*, *5*, 225–230.

Galiano, R. D., Michaels, J. V., Dobryansky, M., Levine, J. P., & Gurtner, G. C. (2004). Murine model of impaired wound healing. *Wound Repair and Regeneration*, *12*(4), 485–492. DOI: 10.1111/j.1067-1927.2004.12404.x PMID: 15260814

Gallelli, G., Cione, E., Serra, R., Leo, A., Citraro, R., Matricardi, P., Di Meo, C., Bisceglia, F., Caroleo, M. C., Basile, S., & Gallelli, L. (2020). Nano-hydrogel embedded with quercetin and oleic acid as a new formulation in the treatment of diabetic foot ulcer: A pilot study. *International Wound Journal*, *17*(2), 485–490. DOI: 10.1111/iwj.13299 PMID: 31876118

Ganesan, S., Parasuraman, S., Maheswaran, S. U., & Gnanasekar, N. (2012). Wound healing activity of *Hemidesmus indicus* formulation. *Journal of Pharmacology & Pharmacotherapeutics*, *3*(1), 66–67. DOI: 10.4103/0976-500X.92516 PMID: 22368424

Gang, G., Gao, R., Zhao, H., Xu, Y., Xing, Y., Jin, X., Hong, L., Yan, S., & Shi, B. (2024). Effects of water extracts of Artemisia annua L. on rumen immune and antioxidative indexes, fermentation parameters and microbials diversity in lambs. *Frontiers in Microbiology*, *15*, 1485882. Advance online publication. DOI: 10.3389/fmicb.2024.1485882 PMID: 39493850

Gantwerker, E. A., & Hom, D. B. (2011). Skin: Histology and physiology of wound healing. *Facial Plastic Surgery Clinics of North America*, *19*(3), 441–453. DOI: 10.1016/j.fsc.2011.06.009 PMID: 21856533

Gao, H., Pei, X., Song, X., Wang, S., Yang, Z., Zhu, J., Lin, Q., Zhu, Q., & Yang, X. (2025). Application and development of CRISPR technology in the secondary metabolic pathway of the active ingredients of phytopharmaceuticals. *Frontiers in Plant Science*, *15*, 1477894. Advance online publication. DOI: 10.3389/fpls.2024.1477894 PMID: 39850214

Gao, K., Zhang, B., & Zhao, F. (2023). Antibiofilm effect of curcumin against Staphylococcus aureus surface wound biofilm–associated infection: In vitro and in silico. *Applied Biochemistry and Biotechnology*, *195*(9), 5329–5337. DOI: 10.1007/s12010-022-03844-5 PMID: 35166997

Garg, S. K., Shukla, A., & Choudhury, S. (2019). Polyphenols and flavonoids. In Gupta, R., Srivastava, A., & Lall, R. (Eds.), *Nutraceuticals in veterinary medicine* (pp. 171–184). Springer., DOI: 10.1007/978-3-030-04624-8_13

Gawali, V. (2021). Antihyperglycemic and antimicrobial activity of Berberis vulgaris in the treatment of diabetic foot ulcers. *Phytotherapy Research*, *35*(5), 1863–1872. DOI: 10.1002/ptr.6907

Geethalakshmi, R., Sakravarthi, C., Kritika, T., Arul Kirubakaran, M., & Sarada, D. V. L. (2013). Evaluation of Antioxidant and Wound Healing Potentials of Sphaeranthus Amaranthoides Burm.f. *BioMed Research International*, *2013*, 607109. DOI: 10.1155/2013/607109 PMID: 23509751

Gethin, G. (2007). The significance of surface pH in chronic wounds. *Wounds UK*, *3*(1), 52–56.

Ghaisas, M. M., Kshirsagar, S. B., & Sahane, R. S. (2014). Evaluation of wound healing activity of ferulic acid in diabetic rats. *International Wound Journal*, *11*(5), 523–532. DOI: 10.1111/j.1742-481X.2012.01119.x PMID: 23236955

Ghanadian, M., Soltani, R., Homayouni, A., Khorvash, F., Jouabadi, S. M., & Abdollahzadeh, M. (2022). The Effect of Plantago major Hydroalcoholic Extract on the Healing of Diabetic Foot and Pressure Ulcers: A Randomized Open-Label Controlled Clinical Trial. *The International Journal of Lower Extremity Wounds*, *23*(3), 475–481. DOI: 10.1177/15347346211070723 PMID: 35044254

Ghosh, S.. (2018). Genetic engineering of plants for enhanced polyphenol production. *Plant Biotechnology Journal*, *16*(6), 1035–1047.

Ghosh, S.. (2020). Molecular farming in plants for therapeutic protein production. *Biomolecules*, *10*(9), 1–14.

Ghosh, T., Maity, T. K., Dash, D. K., & Boss, A. (2007). A study on wound healing activity of *Bacopa monnieri* Linn. aerial parts. *Oriental Pharmacy and Experimental Medicine*, *7*(2), 150–156. DOI: 10.3742/OPEM.2007.7.2.150

Giacco, F., & Brownlee, M. (2010). Oxidative Stress and Diabetic Complications. *Circulation Research*, *107*(9), 1058–1070. DOI: 10.1161/CIRCRESAHA.110.223545 PMID: 21030723

Gigliobianco, M. R., Cortese, M., Peregrina, D. V., Villa, C., Lupidi, G., Pruccoli, L., Angeloni, C., Tarozzi, A., Censi, R., & Di Martino, P. (2021). Development of New Extracts of Crocus sativus L. By-Product from Two Different Italian Regions as New Potential Active Ingredient in Cosmetic Formulations. *Cosmetics*, *8*(2), 51. DOI: 10.3390/cosmetics8020051

Glisan, S. L., Grove, K. A., Yennawar, N. H., & Lambert, J. D. (2017). Inhibition of pancreatic lipase by black tea theaflavins: Comparative enzymology and in silico modeling studies. *Food Chemistry*, *216*, 296–300. DOI: 10.1016/j.foodchem.2016.08.052 PMID: 27596423

Godo, S., & Shimokawa, H. (2017). Endothelial functions. *Arteriosclerosis, Thrombosis, and Vascular Biology*, *37*(9), e108–e114. DOI: 10.1161/ATVBAHA.117.309813 PMID: 28835487

Gohil, K., Patel, J., & Gajjar, A. (2010). Pharmacological review on Centella asiatica: A potential herbal cure-all. *Indian Journal of Pharmaceutical Sciences*, *72*(5), 546. DOI: 10.4103/0250-474X.78519 PMID: 21694984

Golebiewska, E. M., & Poole, A. W. (2015). Platelet secretion: From haemostasis to wound healing and beyond. *Blood Reviews*, *29*(3), 153–162. DOI: 10.1016/j.blre.2014.10.003 PMID: 25468720

Gonfa, Y. H., Tessema, F. B., Bachheti, A., Rai, N., Tadesse, M. G., Nasser Singab, A. N., Chaubey, K. K., & Bachheti, R. K. (2023). Anti-inflammatory activity of phytochemicals from medicinal plants and their nanoparticles: A review. *Current Research in Biotechnology*, *6*, 100152. DOI: 10.1016/j.crbiot.2023.100152

Gonzales, A. M., & Orlando, R. A. (2008). Curcumin and resveratrol inhibit nuclear factor-kappaB-mediated cytokine expression in adipocytes. *Nutrition & Metabolism*, *5*(1), 17. DOI: 10.1186/1743-7075-5-17 PMID: 18549505

Gonzalez, A. C. O., Costa, T. F., Andrade, Z. A., & Medrado, A. R. A. P. (2016). Wound healing – A literature review. *Anais Brasileiros de Dermatologia*, *91*(5), 614–620. DOI: 10.1590/abd1806-4841.20164741 PMID: 27828635

González-Fuentes, J., Selva, J., Moya, C., Castro-Vázquez, L., Lozano, M. V., Marcos, P., Plaza-Oliver, M., Rodríguez-Robledo, V., Santander-Ortega, M. J., Villaseca-González, N., & Arroyo-Jimenez, M. M. (2018). Neuroprotective natural molecules from food to brain. *Frontiers in Neuroscience*, *12*, 721. DOI: 10.3389/fnins.2018.00721 PMID: 30405328

Goto, T., & Saligan, L. N. (2020). Wound pain and wound healing biomarkers from wound exudate: A scoping review. *Journal of Wound, Ostomy, and Continence Nursing*, *47*(6), 559–568. DOI: 10.1097/WON.0000000000000703 PMID: 33201141

Gowder, S. J. (2013). New Insights into Toxicity and Drug Testing. In *InTech eBooks*. DOI: 10.5772/55886

Grice, E. A., & Segre, J. A. (2011). The skin microbiome. *Nature Reviews. Microbiology*, *9*(4), 244–253. DOI: 10.1038/nrmicro2537 PMID: 21407241

Groeber, F., Schober, L., Schmid, F. F., Traube, A., Kolbus-Hernandez, S., Daton, K., Hoffmann, S., Petersohn, D., Schaefer-Korting, M., Walles, H., & Mewes, K. R. (2016). Validation of reconstructed epidermis for skin irritation testing. *Toxicology In Vitro*, *36*, 254–261. DOI: 10.1016/j.tiv.2016.07.008 PMID: 27435616

Grubbs, H., & Manna, B. (2023). Wound physiology. In *StatPearls*. StatPearls Publishing., https://www.ncbi.nlm.nih.gov/books/NBK518964/

Guan, Y., Niu, H., Liu, Z., Dang, Y., Shen, J., Zayed, M. A., Ma, L., & Guan, J. (2021). Sustained oxygenation accelerates diabetic wound healing by promoting epithelialization and angiogenesis and decreasing inflammation. *Science Advances*, *7*(35), eabj0153. Advance online publication. DOI: 10.1126/sciadv.abj0153 PMID: 34452918

Guimarães, I., Baptista-Silva, S., Pintado, M., & Oliveira, A. L. (2021). Polyphenols: A Promising Avenue in Therapeutic Solutions for Wound Care. *Applied Sciences (Basel, Switzerland)*, *11*(3), 1230. DOI: 10.3390/app11031230

Gülçin, İ. (2011). Antioxidant activity of eugenol. *Journal of Medicinal Food*, *14*(9), 975–985. DOI: 10.1089/jmf.2010.0197 PMID: 21554120

Guo, S., & Dipietro, L. A. (2010). Factors affecting wound healing. *Journal of Dental Research*, *89*(3), 219–229. DOI: 10.1177/0022034509359125 PMID: 20139336

Guo, Y., Niu, W., Jiao, H., Li, Y., Xu, C., Zhou, X., & Wang, J. (2025). Crosstalk between oxidative stress and inflammatory pathways: Natural therapeutic approaches for diabetic wound healing. *World Journal of Diabetes*, *16*(11), 111400. DOI: 10.4239/wjd.v16.i11.111400 PMID: 41278434

Gupta, S. C., Patchva, S., & Aggarwal, B. B. (2013). Therapeutic roles of curcumin: Lessons learned from clinical trials. *The AAPS Journal, 15*(1), 195–218. DOI: 10.1208/s12248-012-9432-8 PMID: 23143785

Gurjar, M. S., Ali, S., Akhtar, M., & Singh, K. S. (2012). Efficacy of plant extracts in plant disease management. *Agricultural Reviews (Karnal), 33*(2), 125–134.

Gurley, B. J., Fifer, E. K., & Gardner, Z. (2012). Pharmacokinetic Herb-Drug Interactions (Part 2): Drug Interactions Involving Popular Botanical Dietary Supplements and Their Clinical Relevance [Review of *Pharmacokinetic Herb-Drug Interactions (Part 2): Drug Interactions Involving Popular Botanical Dietary Supplements and Their Clinical Relevance*]. *Planta Medica, 78*(13), 1490. Thieme Medical Publishers (Germany). DOI: 10.1055/s-0031-1298331

Gurmet, P. (2004). "Sowa-Rigpa": Himalayan art of healing. *Indian Journal of Traditional Knowledge, 3*, 212–218.

Gurtner, G. C., Werner, S., Barrandon, Y., & Longaker, M. T. (2008). Wound repair and regeneration. *Nature, 453*(7193), 314–321. DOI: 10.1038/nature07039 PMID: 18480812

Haalboom, M. (2018). Chronic Wounds: Innovations in Diagnostics and Therapeutics. *Current Medicinal Chemistry, 25*(41), 5772–5781. DOI: 10.2174/092986732 4666170710120556 PMID: 28699502

Haddadi, G., Abbaszadeh, A., Mosleh-Shirazi, M. A., Okhovat, M. A., Salajeghe, A., & Ghorbani, Z. (2018). Evaluation of the effect of hesperidin on vascular endothelial growth factor gene expression in rat skin animal models following cobalt-60 gamma irradiation. *Journal of Cancer Research and Therapeutics, 14*(12, Suppl. 5), S1098–S1104. DOI: 10.4103/0973-1482.202892 PMID: 30539852

Haghdoost, F., Baradaran Mahdavi, M. M., Zolfaghari, B., Sanei, M. H., Najafi, S., Zandifar, A., Manouchehri, N., & Javanmard, S. H. (2016). The effect of Quercus brantii gall extract on burn wound healing in rat. *Iranian Journal of Basic Medical Sciences., 19*(10), 1144–1150. PMID: 27872712

Hagiwara, A., Hirose, M., Takahashi, S., Ogawa, K., Shirai, T., & Ito, N. (1991). Forestomach and kidney carcinogenicity of caffeic acid in F344 rats and C57BL/6N × C3H/HeN F1 mice. *Cancer Research, 51*(20), 5655–5660. PMID: 1913684

Hajialyani, M., Tewari, D., Sobarzo-Sánchez, E., Nabavi, S. M., Farzaei, M. H., & Abdollahi, M. (2018). Natural product-based nanomedicines for wound healing purposes: Therapeutic targets and drug delivery systems. *International Journal of Nanomedicine, 13*, 5023–5043. DOI: 10.2147/IJN.S174072 PMID: 30214204

Hakim, R. F., Fakhrurrazi, F., & Dinni, . (2019). Effect of Carica papaya Extract toward Incised Wound Healing Process in Mice (Mus musculus) Clinically and Histologically. *Evidence-Based Complementary and Alternative Medicine : eCAM*, *1*, 1–5. Advance online publication. DOI: 10.1155/2019/8306519 PMID: 31827564

Halim, A. S., Khoo, T. L., & Saad, A. Z. M. (2012). Wound bed preparation from a clinical perspective. *Indian Journal of Plastic Surgery : Official Publication of the Association of Plastic Surgeons of India*, *45*(2), 193–202. DOI: 10.4103/0970-0358.101277 PMID: 23162216

Hamdy, A., & Ibrahem, M. (2010). Management of aphthous ulceration with topical quercetin: A randomized clinical trial. *The Journal of Contemporary Dental Practice*, *11*(4), E009–E016. DOI: 10.5005/jcdp-11-4-9 PMID: 20953559

Hamilton, A. C. (2004). Medicinal plants, conservation and livelihoods. *Biodiversity and Conservation*, *13*(8), 1477–1517. DOI: 10.1023/B:BIOC.0000021333.23413.42

Hancock, V., Dahl, M., Vejborg, R. M., & Klemm, P. (2010). Dietary plant components ellagic acid and tannic acid inhibit Escherichiacoli biofilm formation. *Journal of Medical Microbiology*, *59*(4), 496–498. DOI: 10.1099/jmm.0.013680-0 PMID: 19959627

Handa, S. S., Khanuja, S. P. S., Longo, G., & Rakesh, D. D. (2008). *Extraction technologies for medicinal and aromatic plants*. UNIDO.

Hanna, P., Kochi, S. K., & Collier, R. J. (1992). Biochemical and physiological changes induced by anthrax lethal toxin in J774 macrophage-like cells. *Molecular Biology of the Cell*, *3*(11), 1269–1277. DOI: 10.1091/mbc.3.11.1269 PMID: 1457831

Han, Y. Z., Du, B. X., Zhu, X. Y., Wang, Y. Z., Zheng, H. J., & Liu, W. J. (2024). Lipid metabolism disorder in diabetic kidney disease. *Frontiers in Endocrinology*, *15*, 1336402. DOI: 10.3389/fendo.2024.1336402 PMID: 38742197

Haq, M. M., Chowdhury, M. A. R., Tayara, H., Abdelbaky, I., Islam, M. S., Chong, K. T., & Jeong, S. (2021). A report on multi-target anti-inflammatory properties of phytoconstituents from *Monochoria hastata* (Family: Pontederiaceae). *Molecules (Basel, Switzerland)*, *26*(23), 7397. DOI: 10.3390/molecules26237397 PMID: 34885978

Harding, K. (2015). Wounds and wound healing: New insights, fresh challenges. *British Journal of Dermatology*, *173*(2), 318–319. DOI: 10.1111/bjd.14036 PMID: 26346074

Hashemi-Shahraki, F., Shareghi, B., Farhadian, S., & Yadollahi, E. (2023). A comprehensive insight into the effects of caffeic acid (CA) on pepsin: Multi-spectroscopy and MD simulations methods. *Spectrochimica Acta. Part A: Molecular and Biomolecular Spectroscopy*, *289*, 122240. DOI: 10.1016/j.saa.2022.122240 PMID: 36527971

Hasibuan, P. F., Tambunan, A. Z., & Mutia, M. S. (2023). Anti-inflammatory potential of lemongrass leaves in wound healing. *Jurnal Prima Medika Sains*, *5*(2), 126–129. DOI: 10.34012/jpms.v5i2.4402

Hayakawa, S., Ohishi, T., Miyoshi, N., Oishi, Y., Nakamura, Y., & Isemura, M. (2020). Anti- Cancer Effects of Green Tea Epigallocatchin-3 Gallate and Coffee Chlorogenic Acid. *Molecules (Basel, Switzerland)*, *25*(19), 4553. DOI: 10.3390/molecules25194553 PMID: 33027981

Hayat, K., Iqbal, H., Malik, U., Bilal, U., & Mushtaq, S. (2015). Tea and Its Consumption: Benefits and Risks. *Critical Reviews in Food Science and Nutrition*, *55*(7), 939–954. DOI: 10.1080/10408398.2012.678949 PMID: 24915350

Hazian, N. Q., Mokhtar, N. A., Khairudin, N. A., & Yousif, R. H.Nur Qamarina HazianNur Ainun MokhtarNurulbahiyah Ahmad Khairudin. (2022). In silico molecular docking simulation of *Chromolaena odorata* phytoconstituents against matrix metalloproteinase-9 (MMP-9). *Journal of Research in Nanoscience and Nanotechnology*, *7*(1), 1–5. DOI: 10.37934/jrnn.7.1.16

Hazlett, L. D., Ekanayaka, S. A., McClellan, S. A., & Francis, R. (2019). Glycyrrhizin Use for Multi-Drug Resistant Pseudomonas aeruginosa: In Vitro and In Vivo Studies. *Investigative Ophthalmology & Visual Science*, *60*(8), 2978. DOI: 10.1167/iovs.19-27200 PMID: 31311033

Hegazi, S., Aly, R., Mesilhy, R., & Aljohary, H. (2024). Diabetic Foot Ulcer Wound Healing and Tissue Regeneration: Signaling Pathways and Mechanisms. In *IntechOpen eBooks*. IntechOpen. DOI: 10.5772/intechopen.1004267

Helal, M. A., Elwakeel, H., Sorogy, M. E., Abdelmaksoud, M. A., & Abdelgawad, M. S. (2024). Effectiveness of autologous leucocyte and platelet-rich fibrin (L-PRF) in management of chronic diabetic foot ulcers. *The Egyptian Journal of Surgery*, *43*(4), 1540–1547. DOI: 10.21608/EJSUR.2024.297983.1104

Hemingway, R. W. (1998). Practical polyphenolics: From structure to molecular recognition and physiological action, by Edwin Haslam. (Book review). *Journal of Natural Products*, *61*(11), 1454–1455. DOI: 10.1021/np980243t

Herman, A., Herman, A. P., Herman, A. P., & Herman, A. P. (2023). Herbal Products and Their Active Constituents for Diabetic Wound Healing—Preclinical and Clinical Studies. *Systematic Reviews in Pharmacy*, *15*(1), 281. DOI: 10.3390/pharmaceutics15010281 PMID: 36678910

Hermann, R., & von Richter, O. (2012). Clinical Evidence of Herbal Drugs As Perpetrators of Pharmacokinetic Drug Interactions [Review of *Clinical Evidence of Herbal Drugs As Perpetrators of Pharmacokinetic Drug Interactions*]. *Planta Medica, 78*(13), 1458. Thieme Medical Publishers (Germany). DOI: 10.1055/s-0032-1315117

Herrero-Cervera, A., Soehnlein, O., & Kenne, E. (2022). Neutrophils in chronic inflammatory diseases. *Cellular & Molecular Immunology*, *19*(2), 177–191. DOI: 10.1038/s41423-021-00832-3 PMID: 35039631

Hirose, M., Hoshiya, T., Mizoguchi, Y., Nakamura, A., Akagi, K., & Shirai, T. (2001). Green tea catechins enhance tumor development in the colon without effects in the lung or thyroid after pretreatment with 1,2-dimethylhydrazine or 2,2′-dihydroxy-di-n-propylnitrosamine in male F344 rats. *Cancer Letters*, *168*(1), 23–29. DOI: 10.1016/S0304-3835(01)00502-X PMID: 11368873

Hong, M., Cheng, H., Song, L., Wang, W., Wang, Q., Xu, D., & Xing, W. (2018). Wogonin suppresses the activity of matrix metalloproteinase-9 and inhibits migration and invasion in human hepatocellular carcinoma. *Molecules (Basel, Switzerland)*, *23*(2), 384. DOI: 10.3390/molecules23020384 PMID: 29439451

Hong, Y.-K., Chang, Y.-H., Lin, Y.-C., Chen, B., Guevara, B. E. K., & Hsu, C.-K. (2023). Inflammation in wound healing and pathological scarring. *Advances in Wound Care*, *12*(5), 288–300. DOI: 10.1089/wound.2021.0161 PMID: 36541356

Hormozi, M., Assaei, R., & Boroujeni, M. B. (2017). The Effect of Aloe vera on the Expression of Wound Healing Factors (TGF 1 and BFGF) in Mouse Embryonic Fibroblast Cell: In Vitro Study. *Biomedicine and Pharmacotherapy*, *88*, 610–616. DOI: 10.1016/j.biopha.2017.01.095 PMID: 28142117

Horváth, G., & Ács, K. (2015). Essential oils in respiratory tract diseases: Antibacterial and anti-inflammatory actions. *Flavour and Fragrance Journal*, *30*(5), 331–341. DOI: 10.1002/ffj.3252 PMID: 32313366

Hossain, C. M., Gera, M., & Ali, K. A. (2022). CURRENT STATUS AND CHALLENGES OF HERBAL DRUG DEVELOPMENT AND REGULATORY ASPECT: A GLOBAL PERSPECTIVE. *Asian Journal of Pharmaceutical and Clinical Research*, *31*, 31–41. Advance online publication. DOI: 10.22159/ajpcr.2022.v15i12.46134

Hossain, R., Quispe, C., Khan, R. A., Saikat, A. S. M., Ray, P., Ongalbek, D., Yeskaliyeva, B., Jain, D., Smeriglio, A., Trombetta, D., Kiani, R., Kobarfard, F., Mojgani, N., Saffarian, P., Ayatollahi, S. A., Sarkar, C. K., Islam, M. T., Keriman, D., Uçar, A., & Cho, W. C. (2022). Propolis: An update on its chemistry and pharmacological applications. *Chinese Medicine*, *17*(1), 100. Advance online publication. DOI: 10.1186/s13020-022-00651-2 PMID: 36028892

Hossain, S., Yousaf, M., Liu, Y., Chang, D., & Zhou, X. (2022). An Overview of the Evidence and Mechanism of Drug–Herb Interactions Between Propolis and Pharmaceutical Drugs. *Frontiers in Pharmacology*, *13*, 876183. DOI: 10.3389/fphar.2022.876183 PMID: 35444531

Hou, Q., Li, M., Lu, Y., Liu, D., & Li, C.-C. (2016). Burn wound healing properties of asiaticoside and madecassoside. *Experimental and Therapeutic Medicine*, *12*(3), 1269–1274. DOI: 10.3892/etm.2016.3459 PMID: 27588048

Huang, X., Sun, J., Chen, G., Niu, C., Wang, Y., Congcong, Z., Sun, J., Huang, H., Huang, S., Liang, Y., Shen, Y., Cong, W., Jin, L., & Zhu, Z. (2019). Resveratrol Promotes Diabetic Wound Healing via SIRT1-FOXO1-c-Myc Signaling Pathway-Mediated Angiogenesis. *Frontiers in Pharmacology*, *10*, 421. DOI: 10.3389/fphar.2019.00421 PMID: 31068817

Huang, Y., Dan, N., Dan, W., & Zhao, W. (2019). Reinforced polycaprolactone/chitosan nanocomposites for controlled curcumin release in wound dressing. *ACS Omega*, *4*(27), 22292–22301. DOI: 10.1021/acsomega.9b02217 PMID: 31909312

Hu, M. S., Walmsley, G. G., Barnes, L. A., Weiskopf, K., Rennert, R. C., Duscher, D., Januszyk, M., Maan, Z. N., Hong, W. X., Cheung, A. T., Leavitt, T., Marshall, C. D., Ransom, R. C., Malhotra, S., Moore, A. L., Rajadas, J., Lorenz, H. P., Weissman, I. L., Gurtner, G. C., & Longaker, M. T. (2017). Delivery of monocyte lineage cells in a biomimetic scaffold enhances tissue repair. *JCI Insight*, *2*(19), e96260. DOI: 10.1172/jci.insight.96260 PMID: 28978794

Hung, C., Hsiao, C., Hsieh, W.-H., Li, H., Tsai, Y., Lin, C., Chang, H., & Wu, N. (2017). 18ß-glycyrrhetinic acid derivative promotes proliferation, migration and aquaporin-3 expression in human dermal fibroblasts. *PLoS One*, *12*(8), e0182981. Advance online publication. DOI: 10.1371/journal.pone.0182981 PMID: 28813533

Hunt, M., Torres, M. S., Bachar-Wikström, E., & Wikström, J. D. (2024). Cellular and molecular roles of reactive oxygen species in wound healing. *Communications Biology*, *7*(1), 1534. DOI: 10.1038/s42003-024-07219-w PMID: 39562800

Hutchings, G., Kruszyna, Ł., Nawrocki, M. J., Strauss, E., Bryl, R., Spaczyńska, J., Perek, B., Jemielity, M., Mozdziak, P., Kempisty, B., Nowicki, M., & Krasiński, Z. (2021). Molecular Mechanisms Associated with ROS-Dependent Angiogenesis in Lower Extremity Artery Disease. *Antioxidants*, *10*(5), 735. DOI: 10.3390/antiox10050735 PMID: 34066926

Hu, X., Kong, Q., Li, R., Zhou, C., & Li, Z. (2023). Tannin–thioctic acid network hydrogels with antibacterial and hydrophobic properties. *Fibers and Polymers*, *24*(10), 3373–3384. DOI: 10.1007/s12221-023-00329-y

Hwang, J. S.. (2018). Microbial transformations of polyphenolic compounds. *Biotechnology Letters*, *40*(4), 481–490.

Hwang, S. H., & Lee, H. J. (2021). Stimulation of fibroblast growth and collagen synthesis by flavonoids in skin wound healing. *Journal of Dermatological Science*, *103*(2), 114–121. DOI: 10.1016/j.jdermsci.2021.06.007

Ibrahim, N. I., Wong, S. K., Mohamed, I. N., Mohamed, N., Chin, K. Y., Ima-Nirwana, S., & Shuid, A. N. (2018). Wound healing properties of selected natural products. *International Journal of Environmental Research and Public Health*, *15*(11), 2360. DOI: 10.3390/ijerph15112360 PMID: 30366427

İğci, B. K., & Aytaç, Z. (2020). An investigation on the in vitro wound healing activity and phytochemical composition of *Hypericum pseudolaeve* N. Robson growing in Turkey. *Turkish Journal of Pharmaceutical Sciences*, *17*(6), 610–619. DOI: 10.4274/tjps.galenos.2019.80037 PMID: 33389950

Irham, W. H., & Hardiyanti, R. (2021). Wound healing bioactivity of *Curcuma longa* Linn. *Rasayan Journal of Chemistry*, *14*(4), 2386–2391. DOI: 10.31788/RJC.2021.1446345

Ivanova, E. P., Hasan, J., Webb, H. K., Gervinskas, G., Juodkazis, R., & Truong, V. K.. (2023). Biomaterial-assisted bone regeneration inspired by essential oil bioactives. *Materials (Basel)*, *16*(11), 4012. PMID: 37297146

Jagiełło, K., Uchańska, O., Matyja, K., Jackowski, M., Wiatrak, B., Kubasiewicz–Ross, P., & Karuga-Kuźniewska, E. (2023). Supporting the Wound Healing Process—Curcumin, Resveratrol and Baicalin in In Vitro Wound Healing Studies. *Pharmaceuticals (Basel, Switzerland)*, *16*(1), 82. DOI: 10.3390/ph16010082 PMID: 36678579

Jagwani, D., & Krishna, P. H. (2021). Nature's nano-assets: Green synthesis, characterization techniques and applications. *Materials Today: Proceedings*, *46*, 2307–2317. DOI: 10.1016/j.matpr.2021.04.185

Jain, N., Jain, R., Jain, A., Jain, D. K., & Chandel, H. S. (2010). Evaluation of wound-healing activity of *Acorus calamus* Linn. *Natural Product Research, 24*(6), 534–541. DOI: 10.1080/14786410802531782 PMID: 20182947

Jayakumari, S. (2018). Formulation and evaluation of herbal gel from tannin-enriched fraction of Psidium guajava Linn. leaves for diabetic wound healing. [IJGP]. *International Journal of Green Pharmacy, 12*(03). Advance online publication. DOI: 10.22377/ijgp.v12i03.2009

Jen, M. (2022). Prehistoric Medicine: History, Procedures and Practices. *Anthropology; a Guide to First Degree Courses in UK Universities, 10*, 281.

Jen, M. C., Serrano, M. C., van Lith, R., & Ameer, G. A. (2011). Polymer-Based Nitric Oxide Therapies: Recent Insights for Biomedical Applications. *Advanced Functional Materials, 22*(2), 239–260. DOI: 10.1002/adfm.201101707 PMID: 25067935

Jhanji, R., Bhati, V., Singh, A., & Kumar, A. (2020). Phytomolecules against bacterial biofilm and efflux pump: An in silico and in vitro study. *Journal of Biomolecular Structure & Dynamics, 38*(18), 5500–5512. DOI: 10.1080/07391102.2019.1704884 PMID: 31847713

Jiang, X., Zeng, Y., Li, C., Wang, K., & Yu, D. (2024). Enhancing diabetic wound healing: Advances in electrospun scaffolds from pathogenesis to therapeutic applications. *Frontiers in Bioengineering and Biotechnology, 12*, 1354286. Advance online publication. DOI: 10.3389/fbioe.2024.1354286 PMID: 38375451

Jiang, Z., Zheng, Z., Yu, S., Gao, Y., Ma, J., Huang, L., & Yang, L. (2023). Nanofiber scaffolds as drug delivery systems promoting wound healing. *Pharmaceutics, 15*(7), 1829. Advance online publication. DOI: 10.3390/pharmaceutics15071829 PMID: 37514015

Jiji, S., Udhayakumar, S., Rose, C., Muralidharan, C., & Kadirvelu, K. (2019). Thymol enriched bacterial cellulose hydrogel as effective material for third degree burn wound repair. *International Journal of Biological Macromolecules, 122*, 452–460. DOI: 10.1016/j.ijbiomac.2018.10.192 PMID: 30385344

Jing, W., Xiaolan, C., Yu, C., Feng, Q., & Haifeng, Y. (2022). Pharmacological effects and mechanisms of tannic acid. *Biomedicine and Pharmacotherapy, 154*, 113561. DOI: 10.1016/j.biopha.2022.113561 PMID: 36029537

Jin, W., Xue, Y., Xue, Y., Han, X., Song, Q., Zhang, J., Li, Z., Cheng, J., Guan, S., Sun, S., & Chu, L. (2020). Tannic acid ameliorates arsenic trioxide-induced nephrotoxicity, contribution of NF-κB and Nrf2 pathways. *Biomedicine and Pharmacotherapy*, *126*, 110047. DOI: 10.1016/j.biopha.2020.110047 PMID: 32146384

Jin, X., Wei, C., Li, K., Yin, P., Wu, C., & Zhang, W. (2024). Polyphenol-mediated hyaluronic acid/tannic acid hydrogel with short gelation time and high adhesion strength for accelerating wound healing. *Carbohydrate Polymers*, *342*, 122372. DOI: 10.1016/j.carbpol.2024.122372 PMID: 39048222

Johansen, J. S., Harris, A. K., Rychly, D. J., & Ergul, A. (2005). Oxidative stress and the use of antioxidants in diabetes: Linking basic science to clinical practice [Review of *Oxidative stress and the use of antioxidants in diabetes: Linking basic science to clinical practice*]. *Cardiovascular Diabetology, 4*(1). BioMed Central. DOI: 10.1186/1475-2840-4-5

Johnson, K. E., & Wilgus, T. A. (2014). Vascular Endothelial Growth Factor and Angiogenesis in the Regulation of Cutaneous Wound Repair [Review of *Vascular Endothelial Growth Factor and Angiogenesis in the Regulation of Cutaneous Wound Repair*]. *Advances in Wound Care, 3*(10), 647. Mary Ann Liebert, Inc. DOI: 10.1089/wound.2013.0517

Jomova, K., Alomar, S. Y., Valko, R., Liska, J., Nepovimova, E., Kuca, K., & Valko, M. (2025). Flavonoids and their role in oxidative stress, inflammation, and human diseases. *Chemico-Biological Interactions*, *111489*, 111489. https://www.sciencedirect.com/science/article/pii/S000927972500119X. DOI: 10.1016/j.cbi.2025.111489 PMID: 40147618

Jones, M. L. (2015). A short history of the development of wound care dressings. *British Journal of Healthcare Assistants*, *9*(10), 482–485. DOI: 10.12968/bjha.2015.9.10.482

Kala, C. P. (2017). Traditional health care systems and herbal medicines. *European Journal of Environment and Public Health*, *1*(1), 3. DOI: 10.20897/ejeph.201703

Kaleci, B., & Koyutürk, M. (2020). Efficacy of resveratrol in the wound healing process by reducing oxidative stress and promoting fibroblast cell proliferation and migration. *Dermatologic Therapy*, *33*(6). Advance online publication. DOI: 10.1111/dth.14357 PMID: 32996685

Kaludercic, N., & Lisa, F. D. (2020). Mitochondrial ROS Formation in the Pathogenesis of Diabetic Cardiomyopathy. *Frontiers in Cardiovascular Medicine*, *7*, 12. DOI: 10.3389/fcvm.2020.00012 PMID: 32133373

Kamarudin, N. A., Markom, M., & Latip, J. (2016). Effects of Solvents and Extraction Methods on Herbal Plants *Phyllanthus niruri, Orthosiphon stamineus* and *Labisia pumila. Indian Journal of Science and Technology*, *9*(21). Advance online publication. DOI: 10.17485/ijst/2016/v9i21/95235

Kandhwal, M., Behl, T., Singh, S., Sharma, N., Arora, S., Bhatia, S., Al-Harrasi, A., Sachdeva, M., & Bungau, S. (2022). Role of matrix metalloproteinase in wound healing. *American Journal of Translational Research*, *14*(7), 4391–4405. PMID: 35958464

Kant, V., Gopal, A., Kumar, D., Pathak, N. N., Ram, M., Jangir, B. L., Tandan, S. K., & Kumar, D. (2015). Curcumin-induced angiogenesis hastens wound healing in diabetic rats. *The Journal of Surgical Research*, *193*(2), 978–988. DOI: 10.1016/j.jss.2014.10.019 PMID: 25454972

Kant, V., Jangir, B. L., Kumar, V., Nigam, A., & Sharma, V. (2020). Quercetin accelerated cutaneous wound healing in rats by modulation of different cytokines and growth factors. *Growth Factors (Chur, Switzerland)*, *38*(2), 105–119. DOI: 10.1080/08977194.2020.1822830 PMID: 32957814

Kant, V., Jangir, B. L., Sharma, M., Kumar, V., & Joshi, V. G. (2021). Topical application of quercetin improves wound repair and regeneration in diabetic rats. *Immunopharmacology and Immunotoxicology*, *43*(5), 536–553. DOI: 10.1080/08923973.2021.1950758 PMID: 34278923

Kapoor, M., Howard, R., Hall, I., & Appleton, I. (2004). Effects of epicatechin gallate on wound healing and scar formation in a full thickness incisional wound healing model in rats. *American Journal of Pathology*, *165*(1), 299–307. DOI: 10.1016/S0002-9440(10)63297-X PMID: 15215184

Kar, A. K., Singh, A., Dhiman, N., Purohit, M. P., Jagdale, P., Kamthan, M., Singh, D., Kumar, M., Ghosh, D., & Patnaik, S. (2019). Polymer-assisted in situ synthesis of silver nanoparticles with epigallocatechin gallate (EGCG) impregnated wound patch potentiate controlled inflammatory responses for brisk wound healing. *International Journal of Nanomedicine*, *14*, 9837–9854. DOI: 10.2147/IJN.S228462 PMID: 31849472

Karas, M., Jakubczyk, A., Szymanowska, U., Złotek, U., & Zielińska, E. (2017). Digestion and bioavailability of bioactive phytochemicals. *International Journal of Food Science & Technology*, *52*(2), 291–305. DOI: 10.1111/ijfs.13323

Karimzadeh, S., & Farahpour, M. R. (2017). Topical Application of Salvia officinalis Hydroethanolic Leaf Extract Improves Wound Healing Process. *Indian Journal of Experimental Biology*, *55*, 98–106. PMID: 30183236

Karri, V. V., Kuppusamy, G., Talluri, S. V., Mannemala, S. S., Kollipara, R., Wadhwani, A. D., Mulukutla, S., Raju, K. R., & Malayandi, R. (2016). Curcumin loaded chitosan nanoparticles impregnated into collagen-alginate scaffolds for diabetic wound healing. *International Journal of Biological Macromolecules, 93*, 1519–1529. DOI: 10.1016/j.ijbiomac.2016.05.038 PMID: 27180291

Kartini, K., Wati, N. I., Gustav, R., Wahyuni, R., Anggada, Y. F., Hidayani, R., Raharjo, A., Islamie, R., & Putra, S. E. D. (2021). Wound healing effects of Plantago major extract and its chemical compounds in hyperglycemic rats. *Food Bioscience, 41*, 100937. DOI: 10.1016/j.fbio.2021.100937

Karukonda, S. R., Flynn, T. C., Boh, E. E., McBurney, E. I., Russo, G. G., & Millikan, L. E. (2000). The effects of drugs on wound healing: Part 1. *International Journal of Dermatology, 39*(4), 250–257. DOI: 10.1046/j.1365-4362.2000.00948.x PMID: 10809970

Karunamoorthi, K., Jegajeevanram, K., Xavier, J., Vijayalakshmi, J., & Melita, L. (2012). Tamil traditional medicinal system-siddha: an indigenous health practice in the international perspectives. , *2*(2), 25–35.

Karuppagounder, V., Arumugam, S., Thandavarayan, R. A., Pitchaimani, V., Sreedhar, R., Afrin, R., Harima, M., Suzuki, H., Nomoto, M., Miyashita, S., Suzuki, K., Nakamura, M., Ueno, K., & Watanabe, K. (2015). Tannic acid modulates NFκB signaling pathway and skin inflammation in NC/Nga mice through PPARγ expression. *Cytokine, 76*(2), 206–213. DOI: 10.1016/j.cyto.2015.05.016 PMID: 26049169

Kaur, B.. (2021). The role of microbial fermentation in the transformation of polyphenols. *Critical Reviews in Food Science and Nutrition, 61*(10), 1682–1709.

Kelikian, A., Silva, F., & Viana, G. (2021). Antioxidant role of polyphenols in wound healing: A review. *Antioxidants, 10*(2), 188. Advance online publication. DOI: 10.3390/antiox10020188 PMID: 33525629

Kemppainen, K., Siika-aho, M., Pattathil, S., Giovando, S., & Kruus, K. (2014). Spruce bark as an industrial source of condensed tannins and non-cellulosic sugars. *Industrial Crops and Products, 52*, 158–168. DOI: 10.1016/j.indcrop.2013.10.009

Kenawy, E., Kamoun, E. A., Ghaly, Z. S., Shokr, A. M., El-Meligy, M. A., & Mahmoud, Y. A.-G. (2022). Novel Physically Cross-Linked Curcumin-Loaded PVA/Aloe vera Hydrogel Membranes for Acceleration of Topical Wound Healing: In Vitro and In Vivo Experiments. *Arabian Journal for Science and Engineering, 48*(1), 497–514. DOI: 10.1007/s13369-022-07283-6

Keng, J. W., Lee, S. K., Sang, S. H., Liew, K. B., Teo, S. S., Mossadeq, W. M. S. M., & Chew, Y. L. (2024). *Cassia alata* and its phytochemicals: A promising natural strategy in wound recovery. *Sci Mossadeq, W. M. S. M., & Chew, Y. L., 6*(2), 34.

Khamaisi, M., Katagiri, S., Keenan, H. A., Park, K., Maeda, Y., Li, Q., Qi, W., Thomou, T., Eschuk, D., Tellechea, A., Veves, A., Huang, C., Orgill, D. P., Wagers, A. J., & King, G. L. (2016). PKCδ inhibition normalizes the wound-healing capacity of diabetic human fibroblasts. *The Journal of Clinical Investigation, 126*(3), 837–853. DOI: 10.1172/JCI82788 PMID: 26808499

Khan, H., Ullah, H., Aschner, M., Cheang, W. S., & Akkol, E. K. (2019). Neuroprotective Effects of Quercetin in Alzheimer's Disease [Review of *Neuroprotective Effects of Quercetin in Alzheimer's Disease*]. *Biomolecules, 10*(1), 59. Multidisciplinary Digital Publishing Institute. DOI: 10.3390/biom10010059

Khanbabaee, K., & Van Ree, T. (2001). Tannins: Classification and definition. *Natural Product Reports, 18*(6), 641–649. DOI: 10.1039/b1010611 PMID: 11820762

Khan, K. A., Ansari, M. J., Al-Ghamdi, A., Nuru, A., Harakeh, S., & Iqbal, J. (2017). Investigation of gut microbial communities associated with indigenous honeybee (Apis mellifera jemenitica) from two different eco-regions of Saudi Arabia. *Saudi Journal of Biological Sciences, 24*(5), 1061–1068. DOI: 10.1016/j.sjbs.2017.01.055 PMID: 28663705

Khare, T., Anand, U., Dey, A., Assaraf, Y. G., Chen, Z., Liu, Z., & Kumar, V. (2021). Exploring Phytochemicals for Combating Antibiotic Resistance in Microbial Pathogens [Review of *Exploring Phytochemicals for Combating Antibiotic Resistance in Microbial Pathogens*]. *Frontiers in Pharmacology, 12*. Frontiers Media. DOI: 10.3389/fphar.2021.720726

Khedkar, D. D. (2025). Plant products and essential oil therapy for emerging viral diseases. In *Fighting multidrug resistance with herbal extracts, essential oils and their components* (pp. 391–419). Academic Press. DOI: 10.1016/B978-0-443-29044-2.00017-5

Kim, H., Kawazoe, T., Han, D. W., Matsumara, K., Suzuki, S., Tsutsumi, S., & Hyon, S.-H. (2008). Enhanced wound healing by an epigallocatechin gallate-incorporated collagen sponge in diabetic mice. *Wound Repair and Regeneration, 16*(5), 714–720. DOI: 10.1111/j.1524-475X.2008.00422.x PMID: 19128267

Kim, K.. (2019). Advances in nanotechnology for drug delivery. *Current Drug Targets, 20*(4), 389–400.

Kimmerer, R. W. (2013). *Braiding sweetgrass: Indigenous wisdom, scientific knowledge and the teachings of plants*. Milkweed editions.

Kim, S. H., Shin, H. L., Son, T. H., Lim, S. A., Kim, D., Yoon, J. H., Choi, H., Kim, H. G., & Choi, S. W. (2024). Quercus glauca Acorn Seed Coat Extract Promotes Wound Re-Epithelialization by Facilitating Fibroblast Migration and Inhibiting Dermal Inflammation. *Biology (Basel), 13*(10), 775. DOI: 10.3390/biology13100775 PMID: 39452084

Kim, W. K., Song, S.-Y., Oh, W. K., Kaewsuwan, S., Tran, T. L., Kim, W.-S., & Sung, J.-H. (2013). Wound-healing effect of ginsenoside Rd from leaves of Panax ginseng via cyclic AMP-dependent protein kinase pathway. *European Journal of Pharmacology, 702*(1-3), 1–3, 285–293. DOI: 10.1016/j.ejphar.2013.01.048 PMID: 23399764

Kisseleva, T., & Brenner, D. A. (2008). Mechanisms of fibrogenesis. *Experimental Biology and Medicine, 233*(2), 109–122. DOI: 10.3181/0707-MR-190 PMID: 18222966

Knecht, K. T., Kinder, D. H., & Stockert, A. (2020). Biologically-Based Complementary and Alternative Medicine (CAM) Use in Cancer Patients: The Good, the Bad, the Misunderstood [Review of *Biologically-Based Complementary and Alternative Medicine (CAM) Use in Cancer Patients: The Good, the Bad, the Misunderstood*]. *Frontiers in Nutrition, 6*. Frontiers Media. DOI: 10.3389/fnut.2019.00196

Ko, K. I., Sculean, A., & Graves, D. T. (2021). Diabetic wound healing in soft and hard oral tissues. *Translational Research; the Journal of Laboratory and Clinical Medicine, 236*, 72–86. DOI: 10.1016/j.trsl.2021.05.001 PMID: 33992825

Kokane, D. D., More, R. Y., Kale, M. B., Nehete, M. N., Mehendale, P. C., & Gadgoli, C. H. (2009). Evaluation of wound healing activity of root of *Mimosa pudica. Journal of Ethnopharmacology, 124*(2), 311–315. DOI: 10.1016/j.jep.2009.04.038 PMID: 19397984

Komes, D., Belščak-Cvitanović, A., Jurić, S., Bušić, A., Vojvodić, A., & Durgo, K. (2016). Consumer Acceptability of Liquorice Root (Glycyrrhiza glabra L.) as an Alternative Sweetener and Correlation with Its Bioactive Content and Biological Activity. *International Journal of Food Sciences and Nutrition, 67*(1), 53–66. DOI: 10.3109/09637486.2015.1126563 PMID: 26707111

Kondo, T., & Ishida, Y. (2010). Molecular pathology of wound healing. *Forensic Science International, 203*(1-3), 93–98. DOI: 10.1016/j.forsciint.2010.07.004 PMID: 20739128

Koyama, S., Purk, A., Kaur, M., Soini, H. A., Novotny, M. V., Davis, K., Kao, C. C., Matsunami, H., & Mescher, A. (2019). β-Caryophyllene enhances wound healing through multiple routes. *PLoS One*, *14*(12), e0216104. DOI: 10.1371/journal.pone.0216104 PMID: 31841509

Krausz, A. E., Adler, B. L., Cabral, V., Navati, M., Doerner, J., Charafeddine, R. A., Chandra, D., Liang, H., Gunther, L., Clendaniel, A., Harper, S., Friedman, J. M., Nosanchuk, J. D., & Friedman, A. J. (2015). Curcumin-encapsulated nanoparticles as innovative antimicrobial and wound healing agent. *Nanomedicine; Nanotechnology, Biology, and Medicine*, *11*(1), 195–206. DOI: 10.1016/j.nano.2014.09.004 PMID: 25240595

Krishnaswamy, V. R., Mintz, D., & Sagi, I. (2017). Matrix metalloproteinases: The sculptors of chronic cutaneous wounds. *Biochimica et Biophysica Acta (BBA) –. Biochimica et Biophysica Acta. Molecular Cell Research*, *1864*(11, 11 Pt B), 2220–2227. DOI: 10.1016/j.bbamcr.2017.08.003 PMID: 28797647

Krutmann, J., Bouloc, A., Sore, G., Bernard, B. A., & Passeron, T. (2017). The Skin Aging Exposome. *Journal of Dermatological Science*, *85*(3), 152–161. DOI: 10.1016/j.jdermsci.2016.09.015 PMID: 27720464

Kubat, M., Karabulut, Z., & Şengül, S. (2021). Effect of propolis on wound healing in sacrococcygeal pilonidal disease: A randomized controlled clinical trial. *Pakistan Journal of Pharmaceutical Sciences*, *34*(1). PMID: 34602433

Kujath, P., & Michelsen, A. (2008). Wounds—From physiology to wound dressing. *Deutsches Ärzteblatt International*, *105*(13), 239. DOI: 10.3238/arztebl.2008.0239 PMID: 19629204

Kumar, A., Kumar, B., Singh, S. K., Kaur, B., & Singh, S. (2017). A REVIEW ON PHYTOSOMES: NOVEL APPROACH FOR HERBAL PHYTOCHEMICALS [Review of *A REVIEW ON PHYTOSOMES: NOVEL APPROACH FOR HERBAL PHYTOCHEMICALS*]. *Asian Journal of Pharmaceutical and Clinical Research*, *10*(10), 41. Innovare Academic Sciences. DOI: 10.22159/ajpcr.2017.v10i10.20424

Kumar, M., Kumar, D., Mahmood, S., Singh, V., Chopra, S., Hilles, A. R., & Bhatia, A. (2024). Nanotechnology-driven wound healing potential of asiaticoside: a comprehensive review [Review of *Nanotechnology-driven wound healing potential of asiaticoside: a comprehensive review*]. *RSC Pharmaceutics*, *1*(1), 9. Royal Society of Chemistry. DOI: 10.1039/D3PM00024A

Kumari, A., Raina, N., Wahi, A., Goh, K. W., Sharma, P., Nagpal, R., Jain, A., Ming, L. C., & Gupta, M. (2022). Wound-Healing Effects of Curcumin and Its Nanoformulations: A Comprehensive Review. *Pharmaceutics*, *14*(11), 2288. DOI: 10.3390/pharmaceutics14112288 PMID: 36365107

Kumar, M., Keshwania, P., Chopra, S., Mahmood, S., & Bhatia, A. (2023). Therapeutic Potential of Nanocarrier-Mediated Delivery of Phytoconstituents for Wound Healing: Their Current Status and Future Perspective. *AAPS PharmSciTech*, *24*(6), 155. DOI: 10.1208/s12249-023-02616-6 PMID: 37468691

Kumar, P., & Kothari, V. (2021). *Wound healing research: Current trends and future directions*. Springer. DOI: 10.1007/978-981-16-2677-7

Kumar, P., Kumar, S., Udupa, E. P., Kumar, U., Rao, P., & Honnegowda, T. (2015). Role of angiogenesis and angiogenic factors in acute and chronic wound healing. *Plastic and Aesthetic Research*, *2*(5), 243. DOI: 10.4103/2347-9264.165438

Kumar, S., Balachandran, V., Njoo, L., Anastasopoulos, A., & Tsvetkov, Y. (2023). Language Generation Models Can Cause Harm: So What Can We Do About It? *An Actionable Survey*, *3299*, 3299–3321. Advance online publication. DOI: 10.18653/v1/2023.eacl-main.241

Kumar, S., Kumar, A., Kumar, N., Singh, P., Singh, T. U., Singh, B. R., Gupta, P. K., & Thakur, V. K. (2022). In vivo therapeutic efficacy of *Curcuma longa* extract loaded ethosomes on wound healing. *Veterinary Research Communications*, *46*(4), 1033–1049. DOI: 10.1007/s11259-022-09952-1 PMID: 35796857

Kumar, S., & Pandey, A. K. (2013). Chemistry and biological activities of flavonoids: An overview. *TheScientificWorldJournal*, *2013*(1), 162750. DOI: 10.1155/2013/162750 PMID: 24470791

Kumar, S., Wong, P. F., & Leaper, D. J. (2004). What is new in wound healing? *Turkish Journal of Medical Sciences*, *34*(3), 147–160.

Kung, H. N., Yang, M. J., Chang, C. F., Chau, Y. P., & Lu, K. S. (2008). In vitro and in vivo wound healing-promoting activities of β-lapachone. *American Journal of Physiology. Cell Physiology*, *295*(4), C931–C943. DOI: 10.1152/ajpcell.00266.2008 PMID: 18650264

Kunkemoeller, B., & Kyriakides, T. R. (2017). Redox Signaling in Diabetic Wound Healing Regulates Extracellular Matrix Deposition [Review of *Redox Signaling in Diabetic Wound Healing Regulates Extracellular Matrix Deposition*]. *Antioxidants and Redox Signaling, 27*(12), 823. Mary Ann Liebert, Inc. DOI: 10.1089/ars.2017.7263

Kunle, O. F., Egharevba, H. O., & Ahmadu, P. O. (2012). Standardization of herbal medicines. *International Journal of Biodeversity and Conservation*, *4*(3), 101–112. DOI: 10.5897/IJBC11.163

Küpeli Akkol, E., Kosar, M., Baldemir, A., Şeker Karatoprak, G., Demirpolat, E., Yerer Aycan, M. B., Süntar, I., & Ilgün, S. (2023). The wound-healing potential of the endemic plant *Helianthemum canum* (L.) Baumg.: Preclinical studies supported with phytochemical profiling. *Chemistry & Biodiversity*, *20*(12), e202301529. DOI: 10.1002/cbdv.202301529 PMID: 37955210

Kurahashi, T., & Fujii, J. (2015). Roles of antioxidative enzymes in wound healing. *Journal of Developmental Biology*, *3*(2), 57–70. DOI: 10.3390/jdb3020057

Kwiecień, K., Zegar, A., Jung, J., Brzoza, P., Kwitniewski, M., Godlewska, U., Grygier, B., Kwiecińska, P., Morytko, A., & Cichy, J. (2019). Architecture of antimicrobial skin defense. *Cytokine & Growth Factor Reviews*, *49*, 70–84. DOI: 10.1016/j.cytogfr.2019.08.001 PMID: 31473081

Landén, N. X., Li, D., & Ståhle, M. (2016). Transition from inflammation to proliferation: A critical step during wound healing. *Cellular and Molecular Life Sciences : CMLS*, *73*(20), 3861–3885. DOI: 10.1007/s00018-016-2268-0 PMID: 27180275

Lau, K., Paus, R., Tiede, S., Day, P., & Bayat, A. (2009). Exploring the role of stem cells in cutaneous wound healing. *Experimental Dermatology*, *18*(11), 921–933. DOI: 10.1111/j.1600-0625.2009.00942.x PMID: 19719838

Lazarus, G. S., Cooper, D. M., Knighton, D. R., Margolis, D. J., Pecoraro, R. E., Rodeheaver, G., & Robson, M. C. (1994). Definitions and guidelines for assessment of wounds and evaluation of healing. *Wound Repair and Regeneration*, *2*(3), 165–170. DOI: 10.1046/j.1524-475X.1994.20305.x PMID: 17156107

LeBlanc, K., Woo, K., Gethin, G., Beeckman, D., Heerschap, C., Reddy, M., Parker, C., MacAndrew, M., White-Chu, F., & Wiesenfeld, S. (2025). Adults living with dementia and chronic wounds, wound types, care challenges, and impact across dementia stages: A scoping review protocol. *Journal ofTissue Viability*, *34*(1), 100845. Advance online publication. DOI: 10.1016/j.jtv.2024.12.009 PMID: 39724737

Lee, E. M., Lee, S. S., Chung, B. Y., Cho, J.-Y., Lee, I. C., Ahn, S. R., Jang, S. J., & Kim, T. H. (2010). Pancreatic lipase inhibition by C-glycosidic flavones isolated from Eremochloa ophiuroides. *Molecules (Basel, Switzerland)*, *15*(11), 8251–8259. DOI: 10.3390/molecules15118251 PMID: 21081855

Lee, G. Y., Park, K. G., Namgoong, S., Han, S. K., Jeong, S. H., Dhong, E. S., & Kim, W. K. (2016). Effects of Panax ginseng extract on human dermal fibroblast proliferation and collagen synthesis. *International Wound Journal, 1*(S1), 42–46. DOI: 10.1111/iwj.12530 PMID: 26507878

Lee, J. H., Lee, J. H., & Lee, Y. C. (2020). Polyphenols and their therapeutic role in wound healing: A review. *Antioxidants, 9*(9), 697.

Lee, K. E., Bharadwaj, S., Yadava, U., & Kang, S. G. (2020). Computational and in vitro investigation of (-)-epicatechin and proanthocyanidin B2 as inhibitors of human matrix metalloproteinase 1. *Biomolecules, 10*(10), 1379. DOI: 10.3390/biom10101379 PMID: 32998374

Lee, S., Lee, J., & Lee, H. (2022). The role of Centella Asiatica in the treatment of chronic wounds and its effect on collagen synthesis. *Journal of Wound Care, 31*(6), 350–355. DOI: 10.12968/jowc.2022.31.6.350

Lee, Y., Kim, S., & An, S. (2023). Patient variability in response to wound healing therapies. *International Wound Journal, 20*(5), 1610–1622. DOI: 10.1111/iwj.13986

Li, M., Hou, Q., Zhong, L., Zhao, Y., & Fu, X. (2021). Macrophage Related Chronic Inflammation in Non-Healing Wounds [Review of *Macrophage Related Chronic Inflammation in Non-Healing Wounds*]. *Frontiers in Immunology, 12*. Frontiers Media. DOI: 10.3389/fimmu.2021.681710

Liarte, S., Bernabé-García, A., & Nicolás, F. J. (2020). Role of TGF-β in skin chronic wounds: A keratinocyte perspective. *Cells, 9*(2), 306. DOI: 10.3390/cells9020306 PMID: 32012802

Li, H., Beg, O. U., Rafie, A. R., Kanwal, S., Ovalle-Cisneros, A., Faison, M. O., & Siddiqui, R. A. (2023). Characterization of Green and Yellow Papaya (Carica papaya) for Anti-Diabetic Activity in Liver and Myoblast Cells and Wound-Healing Activity in Fibroblast Cells. *Nutrients, 15*(8), 1929. DOI: 10.3390/nu15081929 PMID: 37111148

Li, H., Tanaka, T., Zhang, Y.-J., Yang, C.-R., & Kouno, I. (2007). Rubusuaviins A–F, monomeric and oligomeric ellagitannins from Chinese sweet tea and their α-amylase inhibitory activity. *Chemical & Pharmaceutical Bulletin, 55*(9), 1325–1331. DOI: 10.1248/cpb.55.1325 PMID: 17827756

Li, J., Chen, J., & Kirsner, R. (2007). Pathophysiology of acute wound healing. *Clinics in Dermatology, 25*(1), 9–18. DOI: 10.1016/j.clindermatol.2006.09.007 PMID: 17276196

Li, J., Zhang, Y.-P., & Kirsner, R. S. (2003). Angiogenesis in wound repair: Angiogenic growth factors and the extracellular matrix. *Microscopy Research and Technique*, *60*(1), 107–114. DOI: 10.1002/jemt.10249 PMID: 12500267

Li, K., Diao, Y., Zhang, H., Wang, S., Zhang, Z., Yu, B., Huang, S., & Yang, H. (2011). Tannin extracts from immature fruits of Terminalia chebula Fructus Retz. promote cutaneous wound healing in rats. *BMC Complementary and Alternative Medicine*, *11*(1), 86. DOI: 10.1186/1472-6882-11-86 PMID: 21982053

Li, M., Xu, J., Shi, T., Yu, H., Bi, J., & Chen, G. (2016). Epigallocatechin-3-gallate augments therapeutic effects of mesenchymal stem cells in skin wound healing. *Clinical and Experimental Pharmacology & Physiology*, *43*(11), 1115–1124. DOI: 10.1111/1440-1681.12652 PMID: 27557877

Li, M., Yu, H., Pan, H., Zhou, X., Ruan, Q., Kong, D., Chu, Z., Li, H., Huang, J., Huang, X., Chau, A., Xie, W., Ding, Y., & Yao, P. (2019). Nrf2 Suppression Delays Diabetic Wound Healing Through Sustained Oxidative Stress and Inflammation. *Frontiers in Pharmacology*, *10*, 1099. Advance online publication. DOI: 10.3389/fphar.2019.01099 PMID: 31616304

Limsitthichaikoon, S., Khampaenjiraroch, B., Damrongrungruang, T., Limphirat, W., Thapphasaraphong, S., & Priprem, A. (2018). Topical oral wound healing potential of anthocyanin complex: Animal and clinical studies. *Therapeutic Delivery*, *9*(5), 359–374. DOI: 10.4155/tde-2017-0123 PMID: 29681233

Lin, C.-M., Chiu, J., Wu, I.-H., Baowei, W., Pan, C.-M., & Chen, Y. (2009). Ferulic acid augments angiogenesis via VEGF, PDGF and HIF-1α. *The Journal of Nutritional Biochemistry*, *21*(7), 627–633. DOI: 10.1016/j.jnutbio.2009.04.001 PMID: 19443196

Linde, K., Berner, M. M., & Kriston, L. (2008). St John's wort for major depression. *Cochrane Database of Systematic Reviews*, *4*. Advance online publication. DOI: 10.1002/14651858.CD000448.pub3 PMID: 18843608

Lindley, L. E., Stojadinovic, O., Pastar, I., & Tomic-Canic, M. (2016). Biology and biomarkers for wound healing. *Plastic and Reconstructive Surgery*, *138*(3, Suppl.), 18S–28S. DOI: 10.1097/PRS.0000000000002682 PMID: 27556760

Lipinski, C. A. (2000). Drug-like properties and the causes of poor solubility and poor permeability. *Journal of Pharmacological and Toxicological Methods*, *44*(1), 235–249. DOI: 10.1016/S1056-8719(00)00107-6 PMID: 11274893

Li, Q., Wei, Q., Yuan, E., Yang, J., & Ning, Z. (2014). Interaction between four flavonoids and trypsin: Effect on the characteristics of trypsin and antioxidant activity of flavonoids. *International Journal of Food Science & Technology*, *49*(4), 1063–1069. DOI: 10.1111/ijfs.12401

Liu, Y., Liu, Y., Deng, J., Li, W., & Nie, X. (2021). Fibroblast Growth Factor in Diabetic Foot Ulcer: Progress and Therapeutic Prospects [Review of *Fibroblast Growth Factor in Diabetic Foot Ulcer: Progress and Therapeutic Prospects*]. *Frontiers in Endocrinology, 12*. Frontiers Media. DOI: 10.3389/fendo.2021.744868

Liu, H., Sun, W., Cai, W., Luo, K., Lu, C., Jin, A., Zhang, J., & Liu, Y. (2025). Current status, challenges, and prospects of artificial intelligence applications in wound repair theranostics. *Theranostics*, *15*(5), 1662–1688. DOI: 10.7150/thno.105109 PMID: 39897550

Liu, H., Yang, Y., Deng, L., Shen, Z., Huang, Q., Shah, N. G., Chen, W., Zhang, Y., Wang, X., Yu, L., & Chen, Z. (2024). Antibacterial and antioxidative hydrogel dressings based on tannic acid-gelatin/oxidized sodium alginate loaded with zinc oxide nanoparticles for promoting wound healing. *International Journal of Biological Macromolecules*, *279*(Pt 2), 135177. DOI: 10.1016/j.ijbiomac.2024.135177 PMID: 39214222

Liu, J., Zhang, Y., & Zhang, M.. (2022). The roles of polyphenols in wound healing: A comprehensive review. *Frontiers in Nutrition*, *9*, 891575.

Liu, Y., Gong, P., Xiao, F., Shao, S., Zhao, D., Yan, M., & Yang, X. (2018). Chemical Constituents and Antioxidant, Anti-Inflammatory and Anti-Tumor Activities of Melilotus officinalis (Linn.) Pall. *Molecules (Basel, Switzerland)*, *23*(2), 271. DOI: 10.3390/molecules23020271 PMID: 29382154

Liu, Y., Xie, J., & Yang, Z. (2021). Asiaticoside promotes angiogenesis and wound healing by upregulating VEGF expression in diabetic rats. *Journal of Ethnopharmacology*, *267*, 113642. DOI: 10.1016/j.jep.2020.113642

Liu, Y.-Y., Wu, J.-Q., Fan, R.-Y., He, Z.-H., Li, C.-Y., & He, M.-F. (2020). Isoliquiritin promote angiogenesis by recruiting macrophages to improve the healing of zebrafish wounds. *Fish & Shellfish Immunology*, *100*, 238–245. DOI: 10.1016/j.fsi.2020.02.071 PMID: 32135341

Liu, Z., & Fang, Y. (2025). Wound healing and signaling pathways. *Open Life Sciences*, *20*(1), 20251166. DOI: 10.1515/biol-2025-1166 PMID: 40917783

Li, W., Kandhare, A. D., Mukherjee, A. A., & Bodhankar, S. L. (2018). Hesperidin, a plant flavonoid accelerated the cutaneous wound healing in streptozotocin-induced diabetic rats: Role of TGF-β/Smads and Ang-1/Tie-2 signaling pathways. *EXCLI Journal*, *17*, 399–415. PMID: 29805347

Li, W.-T., Chuang, Y.-H., & Hsieh, J.-F. (2019). Characterization of maltase and sucrase inhibitory constituents from Rhodiola crenulata. *Foods*, *8*(11), 540. DOI: 10.3390/foods8110540 PMID: 31684079

Li, X., Guan, C., He, Y., Wang, Y., Liu, X., & Zhou, X. (2016). Effects of Total Alkaloids ofSophora alopecuroideson Biofilm Formation inStaphylococcus epidermidis. *BioMed Research International*, *1*, 1–10. Advance online publication. DOI: 10.1155/2016/4020715 PMID: 27413745

Li, Y., Ren, M., Wang, J., Ma, R., Chen, H., Xie, Q., Li, H., Li, J., & Wang, J. (2021). Borneol intervention in ischemic stroke: A systematic review. *Frontiers in Pharmacology*, *12*, 606682. DOI: 10.3389/fphar.2021.606682 PMID: 34017247

Lohray, B. B.. (2006). Pharmaceutical applications of cyclodextrins. *Biopharmaceutics & Drug Disposition*, *27*(4), 145–156.

Long, D. R., Mead, J., Hendricks, J. M., Hardy, M. E., & Voyich, J. M. (2013). 18β-Glycyrrhetinic acid inhibits methicillin-resistant Staphylococcus aureus survival and attenuates virulence gene expression. Antimicrobial agents and chemotherapy, 57(1), 241-247.

López-Díez, R., Egaña-Gorroño, L., Senatus, L. M., Shekhtman, A., Ramasamy, R., & Schmidt, A. M. (2021). Diabetes and Cardiovascular Complications: The Epidemics Continue [Review of *Diabetes and Cardiovascular Complications: The Epidemics Continue*]. *Current Cardiology Reports, 23*(7), 74. Springer Science+Business Media. DOI: 10.1007/s11886-021-01504-4

Lu, L., Ying, K., Wei, S.-M., Fang, Y., Liu, Y., Lin, H., Ma, L., & Mao, Y. (2004). Asiaticoside induction for cell-cycle progression, proliferation and collagen synthesis in human dermal fibroblasts. *International Journal of Dermatology*, *43*(11), 801–807. DOI: 10.1111/j.1365-4632.2004.02047.x PMID: 15533060

Lu, Y., Zhao, D., Cao, G., Yin, S., Liu, C., Song, R., Ma, J., Sun, R., Wu, Z., Liu, J., Wu, P., & Wang, Y. (2024). Research progress on and molecular mechanism of vacuum sealing drainage in the treatment of diabetic foot ulcers. *Frontiers in Surgery*, *11*, 1265360. Advance online publication. DOI: 10.3389/fsurg.2024.1265360 PMID: 38464666

Lyseng-Williamson, K. A., & Perry, C. M. (2003). Micronised purified flavonoid fraction: A review of its use in chronic venous insufficiency, venous ulcers and haemorrhoids. *Drugs*, *63*(1), 71–100. DOI: 10.2165/00003495-200363010-00005 PMID: 12487623

Mace, K. A., Pearson, J. C., & McGinnis, W. (2005). An epidermal barrier wound repair pathway in Drosophila is mediated by grainy head. *Science*, *308*(5720), 381–385. DOI: 10.1126/science.1107573 PMID: 15831751

Machado, B. A., Pereira, C. G., Nunes, S. B., Padilha, F. F., & Umsza-Guez, M. A. (2013). Supercritical CO_2 extraction: Applications and perspectives. *Separation Science and Technology*, *48*(18), 2741–2760. DOI: 10.1080/01496395.2013.811422

Madhyastha, H., Halder, S., Queen Intan, N., Madhyastha, R., Mohanapriya, A., & Sudhakaran, R.. (2020). Surface refined Au quercetin nanoconjugate stimulates dermal cell migration: Possible implication in wound healing. *RSC Advances*, *10*(62), 37683–37694. DOI: 10.1039/D0RA06690G PMID: 35515178

Madkour, M., Aboelenin, M. M., Habashy, W. S., Matter, I. A., Shourrap, M., Elolimy, A. A., & Aboelazab, O. (2024). Oregano and rosemary extracts modulate gut and antioxidant responses in heat-stressed broilers. *Poultry Science*, *103*(9), 103996. DOI: 10.1016/j.psj.2024.103996 PMID: 39024691

Mafuva, C., & Marima-Matarira, H. T. (2014). *Towards professionalization of traditional medicine in Zimbabwe: A comparative analysis to the South African policy on traditional medicine and the Indian Ayurvedic system.*

Mahmoud, G. A. E., Rashed, N. M., El-Ganainy, S. M., & Salem, S. H. (2024). Unveiling the neem (Azadirachta indica) effects on biofilm formation of food-borne bacteria and the potential mechanism using a molecular docking approach. *Plants*, *13*(18), 2669. DOI: 10.3390/plants13182669 PMID: 39339644

Mahmoudi, M., & Gould, L. J. (2020). Opportunities and challenges of the management of chronic wounds: A multidisciplinary viewpoint. *Chronic Wound Care Management and Research*, *7*, 27–36. DOI: 10.2147/CWCMR.S260136

Majtan, J. (2014). Honey: An immunomodulator in wound healing. *Wound Repair and Regeneration*, *22*(2), 187–192. DOI: 10.1111/wrr.12117 PMID: 24612472

Maliar, T., Jedinák, A., Kadrabová, J., & Šturdík, E. (2004). Structural aspects of flavonoids as trypsin inhibitors. *European Journal of Medicinal Chemistry*, *39*(3), 241–248. DOI: 10.1016/j.ejmech.2003.12.003 PMID: 15051172

Malone, M., & Schultz, G. (2022). Challenges in the diagnosis and management of wound infection. *British Journal of Dermatology*, *187*(2), 159–166. DOI: 10.1111/bjd.21612 PMID: 35587707

Mamun, A. A., Shao, C., Geng, P., Wang, S., & Xiao, J. (2024). Recent advances in molecular mechanisms of skin wound healing and its treatments. *Frontiers in Immunology*, *15*, 1395479. Advance online publication. DOI: 10.3389/fimmu.2024.1395479 PMID: 38835782

Manisha, N., Niharika, , Gaur, P., Goel, R., Lata, K., & Mishra, R. (2025, March). Gaur, P., Goel, R., Lata, K., & Mishra, R. (2024). Understanding diabetic wounds: A review of mechanisms, pathophysiology, and multimodal management strategies. *Current Reviews in Clinical and Experimental Pharmacology*, *20*(3), 207–228. Advance online publication. DOI: 10.2174/0127724328326480240927065600 PMID: 40452447

Manosalva, L., Mutis, A., Urzúa, A., Fajardo, V., & Quiróz, A.Forst and Study of Synergism with Ampicillin and Cephalothin. (2016). Antibacterial Activity of Alkaloid Fractions from Berberis microphylla G. Forst and Study of Synergism with Ampicillin and Cephalothin. *Molecules (Basel, Switzerland)*, *21*(1), 76. DOI: 10.3390/molecules21010076 PMID: 26760994

Marcińczyk, N., Gromotowicz-Popławska, A., Tomczyk, M., & Chabielska, E. (2022). Tannins as hemostasis modulators. *Frontiers in Pharmacology*, *12*, 806891. DOI: 10.3389/fphar.2021.806891 PMID: 35095516

Martinengo, L., Olsson, M., Bajpai, R., Soljak, M., Upton, Z., Schmidtchen, A., Car, J., & Järbrink, K. (2019). Prevalence of chronic wounds in the general population: Systematic review and meta-analysis of observational studies. *Annals of Epidemiology*, *29*, 8–15. DOI: 10.1016/j.annepidem.2018.10.005 PMID: 30497932

Martinez-Gonzalez, A. I., Díaz-Sánchez, Á. G., de la Rosa, L. A., Bustos-Jaimes, I., & Alvarez-Parrilla, E. (2021). A novel approach to trypsin inhibition by flavonoids. *Journal of Food Bioactives: An Official Scientific Publication of the International Society of Nutraceuticals and Functional Foods (ISNFF)*, •••, 14. DOI: 10.31665/JFB.2021.14272

Martins, S. G., Zilhão, R., Þorsteinsdóttir, S., & Carlos, A. R. (2021). Linking Oxidative Stress and DNA Damage to Changes in the Expression of Extracellular Matrix Components [Review of *Linking Oxidative Stress and DNA Damage to Changes in the Expression of Extracellular Matrix Components*]. *Frontiers in Genetics, 12*. Frontiers Media. DOI: 10.3389/fgene.2021.673002

Mashamba, T. G., Adeosun, I. J., Baloyi, I. T., Tshikalange, E. T., & Cosa, S. (2022). Quorum sensing modulation and inhibition in biofilm forming foot ulcer pathogens by selected medicinal plants. *Heliyon*, *8*(4), e09303. DOI: 10.1016/j.heliyon.2022.e09303 PMID: 35520625

Matera, R., Lucchi, E., & Valgimigli, L. (2023). Plant essential oils as healthy functional ingredients of nutraceuticals and diet supplements: A review. *Molecules (Basel, Switzerland)*, *28*(2), 901. DOI: 10.3390/molecules28020901 PMID: 36677959

Mathi, P., Bokka, V., & Botlagunta, M. (2015). MEDICINAL USES AND BIOLOGICAL ACTIVITIES OF SOPHORA INTERRUPTA BEDD-A REVIEW. *International Journal of Pharmaceutics*, *5*(15), 265. https://www.pharmascholars.com/articles/medicinal-uses-and-biological-activities-of-sophora-interrupta-bedda-review.pdf

Matta-Gutiérrez, G., García-Morales, E., García-Álvarez, Y., Álvaro-Afonso, F. J., Molines-Barroso, R. J., & Lázaro-Martínez, J. L. (2021). The Influence of Multidrug-Resistant Bacteria on Clinical Outcomes of Diabetic Foot Ulcers: A Systematic Review. *Journal of Clinical Medicine*, *10*(9), 1948. DOI: 10.3390/jcm10091948 PMID: 34062775

Maver, T., Kurečič, M., Smrke, D., Kleinschek, K. S., & Maver, U. (2019). Plant-Derived Medicines with Potential Use in Wound Treatment. In *IntechOpen eBooks*. IntechOpen., DOI: 10.5772/intechopen.72813

McClements, D. J., & Gumus, C. E. (2016). Nanoemulsions: Formation, properties, and applications. *Comprehensive Reviews in Food Science and Food Safety*, *15*(4), 630–646.

McKay, T. B., Emmitte, K. A., German, C., & Karamichos, D. (2023). Quercetin and Related Analogs as Therapeutics to Promote Tissue Repair. *Bioengineering (Basel, Switzerland)*, *10*(10), 1127. DOI: 10.3390/bioengineering10101127 PMID: 37892857

Melrose, J. (2023). The potential of flavonoids and flavonoid metabolites in the treatment of neurodegenerative pathology in disorders of cognitive decline. *Antioxidants*, *12*(3), 663. DOI: 10.3390/antiox12030663 PMID: 36978911

Mi, B., Liu, J., Liu, G., Zhou, W., Liu, Y., Hu, L., Xiong, L., Ye, S., & Wu, Y. (2018). Icariin promotes wound healing by enhancing the migration and proliferation of keratinocytes via the AKT and ERK signaling pathway. *International Journal of Molecular Medicine*, *42*(2), 831–838. DOI: 10.3892/ijmm.2018.3676 PMID: 29767263

Michael Heinrich, A. K. J. (2020). Ethnopharmacology and sustainable development: Reviewing recent progress in holistic research. *Journal of Ethnopharmacology, 257*, 112830.

Middleton, E. J.Jr. (1998). Effect of plant flavonoids on immune and inflammatory cell function. *Advances in Experimental Medicine and Biology, 439*, 175–182. DOI: 10.1007/978-1-4615-5335-9_13 PMID: 9781303

Middleton, E.Jr, Kandaswami, C., & Theoharides, T. C. (2000). The effects of plant flavonoids on mammalian cells: Implications for inflammation, heart disease, and cancer. *Pharmacological Reviews, 52*(4), 673–751. DOI: 10.1016/S0031-6997(24)01472-8 PMID: 11121513

Midwood, K. S., Williams, L. V., & Schwarzbauer, J. E. (2004). Tissue repair and the dynamics of the extracellular matrix. *The International Journal of Biochemistry & Cell Biology, 36*(6), 1031–1037. DOI: 10.1016/j.biocel.2003.12.003 PMID: 15094118

Miguel, M. G., & Antunes, M. D. (2011). Is propolis safe as an alternative medicine? *Journal of Pharmacy & Bioallied Sciences, 3*(4), 479–495. DOI: 10.4103/0975-7406.90101 PMID: 22219581

Miller, A., Barbera, R., & Adriani, W.. (2022). Personalized approaches to wound healing: Evaluating the role of polyphenol delivery in chronic wound management. *Journal of Wound Care, 31*(4), 192–201.

Min, D.-H., Yu, Y.-B., Kim, T., Kim, H., & Lee, S. (2024). Pharmacological effects of pentacyclic triterpenoids isolated from Centella asiatica. *Horticulture, Environment and Biotechnology, 65*(2), 189–197. DOI: 10.1007/s13580-023-00561-8

Minden-Birkenmaier, B. A., & Bowlin, G. L. (2018). Honey-based templates in wound healing and tissue engineering. *Bioengineering (Basel, Switzerland), 5*(2), 46. DOI: 10.3390/bioengineering5020046 PMID: 29903998

Mira, B., Blasco, M., Berna, A., & Subirats, S. (1999). Supercritical CO_2 extraction of orange peel oil. *The Journal of Supercritical Fluids, 14*(2), 95–104. DOI: 10.1016/S0896-8446(98)00111-9

Mirastschijski, U., Schnabel, R., Claes, J., Schneider, W., Agren, M. S., Haaksma, C., & Tomasek, J. J. (2010). Matrix metalloproteinase inhibition delays wound healing and blocks the latent transforming growth factor-beta1-promoted myofibroblast formation and function. *Wound Repair and Regeneration, 18*(2), 223–234. DOI: 10.1111/j.1524-475X.2010.00574.x PMID: 20409148

Miricescu, D., Totan, A., & Stanescu, S.. (2021). Growth factors, reactive oxygen species, and wound healing. *International Journal of Molecular Sciences*, *22*(17), 9512. DOI: 10.3390/ijms22179512 PMID: 34502429

Mittal, G., Prashanth, A., Dhali, A., Prasad, R., Yogesh, S., Nurani, K. M., & Găman, M. A. (2025). Plant extracts with antioxidant and hepatoprotective benefits for liver health: A bibliometric analysis of drug delivery systems. *World Journal of Gastroenterology*, *31*(18), 105836. https://pmc.ncbi.nlm.nih.gov/articles/PMC12146923. DOI: 10.3748/wjg.v31.i18.105836 PMID: 40496363

Mittal, M., Siddiqui, M. R., Tran, K. A., Reddy, S. P., & Malik, A. B. (2013). Reactive Oxygen Species in Inflammation and Tissue Injury. *Antioxidants & Redox Signaling*, *20*(7), 1126–1167. DOI: 10.1089/ars.2012.5149 PMID: 23991888

Mndlovu, H., du Toit, L. C., Kumar, P., & Choonara, Y. E. (2023). Tannic acid-loaded chitosan-RGD-alginate scaffolds for wound healing and skin regeneration. *Biomedical Materials (Bristol, England)*, *18*(4), 045009. DOI: 10.1088/1748-605X/acce88 PMID: 37075773

Mohammadi, Z., Sharif Zak, M., Majdi, H., Seidi, K., Barati, M., Akbarzadeh, A., & Latifi, A. M. (2017). The effect of chrysin-loaded nanofiber on wound healing process in male rat. *Chemical Biology & Drug Design*, *90*(6), 1106–1114. DOI: 10.1111/cbdd.12996 PMID: 28388004

Mohan, Ch., Naresh, B., Kumar, B. K., Reddy, V. D., Manjula, P., Keerthi, B., Sreekanth, D., Manzelat, S. F., & Cherku, P. D. (2017). Micropropagation studies and phytochemical analysis of the endangered tree Commiphora wightii. *Journal of Applied Research on Medicinal and Aromatic Plants*, *6*, 70–79. DOI: 10.1016/j.jarmap.2017.02.004

Mohan, S., Prabhakaran, V. S., & Narayanaswamy, R. (2022). In silico analysis of *Cissus rotundifolia* constituents as human neutrophil elastase (HNE), matrix metalloproteinases (MMP 2 and MMP 9), and tyrosinase inhibitors. *Applied Biochemistry and Biotechnology*, *194*(1), 232–245. DOI: 10.1007/s12010-021-03758-8 PMID: 34800282

Mohsin, F., Javaid, S., Tariq, M. H., & Mustafa, M. (2024). Molecular immunological mechanisms of impaired wound healing in diabetic foot ulcers (DFU), current therapeutic strategies and future directions. *International Immunopharmacology*, *139*, 112713. DOI: 10.1016/j.intimp.2024.112713 PMID: 39047451

Molan, P. C., & White, R. (2009). *Honey in modern wound management*. Wounds UK Ltd.

Molan, P., & Rhodes, T. (2015). Honey: A biologic wound dressing. *Wounds : a Compendium of Clinical Research and Practice*, *27*(6), 141–151. PMID: 26061489

Molin, M. L., Segat, B., Garcia, M. C. F., Pezzin, A. P. T., Silveira, M. L. L., & Schneider, A. L. D. S. (2023). Development and characterization of bacterial cellulose membrane incorporated with Witch hazel extract. Matéria (Rio de Janeiro), 28, e20230008. https://doi.org/ K.N., Reddy, V. K., Veigas, J. M., & Murthy, U. D. (2004). Study on wound healing activity of Punica granatum peel. *Journal of Medicinal Food, 7*(2), 256-259.doi: DOI: 10.1590/1517-7076-RMAT-2023-0008Murthy

Monika, P., Chandraprabha, M. N., Rangarajan, A., Waiker, P. V., & Chidambara Murthy, K. N. (2022). Challenges in healing wound: Role of complementary and alternative medicine. *Frontiers in Nutrition*, *8*, 791899. DOI: 10.3389/fnut.2021.791899 PMID: 35127787

Moore, Z., Butcher, G., Corbett, L. Q., McGuiness, W., Snyder, R. J., & van Acker, K. (2014). Exploring the concept of a team approach to wound care: Managing wounds as a team. *Journal of Wound Care*, *23*(Sup5b, Suppl. 5b), S1–S38. DOI: 10.12968/jowc.2014.23.Sup5b.S1 PMID: 25191792

Morris, G. M., & Lim-Wilby, M. (2008). Molecular docking. In Kukol, A. (Ed.), *Molecular Modeling of Proteins* (pp. 365–382). Humana Press., DOI: 10.1007/978-1-59745-177-2_19

Morton, L. M., & Phillips, T. J. (2016). Wound healing and treating wounds: Differential diagnosis and evaluation of chronic wounds. *Journal of the American Academy of Dermatology*, *74*(4), 589–605. DOI: 10.1016/j.jaad.2015.08.068 PMID: 26979352

Moses, R. L., Prescott, T. A., Mas-Claret, E., Steadman, R., Moseley, R., & Sloan, A. J. (2023). Evidence for natural products as alternative wound-healing therapies. *Biomolecules*, *13*(3), 444. DOI: 10.3390/biom13030444 PMID: 36979379

Moura, L. I., Dias, A. M., Carvalho, E., & de Sousa, H. C. (2013). Wound dressings for diabetic foot ulcers. *Acta Biomaterialia*, *9*(7), 7093–7114. DOI: 10.1016/j.actbio.2013.03.033 PMID: 23542233

Mousaei Ghasroldasht, M., Seok, J., Park, H.-S., Liakath Ali, F. B., & Al-Hendy, A. (2022). Stem cell therapy: From idea to clinical practice. *International Journal of Molecular Sciences*, *23*(5), 2850. Advance online publication. DOI: 10.3390/ijms23052850 PMID: 35269990

Muchlinski, A., Ibdah, M., Ellison, S., Yahyaa, M., Nawade, B., & Laliberte, S.. (2020). Diversity and function of terpene synthases in carrot aroma and flavor biosynthesis. *Scientific Reports*, *10*(1), 9989. DOI: 10.1038/s41598-020-66866-1 PMID: 32561772

Mukhopadhyay, M. (2009). Extraction and processing with supercritical fluids. *Journal of Chemical Technology and Biotechnology*, *84*(1), 6–12. DOI: 10.1002/jctb.2072

Mumtaz, R., Zubair, M., Khan, M. A., Muzammil, S., & Siddique, M. H. (2022). Extracts of *Eucalyptus alba* promote diabetic wound healing by inhibiting α-glucosidase and stimulating cell proliferation. *Evidence-Based Complementary and Alternative Medicine : eCAM*, *4953105*, 1–12. Advance online publication. DOI: 10.1155/2022/4953105 PMID: 35463094

Munro, I. C., Harwood, M., Hlywka, J. J., Stephen, A. M., Doull, J., Flamm, W. G., & Adlercreutz, H. (2003). Soy isoflavones: A safety review. *Nutrition Reviews*, *61*(1), 1–33. DOI: 10.1301/nr.2003.janr.1-33 PMID: 12638461

Muralidhar, A., Babu, K. S., Sankar, T. R., Reddanna, P., & Latha, J. (2013). Wound healing activity of flavonoid fraction isolated from the stem bark of *Butea monosperma* (Lam.) in albino Wistar rats. *European Journal of Experimental Biology*, *3*(6), 1–6.

Murdiana, H. E., Murwanti, R., Fakhrudin, N., & Ikawati, Z. (2024). Multi-target mechanism of polyherbal extract to treat diabetic foot ulcer based on network pharmacology and molecular docking. *Journal of Herbmed Pharmacology*, *13*(2), 289–299. DOI: 10.34172/jhp.2024.49362

Mussbacher, M., Kral-Pointner, J. B., Salzmann, M., Schrottmaier, W. C., & Assinger, A. (2024). Mechanisms of hemostasis: Contributions of platelets, coagulation factors, and the vessel wall. In Geiger, M. (Ed.), *Fundamentals of vascular biology*. Springer., DOI: 10.1007/978-3-031-64591-4_8

Mustafa, I., Chin, N. L., Fakurazi, S., & Arulselvan, P. (2019). Comparison of Phytochemicals, Antioxidant and Anti-Inflammatory Properties of Sun-, Oven- and Freeze-Dried Ginger Extracts. *Foods*, *8*(10), 456. DOI: 10.3390/foods8100456 PMID: 31590464

Mustoe, T. (2004). Understanding chronic wounds: A unifying hypothesis on their pathogenesis and implications for therapy. *American Journal of Surgery*, *187*(5, Suppl.), S65–S70. DOI: 10.1016/S0002-9610(03)00306-4 PMID: 15147994

Mustoe, T. A., O'Shaughnessy, K., & Kloeters, O. (2006). Chronic wound pathogenesis and current treatment strategies: A unifying hypothesis. *Plastic and Reconstructive Surgery, 117*(7, Suppl), 35S–41S. DOI: 10.1097/01.prs.0000225431.63010.1b PMID: 16799373

Mu, X., Chen, J., Zhu, H., Deng, J., Wu, X., He, W., Ye, P., Gu, R., Wu, Y., Han, F. Y., & Nie, X. (2025). Asiaticoside–nitric oxide synergistically accelerate diabetic wound healing by regulating key metabolites and SRC/STAT3 signaling. *Burns and Trauma, 13*, tkaf009. Advance online publication. DOI: 10.1093/burnst/tkaf009 PMID: 40066291

Nafiu, M. O., Hamid, A. A., Muritala, H. F., & Adeyemi, S. B. (2017). Standardization of medicinal plants. In Medicinal spices and vegetables from Africa (pp. 171–204).

Nagase, H., Visse, R., & Murphy, G. (2006). Structure and function of matrix metalloproteinases and TIMPs. *Cardiovascular Research, 69*(3), 562–573. DOI: 10.1016/j.cardiores.2005.12.002 PMID: 16405877

Najafian, Y., Khorasani, Z. M., Najafi, M. N., Hamedi, S. S., Mahjour, M., & Feyz-abadi, Z. (2018). Efficacy of Aloe vera/ Plantago Major Gel in Diabetic Foot Ulcer: A Randomized Double-Blind Clinical Trial. *Current Drug Discovery Technologies, 16*(2), 223–231. DOI: 10.2174/1570163815666180115093007 PMID: 29336265

Nanda, S., & Madan, K. (2021). TheRoleofSafranal and Saffron Stigma Extracts in Oxidative Stress, Diseases and Photoaging: A Systematic Review. *Heliyon, 7*(2), e06117. DOI: 10.1016/j.heliyon.2021.e06117 PMID: 33615006

Nascimento, R., Lima, J., & Oliveira, J.. (2020). Nanotechnology as a tool for enhancing the bioavailability and effectiveness of polyphenols and flavonoids. *Nutrients, 12*(8), 2450. PMID: 32824061

Naswamy, V. (1981). Origin and development of Ayurveda. *Ancient Science of Life, 1*, 1–7. PMID: 22556454

Natarajan, V., Krithica, N., Madhan, B., & Sehgal, P. K. (2013). Preparation and properties of tannic acid cross-linked collagen scaffold and its application in wound healing. *Journal of Biomedical Materials Research. Part B, Applied Biomaterials, 101*(4), 560–567. DOI: 10.1002/jbm.b.32856 PMID: 23255343

Navarrete, P., Pizzi, A., Pasch, H., Rode, K., & Delmotte, L. (2013). Characterization of two maritime pine tannins as wood adhesives. *Journal of Adhesion Science and Technology, 27*(22), 2462–2479. DOI: 10.1080/01694243.2013.787515

Nayak, B. S., Raju, S. S., & Rao, A. V. C. (2007). Wound healing activity of *Matricaria recutita* L. extract. *Journal of Wound Care*, *16*(7), 298–302. DOI: 10.12968/jowc.2007.16.7.27061 PMID: 17708380

Nayak, S. B., Rodrigues, V., Maharaj, S., & Bhogadi, V. S. (2013). Wound healing activity of the fruit skin of *Punica granatum*. *Journal of Medicinal Food*, *16*(9), 857–861. DOI: 10.1089/jmf.2012.0229 PMID: 24044494

Naz, S., Siddiqi, R., Dew, T. P., & Williamson, G. (2011). Epigallocatechin-3-gallate inhibits lactase but is alleviated by salivary proline-rich proteins. *Journal of Agricultural and Food Chemistry*, *59*(6), 2734–2738. DOI: 10.1021/jf103072z PMID: 21348516

Negi, A., Sharma, N., & Singh, M. F. (2012). Formulation and Evaluation of an Herbal Anti-Inflammatory Gel Containing Eupatorium Leaves Extract. *Journal of Pharmacognosy and Phytochemistry*, *1*(4), 112. https://www.phytojournal.com/archives/2012/vol1issue4/PartA/16.1.pdf

Negut, I., Grumezescu, V., & Grumezescu, A. M. (2018). Treatment strategies for infected wounds. *Molecules (Basel, Switzerland)*, *23*(9), 2392. DOI: 10.3390/molecules23092392 PMID: 30231567

Newman, D. J., & Cragg, G. M. (2016). Natural products as sources of new drugs from 1981 to 2014. *Journal of Natural Products*, *79*(3), 629–661. DOI: 10.1021/acs.jnatprod.5b01055 PMID: 26852623

Ngeow, W. C., Tan, C. C., Goh, Y. C., Deliberador, T. M., & Cheah, C. W. (2022). Promoting oxygenation and angiogenesis in oral wound healing. *Bioengineering (Basel, Switzerland)*, *9*(11), 636. DOI: 10.3390/bioengineering9110636 PMID: 36354548

Nguyen, T. T., Le, T. V., Dang, N. N., Nguyen, D. C., Nguyen, P. T., Tran, T. T., Nguyen, Q. V., Bach, L. G., & Pham, D. N. T. (2021). Microencapsulation of essential oils by spray-drying and influencing factors. *Journal of Food Quality*, *2021*, 5525879. DOI: 10.1155/2021/5525879

Nguyen, T., Nioi, P., & Pickett, C. B. (2009). The Nrf2-antioxidant response element signaling pathway and its activation by oxidative stress. *The Journal of Biological Chemistry*, *284*(20), 13291–13295. DOI: 10.1074/jbc.R900010200 PMID: 19182219

Niethammer, P., Grabher, C., Look, A. T., & Mitchison, T. J. (2009). A tissue-scale gradient of hydrogen peroxide mediates rapid wound detection in zebrafish. *Nature*, *459*(7249), 996–999. DOI: 10.1038/nature08119 PMID: 19494811

Nie, X., Zhang, H., Shi, X., Zhao, J., Chen, Y., Wu, F., Yang, J., & Li, X. (2019). Asiaticoside nitric oxide gel accelerates diabetic cutaneous ulcers healing by activating Wnt/β-catenin signaling pathway. *International Immunopharmacology, 79*, 106109. DOI: 10.1016/j.intimp.2019.106109 PMID: 31865242

Ninan, N., Forget, A., Shastri, V. P., Voelcker, N. H., & Blencowe, A. (2016). Antibacterial and anti-inflammatory pH-responsive tannic acid-carboxylated agarose composite hydrogels for wound healing. *ACS Applied Materials & Interfaces, 8*(42), 28511–28521. DOI: 10.1021/acsami.6b10491 PMID: 27704757

Ning, S., Zang, J., Zhang, B., Feng, X., & Qiu, F. (2022). Botanical Drugs in Traditional Chinese Medicine With Wound Healing Properties [Review of *Botanical Drugs in Traditional Chinese Medicine With Wound Healing Properties*]. *Frontiers in Pharmacology, 13*. Frontiers Media. DOI: 10.3389/fphar.2022.885484

Nirenjen, S., Narayanan, J., Tamilanban, T., Subramaniyan, V., Chitra, V., Fuloria, N. K., Wong, L. S., Ramachawolran, G., Sekar, M., Gupta, G., Fuloria, S., Chinni, S. V., & Selvaraj, S. (2023). Exploring the contribution of pro-inflammatory cytokines to impaired wound healing in diabetes. *Frontiers in Immunology, 14*, 1216321. DOI: 10.3389/fimmu.2023.1216321 PMID: 37575261

Nitin Nama, D. R. (2022). Modern Science Meets Traditional Medicine. In M. R. K. A. (Edit. DR. Harshita Jain, Dr. Nitin Nama, Dr. Girish Kumar Vyas (Ed.), *Herbal Remedies: Bridging Traditional Knowledge with Modern Sciences* (pp. 14–22). Career Point Ltd.

Noor, S., Khan, R. U., & Ahmad, J. (2017). Understanding diabetic foot infection and its management. *Diabetes & Metabolic Syndrome, 11*(2), 149–156. DOI: 10.1016/j.dsx.2016.06.023 PMID: 27377687

Noriega, P. (2020). Terpenes in essential oils: Bioactivity and applications. In *Terpenes and terpenoids—Recent advances*. IntechOpen.

Nosbaum, A., Prevel, N., Truong, H.-A., Mehta, P., Ettinger, M., Scharschmidt, T. C., Ali, N. H., Pauli, M. L., Abbas, A. K., & Rosenblum, M. D. (2016). Cutting edge: Regulatory T cells facilitate cutaneous wound healing. *The Journal of Immunology : Official Journal of the American Association of Immunologists, 196*(5), 2010–2014. DOI: 10.4049/jimmunol.1502139 PMID: 26826250

Novaes, R., Costa, T. F. R., Goundry, A. L., Verçoza, B. R. F., Rodrigues, J. C. F., Godinho, J. L. P., Reis, F. C. G., Morrot, A., & Lima, A. P. C. A. (2022). Bone marrow granulocytes downregulate IL-1β and TNF production and the microbicidal activity of inflammatory macrophages. *Biochemistry and Cell Biology, 100*(3), 246–265. DOI: 10.1139/bcb-2021-0460 PMID: 35443139

Nowwarote, N., Osathanon, T., Jitjaturunt, P., Manopattanasoontorn, S., & Pavasant, P. (2012). Asiaticoside Induces Type I Collagen Synthesis and Osteogenic Differentiation in Human Periodontal Ligament Cells. *Phytotherapy Research*, *27*(3), 457–462. DOI: 10.1002/ptr.4742 PMID: 22628285

Nusantoro, A. P., Kuntaman, K., & Perdanakusuma, D. S. (2024). Synergistic Interaction between Topical Application of Allicin and Quercetin Enhances Diabetic Wound Healing Via Inflammatory in Wistar Rats. *Sains Malaysiana*, *53*(2), 397–408. DOI: 10.17576/jsm-2024-5302-13

Nussbaum, S. R., Carter, M. J., Fife, C. E., DaVanzo, J., Haught, R., Nusgart, M., & Cartwright, D. (2018). An economic evaluation of the impact, cost, and Medicare policy implications of chronic nonhealing wounds. *Value in Health*, *21*(1), 27–32. DOI: 10.1016/j.jval.2017.07.007 PMID: 29304937

O'Brien, J., Wang, A., & Miller, D. (2022). Regulatory framework for bioactive compounds in wound care: Challenges and perspectives. *Regulatory Toxicology and Pharmacology*, *129*, 105056. Advance online publication. DOI: 10.1016/j.yrtph.2022.105056

Oboh, G., Ademosun, A., Olasehinde, T., Oyeleye, S., & Ehiakhamen, E. (2015). Effect of processing methods on the antioxidant properties and inhibition of α-amylase and α-glucosidase by African pear (Dacryodes edulis) fruit. *NUTRAfoods : International Journal of Science and Marketing for Nutraceutical Actives, Raw Materials, Finish Products*, *14*(1), 19–26. DOI: 10.1007/s13749-014-0065-7

Ohnishi, N., Kusuhara, M., Yoshioka, M., Kuroda, K., Soga, A., Nishikawa, F., Koishi, T., Nakagawa, M., Hori, S., Matsumoto, T., Yamashita, M., Ohta, S., Takara, K., & Yokoyama, T. (2003). Studies on Interactions between Functional Foods or Dietary Supplements and Medicines. I. Effects of Ginkgo biloba Leaf Extract on the Pharmacokinetics of Diltiazem in Rats. *Biological & Pharmaceutical Bulletin*, *26*(9), 1315–1320. DOI: 10.1248/bpb.26.1315 PMID: 12951478

Okem, A., Henstra, C., Lambert, M., & Hayeshi, R. (2022). A review of the pharmacodynamic effect of chemo-herbal drug combinations therapy for cancer treatment [Review of *A review of the pharmacodynamic effect of chemo-herbal drug combinations therapy for cancer treatment*]. *Medicine in Drug Discovery*, *17*, 100147. Elsevier BV. DOI: 10.1016/j.medidd.2022.100147

Okonkwo, U. A., & DiPietro, L. A. (2017). Diabetes and Wound Angiogenesis [Review of *Diabetes and Wound Angiogenesis*]. *International Journal of Molecular Sciences*, *18*(7), 1419. Multidisciplinary Digital Publishing Institute. DOI: 10.3390/ijms18071419

Okonkwo, U. A., Chen, L., Ma, D., Haywood, V. A., Barakat, M., Urao, N., & DiPietro, L. A. (2020). Compromised angiogenesis and vascular Integrity in impaired diabetic wound healing. *PLoS One*, *15*(4), e0231962. Advance online publication. DOI: 10.1371/journal.pone.0231962 PMID: 32324828

Oliveira, R.N., Mancini, M.C., de Oliveira, F.C.S., Passos, T.M., Quilty, B., Thiré, R.M.d.S.M., & McGuinness, G.B. (2016). FTIR Analysis and Quantification of Phenols and Flavonoids of Five Commercially Available Plants Extracts Used in Wound Healing. Matér. Rio Jan. 21, 767–779.

Oliveira, D., Borges, A., Saavedra, M. J., Borges, F., & Simões, M. (2022). Screening of Natural Molecules as Adjuvants to Topical Antibiotics to Treat Staphylococcus aureus from Diabetic Foot Ulcer Infections. *Antibiotics (Basel, Switzerland)*, *11*(5), 620. DOI: 10.3390/antibiotics11050620 PMID: 35625264

Olson, M. E., Ceri, H., Morck, D. W., Buret, A. G., & Read, R. R. (2002). Biofilm bacteria. *Canadian Journal of Veterinary Research = Revue Canadienne de Recherche Vétérinaire*, *66*(2), 86–92. PMID: 11989739

Olsson, M., Järbrink, K., Divakar, U., Bajpai, R., Upton, Z., Schmidtchen, A., & Car, J. (2019). The humanistic and economic burden of chronic wounds: A systematic review. *Wound Repair and Regeneration*, *27*(1), 114–125. DOI: 10.1111/wrr.12683 PMID: 30362646

Omidian, H., Wilson, R. L., & Chowdhury, S. D. (2023). Enhancing Therapeutic Efficacy of Curcumin: Advances in Delivery Systems and Clinical Applications. *Gels (Basel, Switzerland)*, *9*(8), 596. DOI: 10.3390/gels9080596 PMID: 37623051

Organization, W. H. (2004). The world medicines situation. In *The world medicines situation.*

Organization, W. H. (2018). *WHO traditional medicine strategy 2014–2023. Geneva: WHO; 2013.*

Orlowski, P., Zmigrodzka, M., Tomaszewska, E., Ranoszek-Soliwoda, K., Czupryn, M., Antos-Bielska, M., Szemraj, J., Celichowski, G., Grobelny, J., & Krzyzowska, M. (2018). Tannic acid-modified silver nanoparticles for wound healing: The importance of size. *International Journal of Nanomedicine*, *13*, 991–1007. DOI: 10.2147/IJN.S154797 PMID: 29497293

Oryan, A., Mohammadalipour, A., Moshiri, A., & Tabandeh, M. R. (2016). Topical Application of Aloe vera Accelerated Wound Healing, Modeling, and Remodeling: An Experimental Study. *Annals of Plastic Surgery*, *77*(1), 37–46. DOI: 10.1097/SAP.0000000000000239 PMID: 25003428

Ousey, K., Gilchrist, B., & James, H. (2018). Understanding clinical practice challenges: A survey performed with wound care clinicians to explore wound assessment frameworks. *Wounds International*, *9*(4), 58–62.

Owens, B. D., Kragh, J. F.Jr, Macaitis, J., Svoboda, S. J., & Wenke, J. C. (2007). Characterization of extremity wounds in Operation Iraqi Freedom and Operation Enduring Freedom. *Journal of Orthopaedic Trauma*, *21*(4), 254–257. DOI: 10.1097/BOT.0b013e31802f78fb PMID: 17414553

Oyama, K., Kawada-Matsuo, M., Oogai, Y., Hayashi, T., Nakamura, N., & Komatsuzawa, H. (2016). Antibacterial Effects of Glycyrrhetinic Acid and Its Derivatives on Staphylococcus aureus. *PLoS One*, *11*(11), e0165831. Advance online publication. DOI: 10.1371/journal.pone.0165831 PMID: 27820854

Oyebode, O., Kandala, N. B., Chilton, P. J., & Lilford, R. J. (2016). Use of traditional medicine in middle-income countries: A WHO-SAGE study. *Health Policy and Planning*, *31*(8), 984–991. DOI: 10.1093/heapol/czw022 PMID: 27033366

Özay, Y., Guzel, S., Erdoğdu, İ., Yıldırım, Z., Pehlivanoğlu, B., Turk, B., & Darcan, S. (2018). Evaluation of the wound healing properties of luteolin ointments on excision and incision wound models in diabetic and non-diabetic rats. *Records of Natural Products*, *12*(4), 350–366. DOI: 10.25135/rnp.38.17.08.135

Özay, Y., Güzel, S., Yumrutaş, Ö., Pehlivanoğlu, B., Erdoğdu, İ. H., Yildirim, Z., Türk, B. A., & Darcan, S. (2019). Wound healing effect of kaempferol in diabetic and nondiabetic rats. *The Journal of Surgical Research*, *233*, 284–296. DOI: 10.1016/j.jss.2018.08.009 PMID: 30502261

Pacyga, K., Pacyga, P., Topola, E., Viscardi, S., & Duda-Madej, A. (2024). Bioactive compounds from plant origin as natural antimicrobial agents for the treatment of wound infections. *International Journal of Molecular Sciences*, *25*(4), 2100. DOI: 10.3390/ijms25042100 PMID: 38396777

Pagadala, N. S., Syed, K., & Tuszynski, J. (2017). Software for molecular docking: A review. *Biophysical Reviews*, *9*(2), 91–102. DOI: 10.1007/s12551-016-0247-1 PMID: 28510083

Page-McCaw, A., Ewald, A. J., & Werb, Z. (2007). Matrix metalloproteinases and the regulation of tissue remodelling. *Nature Reviews. Molecular Cell Biology*, *8*(3), 221–233. DOI: 10.1038/nrm2125 PMID: 17318226

Paggi, J. M., Pandit, A., & Dror, R. O. (2024). The art and science of molecular docking. *Annual Review of Biochemistry*, *93*(1), 389–410. DOI: 10.1146/annurev-biochem-030222-120000 PMID: 38594926

Pai, L., Patil, S., Liu, S., & Wen, F. (2023). A growing battlefield in the war against biofilm-induced antimicrobial resistance: Insights from reviews on antibiotic resistance. *Frontiers in Cellular and Infection Microbiology*, *13*, 1327069. Advance online publication. DOI: 10.3389/fcimb.2023.1327069 PMID: 38188636

Palmieri, B., Vadalà, M., & Laurino, C. (2017). Review of the molecular mechanisms in wound healing: New therapeutic targets? *Journal of Wound Care*, *26*(12), 765–775. DOI: 10.12968/jowc.2017.26.12.765 PMID: 29244975

Panares Arcaya, P. J., Claro Capilitan, L. M. D., Decano, Z. R., Guyo Dugaduga, C. D., Lamasan Hernane, A. G., Calo, J. R. D., & Paspe Saldo, I. J. (2025). Phytochemical analysis and wound healing potential of *Aloe vera* (*Aloe massawana*) peel extract. *Asian Journal of Biological and Life Sciences*, *14*(1), 147–152. DOI: 10.5530/ajbls.20251440

Pandey, M. M., Rastogi, S., & Rawat, A. K. S. (2013). Indian traditional ayurvedic system of medicine and nutritional supplementation. *Evidence-Based Complementary and Alternative Medicine : eCAM*, *2013*(1), 376327. DOI: 10.1155/2013/376327 PMID: 23864888

Pandith, H., Zhang, X., Liggett, J., Min, K. W., Gritsanapan, W., & Baek, S. J. (2013). Hemostatic and wound healing properties of Chromolaena odorata leaf extract. *ISRN Dermatology*, *2013*, 168269. DOI: 10.1155/2013/168269 PMID: 23984087

Pan, S.-Y., Litscher, G., Gao, S.-H., Zhou, S.-F., Yu, Z.-L., Chen, H.-Q., Zhang, S.-F., Tang, M.-K., Sun, J.-N., & Ko, K.-M. (2014). Historical perspective of traditional indigenous medical practices: The current renaissance and conservation of herbal resources. *Evidence-Based Complementary and Alternative Medicine : eCAM*, *2014*(1), 525340. DOI: 10.1155/2014/525340 PMID: 24872833

Pan, S.-Y., Zhou, S.-F., Gao, S.-H., Yu, Z.-L., Zhang, S.-F., Tang, M.-K., Sun, J.-N., Ma, D.-L., Han, Y.-F., Fong, W.-F., & Ko, K.-M. (2013). New perspectives on how to discover drugs from herbal medicines: CAM′ S outstanding contribution to modern therapeutics. *Evidence-Based Complementary and Alternative Medicine : eCAM*, *2013*(1), 627375. DOI: 10.1155/2013/627375 PMID: 23634172

Papadopoulou, A., Green, R., & Frazier, R. A. (2004). Interaction of Flavonoids with Bovine Serum Albumin: A Fluorescence Quenching Study. *Journal of Agricultural and Food Chemistry*, *53*(1), 158–163. DOI: 10.1021/jf048693g PMID: 15631523

Paredes, A. J.. (2021). Smart polymeric systems for drug delivery of bioactive compounds. *Journal of Drug Targeting*, *29*(3), 224–239.

Park, E., Lee, S. M., Jung, I.-K., Lim, Y., & Kim, J.-H. (2011). Effects of genistein on early-stage cutaneous wound healing. *Biochemical and Biophysical Research Communications*, *410*(3), 514–519. DOI: 10.1016/j.bbrc.2011.06.013 PMID: 21679688

Park, J. S., An, S. J., Jeong, S. I., Gwon, H. J., Lim, Y. M., & Nho, Y. C. (2017). Chestnut honey impregnated carboxymethyl cellulose hydrogel for diabetic ulcer healing. *Polymers*, *9*(7), 248. DOI: 10.3390/polym9070248 PMID: 30970925

Parks, W. C., Wilson, C. L., & López-Boado, Y. S. (2004). Matrix metalloproteinases as modulators of inflammation and innate immunity. *Nature Reviews. Immunology*, *4*(8), 617–629. DOI: 10.1038/nri1418 PMID: 15286728

Parveen, B., Rajinikanth, V., & Narayanan, M. (2025). Natural plant antioxidants for food preservation and emerging trends in nutraceutical applications. *Discover Applied Sciences*, *7*(8), 845. Advance online publication. DOI: 10.1007/s42452-025-07464-6

Pase, L., Layton, J. E., Wittmann, C., Ellett, F., Nowell, C. J., Reyes-Aldasoro, C. C., Varma, S., Rogers, K. L., Hall, C. J., Keightley, M. C., Crosier, P. S., Grabher, C., Heath, J. K., Renshaw, S. A., & Lieschke, G. J. (2012). Neutrophil-delivered myeloperoxidase dampens the hydrogen peroxide burst after tissue wounding in zebrafish. *Current Biology*, *22*(19), 1818–1824. DOI: 10.1016/j.cub.2012.07.060 PMID: 22940471

Pastar, I., Stojadinovic, O., Krzyzanowska, A., Barrientos, S., Stuelten, C., Zimmerman, K., Blumenberg, M., Brem, H., & Tomic-Canic, M. (2010). Attenuation of the transforming growth factor β-signaling pathway in chronic venous ulcers. *Molecular Medicine (Cambridge, Mass.)*, *16*(3–4), 92–101. DOI: 10.2119/molmed.2009.00149 PMID: 20069132

Pastorino, G., Cornara, L., Soares, S., Rodrigues, F., & Oliveira, M. B. P. P. (2018). Liquorice (Glycyrrhiza glabra): A Phytochemical and Pharmacological Review. *Phytotherapy Research*, *32*(12), 2323–2339. DOI: 10.1002/ptr.6178 PMID: 30117204

Patel, S., Maheshwari, A., & Chandra, A. (2016). Biomarkers for wound healing and their evaluation. *Journal of Wound Care*, *25*(1), 46–55. DOI: 10.12968/jowc.2016.25.1.46 PMID: 26762498

Patil, K. R., Mahajan, U. B., Unger, B. S., Goyal, S. N., Belemkar, S., Surana, S. J., Ojha, S., & Patil, C. R. (2019). Animal Models of Inflammation for Screening of Anti-inflammatory Drugs: Implications for the Discovery and Development of Phytopharmaceuticals [Review of *Animal Models of Inflammation for Screening of Anti-inflammatory Drugs: Implications for the Discovery and Development of Phytopharmaceuticals*]. *International Journal of Molecular Sciences, 20*(18), 4367. Multidisciplinary Digital Publishing Institute. DOI: 10.3390/ijms20184367

Pattarayan, D., Sivanantham, A., Bethunaickan, R., Palanichamy, R., & Rajasekaran, S. (2018). Tannic acid modulates fibroblast proliferation and differentiation in response to pro-fibrotic stimuli. *Journal of Cellular Biochemistry, 119*(8), 6732–6742. DOI: 10.1002/jcb.26866 PMID: 29665059

Patwardhan, B., Warude, D., Pushpangadan, P., & Bhatt, N. (2005). Ayurveda and traditional Chinese medicine: A comparative overview. *Evidence-Based Complementary and Alternative Medicine : eCAM*, *2*(4), 465–473. DOI: 10.1093/ecam/neh140 PMID: 16322803

Pavela, R., & Benelli, G. (2016). Essential oils as ecofriendly biopesticides: Challenges and constraints. *Trends in Plant Science, 21*(12), 1000–1007. DOI: 10.1016/j.tplants.2016.10.005 PMID: 27789158

Pawar, S., Pawade, K., Nipate, S., Balap, A., Pimple, B., Wagh, V., Kachave, R., & Gaikwad, A. (2023). Preclinical evaluation of the diabetic wound healing activity of phytoconstituents extracted from Ficus racemosa Linn. leaves. *Int. J. Exp. Res. Rev*, *32*, 365–377. DOI: 10.52756/ijerr.2023.v32.032

Payne, D. E., Martin, N. R., Parzych, K. R., Rickard, A. H., Underwood, A., & Boles, B. R. (2013). Tannic acid inhibits Staphylococcus aureus surface colonization in an IsaA-dependent manner. *Infection and Immunity*, *81*(2), 496–504. DOI: 10.1128/IAI.00877-12 PMID: 23208606

Percival, N. J. (2002). Classification of wounds and their management. *Surgery (Oxford)*, *20*(5), 114–117. DOI: 10.1383/surg.20.5.114.14626

Percival, N. J. (2002). Classification of wounds and their management. *Surgical Infections*, *20*(5), 114–117.

Percival, S. L., McCarty, S. M., & Lipsky, B. (2015). Biofilms and wounds: An overview of the evidence. *Advances in Wound Care*, *4*(7), 373–381. DOI: 10.1089/wound.2014.0557 PMID: 26155379

Pereira, S. G., Moura, J., Carvalho, E., & Empadinhas, N. (2017). Microbiota of Chronic Diabetic Wounds: Ecology, Impact, and Potential for Innovative Treatment Strategies [Review of *Microbiota of Chronic Diabetic Wounds: Ecology, Impact, and Potential for Innovative Treatment Strategies*]. *Frontiers in Microbiology, 8*. Frontiers Media. DOI: 10.3389/fmicb.2017.01791

Pereira, J. A., Casado, N., Porto-Figueira, P., & Câmara, J. S. (2022). The potential of microextraction techniques for the analysis of bioactive compounds in food. *Frontiers in Nutrition*, *9*, 825519. DOI: 10.3389/fnut.2022.825519 PMID: 35257008

Pereira, R. V. S., EzEldeen, M., Ugarte-Berzal, E., Martens, E., Malengier-Devlies, B., Vandooren, J., Vranckx, J. J., Matthys, P., & Opdenakker, G. (2023). Physiological fibrin hydrogel modulates immune cells and molecules and accelerates mouse skin wound healing. *Frontiers in Immunology*, *14*, 1170153. Advance online publication. DOI: 10.3389/fimmu.2023.1170153 PMID: 37168862

Pérez-Recalde, M., Arias, I. E., & Hermida, É. B. (2018). Essential oils in biopolymer wound dressings: A systematic review. *Phytomedicine*, *38*, 57–65. DOI: 10.1016/j.phymed.2017.09.024 PMID: 29425655

Perović, A. B., Karabegović, I. T., Krstić, M. S., Veličković, A. V., Avramović, J. M., Danilović, B. R., & Veljković, V. B. (2024). Lavender essential oil recovery by novel hydrodistillation methods. *Industrial Crops and Products*, *211*, 118244. DOI: 10.1016/j.indcrop.2024.118244

Peter, A. E., Sandeep, B. V., Rao, B. G., & Kalpana, V. L. (2021). Natural immunosuppressants as adjuvants to target cytokine storm in COVID-19. *Frontiers in Pharmacology*, *11*, 583777. DOI: 10.3389/fphar.2020.583777 PMID: 33708109

Phelps, E. A., & García, A. J. (2008). Update on Therapeutic Vascularization Strategies [Review of *Update on Therapeutic Vascularization Strategies*]. *Regenerative Medicine, 4*(1), 65. Future Medicine. DOI: 10.2217/17460751.4.1.65

Piazza, S., Martinelli, G., Vrhovsek, U., Masuero, D., Fumagalli, M., Magnavacca, A., Pozzoli, C., Canilli, L., Terno, M., Angarano, M., Dell'Agli, M., & Sangiovanni, E. (2022). Anti-Inflammatory and Anti-Acne Effects of Hamamelis virginiana Bark in Human Keratinocytes. *Antioxidants*, *11*(6), 1119. DOI: 10.3390/antiox11061119 PMID: 35740016

Pignet, A.-L., Schellnegger, M., Hecker, A., Kohlhauser, M., Kotzbeck, P., & Kamolz, L. (2021). Resveratrol-Induced Signal Transduction in Wound Healing. *International Journal of Molecular Sciences*, *22*(23), 12614. DOI: 10.3390/ijms222312614 PMID: 34884419

Pisarik, P. (2016). Choosing tap water vs. sterile saline for wound irrigation. *American Family Physician, 94*(2), 83–84. https://www.aafp.org/pubs/afp/issues/2016/0715/p83.html PMID: 27419322

Pizzi, A. (2021). Tannins medical/pharmacological and related applications: A critical review. *Sustainable Chemistry and Pharmacy, 22*, 100481. DOI: 10.1016/j.scp.2021.100481

Polaka, S., Katare, P., Pawar, B., Vasdev, N., Gupta, T., Rajpoot, K., Sengupta, P., & Tekade, R. K. (2022). Emerging ROS-Modulating Technologies for Augmentation of the Wound Healing Process. *ACS Omega, 7*(35), 30657–30672. DOI: 10.1021/acsomega.2c02675 PMID: 36092613

Ponrasu, T., Veerasubramanian, P. K., Kannan, R., Gopika, S., Suguna, L., & Muthuvijayan, V. (2018). Morin incorporated polysaccharide–protein (psyllium–keratin) hydrogel scaffolds accelerate diabetic wound healing in Wistar rats. *RSC Advances, 8*(5), 2305–2314. DOI: 10.1039/C7RA10334D PMID: 35541447

Porel, P., Kaur, M., Sharma, V., & Aran, K. R. (2025). Understanding molecular mechanism of diabetic wound healing: Addressing recent advancements in therapeutic managements. *Journal of Diabetes and Metabolic Disorders, 24*(1), 76. DOI: 10.1007/s40200-025-01588-7 PMID: 40060271

Portou, M. J., Baker, D., Abraham, D., & Tsui, J. (2015). The innate immune system, toll-like receptors and dermal woundhealing: A review. *Vascular Pharmacology, 71*, 31–36. DOI: 10.1016/j.vph.2015.02.007 PMID: 25869514

Possemiers, S.. (2013). The role of biotransformation in the bioavailability of dietary polyphenols. *Food Research International, 54*(1), 213–229.

Potekaev, N. N., Borzykh, O. B., Medvedev, G. V., Pushkin, D. V., Petrova, M. M., Petrov, A. V., Dmitrenko, D. V., Karpova, E. I., Demina, O. M., & Shnayder, N. A. (2021). The role of extracellular matrix in skin wound healing. *Journal of Clinical Medicine, 10*(24), 5947. DOI: 10.3390/jcm10245947 PMID: 34945243

Pouget, C., Dunyach-Remy, C., Pantel, A., Schuldiner, S., Sotto, A., & Lavigne, J. (2020). Biofilms in Diabetic Foot Ulcers: Significance and Clinical Relevance. *Microorganisms, 8*(10), 1580. DOI: 10.3390/microorganisms8101580 PMID: 33066595

Pourmortazavi, S. M., & Hajimirsadeghi, S. S. (2007). Supercritical fluid extraction in essential oil analysis. *Journal of Chromatography. A, 1163*(1–2), 2–4. DOI: 10.1016/j.chroma.2007.06.021 PMID: 17624357

Powanda, M. C., & Moyer, E. D. (1981). Plasma proteins and wound healing. *Surgery, Gynecology & Obstetrics, 153*(5), 749–755. PMID: 7027503

Powers, J. G., Higham, C., Broussard, K., & Phillips, T. J. (2016). Wound healing and treating wounds: Chronic wound care and management. *Journal of the American Academy of Dermatology, 74*(4), 607–625. DOI: 10.1016/j.jaad.2015.08.070 PMID: 26979353

Przekora, A. (2020). Tissue engineered artificial skin grafts for chronic wound treatment: Can functional skin be reconstructed in vitro? *Cells, 9*(7), 1622. DOI: 10.3390/cells9071622 PMID: 32640572

Putri, R. F., Abdjan, M. I., Santosaningsih, D., Ni, Y. L., Putri, N., Khairunnisa, F., & Baktir, A. (2025). molecular dynamic and docking simulation in exploring a potential extracellular matrix-based antibiofilm: Fighting the resistant Pseudomonas aeruginosa-Candida albicans. *Engineering and Science, 34*, 1386. DOI: 10.30919/es1386

Pyner, A., Nyambe-Silavwe, H., & Williamson, G. (2017). Inhibition of human and rat sucrase and maltase activities to assess antiglycemic potential: Optimization of the assay using acarbose and polyphenols. *Journal of Agricultural and Food Chemistry, 65*(39), 8643–8651. DOI: 10.1021/acs.jafc.7b03678 PMID: 28914528

Qiao, W., Niu, L., Jiang, W., Lu, L., & Liu, J. (2024). Berberine ameliorates endothelial progenitor cell function and wound healing in vitro and in vivo via the miR-21-3p/RRAGB axis for venous leg ulcers. *Regenerative Therapy, 26*, 458–468. DOI: 10.1016/j.reth.2024.06.011 PMID: 39100534

Quave, C. L. (2018). Wound healing with botanicals: A review and future perspectives. *Current Dermatology Reports, 7*(4), 287–295. DOI: 10.1007/s13671-018-0247-4 PMID: 31106027

Quintans, J. S., Shanmugam, S., Heimfarth, L., Araújo, A. A., Almeida, J. R., Picot, L., & Quintans-Júnior, L. J. (2019). Monoterpenes modulating cytokines. *Food and Chemical Toxicology, 123*, 233–257. DOI: 10.1016/j.fct.2018.10.058 PMID: 30389585

Rafey, H. A., Amin, A., Ross, S. A., El-Shazly, M., Zahid, M. A., Niaz, S. I., Ul Mahmood, F., & Ullah, H. (2024). Multiple integrated computational approach to analyse wound healing potential of *Symplocos racemosa* bark as matrix metalloproteinase inhibitors. *Natural Product Research*, •••, 1–10. DOI: 10.1080/14786419.2024.2345678 PMID: 38497294

Raghavendra, G., & Mohan, B. (2018). Nanostructured materials for drug delivery. *Nanomedicine; Nanotechnology, Biology, and Medicine, 14*(1), 215–230.

Raguraman, R., Srivastava, A., Munshi, A., & Ramesh, R. (2021). Therapeutic approaches targeting molecular signaling pathways common to diabetes, lung diseases and cancer [Review of *Therapeutic approaches targeting molecular signaling pathways common to diabetes, lung diseases and cancer*]. *Advanced Drug Delivery Reviews, 178*, 113918. Elsevier BV. DOI: 10.1016/j.addr.2021.113918

Rai, V., Moellmer, R., & Agrawal, D. K. (2022). Stem Cells and Angiogenesis: Implications and Limitations in Enhancing Chronic Diabetic Foot Ulcer Healing. *Cells, 11*(15), 2287. DOI: 10.3390/cells11152287 PMID: 35892584

Rajasekaran, S., Rajasekar, N., & Sivanantham, A. (2021). Therapeutic potential of plant-derived tannins in non-malignant respiratory diseases. *The Journal of Nutritional Biochemistry, 94*, 108632. DOI: 10.1016/j.jnutbio.2021.108632 PMID: 33794331

Rajendran, P., Rengarajan, T., Thangavel, J., Nishigaki, Y., Sakthisekaran, D., Sethi, G., & Nishigaki, I. (2013). The Vascular Endothelium and Human Diseases [Review of *The Vascular Endothelium and Human Diseases*]. *International Journal of Biological Sciences, 9*(10), 1057. Ivyspring International Publisher. DOI: 10.7150/ijbs.7502

Rajput, S., Kaur, S., Panesar, P. S., & Thakur, A. (2023). Supercritical fluid extraction of essential oils from Citrus reticulata. *Biomass Conversion and Biorefinery, 13*(16), 14605–14614. DOI: 10.1007/s13399-022-02807-4

Ramirez-Acuña, J. M., Cardenas-Cadena, S. A., Marquez-Salas, P. A., Garza-Veloz, I., Perez-Favila, A., Cid-Báez, M. A., Flores-Morales, V., & Martínez-Fierro, M. L. (2019). Diabetic Foot Ulcers: Current Advances in Antimicrobial Therapies and Emerging Treatments. *Antibiotics (Basel, Switzerland), 8*(4), 193. DOI: 10.3390/antibiotics8040193 PMID: 31652990

Rani, M., Jindal, S., Anand, R., Sharma, N., Ranjan, K. R., Mukherjee, M. D., & Mishra, V. (2023). Essential oils for therapeutic benefits. In Essential oils: Extraction methods and applications (pp. 977–1008).

Rani, M., Jindal, S., Anand, R., Sharma, N., Ranjan, K. R., Mukherjee, M. D., & Mishra, V. (2023). Plant essential oils and their constituents for therapeutic benefits. In Essential oils: Extraction methods and applications (pp. 977–1008). DOI: 10.1002/9781119829614.ch42

Ranjha, M.. (2022). Scalability of nanomaterial production: Issues and strategies. *Journal of Nanomedicine & Nanotechnology, 13*(2), e1234.

Ranzato, E., Martinotti, S., & Burlando, B. (2011). Wound healing properties of jojoba liquid wax: An in vitro study. *Journal of Ethnopharmacology, 134*(2), 443–449. DOI: 10.1016/j.jep.2010.12.042 PMID: 21211559

Rao, K. M., Uthappa, U. T., Kim, H. J., & Han, S. S. (2023). Tissue adhesive, biocompatible, antioxidant, and antibacterial hydrogels based on tannic acid and fungal-derived carboxymethyl chitosan for wound-dressing applications. *Gels (Basel, Switzerland), 9*(5), 354. DOI: 10.3390/gels9050354 PMID: 37232946

Rautela, D. S., Das, S., Mazumder, A., Sharma, B., Khan, F., & Chaitanya, M. V. (2024). Understanding the influence of diabetes on wound healing and exploring effective management study. *Current Science, 126*(12), 1438. Advance online publication. DOI: 10.18520/cs/v126/i12/1438-1445

Raza, A., Chohan, T. A., Zaidi, S. H., Hai, A., Alzahrani, A. R., Imran, M., & Saleem, H. (2024). Biochemical perspectives on natural products in wound healing: Phytochemicals in tissue repair and scar prevention. *Chemistry & Biodiversity, 21*(10), e202400615. DOI: 10.1002/cbdv.202400615 PMID: 38958197

Razavi, Z. S., Sharafshadehi, S. A., Yousefi, M. H., Javaheri, F., Barghani, M. R. R., Afkhami, H., & Heidari, F. (2025). Application of novel strategies in chronic wound management with focusing on pressure ulcers: New perspective. *Archives of Dermatological Research, 317*(1), 320. DOI: 10.1007/s00403-024-03790-8 PMID: 39888392

Razif, R., Fadilah, N. I. M., Ahmad, H., Hao, D. L. Q., Maarof, M., & Fauzi, M. B. (2025). Asiaticoside-Loaded Multifunctional Bioscaffolds for Enhanced Hyperglycemic Wound Healing. *Biomedicines, 13*(2), 277. DOI: 10.3390/biomedicines13020277 PMID: 40002691

Raziyeva, K., Kim, Y., Zharkinbekov, Z., Kassymbek, K., Jimi, S., & Saparov, A. (2021). Immunology of Acute and Chronic Wound Healing. *Biomolecules, 11*(5), 700. DOI: 10.3390/biom11050700 PMID: 34066746

Reinke, J. M., & Sorg, H. (2012). Wound repair and regeneration. *European Surgical Research, 49*(1), 35–43. DOI: 10.1159/000339613 PMID: 22797712

Rendra, E., Riabov, V., Mossel, D. M., Sevastyanova, T., Harmsen, M. C., & Kzhyshkowska, J. (2018). Reactive oxygen species (ROS) in macrophage activation and function in diabetes [Review of *Reactive oxygen species (ROS) in macrophage activation and function in diabetes*]. *Immunobiology, 224*(2), 242. Elsevier BV. DOI: 10.1016/j.imbio.2018.11.010

Ren, S., Chen, J., Guo, J., Liu, Y., Xiong, H., Jing, B., Yang, X., Li, G., Kang, Y., Wang, C., Xu, X., Liu, Z., Zhang, M., Xiang, K., Li, C., Li, Q., Machens, H., & Chen, Z. (2022). Exosomes from Adipose Stem Cells Promote Diabetic Wound Healing through the eHSP90/LRP1/AKT Axis. *Cells*, *11*(20), 3229. DOI: 10.3390/cells11203229 PMID: 36291096

Reyes-Jurado, F., Franco-Vega, A., Ramírez-Corona, N., Palou, E., & López-Malo, A. (2015). Essential oils: Antimicrobial activity and extraction modeling. *Food Engineering Reviews*, *7*(3), 275–297. DOI: 10.1007/s12393-014-9099-2

Ribeiro, A.. (2021). Safety and toxicological assessment of nanomaterials in drug delivery systems. *Nano Today*, *36*, 100969.

Richard, S. A. (2021). Exploring the Pivotal Immunomodulatory and Anti-Inflammatory Potentials of Glycyrrhizic and Glycyrrhetinic Acids. *Mediators of Inflammation*, *2021*(1), 6699560. DOI: 10.1155/2021/6699560 PMID: 33505216

Richardson, R., Slanchev, K., Kraus, C., Knyphausen, P., Eming, S., & Hammerschmidt, M. (2013). Adult zebrafish as a model system for cutaneous wound-healing research. [Note: Original DOI corrected to match article]. *The Journal of Investigative Dermatology*, *133*(6), 1655–1665. DOI: 10.1038/jid.2013.16 PMID: 23325040

Richmond, J. M., & Harris, J. E. (2014). Immunology and skin in health and disease. *Cold Spring Harbor Perspectives in Medicine*, *4*(12), a015339. DOI: 10.1101/cshperspect.a015339 PMID: 25452424

Rippa, A. L., Kalabusheva, E. P., & Vorotelyak, E. A. (2019). Regeneration of Dermis: Scarring and Cells Involved. *Cells*, *8*(6), 607. DOI: 10.3390/cells8060607 PMID: 31216669

Rithichai, P., Jirakiattikul, Y., Nambuddee, R., & Itharat, A. (2024). Effect of salicylic acid foliar application on bioactive compounds and antioxidant activity in holy basil (Ocimum sanctum L.). *International Journal of Agronomy*, *2024*, 8159886. Advance online publication. DOI: 10.1155/2024/8159886

Robson, M. C., & Barbul, A. (2006). Guidelines for the best care of chronic wounds. *Wound Repair and Regeneration*, *14*(6), 647–648. DOI: 10.1111/j.1524-475X.2006.00173.x PMID: 17199830

Rodero, M. P., & Khosrotehrani, K. (2010). Skin wound healing modulation by macrophages. *International Journal of Clinical and Experimental Pathology*, *3*(7), 643–653. PMID: 20830235

Rodrigues, M., Kosaric, N., Bonham, C. A., & Gurtner, G. C. (2018). Wound healing: A cellular perspective. *Physiological Reviews.* Advance online publication. DOI: 10.1152/physrev.00067.2017 PMID: 30475656

Rodriguez, P. G., Felix, F. N., Woodley, D. T., & Shim, E. K. (2008). The role of oxygen in wound healing: A review of the literature. *Dermatologic Surgery*, *34*(9), 1159–1169. DOI: 10.1097/00042728-200809000-00001 PMID: 18513296

Rodríguez-Rodríguez, N., Martínez-Jiménez, I., García-Ojalvo, A., Mendoza-Marí, Y., Guillén, G., Armstrong, D., & Berlanga-Acosta, J. (2021). Wound Chronicity, Impaired Immunity and Infection in Diabetic Patients [Review of Wound Chronicity, Impaired Immunity and Infection in Diabetic Patients]. *MEDICC Review*, *24*(1), 44. DOI: 10.37757/MR2021.V23.N3.8 PMID: 34653116

Rubeena, P., Kumari, G., & Siqun, M. L. (2021). Polyphenolic ointment for wound healing: An innovative approach. *Journal of Pharmacy & Bioallied Sciences*, *13*(1), 67–73. DOI: 10.4103/jpbs.JPBS_84_20

Rudrapal, M., Rakshit, G., Singh, R. P., Garse, S., Khan, J., & Chakraborty, S. (2024). Dietary polyphenols: Review on chemistry/sources, bioavailability/metabolism, antioxidant effects, and their role in disease management. *Antioxidants*, *13*(4), 429. https://www.mdpi.com/2076-3921/13/4/429. DOI: 10.3390/antiox13040429 PMID: 38671877

Rumbaut, R. E. (2010). *Platelet-vessel wall interactions in hemostasis and thrombosis*. Morgan & Claypool Life Sciences. DOI: 10.4199/C00007ED1V01Y201002ISP004

Ruthenborg, R. J., Ban, J.-J., Wazir, A., Takeda, N., & Kim, J. (2014). Regulation of Wound Healing and Fibrosis by Hypoxia and Hypoxia-Inducible Factor-1. *Molecules and Cells*, *37*(9), 637–643. DOI: 10.14348/molcells.2014.0150 PMID: 24957212

Saad, H., Khoukh, A., Ayed, N., Charrier, B., & Charrier-El Bouhtoury, F. (2014). Characterization of Tunisian Aleppo pine tannins for a potential use in wood adhesive formulation. *Industrial Crops and Products*, *61*, 517–525. DOI: 10.1016/j.indcrop.2014.07.035

Saeed Anowar and Somnath Das. (2024). The role of Folk medicine in achieving the traditional goals through IKS: A Review. In D. M. D. and D. V. A. (eds. Dr. Somnath Das, Dr. Jayanta Kumar Das (Ed.), *Life as Basic Science: An Overview and Prospects for the Future Volume: 3* (pp. 164–179). International Academic Publishing House (IAPH), https://doi.org/https://doi.org/10.52756/lbsopf.2024.e03.007

Safarzadeh, E., Shotorbani, S. S., & Baradaran, B. (2014). Herbal medicine as inducers of apoptosis in cancer therapy. *Advanced Pharmaceutical Bulletin*, *4*(Suppl 1), 421. PMID: 25364657

Sahu, P., Bhowmick, A. K., & Kali, G. (2020). Terpene-based elastomers: Synthesis, properties, and applications. *Processes (Basel, Switzerland)*, *8*(5), 553. DOI: 10.3390/pr8050553

Sakihama, Y., Cohen, M. F., Grace, S. C., & Yamasaki, H. (2002). Plant phenolic antioxidant and prooxidant activities: Phenolics-induced oxidative damage mediated by metals in plants. *Toxicology*, *177*(1), 67–80. DOI: 10.1016/S0300-483X(02)00196-8 PMID: 12126796

Salama, A., El-Kassaby, M. I., Refaat, A., & Mohasib, R. M. (2024). GC–MS and molecular docking analyses of phytochemicals from *Calendula officinalis* L. hexane extract and evaluation of its antioxidant and wound healing properties in rats. *Egyptian Journal of Chemistry*, *67*(13), 1037–1058. DOI: 10.21608/ejchem.2024.328645.10642

Salazar-Gómez, A., & Alonso-Castro, A. J. (2022). Medicinal plants from Latin America with wound healing activity: Ethnomedicine, phytochemistry, preclinical and clinical studies—A review. *Pharmaceuticals (Basel, Switzerland)*, *15*(9), 1095. DOI: 10.3390/ph15091095 PMID: 36145316

Salehi, M., Ehterami, A., Farzamfar, S., Vaez, A., & Ebrahimi-Barough, S. (2021). Accelerating healing of excisional wound with alginate hydrogel containing naringenin in rat model. *Drug Delivery and Translational Research*, *11*(1), 142–153. DOI: 10.1007/s13346-020-00731-6 PMID: 32086788

Salehi, M., Vaez, A., Naseri-Nosar, M., Farzamfar, S., Ai, A., Ai, J., Tavakol, S., Khakbiz, M., & Ebrahimi-Barough, S. (2018). Naringin-loaded poly(ε-caprolactone)/gelatin electrospun mat as a potential wound dressing: In vitro and in vivo evaluation. *Fibers and Polymers*, *19*(1), 125–134. DOI: 10.1007/s12221-018-7528-6

Salem, M. A., Mohamed, O. G., Mosalam, E. M., Elberri, A. I., Abdel-Bar, H. M., Hassan, M., Al-Karmalawy, A. A., Tripathi, A., Ezzat, S. M., & Abo Mansour, H. E. (2023). Investigation of the phytochemical composition, antioxidant, antibacterial, anti-osteoarthritis, and wound healing activities of selected vegetable waste. *Scientific Reports*, *13*(1), 13034. DOI: 10.1038/s41598-023-38591-y PMID: 37563154

Sallam, I. E., Abdelwareth, A., Attia, H., Aziz, R. K., Homsi, M. N., von Bergen, M., & Farag, M. A. (2021). Effect of Gut Microbiota Biotransformation on Dietary Tannins and Human Health Implications. *Microorganisms*, *9*(5), 965. DOI: 10.3390/microorganisms9050965 PMID: 33947064

Salman, M., Tabassum, H., & Parvez, S. (2020). Tannic acid provides neuroprotective effects against traumatic brain injury through the PGC-1α/Nrf2/HO-1 pathway. *Molecular Neurobiology*, *57*(6), 2870–2885. DOI: 10.1007/s12035-020-01924-3 PMID: 32399817

Samanta, R., Pattnaik, A. K., Pradhan, K. K., Mehta, B. K., Pattanayak, S. P., & Banerjee, S. (2016). Wound healing activity of silibinin in mice. *Pharmacognosy Research*, *8*(4), 298. DOI: 10.4103/0974-8490.188880 PMID: 27695272

Samavati, S. S., Kashanian, S., Derakhshankhah, H., Rabiei, M., Sajadimajd, S., Fakhri, S., & Rashidi, Z. (2025). Accelerated wound healing through tannin-rich Jaft extract: Concentration-dependent efficacy and mechanistic insights from Quercus brantii ointment formulations. *Scientific Reports*, *15*(1), 29004. DOI: 10.1038/s41598-025-13832-4 PMID: 40781116

Samuelson, R., Lobl, M., Higgins, S., Clarey, D., & Wysong, A. (2020). Effects of lavender essential oil on wound healing: A review. *Journal of Alternative and Complementary Medicine (New York, N.Y.)*, *26*(8), 680–690. DOI: 10.1089/acm.2019.0286 PMID: 32589447

Sánchez-Machado, D. I., López-Cervantes, J., Sendón, R., & Sanches-Silva, A. (2017). Aloe vera: Ancient Knowledge with New Frontiers. *Trends in Food Science & Technology*, *61*, 94–102. DOI: 10.1016/j.tifs.2016.12.005

Sanyal, R. (2023). Green Pharmacy : Unveiling the Healing Potential of Medicinal Plants. In D. A. C. S. & D. A. P. (eds. Mrs. Bhanumati Sarkar, Dr. (Professor) Surjyo Jyoti Biswas (Ed.), *The Basic Handbook of Indian Ethnobotany and Traditional Medicine [Volume: 2]* (pp. 104–116). International Academic Publishing House (IAPH). https://doi.org/https://doi.org/10.52756/bhietm.2023.e02.009

Sapienza, P., Mingoli, A., Borrelli, V., Brachini, G., Biacchi, D., Sterpetti, A. V., Grande, R., Serra, R., & Tartaglia, E. (2019). Inflammatory biomarkers, vascular procedures of lower limbs, and wound healing. *International Wound Journal*, *16*(3), 716–723. DOI: 10.1111/iwj.13086 PMID: 30773823

Sarkar, B., & Madhu, N. R. (2016). *Knowledge of traditional medicines and practices used for reproductive health problems. Review article. UGC-sponsored National Seminar on "Medicinal Plants, Health and Environment*. Nectar Publisher, Kolkata.

Satpathy, R. (2024b). Computational approaches to discover novel phytochemical inhibitors against novel coronavirus (SARS-CoV-2). In *Computational Biology in Drug Discovery and Repurposing* (pp. 323–336). Apple Academic Press. DOI: 10.1201/9781003455424-15

Satpathy, R. (2020a). Therapeutic importance and application of boswellic acid from the plant *Boswellia serrata*. In Singh, A., Singh, P., & Bithel, N. (Eds.), *Advanced Pharmacological Uses of Medicinal Plants and Natural Products* (pp. 302–315). IGI Global., DOI: 10.4018/978-1-7998-2094-9.ch016

Satpathy, R. (2020b). Application of molecular docking methods on endocrine disrupting chemicals: A review. *Journal of Applied Biotechnology Reports*, *7*(2), 74–80.

Satpathy, R. (2021). Application of some medicinal plants and their constituents in the treatment of diabetes mellitus. In Hussain, A., & Behl, S. (Eds.), *Treating Endocrine and Metabolic Disorders with Herbal Medicines* (pp. 32–47). IGI Global., DOI: 10.4018/978-1-7998-4808-0.ch002

Satpathy, R. (2024a). Application of molecular modeling techniques to investigate phytochemicals as prospective anti-malarial agents. In Tripathi, R. K. P., & Tiwari, S. (Eds.), *Converging Pharmacy Science and Engineering in Computational Drug Discovery* (pp. 116–139). IGI Global., DOI: 10.4018/979-8-3693-2897-2.ch006

Saxena, A., Tripathi, K. P., Roy, S., Khan, F., & Sharma, A. (2008). Pharmacovigilance: Effects of herbal components on human drugs interactions involving Cytochrome P450. *Bioinformation, 3*(5), 198–204. DOI: 10.6026/97320630003198 PMID: 19255634

Scherr, T. D., Heim, C. E., Morrison, J. M., & Kielian, T. (2014). Hiding in Plain Sight: Interplay between Staphylococcal Biofilms and Host Immunity [Review of *Hiding in Plain Sight: Interplay between Staphylococcal Biofilms and Host Immunity*]. *Frontiers in Immunology, 5*. Frontiers Media. DOI: 10.3389/fimmu.2014.00037

Schippmann, U., Leaman, D. J., & Cunningham, A. B. (2002). *Impact of cultivation and gathering of medicinal plants on biodiversity: global trends and issues*. Biodiversity and the Ecosystem Approach in Agriculture, Forestry and Fisheries.

Schreml, S., Szeimies, R. M., Prantl, L., Landthaler, M., & Babilas, P. (2010). Wound healing in the 21st century. *Journal of the American Academy of Dermatology, 63*(5), 866–881. DOI: 10.1016/j.jaad.2009.10.048 PMID: 20576319

Schultz, G. S., Davidson, J. M., Kirsner, R. S., Bornstein, P., & Herman, I. M. (2011). Dynamic reciprocity in the wound microenvironment. *Wound Repair and Regeneration, 19*(2), 134–148. DOI: 10.1111/j.1524-475X.2011.00673.x PMID: 21362080

Schultz, G. S., & Wysocki, A. (2009). Interactions between extracellular matrix and growth factors in wound healing. *Wound Repair and Regeneration, 17*(2), 153–162. DOI: 10.1111/j.1524-475X.2009.00466.x PMID: 19320882

Scialò, F., Fernández-Ayala, D. J., & Sanz, A. (2017). Role of mitochondrial reverse electron transport in ROS signaling: Potential roles in health and disease. *Frontiers in Physiology*, *8*, 428. DOI: 10.3389/fphys.2017.00428 PMID: 28701960

Scrima, M., Melito, C., Merola, F., Iorio, A., Vito, N., Giori, A. M., & Ferravante, A. (2020). Evaluation of Wound Healing Activity of Salvia Haenkei Hydroalcoholic Aerial Part Extract on in Vitro and in vivo Experimental Models. *Clinical, Cosmetic and Investigational Dermatology*, *13*, 627–637. DOI: 10.2147/CCID.S224418 PMID: 32922060

Sebestyén, Z., Jakab, E., Badea, E., Barta-Rajnai, E., Şendrea, C., & Czégény, Z. S. (2019). Thermal degradation study of vegetable tannins and vegetable tanned leathers. *Journal of Analytical and Applied Pyrolysis*, *138*, 178–187. DOI: 10.1016/j.jaap.2018.12.022

Sęczyk, Ł., Świeca, M., Kapusta, I., & Gawlik-Dziki, U. (2019). Protein–phenolic interactions as a factor affecting the physicochemical properties of white bean proteins. *Molecules (Basel, Switzerland)*, *24*(3), 408. DOI: 10.3390/molecules24030408 PMID: 30678067

Seema, B. R., Jyothi, Y., Bhovi, C. N., Aradhya, M. V., Lekhak, M., & Mane, S. R. (2023). Advanced up and down methodology for acute toxicity assessment with reliable LD50 verified by aqueous extract of curly kale using Wistar rats. *Research Journal of Pharmacy and Technology*, *16*(10), 4519–4524. https://www.indianjournals.com/article/rjpt-16-10-006

Selvaraj, K., Theivendren, P., Pavadai, P., Sureshbabu, R. K., Ravishankar, V., & Palanisamy, P.. (2022). Physicochemical parameters affecting extraction of bioactives. *Current Bioactive Compounds*, *18*(4), 11–27.

Sen, C. K. (2019). Human wounds and its burden: An updated compendium of estimates. *Advances in Wound Care*, *8*(2), 39–48. DOI: 10.1089/wound.2019.0946 PMID: 30809421

Sen, C. K. (2021). Human wound and its burden: Updated compendium of estimates. *Advances in Wound Care*, *10*(5), 281–292. DOI: 10.1089/wound.2021.0026 PMID: 33733885

Sen, C. K., Gordillo, G. M., Roy, S., Kirsner, R., Lambert, L., Hunt, T. K., Gottrup, F., Gurtner, G. C., & Longaker, M. T. (2009). Human skin wounds: A major and snowballing threat to public health and the economy. *Wound Repair and Regeneration*, *17*(6), 763–771. DOI: 10.1111/j.1524-475X.2009.00543.x PMID: 19903300

Şenol, F. S., Tuğcu-Demiröz, F., Ulutaş, O. K., Erdoğan, İ., & Orhan, I. E. (2025). Phytosomes-Unraveling the Unique Properties of Plant-Derived Nanotechnological Drug Delivery Systems: A Review [Review of *Phytosomes-Unraveling the Unique Properties of Plant-Derived Nanotechnological Drug Delivery Systems: A Review*]. *Current Medicinal Chemistry, 32*(15), 3088. Bentham Science Publishers. DOI: 10.2174/0109298673319759250116104648

Sen, S., & Chakraborty, R. (2015). Toward the integration and advancement of herbal medicine: A focus on traditional Indian medicine. *Botanics : Targets and Therapy*, •••, 33–44. DOI: 10.2147/BTAT.S66308

Sen, S., & Chakraborty, R. (2017). Revival, modernization and integration of Indian traditional herbal medicine in clinical practice: Importance, challenges and future. *Journal of Traditional and Complementary Medicine*, *7*(2), 234–244. DOI: 10.1016/j.jtcme.2016.05.006 PMID: 28417092

Sen, S., Chakraborty, R., & De, B. (2011). Challenges and opportunities in the advancement of herbal medicine: India's position and role in a global context. *Journal of Herbal Medicine, 1*(3–4), 67–75. DOI: 10.1016/j.hermed.2011.11.001

Sen, S., Chakraborty, R., De, B., Ganesh, T., Raghavendra, H. G., & Debnath, S. (2010). Analgesic and anti-inflammatory herbs: A potential source of modern medicine. *International Journal of Pharmaceutical Sciences and Research*, *1*(11), 32.

Sen, S., Chakraborty, R., De, B., & Mazumder, J. (2009). Plants and phytochemicals for peptic ulcer: An overview. *Pharmacognosy Reviews*, *3*(6), 270.

Seo, G. Y., Lim, Y., Koh, D., Huh, J. S., Hyun, C., Kim, Y. M., & Cho, M. (2017). TMF and glycitin act synergistically on keratinocytes and fibroblasts to promote wound healing and anti-scarring activity. *Experimental & Molecular Medicine*, *49*(3), e302. DOI: 10.1038/emm.2016.167 PMID: 28303029

Sethunga, M., Ranaweera, K. K., Gunathilake, K. D., & Munaweera, I. (2022). Extraction methods of essential oils and oleoresins. *Journal of Food Bioprocess Engineering*, *5*(2), 151–167.

Setia, A., Vallamkonda, B., Challa, R. R., Mehata, A. K., Badgujar, P., & Muthu, M. S. (2024). Herbal Theranostics: Controlled, Targeted Delivery and Imaging of Herbal Molecules [Review of *Herbal Theranostics: Controlled, Targeted Delivery and Imaging of Herbal Molecules*]. *Nanotheranostics, 8*(3), 344. Ivyspring International Publisher. DOI: 10.7150/ntno.94987

Setiawati, A., Maharani, B. A., Sari, P. A. P., Widyantara, K. A., Saputra, B. W., Febriansah, R., & Dwiastuti, R. (2024). Deciphering the molecular pathway of an asiaticosiderich fraction of Centella asiatica as an anti-melanogenesis agent. *Journal of Herbmed Pharmacology*, *13*(2), 269–279. DOI: 10.34172/jhp.2024.49332

Sevgi, M., Toklu, A., Vecchio, D., & Hamblin, M. R. (2014). Topical Antimicrobials for Burn Infections – An Update [Review of *Topical Antimicrobials for Burn Infections – An Update*]. *Recent Patents on Anti-Infective Drug Discovery, 8*(3), 161. Bentham Science Publishers. DOI: 10.2174/1574891X08666131112143447

Shady, N. H., Mostafa, N. M., Fayez, S., Abdel-Rahman, I. M., Maher, S. A., Zayed, A., Saber, E. A., Khowdiary, M. M., Elrehany, M. A., Alzubaidi, M. A., Altemani, F. H., Shawky, A. M., & Abdelmohsen, U. R. (2022). Mechanistic wound healing and antioxidant potential of *Moringa oleifera* seeds extract supported by metabolic profiling, in silico network design, molecular docking, and in vivo studies. *Antioxidants*, *11*(9), 1743. DOI: 10.3390/antiox11091743 PMID: 36139817

Shah, A., & Amini-Nik, S. (2017). The role of phytochemicals in the inflammatory phase of wound healing. *International Journal of Molecular Sciences*, *18*(5), 1068. DOI: 10.3390/ijms18051068 PMID: 28509885

Shahidi, F., & Naczk, M. (1995). *Food phenolics: Sources, chemistry, effects, applications*. Technomic Publishing.

Shahrahmani, H., Kariman, N., Jannesari, S., Rafieian-Kopaei, M., Mirzaei, M., Ghalandari, S., Shahrahmani, N., & Mardani, G. (2018). The effect of green tea ointment on episiotomy pain and wound healing in primiparous women: A randomized, double-blind, placebo-controlled clinical trial. *Phytotherapy Research*, *32*(3), 522–530. DOI: 10.1002/ptr.5999 PMID: 29235159

Shalaby, E. A., Aboul-Enein, A., El-Shemy, H., & Nassrallah, A. (2025). Active ingredients of some Egyptian plants species with anticancer activity and its mode of action. *Arab Journal of Biotechnology*, *24*(1), 18–40. https://journals.ekb.eg/article_432494_09dce9501746bb2f9536fa511639e7fc.pdf. DOI: 10.21608/arjb.2025.365025.1000

Shalini, R., Chandrasekar, M. J. N., Nanjan, M. J., Madhunapantula, S. V., Karnik, M., Selvaraj, J., & Ganesh, G. N. K. (2023). 1-Tetracosanol isolated from the leaves of *Eupatorium glandulosum* accelerates wound healing by expressing inflammatory cytokines and matrix metalloproteinase. *Journal of Ethnopharmacology*, *315*, 116654. DOI: 10.1016/j.jep.2023.116654 PMID: 37225028

Shams, F., Moravvej, H., Hosseinzadeh, S., Mostafavi, E., Bayat, H., Kazemi, B., Bandehpour, M., Rostami, E., Rahimpour, A., & Moosavian, H. (2022). Overexpression of VEGF in dermal fibroblast cells accelerates the angiogenesis and wound healing function: In vitro and in vivo studies. *Scientific Reports, 12*(1), 18529. Advance online publication. DOI: 10.1038/s41598-022-23304-8 PMID: 36323953

Shankar, D., & Majumdar, B. (1997). Beyond the biodiversity convention: The challenges facing the biocultural heritage of India's medicinal plants. *Medicinal Plants for Forest Conservation and Health Care, 11*, 87–99.

Sharifi-Rad, J., Sureda, A., Tenore, G. C., Daglia, M., Valussi, M., & Tundis, R.. (2017). Biological activities of essential oils. *Molecules (Basel, Switzerland), 22*(1), 70. DOI: 10.3390/molecules22010070 PMID: 28045446

Sharma, V., Moshage, H., & Haznadar, M. (2021). Polyphenols: Potential role in cutaneous wound healing processes. *Foods, 10*(5), 992. PMID: 34062860

Sharma, Y., Bhardwaj, R., Kaur, A., Singh, G., Kulkarni, S., & Bhatia, A.. (2024). Nano-gel formulation of polyphenolic fraction of tobacco stem for wound healing and its inhibitory efficacies against the receptors of chronic wound development. *Journal of Biomolecular Structure & Dynamics*. Advance online publication. PMID: 38433407

Shedoeva, A., Leavesley, D., Upton, Z., & Fan, C. (2019). Wound healing and the use of medicinal plants. *Evidence-Based Complementary and Alternative Medicine : eCAM, 2684108*, 1–30. Advance online publication. DOI: 10.1155/2019/2684108 PMID: 31662773

Sheppard, A. J., Barfield, A. M., Barton, S., & Dong, Y. (2022). Reactive oxygen species in bone regeneration. *Frontiers in Bioengineering and Biotechnology, 10*, 836764. DOI: 10.3389/fbioe.2022.836764 PMID: 35198545

Shevelev, A. B., La Porta, N., Isakova, E. P., Martens, S., Biryukova, Y. K., Belous, A. S., Sivokhin, D. A., Trubnikova, E. V., Zylkova, M. V., Belyakova, A. V., Smirnova, M. S., & Deryabina, Y. I. (2020). In vivo antimicrobial and wound-healing activity of resveratrol, dihydroquercetin, and dihydromyricetin against Staphylococcus aureus, Pseudomonas aeruginosa, and Candida albicans. *Pathogens (Basel, Switzerland), 9*(4), 296. DOI: 10.3390/pathogens9040296 PMID: 32316572

Shi, Z., Yao, C., Shui, Y., Li, S., & Yan, H. (2023). Research progress on the mechanism of angiogenesis in wound repair and regeneration [Review of *Research progress on the mechanism of angiogenesis in wound repair and regeneration*]. *Frontiers in Physiology, 14*, 1284981. Frontiers Media. DOI: 10.3389/fphys.2023.1284981

Shi, Y., Shu, B., Yang, R., Xu, Y., Xing, B., & Liu, J.. (2015). Wnt and Notch signaling pathway involved in wound healing by targeting separately c-Myc and Hes1. *Stem Cell Research & Therapy*, *6*, 120. Advance online publication. DOI: 10.1186/s13287-015-0103-4 PMID: 26076648

Shoham, Y., Snyder, R. J., Levy, Y. K., Zarbiv, K. D., Klinger, E., & Kramer, M.. (2024). Bromelain-based enzymatic debridement. *EClinicalMedicine*, *75*, 102655. PMID: 39763591

Shubina, V., & Shatalin, Y. V. (2012). Skin regeneration after chemical burn under the effect of taxifolin-based preparations. *Bulletin of Experimental Biology and Medicine*, *154*(2), 152–157. DOI: 10.1007/s10517-012-1897-z PMID: 23330113

Shukla, L., Yuan, Y., Shayan, R., Greening, D. W., & Karnezis, T. (2020). Fat Therapeutics: The Clinical Capacity of Adipose-Derived Stem Cells and Exosomes for Human Disease and Tissue Regeneration [Review of *Fat Therapeutics: The Clinical Capacity of Adipose-Derived Stem Cells and Exosomes for Human Disease and Tissue Regeneration*]. *Frontiers in Pharmacology, 11*. Frontiers Media. DOI: 10.3389/fphar.2020.00158

Shukla, A., Rasik, A. M., & Dhawan, B. N. (1999). Asiaticoside-induced elevation of antioxidant levels in healing wounds. *Phytotherapy Research*, *13*(1), 50–54. DOI: 10.1002/(SICI)1099-1573(199902)13:1<50::AID-PTR368>3.0.CO;2-V PMID: 10189951

Shukla, R., Kashaw, S. K., Jain, A. P., & Lodhi, S. (2016). Fabrication of apigenin loaded gellan gum–chitosan hydrogels (GGCH-HGs) for effective diabetic wound healing. *International Journal of Biological Macromolecules*, *91*, 1110–1119. DOI: 10.1016/j.ijbiomac.2016.06.075 PMID: 27344952

Shukla, S. D., Bhatnagar, M., & Khurana, S. (2012). Critical Evaluation of Ayurvedic Plants for Stimulating Intrinsic Antioxidant Response. *Frontiers in Neuroscience*, *6*. Advance online publication. DOI: 10.3389/fnins.2012.00112 PMID: 22855669

Sideek, S. A., El-Nassan, H. B., Fares, A., ElMeshad, A. N., & Elkasabgy, N. A. (2022). Different Curcumin-Loaded Delivery Systems for Wound Healing Applications: A Comprehensive Review. *Pharmaceutics*, *15*(1), 38. DOI: 10.3390/pharmaceutics15010038 PMID: 36678665

Silva, L. P., Santos, E. C., Borges, B. A., Veloso, M. P., Chagas-Paula, D. A., Gonçalves, R. V., & Novaes, R. D. (2021). Tagitinin F has anti-inflammatory, anti-nociceptive and anti-matrix metalloproteinase properties: An in silico, in vitro and in vivo study. *Pharmacological Research*, *164*, 105303. DOI: 10.1016/j.phrs.2020.105303 PMID: 33212290

Singer, A. J. (2022). Healing mechanisms in cutaneous wounds: Tipping the balance. *Tissue Engineering. Part B, Reviews, 28*(6), 1151–1167. DOI: 10.1089/ten.teb.2021.0114 PMID: 34915757

Singh, B., Yadav, A., & Sharma, N. (2020). The efficacy of aloe vera in wound healing and its application in the management of burns. *Burns and Trauma, 8*, 47. DOI: 10.1186/s41038-020-00222-x

Siriwattanasatorn, M., Itharat, A., Thongdeeying, P., & Ooraikul, B. (2020). In Vitro Wound Healing Activities of Three Most Commonly Used Thai Medicinal Plants and Their Three Markers. *Evidence-Based Complementary and Alternative Medicine : eCAM, 2020*(1), e6795383. DOI: 10.1155/2020/6795383 PMID: 32714411

Sivamani, R. K., Brian, R., Wehrli, L. N., & Maverakis, E. (2012). Phytochemicals and Naturally Derived Substances for Wound Healing [Review of *Phytochemicals and Naturally Derived Substances for Wound Healing*]. *Advances in Wound Care, 1*(5), 213. Mary Ann Liebert, Inc. DOI: 10.1089/wound.2011.0330

Smeriglio, A., Barreca, D., Bellocco, E., & Trombetta, D. (2017). Proanthocyanidins and hydrolysable tannins: Occurrence, dietary intake and pharmacological effects. *British Journal of Pharmacology, 174*(11), 1244–1262. DOI: 10.1111/bph.13630 PMID: 27646690

Smith, J., & Kaur, J. (2022). Advanced dressings infused with bioactive compounds for enhanced wound healing. *Wound Repair and Regeneration, 30*(2), 174–185. DOI: 10.1111/wrr.12952

Somwong, K., Lertpatipanpong, P., Nimlamool, W., Panya, A., Tragoolpua, Y., Yongsawas, R., Gritsanapan, W., Pandith, H., & Baek, S. J. (2022). Effect of *Holoptelea integrifolia* (Roxb.) Planch. N-hexane extract and its bioactive compounds on wound healing and anti-inflammatory activity. *Molecules (Basel, Switzerland), 27*(23), 8540. DOI: 10.3390/molecules27238540 PMID: 36500633

Son, H. Y., Jun, H., Kim, K. R., Hong, C. A., & Nam, Y. S. (2018). Tannin-mediated assembly of gold–titanium oxide hybrid nanoparticles for plasmonic photochemical applications. *Journal of Industrial and Engineering Chemistry, 63*, 420–425. DOI: 10.1016/j.jiec.2018.03.002

Soni, H., Khare, S., & Singh, G. (2020). Silymarin for diabetic foot ulcer healing: Mechanisms and therapeutic application. *Journal of Wound Care, 29*(5), 270–278. DOI: 10.12968/jowc.2020.29.5.270 PMID: 32421483

Sorg, H., & Sorg, C. G. G. (2023). Skin wound healing: Of players, patterns, and paradigms. *European Surgical Research, 64*(2), 141–157. DOI: 10.1159/000528271 PMID: 36417847

Sorrentino, S., Studt, J. D., Medalia, O., & Tanuj Sapra, K. (2015). Roll, adhere, spread and contract: Structural mechanics of platelet function. *European Journal of Cell Biology, 94*(3–4), 129–138. DOI: 10.1016/j.ejcb.2015.01.001 PMID: 25655000

Souiy, Z. (2023). Essential oil extraction process. In *Essential oils—Recent advances*. IntechOpen.

Spinetti, G., Kraenkel, N., Emanueli, C., & Madeddu, P. (2008). Diabetes and vessel wall remodelling: from mechanistic insights to regenerative therapies [Review of *Diabetes and vessel wall remodelling: from mechanistic insights to regenerative therapies*]. *Cardiovascular Research, 78*(2), 265. Oxford University Press. DOI: 10.1093/cvr/cvn039

Sriwatanametanon, S.. (2015). Nanoparticle-assisted pharmacotherapy for cancer. *Asian Pacific Journal of Cancer Prevention : APJCP, 16*(4), 1477–1485.

Stacey, M. C., Wysocki, S. J., Wallace, H. J., Yeoh, S., & Davies, M. J. (2002). Matrix metalloproteinases in wound healing. *ANZ Journal of Surgery, 72*(S1), A118.

Stephane, F. F. (2020). Terpenoids as important bioactives. In *Essential oils: Bioactive compounds, new perspectives and applications*. IntechOpen. DOI: 10.5772/intechopen.91426

Stevens Barrón, J. C., Chapa González, C., Álvarez Parrilla, E., & De la Rosa, L. A. (2023). Nanoparticle-mediated delivery of flavonoids: Impact on proinflammatory cytokine production: A systematic review. *Biomolecules, 13*(7), 1158. Advance online publication. DOI: 10.3390/biom13071158 PMID: 37509193

Strbo, N., Yin, N., & Stojadinovic, O. (2014). Innate and adaptive immune responses in wound epithelialization. *Advances in Wound Care, 3*(7), 492–501. DOI: 10.1089/wound.2012.0435 PMID: 25032069

Strodtbeck, F. (2001). Physiology of wound healing. *Newborn and Infant Nursing Reviews; NAINR, 1*(1), 43–52. DOI: 10.1053/nbin.2001.23176

Subbarayappa, B. V. (1997). Siddha medicine: An overview. *Lancet, 350*(9094), 1841–1844. DOI: 10.1016/S0140-6736(97)04223-2 PMID: 9428267

Subbarayappa, B. V. (2001). The roots of ancient medicine: An historical outline. *Journal of Biosciences, 26*(2), 135–143. DOI: 10.1007/BF02703637 PMID: 11426049

Suganya, T., Packiavathy, I. A. S. V., Aseervatham, G. S. B., Carmona, A., Rashmi, V., Mariappan, S., Devi, N. R., & Ananth, D. A. (2022). Tackling Multiple-Drug-Resistant Bacteria With Conventional and Complex Phytochemicals [Review of *Tackling Multiple-Drug-Resistant Bacteria With Conventional and Complex Phytochemicals*]. *Frontiers in Cellular and Infection Microbiology, 12*. Frontiers Media. DOI: 10.3389/fcimb.2022.883839

Suguna, L., Singh, S., Sivakumar, P., Sampath, P., & Chandrakasan, G. (2002). Influence of Terminalia chebula on dermal wound healing in rats. *Phytotherapy Research, 16*(3), 227–231. DOI: 10.1002/ptr.827 PMID: 12164266

Sun, B. K., Siprashvili, Z., & Khavari, P. A. (2014). Advances in skin grafting and treatment of cutaneous wounds. *Science, 346*(6212), 941–945. DOI: 10.1126/science.1253836 PMID: 25414301

Sundoro, A., Nadia, K., Nur, A., Sudjatmiko, G., & Tedjo, A. (2012). Comparison of physical–chemical characteristic and antibacterial effect between Manuka honey and local honey. *J Plastik Rekonstruksi., 2012*(1), 3.

Sun, L., Gidley, M. J., & Warren, F. J. (2018). Tea polyphenols enhance binding of porcine pancreatic α-amylase with starch granules but reduce catalytic activity. *Food Chemistry, 258*, 164–173. DOI: 10.1016/j.foodchem.2018.03.017 PMID: 29655719

Sun, L., Warren, F. J., Netzel, G., & Gidley, M. J. (2016). 3 or 3′-galloyl substitution plays an important role in association of catechins and theaflavins with porcine pancreatic α-amylase: The kinetics of inhibition of α-amylase by tea polyphenols. *Journal of Functional Foods, 26*, 144–156. DOI: 10.1016/j.jff.2016.07.012

Sun, S., Yu, Y., Jo, Y., Han, J. H., Xue, Y., Cho, M., Bae, S.-J., Ryu, D., Park, M.-J., Ha, K., & Zhuang, S. (2025). Impact of extraction techniques on phytochemical composition and bioactivity of natural product mixtures. *Frontiers in Pharmacology, 16*, 1615338. DOI: 10.3389/fphar.2025.1615338 PMID: 40808686

Suntar, I., Çetinkaya, S., Panieri, E., Saha, S., Buttari, B., Profumo, E., & Saso, L. (2021). Regulatory Role of Nrf2 Signaling Pathway in Wound Healing Process. *Molecules (Basel, Switzerland), 26*(9), 2424. DOI: 10.3390/molecules26092424 PMID: 33919399

Su, X., Liu, X., Wang, S., Li, B., Pan, T., Liu, D., Wang, F., Diao, Y., & Li, K. (2017). Wound-healing promoting effect of total tannins from Entada phaseoloides (L.) Merr. in rats. *Burns, 43*(4), 830–838. DOI: 10.1016/j.burns.2016.10.010 PMID: 28040363

Su, Y., & Richmond, A. (2015). Chemokine regulation of neutrophil infiltration of skin wounds. *Advances in Wound Care*, *4*(11), 631–640. DOI: 10.1089/wound.2014.0559 PMID: 26543677

Swann, G. (2010). The skin is the body's largest organ. *Journal of Visual Communication in Medicine*, *33*(4), 148–149. DOI: 10.3109/17453054.2010.525439 PMID: 21087182

Sysak, S., Czarczynska-Goslinska, B., Szyk, P., Koczorowski, T., Mlynarczyk, D. T., Szczolko, W., Lesyk, R., & Goslinski, T. (2023). Metal nanoparticle-flavonoid connections: Synthesis, physicochemical and biological properties, as well as potential applications in medicine. *Nanomaterials (Basel, Switzerland)*, *13*(9), 1531. Advance online publication. DOI: 10.3390/nano13091531 PMID: 37177076

Szpaderska, A. M., Egozi, E. I., Gamelli, R. L., & DiPietro, L. A. (2003). The effect of thrombocytopenia on dermal wound healing. *The Journal of Investigative Dermatology*, *120*(6), 1130–1137. DOI: 10.1016/S0022-202X(18)32320-0 PMID: 12787144

Tabassum, A., & Raj, M. (2020). Plant extracts as potential antibacterial agents against pathogens causing diabetic foot ulcers. *Journal of Ethnopharmacology*, *262*, 113145. DOI: 10.1016/j.jep.2020.113145

Takeuchi, O., & Akira, S. (2010). Pattern recognition receptors and inflammation. *Cell*, *140*(6), 805–820. DOI: 10.1016/j.cell.2010.01.022 PMID: 20303872

Tan, C., Aziz, M., & Wang, P. (2021). The vitals of NETs. *Journal of Leukocyte Biology*, *110*(4), 797–808. DOI: 10.1002/JLB.3RU0620-375R PMID: 33378572

Tanveer, M., Wagner, C., ul Haq, M. I., Ribeiro, N. C., Rathinasabapathy, T., Butt, M. S., Shehzad, A., & Komarnytsky, S. (2020). Spicing up gastrointestinal health with dietary essential oils. *Phytochemistry Reviews*, *19*(2), 243–263. DOI: 10.1007/s11101-020-09664-x

Tan, Y., & Chang, S. K. (2017). Digestive enzyme inhibition activity of the phenolic substances in selected fruits, vegetables and tea as compared to black legumes. *Journal of Functional Foods*, *38*, 644–655. DOI: 10.1016/j.jff.2017.04.005

Tatarusanu, S. M., Lupascu, F. G., Profire, B. S., Szilagyi, A., Gardikiotis, I., Iacob, A. T., Caluian, I., Herciu, L., Giscă, T. C., Baican, M. C., Crivoi, F., & Profire, L. (2023). Modern approaches in wounds management. *Polymers*, *15*(17), 3648. DOI: 10.3390/polym15173648 PMID: 37688274

Tecilazich, F., Dinh, T. L., & Veves, A. (2013). Emerging drugs for the treatment of diabetic ulcers [Review of *Emerging drugs for the treatment of diabetic ulcers*]. *Expert Opinion on Emerging Drugs, 18*(2), 207. Taylor & Francis. DOI: 10.1517/14728214.2013.802305

Teixeira, M. A., Paiva, M. C., Amorim, M. T., & Felgueiras, H. P. (2020). Electrospun nanocomposites containing cellulose derivatives modified with specialized biomolecules for enhanced wound healing. *Nanomaterials (Basel, Switzerland), 10*(3), 557. DOI: 10.3390/nano10030557 PMID: 32204521

Teller, P., & White, T. K. (2009). The physiology of wound healing: Injury through maturation. *The Surgical Clinics of North America, 89*(3), 599–610. DOI: 10.1016/j.suc.2009.03.006 PMID: 19465199

Tenci, M., Rossi, S., Aguzzi, C., Carazo, E., Sandri, G., Bonferoni, M., Grisoli, P., Viseras, C., Caramella, C. M., & Ferrari, F. (2017). Carvacrol/clay hybrids loaded into in situ gelling films. *International Journal of Pharmaceutics, 531*(2), 676–688. DOI: 10.1016/j.ijpharm.2017.06.024 PMID: 28619454

Teng, H., & Chen, L. (2019). Polyphenols and bioavailability: An update. *Critical Reviews in Food Science and Nutrition, 59*(13), 2040–2051. DOI: 10.1080/10408398.2018.1437023 PMID: 29405736

Teodoro, J. S., Nunes, S., Rolo, A. P., Reis, F., & Palmeira, C. M. (2019). Therapeutic Options Targeting Oxidative Stress, Mitochondrial Dysfunction and Inflammation to Hinder the Progression of Vascular Complications of Diabetes. *Frontiers in Physiology, 9*, 1857. DOI: 10.3389/fphys.2018.01857 PMID: 30705633

Tetali, S. D. (2019). Terpenes and isoprenoids: A wealth of compounds for global use. *Planta, 249*(1), 1–8. DOI: 10.1007/s00425-018-3056-x PMID: 30467631

Teymoorian, S. K., Nouri, H., & Moghimi, H. (2024). In-vivo and in-vitro wound healing and tissue repair effect of *Trametes versicolor* polysaccharide extract. *Scientific Reports, 14*(1), 3796. DOI: 10.1038/s41598-024-54565-0 PMID: 38360911

Thandavarayan, R. A., Garikipati, V. N. S., Joladarashi, D., Babu, S. S., Jeyabal, P., Verma, S. K., Mackie, A. R., Khan, M., Arumugam, S., Watanabe, K., Kishore, R., & Krishnamurthy, P. (2015). Sirtuin-6 deficiency exacerbates diabetes-induced impairment of wound healing. *Experimental Dermatology, 24*(10), 773–778. DOI: 10.1111/exd.12762 PMID: 26010430

Thangapazham, R. L., Sharad, S., & Maheshwari, R. K. (2016). Phytochemicals in wound healing. *Advances in Wound Care, 5*(5), 230–241. DOI: 10.1089/wound.2013.0505 PMID: 27134766

Thawabteh, A. M., Ghanem, A. W., AbuMadi, S., Thaher, D., Jaghama, W., Karaman, R., Scrano, L., & Bufo, S. A. (2024). Antibacterial Activity and Antifungal Activity of Monomeric Alkaloids. *Toxins*, *16*(11), 489. DOI: 10.3390/toxins16110489 PMID: 39591244

Thomas, A., Shinde, S., Wavhale, R., Jadhav, P., Tambe, S., Lokhande, K., & Chitlange, S. (2024). In-silico screening of phytomolecules against multiple targets for wound management. *In Silico Pharmacology*, *12*(1), 19. DOI: 10.1007/s40203-024-00194-4 PMID: 38550524

Thomas, L., Zakir, F., Mirza, M. A., Anwer, M. K., Ahmad, F. J., & Iqbal, Z. (2017). Development of curcumin loaded chitosan polymer based nanoemulsion gel: In vitro, ex vivo evaluation and in vivo wound healing studies. *International Journal of Biological Macromolecules*, *101*, 569–579. DOI: 10.1016/j.ijbiomac.2017.03.066 PMID: 28322948

Tian, J.-L., Si, X., Wang, Y.-H., Gong, E.-S., Xie, X., Zhang, Y., Li, B., & Shu, C. (2021). Bioactive flavonoids from Rubus corchorifolius inhibit α-glucosidase and α-amylase to improve postprandial hyperglycemia. *Food Chemistry*, *341*, 128149. DOI: 10.1016/j.foodchem.2020.128149 PMID: 33039745

Tie, Y., Liu, H., Zhang, T., Meng, T., & Liang, Q. (2025). Natural products alleviate viral pneumonia by modulating inflammatory and oxidative-stress pathways [Review of *Natural products alleviate viral pneumonia by modulating inflammatory and oxidative-stress pathways*]. *Frontiers in Pharmacology, 16*. Frontiers Media. DOI: 10.3389/fphar.2025.1657829

Tie, L., An, Y., Han, J., Xiao, Y., Xiaokaiti, Y., Fan, S., Liu, S., Chen, A. F., & Li, X. (2013). Genistein accelerates refractory wound healing by suppressing superoxide and FoxO1/iNOS pathway in type 1 diabetes. *The Journal of Nutritional Biochemistry*, *24*(1), 88–96. DOI: 10.1016/j.jnutbio.2012.02.011 PMID: 22819564

Tirant, M., Tirant, H., & Wollina, U. (2024). Herbal Bioactive Compounds for Skin Infections and Inflammatory Conditions. *Open Access Macedonian Journal of Medical Sciences*, *12*, 1–44. DOI: 10.3889/oamjms.2024.11888

Tirona, R. G., & Bailey, D. G. (2006). Herbal product–drug interactions mediated by induction. *British Journal of Clinical Pharmacology*, *61*(6), 677–681. DOI: 10.1111/j.1365-2125.2006.02684.x PMID: 16722828

Tomasek, J. J., Gabbiani, G., Hinz, B., Chaponnier, C., & Brown, R. A. (2002). Myofibroblasts and mechano-regulation of connective tissue remodelling. *Nature Reviews. Molecular Cell Biology*, *3*(5), 349–363. DOI: 10.1038/nrm809 PMID: 11988769

Tomasek, J. J., Haaksma, C. J., Schwartz, R. J., & Howard, E. W. (2013). Whole animal knockout of smooth muscle alpha-actin does not alter excisional wound healing or the fibroblast-to-myofibroblast transition. *Wound Repair and Regeneration, 21*(1), 166–176. DOI: 10.1111/wrr.12001 PMID: 23253249

Tome-Garcia, J., Li, D., Ghazaryan, S., Shu, L., & Wu, L. (2014). ERBB2 increases metastatic potentials specifically in androgen-insensitive prostate cancer cells. *PLoS One*, *9*(6), e99525. Advance online publication. DOI: 10.1371/journal.pone.0099525 PMID: 24937171

Trinh, X.-T., & Nguyen-Van Long, L. T. V. A. (2022). A comprehensive review of natural compounds for wound healing: Targeting bioactivity perspective. *International Journal of Molecular Sciences*, *23*(17), 9573. DOI: 10.3390/ijms23179573 PMID: 36076971

Tripathi, S., Maier, K. G., Bruch, D., & Kittur, D. S. (2007). Effect of 6-Gingerol on Pro- Inflammatory Cytokine Production and Costim-ulatory Molecule Expression in Murine Peritoneal Macrophages. *The Journal of Surgical Research, 138*(2), 209–213. DOI: 10.1016/j.jss.2006.07.051 PMID: 17291534

Trottier, L.. (2018). Regulatory challenges for nanomedicine: Perspectives and future directions. *Nanomedicine (London), 14*(8), 1129–1138.

Tu, Y. (2011). The discovery of artemisinin (qinghaosu) and gifts from Chinese medicine. *Nature Medicine*, *17*(10), 1217–1220. DOI: 10.1038/nm.2471 PMID: 21989013

Tu, Y., Zheng, W., Ding, Z., Xiang, J., Yang, Q., Liu, Y., Cao, J., Shen, Y., Tang, Z., Lin, S., Fan, L., Xu, Y., & Chen, B. (2025). Exosome-loaded tannic acid-thioctic acid hydrogel enhances wound healing in coagulation disorders. *Materials Today. Bio*, *31*, 101496. DOI: 10.1016/j.mtbio.2025.101496 PMID: 39990738

U.S. FDA. (1994). *Dietary Supplement Health and Education Act (DSHEA).*

Ukaegbu, K., Allen, E. P., & Svoboda, K. K. H. (2025). Reactive Oxygen Species and Antioxidants in Wound Healing: Mechanisms and Therapeutic Potential [Review of *Reactive Oxygen Species and Antioxidants in Wound Healing: Mechanisms and Therapeutic Potential*]. *International Wound Journal, 22*(5). Wiley. DOI: 10.1111/iwj.70330

Unalan, I., & Boccaccini, A. R. (2021). Essential oils in biomedical applications. *Current Opinion in Biomedical Engineering*, *17*, 100261. DOI: 10.1016/j.cobme.2021.100261

Upadhyay, R., & Bano, S. (2023). A review on terpenoid synthesized nanoparticles and their antimicrobial activity. *Oriental Journal of Chemistry*, *39*(2), 1–10. DOI: 10.13005/ojc/390226

Uzun, M., Guvenalp, Z., Kazaz, C., & Demirezer, L. O. (2020). Matrix metalloproteinase inhibitor and sunscreen effective compounds from *Rumex crispus* L.: Isolation, identification, bioactivity and molecular docking study. *Phytochemical Analysis*, *31*(6), 818–834. DOI: 10.1002/pca.2948 PMID: 32488908

Vandamme, L., Heyneman, A., Hoeksema, H., Verbelen, J., & Monstrey, S. (2013). Honey in modern wound care: A systematic review. *Burns*, *39*(8), 1514–1525. DOI: 10.1016/j.burns.2013.06.014 PMID: 23896128

Varzakas, T. (2016). Food additives and contaminants. In Varzakas, T., & Tzia, C. (Eds.), *Handbook of food processing and preservation* (pp. 319–388).

Veith, A. P., Henderson, K., Spencer, A., Sligar, A. D., & Baker, A. B. (2020). Therapeutic strategies for enhancing angiogenesis in wound healing. *Advanced Drug Delivery Reviews*, *146*, 97–125. DOI: 10.1016/j.addr.2018.09.010 PMID: 30267742

Velnar, T., Bailey, T., & Smrkolj, V. (2009). The wound healing process: An overview of the cellular and molecular mechanisms. *The Journal of International Medical Research*, *37*(5), 1528–1542. DOI: 10.1177/147323000903700531 PMID: 19930861

Vestita, M., Tedeschi, P., & Bonamonte, D. (2022). Anatomy and physiology of the skin. In Maruccia, M., & Giudice, G. (Eds.), *Textbook of plastic and reconstructive surgery*. Springer., DOI: 10.1007/978-3-030-82335-1_1

Vestweber, D. (2015). How leukocytes cross the vascular endothelium. *Nature Reviews. Immunology*, *15*(11), 692–704. DOI: 10.1038/nri3908 PMID: 26471775

Vignesh, K., Thangavelu, A., & Suresh, S. (2020). Nanofibers: Properties and applications. *Nanotechnology*, *31*(30), 305603.

Vijaya, V. T., Srinivasan, D., & Sengottuvelu, S. (2012). Wound healing potential of *Melia azedarach* L. leaves in alloxan induced diabetic rats. *Global Journal of Research on Medicinal Plants and Indigenous Medicine*, *1*(7), 265.

Vitale, S., Colanero, S., Placidi, M., Di Emidio, G., Tatone, C., Amicarelli, F., & D'Alessandro, A. M. (2022). Phytochemistry and biological activity of medicinal plants in wound healing: An overview of current research. *Molecules (Basel, Switzerland)*, *27*(11), 3566. DOI: 10.3390/molecules27113566 PMID: 35684503

Volpe, C. M. O., Villar-Delfino, P. H., dos Anjos, P. M. F., & Nogueira-Machado, J. A. (2018). Cellular death, reactive oxygen species (ROS) and diabetic complications. *Cell Death & Disease*, *9*(2), 119. Advance online publication. DOI: 10.1038/s41419-017-0135-z PMID: 29371661

Vora, L. K., Gholap, A. D., Hatvate, N. T., Naren, P., Khan, S., Chavda, V. P., Balar, P. C., Gandhi, J., & Khatri, D. K. (2024). Essential oils in clinical aromatherapy. *Journal of Ethnopharmacology*, *318*, 118180. DOI: 10.1016/j.jep.2024.118180 PMID: 38614262

Wachtel-Galor, S., & Benzie, I. F. F. (2012). *Herbal medicine: an introduction to its history, usage, regulation, current trends, and research needs.*

Walach, H., Huber, B., Bassler, D., Naumann, J., Bismarck, D., Schneider, M., Müller, E., Walach, H., Mangelsdorf, I., & Mutter, J. (2017). Journal Club. *Complementary Medicine Research*, *24*(3), 135–138. DOI: 10.1159/000477783 PMID: 28641284

Wang, C., Lou, Y., Tong, M., Zhang, L., Zhang, Z., Feng, Y., Shi, L., Xu, H., & Mao, C. (2017). Asperosaponin VI promotes angiogenesis and accelerates wound healing in rats via up-regulating HIF-1α/VEGF signaling. *Acta Pharmacologica Sinica*, *39*(3), 393–404. DOI: 10.1038/aps.2017.161 PMID: 29219948

Wang, G., Yang, F., Zhou, W., Xiao, N., Luo, M., & Tang, Z. (2022). The initiation of oxidative stress and therapeutic strategies in wound healing. *Biomedicine and Pharmacotherapy*, *157*, 114004. DOI: 10.1016/j.biopha.2022.114004 PMID: 36375308

Wang, H., Chen, Y., Wang, L., Liu, Q., Yang, S., & Wang, C. (2023). Advancing herbal medicine through robust quality control practices. *Frontiers in Pharmacology*, *14*, 1265178. DOI: 10.3389/fphar.2023.1265178 PMID: 37818188

Wang, H., Du, Y.-J., & Song, H.-C. (2010). α-Glucosidase and α-amylase inhibitory activities of guava leaves. *Food Chemistry*, *123*(1), 6–13. DOI: 10.1016/j.foodchem.2010.03.088

Wang, L., Xia, K., Han, L., Zhang, M., Fan, J., Liu, S., Liao, A., Wang, W., & Guo, J. (2022). Local Administration of Ginkgolide B Using a Hyaluronan-Based Hydrogel Improves Wound Healing in Diabetic Mice. *Frontiers in Bioengineering and Biotechnology*, *10*, 898231. Advance online publication. DOI: 10.3389/fbioe.2022.898231 PMID: 35694224

Wang, M., Huang, H., Wang, L., Yin, L., Yang, H., Chen, C., Zheng, Q., & He, S. (2022). Tannic acid attenuates intestinal oxidative damage by improving antioxidant capacity and intestinal barrier in weaned piglets and IPEC-J2 cells. *Frontiers in Nutrition*, *9*, 1012207. DOI: 10.3389/fnut.2022.1012207 PMID: 36407512

Wang, P. H., Huang, B. S., Horng, H. C., Yeh, C. C., & Chen, Y. J. (2018). Wound healing. *Journal of the Chinese Medical Association, 81*(2), 94–101. DOI: 10.1016/j.jcma.2017.11.002 PMID: 29169897

Wang, Q., Liu, C., An, J., Liu, C., Wang, Y., & Cai, Y. (2025). Mechanisms of microbial infection and wound healing in diabetic foot ulcer: Pathogenicity in the inflammatory-proliferative phase, chronicity, and treatment strategies. *Frontiers in Endocrinology, 16*, 1657928. DOI: 10.3389/fendo.2025.1657928 PMID: 41103650

Wang, T., Zheng, Y., Zhang, J., & Wu, Z. (2024). Targeting ferroptosis promotes diabetic wound healing via Nrf2 activation. *Heliyon, 10*(19), e37477. DOI: 10.1016/j.heliyon.2024.e37477 PMID: 39421383

Wang, W., & Zhang, T. (2017). Integration of traditional Chinese medicine and Western medicine in the era of precision medicine. *Journal of Integrative Medicine, 15*(1), 1–7. https://doi.org/https://doi.org/10.1016/S2095-4964(17)60314-5. DOI: 10.1016/S2095-4964(17)60314-5 PMID: 28088253

Wang, X., Yang, J., Li, H., Shi, S., & Peng, X. (2022). Mechanistic study and synergistic effect on inhibition of α-amylase by structurally similar flavonoids. *Journal of Molecular Liquids, 360*, 119485. DOI: 10.1016/j.molliq.2022.119485

Wang, Y., Zhang, Z., & Li, Y. (2024). Glycyrrhizic acid-based hydrogel for improving wound healing in diabetic ulcers. *Journal of Materials Chemistry. B, 12*(7), 2125–2132. DOI: 10.1039/D4TB01572J PMID: 38230414

Wei, Q., Yang, Q., Wang, Q., Sun, C., Zhu, Y., Niu, Y., Yu, J., & Xu, X. (2018). Formulation, Characterization, and Pharmacokinetic Studies of 6-Gingerol-Loaded Nanostructured Lipid Carriers. *AAPS PharmSciTech, 19*(8), 3661–3669. DOI: 10.1208/s12249-018-1165-2 PMID: 30324361

Wei, Z., Zuo, Y., Wu, E., Huang, L., Qian, Y., Wang, J., & Chen, Z. (2024). Highly biocompatible, antioxidant and antibacterial gelatin methacrylate/alginate-tannin hydrogels for wound healing. *International Journal of Biological Macromolecules, 279*, 135417. DOI: 10.1016/j.ijbiomac.2024.135417 PMID: 39265900

Wendlocha, D., Krzykawski, K., Mielczarek-Palacz, A., & Kubina, R. (2023). Selected flavonols in breast and gynecological cancer: A systematic review. *Nutrients, 15*(13), 2938. Advance online publication. DOI: 10.3390/nu15132938 PMID: 37447264

Werdin, F., Tennenhaus, M., Schaller, H. E., & Rennekampff, H. O. (2009). Evidence-based management strategies for treatment of chronic wounds. *Eplasty, 9*, e19. PMID: 19578487

Whitney, J. D. (2005). Overview: Acute and chronic wounds. *Nursing Clinics*, *40*(2), 191–205. PMID: 15924889

Whittam, A. J., Maan, Z. N., Duscher, D., Wong, V. W., Barrera, J. A., Januszyk, M., & Gurtner, G. C. (2016). Challenges and opportunities in drug delivery for wound healing. *Advances in Wound Care*, *5*(2), 79–88. DOI: 10.1089/wound.2014.0600 PMID: 26862465

Wilgus, T. A., Roy, S., & McDaniel, J. C. (2013). Neutrophils and wound repair: Positive actions and negative reactions. *Advances in Wound Care*, *2*(7), 379–388. DOI: 10.1089/wound.2012.0383 PMID: 24527354

Wilkinson, H. N., & Hardman, M. J. (2020). Wound healing: Cellular mechanisms and pathological outcomes. *Open Biology*, *10*(9), 200223. Advance online publication. DOI: 10.1098/rsob.200223 PMID: 32993416

Williamson, G. (2013). Possible effects of dietary polyphenols on sugar absorption and digestion. *Molecular Nutrition & Food Research*, *57*(1), 48–57. DOI: 10.1002/mnfr.201200511 PMID: 23180627

Wong, C. M., & Ling, J. J. (2014). In vitro study of wound healing potential in black pepper (*Piper nigrum* L.). *Pharmaceutical and Biosciences Journal*, 5–9.

Woo, Y. C., Park, S. S., Subieta, A. R., & Brennan, T. J. (2004). Changes in tissue pH and temperature after incision indicate acidosis may contribute to postoperative pain. *Anesthesiology*, *101*(2), 468–475. DOI: 10.1097/00000542-200408000-00029 PMID: 15277931

Wu, H., Estevez-Santiago, R., & Angelino, D.. (2021). The influence of gut microbiota on polyphenol metabolism and the resulting bioactivity. *Frontiers in Nutrition*, *8*, 36.

Wunnoo, S., Sermwittayawong, D., Praparatana, R., Voravuthikunchai, S. P., & Jakkawanpitak, C. (2024). Quercus infectoria Gall Ethanolic Extract Accelerates Wound Healing through Attenuating Inflammation and Oxidative Injuries in Skin Fibroblasts. *Antioxidants*, *13*(9), 1094. DOI: 10.3390/antiox13091094 PMID: 39334753

Wu, W.-S., Chen, L.-R., & Chen, K.-H. (2025). Platelet-rich plasma (PRP): Molecular mechanisms, actions and clinical applications in human body. *International Journal of Molecular Sciences*, *26*(21), 10804. Advance online publication. DOI: 10.3390/ijms262110804 PMID: 41226837

Wu, X., Ding, H., Hu, X., Pan, J., Liao, Y., Gong, D., & Zhang, G. (2018). Exploring inhibitory mechanism of gallocatechin gallate on α-amylase and α-glucosidase relevant to postprandial hyperglycemia. *Journal of Functional Foods*, *48*, 200–209. DOI: 10.1016/j.jff.2018.07.022

Wu, X., Hu, M., Hu, X., Ding, H., Gong, D., & Zhang, G. (2019). Inhibitory mechanism of epicatechin gallate on α-amylase and α-glucosidase and its combinational effect with acarbose or epigallocatechin gallate. *Journal of Molecular Liquids*, *290*, 111202. DOI: 10.1016/j.molliq.2019.111202

Xiong, Y., Knoedler, S., Alfertshofer, M., Kim, B. S., Jiang, D., Liu, G., Rinkevich, Y., & Mi, B. (2025). Mechanisms and therapeutic opportunities in metabolic aberrations of diabetic wounds: A narrative review. *Cell Death & Disease*, *16*(1), 341. DOI: 10.1038/s41419-025-07583-3 PMID: 40280905

Xu, B., Bai, L., Chen, L., Tong, R., Feng, Y., & Shi, J. (2022). Terpenoid natural products exert neuroprotection via PI3K/Akt signaling. *Frontiers in Pharmacology*, *13*, 1036506. DOI: 10.3389/fphar.2022.1036506 PMID: 36313360

Xue, M., Le, N. T. V., & Jackson, C. J. (2006). Targeting matrix metalloproteases to improve cutaneous wound healing. *Expert Opinion on Therapeutic Targets*, *10*(1), 143–155. DOI: 10.1517/14728222.10.1.143 PMID: 16441234

Xue, R., He, L., Wu, J., Kong, X., Wang, Q., Chi, Y., Liu, J., Wang, Z., Zeng, K., Chen, W., Ren, H., & Han, B. (2024). Multifunctional sprayable carboxymethyl chitosan/polyphenol hydrogel for wound healing. *International Journal of Biological Macromolecules*, *275*, 133303. DOI: 10.1016/j.ijbiomac.2024.133303 PMID: 38917923

Xu, F. W., Lv, Y. L., Zhong, Y. F., Xue, Y. N., Wang, Y., Zhang, L. Y., Hu, X., & Tan, W. Q. (2021). Beneficial effects of green tea EGCG on skin woundhealing: A comprehensive review. *Molecules (Basel, Switzerland)*, *26*(20), 6123. DOI: 10.3390/molecules26206123 PMID: 34684703

Xu, F., Weng, B., Gilkerson, R., Materon, L. A., & Lozano, K. (2015). Development of tannic acid/chitosan/pullulan composite nanofibers from aqueous solution for potential applications as wound dressing. *Carbohydrate Polymers*, *115*, 16–24. DOI: 10.1016/j.carbpol.2014.08.081 PMID: 25439862

Xu, Z., Liu, G., Liu, P., Hu, Y., Chen, Y., Fang, Y., Sun, G., Huang, H., & Wu, J. (2022). Hyaluronic acid-based glucose-responsive antioxidant hydrogel platform for enhanced diabetic wound repair. *Acta Biomaterialia*, *147*, 147–157. DOI: 10.1016/j.actbio.2022.05.047 PMID: 35649507

Yadav, A., Pandey, A. K., & Jaiswal, H. (2025). History of Ancient Herbal Medicine. *International Journal of Innovative Science and Research Technology, 10*(4), 2344–2347. DOI: 10.38124/ijisrt/25apr1442

Yadav, P., Singh, M. P., Vinayagam, R., & Shukla, P. (2025). Therapies and delivery systems for diabetic wound care: Current insights and future directions. *Frontiers in Pharmacology, 16*, 1628252. DOI: 10.3389/fphar.2025.1628252 PMID: 40766758

Yamaguchi, K., Urai, T., Okuwa, M., & Nakatani, T. (2012). The effect of isoflavone oral medication on cutaneous wound healing in ovariectomized female mice. *Wound Repair and Regeneration, 20*(1), A5.

Yang, B., Dong, Y., Wang, F., & Zhang, Y. (2020). Nanoformulations to Enhance the Bioavailability and Physiological Functions of Polyphenols. *Molecules (Basel, Switzerland), 25*(20), 4613. DOI: 10.3390/molecules25204613 PMID: 33050462

Yang, D., Wang, M., Zhang, C., & Wang, Y. (2024). Endothelial dysfunction in vascular complications of diabetes: A comprehensive review of mechanisms and implications. *Frontiers in Endocrinology, 15*, 1359255. DOI: 10.3389/fendo.2024.1359255 PMID: 38645427

Yang, F., Fang, Z., Yang, M., Cheng, Z., Tian, Y., Liang, J., & Li, T. (2025). Research hotspots in natural products for wound healing: A bibliometric analysis and literature review. *Clinical, Cosmetic and Investigational Dermatology, 18*, 15. Advance online publication. DOI: 10.2147/CCID.S544826 PMID: 40873990

Yang, H. L., Tsai, Y. C., Korivi, M., Chang, C. T., & Hseu, Y. C. (2017). Lucidone promotes the cutaneous wound healing process via activation of the PI3K/AKT, Wnt/β-catenin and NF-κB signaling pathways. *Biochimica et Biophysica Acta (BBA)-. Molecular Cell Research, 1864*(1), 151–168. PMID: 27816443

Yang, K., Zhou, X., Li, Z., Wang, Z., Luo, Y., Deng, L., & He, D. (2022). Ultrastretchable, self-healable, and tissue-adhesive hydrogel dressings involving nanoscale tannic acid/ferric ion complexes for combating bacterial infection and promoting wound healing. *ACS Applied Materials & Interfaces, 14*(38), 43010–43025. DOI: 10.1021/acsami.2c13283 PMID: 36108772

Yang, X., & Kong, F. (2016). Effects of tea polyphenols and different teas on pancreatic α-amylase activity in vitro. *Lebensmittel-Wissenschaft + Technologie, 66*, 232–238. DOI: 10.1016/j.lwt.2015.10.035

Yazarlu, O., Iranshahi, M., Kashani, H. R. K., Reshadat, S., Habtemariam, S., Iranshahy, M., & Hasanpour, M. (2021). Perspective on the application of medicinal plants and natural products in wound healing: A mechanistic review [Review of *Perspective on the application of medicinal plants and natural products in wound healing: A mechanistic review*]. *Pharmacological Research, 174*, 105841. Elsevier BV. DOI: 10.1016/j.phrs.2021.105841

Yeshi, K., Turpin, G., Jamtsho, T., & Wangchuk, P. (2022). Indigenous Uses, Phytochemical Analysis, and Anti-Inflammatory Properties of Australian Tropical Medicinal Plants. *Molecules (Basel, Switzerland)*, *27*(12), 3849. DOI: 10.3390/molecules27123849 PMID: 35744969

Yilmaz, A. C., & Aygin, D. (2020). Honey dressing in wound treatment: A systematic review. *Complementary Therapies in Medicine*, *51*, 102388. DOI: 10.1016/j.ctim.2020.102388 PMID: 32507418

Young, A., & McNaught, C. E. (2011). The physiology of wound healing. *Surgery (Oxford)*, *29*(10), 475–479. DOI: 10.1016/j.mpsur.2011.06.011

Yousefi, M., Rahimi-Nasrabadi, M., Pourmortazavi, S. M., Wysokowski, M., Jesionowski, T., Ehrlich, H., & Mirsadeghi, S. (2019). Supercritical fluid extraction of essential oils. *Trends in Analytical Chemistry*, *118*, 182–193. DOI: 10.1016/j.trac.2019.05.038

Yuan, L., He, Q., Zhang, Y., Luo, H., Xiang, W., Deng, C., Li, C., Li, X., Yao, L. et al. (2025). 6-Gingerol microneedle promotes diabetic wound healing by regulating macrophage polarization. International Immunopharmacology, 151: 4, 114288

Yuan, H., Ma, Q., Ye, L., & Piao, G. (2016). The traditional medicine and modern medicine from natural products. *Molecules (Basel, Switzerland)*, *21*(5), 559. DOI: 10.3390/molecules21050559 PMID: 27136524

Yuliati, L., Mardliyati, E., Bramono, K., & Freisleben, H. J. (2015). Asiaticoside induces cell proliferation and collagen synthesis in human dermal fibroblasts. *Universa Medicina*, *34*(2), 96. DOI: 10.18051/UnivMed.2015.v34.96-103

Zahra, M., Abrahamse, H., & George, B. P. (2024). Flavonoids: Antioxidant powerhouses and their role in nanomedicine. *Antioxidants*, *13*(8), 922. DOI: 10.3390/antiox13080922 PMID: 39199168

Zahra, M., Abrahamse, H., & George, B. P. (2024). Green nanotech paradigm for enhancing sesquiterpene lactone therapeutics in cancer. *Biomedicine and Pharmacotherapy*, *173*, 116426. Advance online publication. DOI: 10.1016/j.biopha.2024.116426 PMID: 38471274

Zangeneh, A., Pooyanmehr, M., Zangeneh, M. M., Moradi, R., Rasad, R., & Kazemi, N. (2019). Therapeutic Effects of Glycyrrhiza glabra Aqueous Extract Ointment on Cutaneous Wound Healing in Sprague Dawley Male Rats. *Comparative Clinical Pathology*, *28*(5), 1507–1514. DOI: 10.1007/s00580-019-03007-9

Zayed, A., Abdelkareem, S., Talaat, N., Dayem, D. A., & Farag, M. A. (2025). Tannin in foods: Classification, Dietary Sources, and Processing Strategies to Minimize Anti-Nutrient Effects. *Food and Bioprocess Technology*, *18*(11), 9221–9249. DOI: 10.1007/s11947-025-04020-3

Zeka, K., Ruparelia, K. C., Sansone, C., Macchiarelli, G., Continenza, M. A., & Arroo, R. R. J. (2018). New Hydrogels Enriched with Antioxidants from Saffron Crocus Can Find Applications in Wound Treatment and/or Beautification. *Skin Pharmacology and Physiology*, *31*(2), 95–98. DOI: 10.1159/000486135 PMID: 29393263

Zekavat, O., Amanat, A., Karami, M., Paydar, S., Gramizadeh, B., & Zareian-Jahromi, M. (2016). Wound healing studies using Punica granatum peel: An animal experimental study. *Advances in Skin & Wound Care*, *29*(5), 217–225. DOI: 10.1097/01.ASW.0000481116.16998.55 PMID: 27089150

Zeng, H.-J., Yang, R., Liang, H., & Qu, L.-B. (2015). Molecular interactions of flavonoids to pepsin: Insights from spectroscopic and molecular docking studies. *Spectrochimica Acta. Part A: Molecular and Biomolecular Spectroscopy*, *151*, 576–590. DOI: 10.1016/j.saa.2015.06.059 PMID: 26162346

Zhang, G., Xie, F., Sun, Y., Yu, X., Xiao, Z., Fang, R., Li, J., Li, Q., Du, L., & Jin, Y. (2021). Inhalable jojoba oil dry nanoemulsion powders for the treatment of lipopolysaccharide-or H2O2-induced acute lung injury. *Pharmaceutics*, *13*(4), 486. DOI: 10.3390/pharmaceutics13040486 PMID: 33918471

Zhang, H., Chen, J., & Cen, Y. (2018). Burn wound healing potential of a polysaccharide from Sanguisorba officinalis L. in mice. *International Journal of Biological Macromolecules*, *112*, 862–867. DOI: 10.1016/j.ijbiomac.2018.01.214 PMID: 29425875

Zhang, H., Lin, X., Cao, X., Wang, Y., Wang, J., & Zhao, Y. (2024). Developing natural polymers for skin wound healing. *Bioactive Materials, 33*, 355-376. Zhang, J., Song, Q., Han, X., Zhang, Y., Zhang, Y., Zhang, X., Chu, X., Zhang, F., & Chu, L. (2017). Multi-targeted protection of acetaminophen-induced hepatotoxicity in mice by tannic acid. *International Immunopharmacology*, *47*, 95–105. DOI: 10.1016/j.intimp.2017.03.027 PMID: 28376392

Zhang, L., Wang, X., & Si, H. (2022). Synergistic anti-inflammatory effects and mechanisms of the combination of resveratrol and curcumin in human vascular endothelial cells and rodent aorta. *The Journal of Nutritional Biochemistry, 108*, 109083. DOI: 10.1016/j.jnutbio.2022.109083 PMID: 35691595

Zhang, P., Lu, J., Jing, Y., Tang, S., Zhu, D., & Bi, Y. (2017). Global epidemiology of diabetic foot ulceration. *Annals of Medicine, 49*(2), 106–116. DOI: 10.1080/07853890.2016.1231932 PMID: 27585063

Zhang, S., Ou, Q., Xin, P., Yuan, Q., Wang, Y., & Wu, J. (2019). Polydopamine/puerarin nanoparticle-incorporated hybrid hydrogels for enhanced wound healing. *Biomaterials Science, 7*(10), 4230–4236. DOI: 10.1039/C9BM00991D PMID: 31393463

Zhang, S., Yang, K., & Zhang, M. (2022). Stability of polyphenols in various topical formulations: A review. *Pharmaceutics, 14*(9), 1818. Advance online publication. DOI: 10.3390/pharmaceutics14091818 PMID: 36145565

Zhang, W., Chen, L., Xiong, Y., Panayi, A. C., Abududilibaier, A., Hu, Y., Yu, C., Zhou, W., Sun, Y., Liu, M., Xue, H., Hu, L., Yan, C., Xie, X., Lin, Z., Cao, F., Mi, B., & Liu, G. (2021). Antioxidant Therapy and Antioxidant-Related Bionanomaterials in Diabetic Wound Healing. *Frontiers in Bioengineering and Biotechnology, 9*, 707479. DOI: 10.3389/fbioe.2021.707479 PMID: 34249895

Zhang, W., Liu, L., Cheng, H., Zhu, J., Li, X., Ye, S., & Li, X. (2023). Hydrogel-based dressings designed to facilitate wound healing. *Materials Advances, 5*(4), 1364–1394. DOI: 10.1039/D3MA00682D

Zhang, X., Liu, S., & Wang, J. (2023). Curcumin and its role in wound healing and inflammation modulation. *Journal of Ethnopharmacology, 289*, 115131. DOI: 10.1016/j.jep.2023.115131

Zhang, Y.. (2019). Polymeric nanoparticles for targeted delivery of therapeutics. *Journal of Controlled Release, 301*, 56–69.

Zhang, Y.. (2020). Nanocarriers for the delivery of bioactive compounds: A review. *Trends in Food Science & Technology, 99*, 61–70.

Zhang, Y., Bai, X., Shen, K., Luo, L., Zhao, M., Xu, C., Jia, Y., Xiao, D., Li, Y., Gao, X., Tian, C., Wang, Y., & Hu, D. (2022). Exosomes Derived from Adipose Mesenchymal Stem Cells Promote Diabetic Chronic Wound Healing through SIRT3/SOD2. *Cells, 11*(16), 2568. DOI: 10.3390/cells11162568 PMID: 36010644

Zhang, Y., Liu, Y., & Zhang, W. (2015). Nanosheets in electronics. *Advanced Materials, 27*(44), 6833–6848. PMID: 26449732

Zhang, Y., Xie, C., Wang, L., Huang, L., & Wei, Y. (2020). The role of catechol-O-methyltransferase (COMT) polymorphisms in the metabolism and bioavailability of flavonoids: Implications for their protective effects on health. *Phytotherapy Research*, *34*(6), 1303–1312. PMID: 31840883

Zhang, Z., Huang, Q., Zhao, D., Lian, F., Li, X., & Qi, W. (2023). The impact of oxidative stress-induced mitochondrial dysfunction on diabetic microvascular complications. *Frontiers in Endocrinology*, *14*, 1112363. Advance online publication. DOI: 10.3389/fendo.2023.1112363 PMID: 36824356

Zhao, L., Wang, F., Lu, Q., Liu, R., Tian, J., & Huang, Y. (2018). Study on interaction between human salivary α-amylase and sorghum procyanidin tetramer: Binding characteristics and structural analysis. *International Journal of Biological Macromolecules*, *118*, 1136–1141. DOI: 10.1016/j.ijbiomac.2018.06.189 PMID: 30001600

Zhao, R., Liang, H., Clarke, E., Jackson, C., & Xue, M. (2016). Inflammation in chronic wounds. *International Journal of Molecular Sciences*, *17*(12), 2085. DOI: 10.3390/ijms17122085 PMID: 27973441

Zheng, S. Y., Wan, X. X., Kambey, P. A., Luo, Y., Hu, X. M., Liu, Y. F., Shan, J. Q., Chen, Y. W., & Xiong, K. (2023). Therapeutic role of growth factors in treating diabetic wound. *World Journal of Diabetes*, *14*(4), 364–395. DOI: 10.4239/wjd.v14.i4.364 PMID: 37122434

Zheng, Y., Tian, J., Yang, W., Chen, S., Liu, D., Fang, H., Zhang, H., & Ye, X. (2020). Inhibition mechanism of ferulic acid against α-amylase and α-glucosidase. *Food Chemistry*, *317*, 126346. DOI: 10.1016/j.foodchem.2020.126346 PMID: 32070843

Zheng, Y., Yang, W., Sun, W., Chen, S., Liu, D., Kong, X., Tian, J., & Ye, X. (2020). Inhibition of porcine pancreatic α-amylase activity by chlorogenic acid. *Journal of Functional Foods*, *64*, 103587. DOI: 10.1016/j.jff.2019.103587

Zhou, F., Li, G., Tan, R., Wu, G., & Deng, C. (2025). Artemisinins in autoimmune diseases: Effects and mechanisms in systemic lupus erythematosus and rheumatoid arthritis. *British Journal of Pharmacology*, *182*(15), 3411–3427. Advance online publication. DOI: 10.1111/bph.70086 PMID: 40474495

Zhou, P., Chrepa, V., Karoussis, I. K., Pikos, M. A., & Kotsakis, G. A. (2021). Cytocompatibility Properties of an Herbal Compound Solution Support In vitro Wound Healing. *Frontiers in Physiology*, *12*, 653661. Advance online publication. DOI: 10.3389/fphys.2021.653661 PMID: 33868021

Zhuang, W. B., Li, Y. H., Shu, X. C., Pu, Y. T., Wang, X. J., Wang, T., & Wang, Z. (2023). The classification, molecular structure and biological biosynthesis of flavonoids, and their roles in biotic and abiotic stresses. *Molecules (Basel, Switzerland), 28*(8), 3599. DOI: 10.3390/molecules28083599 PMID: 37110833

Zhuang, Y., Hu, K., & Wei, Y.. (2020). Quercetin promotes angiogenesis via upregulation of VEGF and bFGF expression in the wound healing process. *Molecules (Basel, Switzerland), 25*(12), 2903. Advance online publication. DOI: 10.3390/molecules25122903 PMID: 32599753

Zhu, Y., Lu, J., Wang, S., Xu, D., Wu, M., Xian, S., Zhang, W., Tong, X., Liu, Y., Huang, J., Jiang, L., Guo, X., Xie, S., Gu, M., Jin, S., Ma, Y., Huang, R., Xiao, S., & Ji, S. (2023). Mapping intellectual structure and research hotspots in the field of fibroblast-associated DFUs: A bibliometric analysis. *Frontiers in Endocrinology, 14*, 1109456. Advance online publication. DOI: 10.3389/fendo.2023.1109456 PMID: 37124747

Zielins, E. R., Brett, E. A., Luan, A., Hu, M. S., Walmsley, G. G., Paik, K., Senarath-Yapa, K., Atashroo, D. A., Wearda, T., Lorenz, H. P., Wan, D. C., & Longaker, M. T. (2015). Emerging drugs for the treatment of wound healing. *Expert Opinion on Emerging Drugs, 20*(2), 235–246. DOI: 10.1517/14728214.2015.1018176 PMID: 25704608

Zielińska-Błajet, M., & Feder-Kubis, J. (2020). Monoterpenes and their derivatives—Recent development in biological and medical applications. *International Journal of Molecular Sciences, 21*(19), 7078. DOI: 10.3390/ijms21197078 PMID: 32992914

Zio, S., Cisse, H., Zongo, O., Guira, F., Tapsoba, F., & Somda, N. S.. (2020). Edible oil refining and contaminants. *Journal of Food Technology Research, 7*(1), 9–47. DOI: 10.18488/journal.58.2020.71.9.47

Zulkefli, N., Che Zahari, C. N. M., Sayuti, N. H., Kamarudin, A. A., Saad, N., Hamezah, H. S., Bunawan, H., Baharum, S. N., Mediani, A., Ahmed, Q. U., Ismail, A. F. H., & Sarian, M. N. (2023). Flavonoids as potential wound-healing molecules: Emphasis on the pathways perspective. *International Journal of Molecular Sciences, 24*(5), 4607. DOI: 10.3390/ijms24054607 PMID: 36902038

Zuo, L., Wang, X., Cao, X., Chen, B., Shao, M., Yang, G., Fu, S., & Wang, L. (2023). Preparation and characterization of tannin-maltodextrin-polyvinyl alcohol hydrogel based on hydrogen bonding for wound healing. *Journal of the Mechanical Behavior of Biomedical Materials, 145*, 105942. DOI: 10.1016/j.jmbbm.2023.105942 PMID: 37423009

About the Contributors

Geeta Deswal holds a Ph.D. in Pharmaceutical Sciences from MM University, Mullana, Ambala, Haryana, awarded in 2017 under the guidance of Prof. (Dr.) Vipin Saini. She completed her M.Pharm in Pharmacognosy in 2011 and B.Pharm in 2009, both from the same institution. Currently, Dr. Deswal serves as a Professor and Head of the Department of Pharmacognosy at Guru Gobind Singh College of Pharmacy, Yamuna Nagar. In addition, she functions as the Internal Quality Assurance Cell (IQAC) and NBA Coordinator for the institution. Dr. Deswal has made significant academic contributions through her extensive list of 39 publications in reputed journals, focusing on phytochemical screening, plant tissue culture, herbal formulation quality audits, and therapeutic applications of natural compounds. Her research also extends to molecular docking, nanotechnological strategies, and biodegradable polymer applications in cancer therapy. She has co-authored and edited 10 books, including: A Practical Book of Pharmaceutical Organic Chemistry (2024, ISBN: 978-9-36109-206-0). Nature's Arsenal Against Cancer: Harnessing the Power of Natural Compounds for Treatment and Prevention (2024, ISBN: 979-8-89530-202-6). Naturally Occurring Chemicals for the Treatment and Management of Diabetes Mellitus and Related Disorders (2024, ISBN: 979-8-89113-491-1). Dr. Deswal has also contributed to numerous book chapters, such as: Nanoparticles for the Treatment of Cancer in Current Therapeutic Strategies for the Management of Cancer (2023, ISBN: 979-8-89113-133-0). Resveratrol: A Phytocentric Overview of Potential in Metabolic Syndrome in Natural Products for Treatment of Metabolic Syndrome (2022, ISBN: 979-8-88697-152-1). Her work has been supported by research grants and consultancy projects, and she has received several awards, including the "Young Achiever Award." Dr. Deswal continues to excel in pharmaceutical education and research, particularly in advancing the therapeutic use of natural compounds.

Ajmer Singh Grewal is working as Professor at the Department of Pharmaceutical Chemistry, Guru Gobind Singh College of Pharmacy, Yamuna Nagar, Haryana (India) since Mar. 02, 2021. He had worked as Assistant Professor at the School College of Pharmacy, Chitkara University (Himachal Pradesh) from July 2019 to February 2021 and at Chitkara College of Pharmacy, Chitkara University (Punjab) from March 2017 to June 2019. Before that, he had worked as Assistant Professor at Jan Nayak Ch. Devi Lal Memorial College of Pharmacy, Sirsa (Haryana, India) from November 2011 to February 2017. He is a dedicated academic professional with experience of 14+ years committed to pharmaceutical education. He had taught B. Pharm., M. Pharm. and Pharm. D. students and supervised 08 M. Pharm. (Pharmaceutical Chemistry) students in the research area of Drug Design and Medicinal Chemistry. He is currently guiding 03 PhD students in the area of Drug Design and Discovery. He had completed PhD in Pharmaceutical Sciences in the subject area of Computational and Medicinal Chemistry on the project entitled "Design, Synthesis and Biological Evaluation of Newer Glucokinase Activators" from I.K. Gujral Punjab Technical University, Jalandhar (Punjab) in 2019. Earlier, He had seasoned Master of Pharmacy in Pharmaceutical Chemistry on the project entitled "Development and Validation of New Analytical Methods for Estimation of Selected Drugs in Bulk & Pharmaceutical Formulations" from the Institute of Pharmacy and Technology, Salipur, Cuttack in 2011. He had qualified for GATE-2009 (87.09 percentile). His present areas of research include rational drug design and medicinal chemistry of matrix metallo-proteinase (MMP2 and MMP9) inhibitors as anticancer agents, antibacterial drug resistance (NDM-1 inhibitors), glucokinase activators as anti-diabetic agents, aldose reductase inhibitors for diabetic complications, PPAR agonists for the treatment of metabolic syndrome, multi-potent anti-Alzheimer's agents, selective PDE-4 & PDE-7 inhibitors as anti-inflammatory agents and environmental chemistry. He owns to his credit more than 100 articles in national and international journals of repute, 22 book chapters, 16 books, 6 patents (3 granted & 3 published), 2 design patents and various poster presentations (16) at several conferences (both national and international). Currently, he is editing 7 books with Nova Science, Elsevier and IGI Global.

Kumar Guarve is an accomplished education professional with over 18 years of experience in the field. He has been serving as the Principal at Guru Gobind Singh College of Pharmacy, Yamuna Nagar, Haryana (India) since 2017. He has specialized expertise in maintaining and achieving prestigious accreditations like NAAC and NBA, demonstrating a commitment to top-tier educational standards. Throughout his career, he has been instrumental in creating robust policies for revenue generation and successfully developing incubation centres within educational institutions. As an innovative leader, he has a proven track record of leveraging strategic planning and decision-making to drive institutional growth and academic excellence. He is renowned for his ability to transform colleges into thriving educational centres by implementing a visionary approach and ensuring its execution. He excels in fostering a culture of innovation, driving profitability, and launching start-up initiatives within colleges to enhance revenue and improve the overall educational experience. He has made significant contributions to academic research, with faculty under his guidance publishing numerous Scopus-indexed papers. He has also played a key role in patenting innovative products and securing copyrights for various educational processes and curricular frameworks. He is an active member of the Association of Pharmaceutical Teachers of India, the Indian Pharmacy Graduates Association, the Indian Society for Technical Education (ISTE), and the Association of Registered Pharmacists (ARP).

***

Gayathiri E is an Assistant Professor in the Department of Plant Biology and Plant Biotechnology at Guru Nanak College (Autonomous) in Chennai, Tamil Nadu. She holds a Ph.D. from Guru Nanak College (2015), an M.Sc. in Botany from Madras University (2010), and a B.Ed. from Thiruvalluvar University (2008). With over 341 citations and an H-index of 9, her academic contributions are notable. She has been part of the academic community since 2017.

Satyavani Kaliamurthi is a Research Associate in Medical and Bioresearch at AVMCH. She is a biotechnologist specializing in data-driven human therapeutics, nano-drug delivery, and phytomedicine, with a vision to develop innovative therapeutics that protect communities from emerging health threats. She actively contributes to the scientific publishing community as a Guest Editor for journals from Bentham Science, Frontiers, and Elsevier, and serves as an IEEE program committee expert member, reviewer for more than 12 international journals, Advisory Board Member for Heliyon (Immunology), and Associate Editor for Scientific Reports. She is also a distinguished Adjunct Faculty at SIMATS and AMET University. Dr. Kaliamurthi received the Concordia University Conference and Exposition Award twice (2022 and 2023) for her work on peptide vaccine design for COVID-19. With over eight years of postdoctoral research experience, she has been awarded several prestigious international fellowships, including the TÜBİTAK 2216, Henan Province Postdoctoral Science Foundation (China), MITACS Globalink Fellowship (Canada), and the Horizon Postdoctoral Fellowship (Canada). She has authored more than 80 research publications with over 2,900 citations and has presented at 23 international conferences, including the International Society for Vaccines (ISV) 2022 Congress. She also completed a 35-hour graduate course in University Teaching at Concordia

University. “Be always confident, compassionate, and true to yourself; surely you will achieve your goals.”

Dharmendra Kumar Khatri, working as an Associate Professor & HoD at the Department of Pharmacology, Nims Institute of Pharmacy at NIMS University Rajasthan, Jaipur, India. He worked almost 5 Years at the National Institute of Pharmaceutical Education and Research (NIPER) Hyderabad, India. Dr. Khatri obtained his Master of Pharmacy with AA outstanding grade and PhD from India's one of the prestigious institutes “Institute of Chemical Technology, Mumbai (Formerly known as UDCT). He is the proud recipient of the University Grant Fellowship (UGC) of Junior Research Fellow two times and Senior Research Fellow for his Master and PhD degree. He is a young neuroscientist in an early career stage, and his major research is focused on leveraging multi-omics (Bioinformatics, Genomics, proteomics, metabolomics and lipidomics) approaches for the precision target and refined hit molecules. In particular his research quest is to understand the α-synuclein dysregulation at post-transcriptional (miRNA's) and post-translational levels (Phosphorylation), leading to its overexpression and its transition to toxic oligomeric form in Parkinson's Disease. He has published more than 130 articles in peer reviewed high impact factors journals and authored 2 books, several book chapters, and has 09 patents granted. He has supervised more than 30 Master, and 6 PhD degree students. Based on his scientific contribution in the neuroscience field he has been conferred with many prestigious fellowships and awards at the national and international level, naming few: ¬ Listed in World top 2% Scientist list released by Stanford 2024. ¬ Admitted as Associate Fellows of Telangana Academy of Sciences, India for the years 2022 for the recognition of contributions to Science & Technology ¬ Member of Royal Society of Biology, London UK. ¬ Awarded Excellent Oral Presentation award at 43rd Annual Scientific Meeting of the Korean Endocrine Society (KES) Seoul in Korea, 2024. ¬ Awarded First position for Dr. P. D. Patil Best Thesis Award 2023, in the subject category M. Pharmacy-Pharmacology as mentor. ¬ Young Neuroscientist award from Japanese Neuroscience Society ¬ He is an Indian Ambassador on behalf of International Society of Neurochemistry (ISN) for the consequently for the year 2022-24 and 2019-21. ¬ Next Generation Young Neuroscientist award from Osaka University, Japan, 2018. ¬ Received Lalitha and Ravindra Nath International Travel Award to present his research work in the American Society for Neurochemistry (ASN) Meeting in held in Montréal, Canada 2019. ¬ Best Poster Award for the poster at 4th International Conference of Korean Movement Disorder Society (ICKMDS 2023) held from 2023, Seoul, Korea He also serves Govt. of India's regulatory bodies like CCSEA ethics committee as Scientist and NABL as assessor. He serves various Editorial board of prestigious journals including Molecular Neurobiology as editor, Journal of Alzheimer disease

(As Associate Editor), Annals of Neuroscience, Frontiers in Neuroscience, Wyno academics journals as board member He is also a member of various prestigious bodies: • Member of Asia Pacific Society for Neurochemistry (17-3-IND-S32) • Member of Japanese Neuroscience Society (JNS): 2212266471 • Member of The International Society for Neurochemistry (ISN)-(21741) • Member of International Parkinson and Movement Disorder Society (MDS)- (39982) • Lifetime member of Indian Pharmacological Society (LD-292) • Lifetime Member of Indian Academy of Neuroscience (IAN)-(LK-143) • Lifetime Member of Society for Neurochemistry, India (SNCI)- (LM-I-662) • Life Member of Association of Physiologist & Pharmacologist of India. (3187) • Life Member Association of Pharmaceutical Teachers of India (TL/LM-916)

N. Prithivi Raj is an accomplished Biotechnologist and AssociateProfessor at Aarupadai Medical College and Hospital, with a rich academic background including a Ph.D. in Marine Biotechnology from Annamalai University, India. He has held post-doctoral positions at Shanghai Jiao Tong University and Shanghai Ocean University, China, focusing on tissue regeneration and bioactive natural products. His research contributions include work on Candida albicans and tissue regeneration-related chemotactic active peptides. He is a recipient of several awards and fellowships, including the Dr. APJ Abdul Kalam Young Scientist and Distinguished Fellow Award, 2020.

Nagarajan Sankaranarayanan currently works at the Department of Chemistry, National Institute of Technology, Manipur. Dr Nagarajan does research in Nanotechnology, Materials Chemistry and Catalysis. Their current project is 'Subsurface Oxygen and its effect on catalytic process towards application in Environmental pollution'.

Rajat Srivastava is an academician and researcher in the field of Pharmaceutical Sciences, with teaching and research experience across reputed pharmacy institutions in Uttar Pradesh. He is currently working as an Assistant Professor at BBS College of Pharmacy & Polytechnic, Prayagraj. He is pursuing his Ph.D. from Maharishi University of Information Technology, Lucknow, and holds an M. Pharm. and B. Pharm. from Shambhunath Institute of Pharmacy (AKTU, Lucknow). His research interests include novel drug delivery systems, nanotechnology-based formulations, and pharmaceutical formulation development. Mr. Srivastava has published numerous research and review articles in national and international journals and has co-authored textbooks and book chapters in pharmaceutics and oncology. He is also a co-inventor of an Indian patent related to ocular nano-emulsion drug delivery. He actively participates in academic administration, examinations, and professional development programs

Priya Tiwari is an experienced academician and researcher in the field of Pharmaceutics and Pharmaceutical Technology, with over seven and a half years of teaching experience at undergraduate and postgraduate levels. She is currently working as an Associate Professor at the Moradabad Educational Trust Group of Institutions, Faculty of Pharmacy. She is pursuing her Ph.D. in Pharmaceutics from Integral University, Lucknow, and holds an M. Pharm. and B. Pharm. from Bundelkhand University, Jhansi, both with first-class distinctions. Her areas of expertise include Pharmaceutics, Industrial Pharmacy, Novel Drug Delivery Systems, and Biopharmaceutics. Dr. Tiwari has published several research and review articles in reputed journals, authored multiple textbooks, contributed book chapters, and holds an Indian patent in pharmaceutical nano-emulsion technology. She is a life member of the Association of Pharmaceutical Teachers of India (APTI) and actively participates in academic administration, research coordination, and professional development activities.

Index

A

B

C

D

E

F

H

I

M

N

O

P

R

S

T

W

www.ingramcontent.com/pod-product-compliance
Lightning Source LLC
LaVergne TN
LVHW061527290926
849162LV00004B/700